Semigroups, Algorithms, Automata and Languages

Semigroups, Algorithms, Automata and Languages

Coimbra, Portugal   May – July 2001

editors

Gracinda M. S. Gomes
University of Lisbon, Portugal

Jean-Éric Pin
University of Paris VII and CNRS, France

Pedro V. Silva
University of Porto, Portugal

World Scientific
New Jersey • London • Singapore • Hong Kong

Published by
World Scientific Publishing Co. Pte. Ltd.
P O Box 128, Farrer Road, Singapore 912805
USA office: Suite 1B, 1060 Main Street, River Edge, NJ 07661
UK office: 57 Shelton Street, Covent Garden, London WC2H 9HE

British Library Cataloguing-in-Publication Data
A catalogue record for this book is available from the British Library.

SEMIGROUPS, ALGORITHMS, AUTOMATA AND LANGUAGES
Copyright © 2002 by World Scientific Publishing Co. Pte. Ltd.

All rights reserved. This book, or parts thereof, may not be reproduced in any form or by any means, electronic or mechanical, including photocopying, recording or any information storage and retrieval system now known or to be invented, without written permission from the Publisher.

For photocopying of material in this volume, please pay a copying fee through the Copyright Clearance Center, Inc., 222 Rosewood Drive, Danvers, MA 01923, USA. In this case permission to photocopy is not required from the publisher.

ISBN 981-238-099-X

This book is printed on acid-free paper.
Printed in Singapore by Mainland Press

Preface

The thematic term on Semigroups, Algorithms, Automata and Languages was held at the International Centre of Mathematics, CIM, in Coimbra during the months of May, June and July 2001. It was designed to make Coimbra the gathering point of researchers in the subjects of semigroup theory and automata theory.

The programme included three schools and two workshops dedicated to specific areas considered presently to have great importance to the study of semigroups, algorithms, automata and languages. These areas were selected considering their huge recent development, motivation from other fields of mathematics and computer science, and their potential applications.

- First School (May 2 to 11, 2001): Algorithmic aspects of the theory of semigroups and its applications

- Second School (June 4 to 8, 2001): Automata and languages

- Third School (July 2 to 6, 2001): Semigroups and applications

- First Workshop (June 11 to 13, 2001): Logic, profinite topology and semigroups

- Second Workshop (July 9 to 11, 2001): Presentations and geometry

Each school consisted of several 5 hour courses held by prominent researchers. The workshops included 50 minute invited lectures and a limited number of 20 minute talks on the specific topics of the workshop, proposed by the participants.

The thematic term was a great success and many participants expressed their deep contentment to the organizers. The insiders enjoyed the beautiful surroundings of Coimbra, the pleasing and relaxed atmosphere and the high scientific quality of the meetings. Altogether, about 90 people from 20 different countries attended the thematic term.

The organizers would like to express their thanks to the sponsors of the thematic term namely Centre of Algebra of the University of Lisbon, Centre of Mathematics of the University of Porto, International Centre of Mathematics, Faculty of Sciences of the University of Lisbon, Foundation Calouste Gulbenkian, Luso-American Foundation for Development, Portuguese Foundation for Science and Technology and the University of Porto. We would also like to acknowledge the township of Coimbra who very kindly organized three receptions in the town hall.

Lisboa, Paris and Porto, 27 June 2002

Gracinda M. S. Gomes
Jean-Éric Pin
Pedro V. Silva

Contents

Preface   v

Advanced courses   1

1. Finite semigroups: An introduction to a unified theory of pseudovarieties
   Jorge Almeida 3

2. On existence varieties of regular semigroups
   Karl Auinger 65

3. Varieties of languages
   Mário J. J. Branco 91

4. A short introduction to automatic group theory
   Christian Choffrut 133

5. An introduction to covers for semigroups
   John Fountain 155

6. E^*-unitary inverse semigroups
   Mark V. Lawson 195

7. Some results on semigroup-graded rings
   W. D. Munn 215

8. Profinite groups and applications to finite semigroups
   Luis Ribes 235

Research articles   267

1 **Dynamics of finite semigroups**
   Jorge Almeida   269

2 **Group presentations for a class of radical rings of matrices**
   Noelle Antony, Clare Coleman and David Easdown   293

3 **Finite semigroups imposing tractable constraints**
   Andrei Bulatov, Peter Jeavons, Mikhail Volkov   313

4 **On the efficiency and deficiency of Rees matrix semigroups**
   C.M. Campbell, J.D. Mitchell and N. Ruškuc   331

5 **Some pseudovariety joins involving groups and locally trivial semigroups**
   José Carlos Costa   341

6 **Partial action of groups on relational structures: A connection between model theory and profinite topology**
   Thierry Coulbois   349

7 **Presentations for some monoids of partial transformations on a finite chain: A survey**
   Vítor H. Fernandes   363

8 **Some relatives of automatic and hyperbolic groups**
   Michael Hoffmann, Dietrich Kuske, Friedrich Otto and Richard M. Thomas   379

9 **Operators on classes of regular languages**
   Libor Polák   407

10 **Automata in autonomous varieties**
   Olga Sokratova   423

11 **A sampler of a topological approach to inverse semigroups**
   Benjamin Steinberg   437

12 **Finite semigroups and the logical description of regular languages**
   Howard Straubing   463

13 Diamonds are forever: The variety DA
    Pascal Tesson and Denis Thérien   475

14 Decidability problems in finite semigroups
    Peter G. Trotter   501

Advanced courses

FINITE SEMIGROUPS: AN INTRODUCTION TO A UNIFIED THEORY OF PSEUDOVARIETIES

JORGE ALMEIDA[*]

Dep. Matemática Pura, Faculdade de Ciências, Universidade do Porto
Rua do Campo Alegre 687, 4169-007 Porto, Portugal
E-mail: jalmeida@fc.up.pt

There are some remarkable similarities between the approaches to the calculation of Mal'cev and semidirect products of pseudovarieties of semigroups using relatively free profinite semigroups and pseudoidentities. These lecture notes serve to extract a common framework to deal with such operators on pseudovarieties as well as to introduce the non-specialist reader to the techniques and most important results in the area. Central to the main results is a new very general compactness theorem which extends and unifies several such results. As a new application of these results a decidability theorem for pseudovarieties of the form $U^{-1}W$ is also presented.

1 Introduction

The aim of these lecture notes is to introduce the reader to the theory of finite semigroups understood as a collection of tools and results concerning the classification of these structures in pseudovarieties [21, 22, 33, 42, 4]. As an introduction to the subject, this is not meant to be covering or surveying exhaustively every particular area in the theory. On the other hand, since the theory is currently quite lively and is likely to remain so for years to come, one cannot expect to prepare the reader for every possible future development. Yet, it is hoped that these notes will provide the present basic background and tools for some of the main aspects of the theory as well as open up new research paths.

---

[*]WORK SUPPORTED, IN PART, BY *FUNDAÇÃO PARA A CIÊNCIA E A TECNOLOGIA* (FCT) THROUGH THE *CENTRO DE MATEMÁTICA DA UNIVERSIDADE DO PORTO*, AND BY THE FCT AND POCTI APPROVED PROJECT POCTI/32817/MAT/2000 WHICH IS COMPARTICIPATED BY THE EUROPEAN COMMUNITY FUND FEDER. THIS WORK WAS DONE IN PART WHILE THE AUTHOR WAS VISITING THE *CENTRO INTERNACIONAL DE MATEMÁTICA*, IN COIMBRA, PORTUGAL. FINANCIAL SUPPORT OF *FUNDAÇÃO CALOUSTE GULBENKIAN* (FCG), FCT, *FACULDADE DE CIÊNCIAS DA UNIVERSIDADE DE LISBOA* (FCUL) AND *REITORIA DA UNIVERSIDADE DO PORTO* IS GRATEFULLY ACKNOWLEDGED.

One of the most fruitful tools in the theory of finite semigroups is relatively free profinite semigroups. Although such semigroups are usually uncountable, their consideration leads to theoretical and even practical algorithms. It is also particularly useful to view their elements as operations, so called *implicit operations*.

An example of application of these ideas is found in the computation of semidirect products of pseudovarieties. By extending to the profinite world some of the ideas and results of Tilson [61], Weil and the author [13] obtained a "basis theorem" for semidirect products of pseudovarieties. Unfortunately, as has been observed recently by Rhodes and Steinberg, the proof of the theorem is faulty and it does not apply in full generality although the validity of the theorem remains open at present. Before this disturbing discovery there had been several related developments whose importance remains as it goes beyond the original problem.

Although on its own the "basis theorem" gives in general no algorithms to compute semidirect products, it was later explored by the author [5] for the systematic development of such algorithms through the notion of *hyperdecidability* in a way which does not depend on the faulty case of the theorem and which has been successfully applied. Shortly after Steinberg and the author [11, 12] further refined the ideas by replacing free profinite semigroups by suitable subalgebras, from which surfaced the notion of *tameness*. This led to a dramatic reduction of the celebrated Krohn-Rhodes complexity problem to solving a natural word problem (done independently by McCammond [37] and Zhil'tsov [65]) and proving an abstract property of the pseudovariety of finite aperiodic semigroups which has been submitted by Rhodes [50] based on further work of McCammond and himself [38, 51]. Unfortunately, the reduction depends on the full strength of the faulty "basis theorem" and so it is at present uncertain whether the reduction holds.

Another example in which similar ideas have been applied is provided by the classical Mal'cev product of pseudovarieties [47]. Here the situation is considerably simpler yet the tools are basically the same and the approach is successful. Moreover, the tameness approach, suitably adapted, also applies in this setting.

One of the main aims of these lecture notes is to put into perspective these two examples, avoiding the accessory and portraying the essential in the hope that the fundamental ideas will emerge and perhaps lead to means to systematically handle similar situations while clarifying the difficulties encountered in the "basis theorem". Besides a short introduction to pseudovarieties of semigroups in Sections 2 and 3, as part of this programme we present a general introduction to profinite algebras and more specifically to pro-V algebras for a pseudovariety V in Sections 4 to 7.

Section 8 presents a new compactness theorem which extends and unifies several known results of the same sort which underlie the above applications. It also generalizes an earlier result from [4] which is frequently used.

Section 9 discusses decidability issues and guides the reader through a wealth of examples in which a general framework is recognized. A strategy to prove such decidability results without caring to exhibit efficient algorithms on a first approximation is discussed in Section 10 where a generalization of the notion of tameness is presented.

Section 11 brings us back to the computation of pseudovarieties defined by operators putting into a unified framework results on semidirect and Mal'cev products through the definition of what we call *generalized Mal'cev* and *wreath products*. There is a delicate relationship between these two types of products, the Mal'cev version being more suitable to apply the tameness approach to prove decidability under reasonable hypotheses. As an application, a general decidability result is obtained for pseudovarieties of the form $U^{-1}W$.

Finally, Section 12 gives some simple examples of application of the main results which are meant to illustrate that a substantial ingredient in future developments of the decidability aspects of the theory of pseudovarieties is likely to be the search for proofs of tameness.

These notes are meant to be self-contained except for some background in general topology, for which we adopt [64] as a standard reference. Some familiarity with (pro)finite semigroup theory will certainly facilitate reading but it is not a requirement. Whatever background is needed in this area is introduced along the way. The reader familiar with pro-V algebras and implicit operations might wish to jump to Section 8, referring back to earlier results when needed.

2 Automata and languages

The motivation to study finite semigroups appeared in the 1950's as a result of work on linguistics and models of computation and reasoning. From such works emerged the notion of a *finite automaton* of which several variants can be found in the literature. The simplest version of a (finite) automaton $\mathcal{A} = \langle Q, A, \delta, I, F \rangle$ is given by

- a (resp. finite) set Q of *states*,

- a (resp. finite) set A of acting *symbols* (or *letters*) a to each of which is associated a binary relation $\delta(a)$ on Q,

- a subset $I \subseteq Q$ consisting of *initial states*, and
- a subset $F \subseteq Q$ of *final states*.

Using the associative operation of composition of binary relations, each finite non-empty sequence of letters (or *word*) $w = a_1 a_2 \cdots a_n$ acts in turn on the states as the composite binary relation $\hat{\delta}(w) = \delta(a_1)\delta(a_2) \cdots \delta(a_n)$ of the actions of its individual letters. Denoting by A^+ the set of all words on the *alphabet* A, the operation of concatenation turns A^+ into a semigroup. Then the extended action $\hat{\delta}$ is in fact a homomorphism $A^+ \to \mathcal{B}_Q$ into the semigroup of binary relations on Q. The image of this homomorphism is called the *transition semigroup* of the automaton $\mathcal{A}$ and is denoted $T(\mathcal{A})$. The congruence

$$\ker \hat{\delta} = \{(u, v) \in A^+ \times A^+ : \hat{\delta}(u) = \hat{\delta}(v)\}$$

is called the *transition congruence* of $\mathcal{A}$. Note that the transition semigroup of a finite automaton is finite.

In an automaton $\mathcal{A}$, the binary relations describing the action of the letters on the states may be actually partial or even full transformations of Q. In case the action of every letter has domain Q, the automaton is called *complete*. If the set I is a singleton $\{q_0\}$ and the action of the letters is given by partial transformations, then the automaton $\mathcal{A}$ is said to be *deterministic*.

Automata are often described by suggestive graphical descriptions which are essentially self explanatory. Figure 2 gives an example of an incomplete non-deterministic automaton with states $0, 1, 2$, letters a, b, initial state 0, and final states $1, 2$. The action of the letter b is given by the binary relation $\{(1,1), (1,2)\}$.

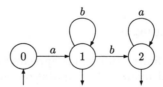

Figure 1. A finite automaton

Automata are used as language recognition devices. A word $w \in A^+$ is said to be *recognized* by the automaton $\mathcal{A} = \langle Q, A, \delta, I, F \rangle$ if $\hat{\delta}(w) \cap (I \times F) \neq \emptyset$. In the graphical representation of the automaton this condition means that there is some directed path from an initial state to a final state *labeled* w in the sense that w is the concatenation of the succesive labels of the edges in the path. The *language recognized* by an automaton $\mathcal{A}$ is the set $L(\mathcal{A})$ of all words recognized by $\mathcal{A}$. A language is said to be *recognizable* if it is recognized by some finite automaton. For instance, for the automaton $\mathcal{A}$ of Figure 2, the language $L(\mathcal{A})$ is $ab^* \cup ab^+a^*$. Here the notation is the following: adding to A^+ the empty word 1, we obtain a monoid A^*; the submonoid (resp. subsemigroup) of A^* generated by a language L is denoted L^* (resp. L^+).

When the action of a word w on the states is given by a transformation, we write qw for the state q' if there is a w-labeled path $q \to q'$.

Say that a state q of an automaton is *accessible* from another state q' if there is some path from q' to q. Any state which is not accessible from an initial state does not intervene in the recognition of the language recognized by the automaton and may therefore be suppressed for recognition purposes. An automaton which has no such states is called *trim*.[a]

In terms of the transition function $\hat{\delta}$ into the transition semigroup $T(\mathcal{A})$ of the automaton $\mathcal{A}$, the language $L(\mathcal{A})$ is the inverse image of the set

$$\{\tau \in \mathcal{B}_Q : \tau \cap (I \times F) \neq \emptyset\}.$$

More generally, we say that a subset L of a semigroup S is *recognized* by a homomorphism $\varphi : S \to T$ into a semigroup T if $L = \varphi^{-1}\varphi(L)$.

For a semigroup S, denote by S^1 a smallest monoid $S \cup \{1\}$ containing S which is obtained by adjoining a neutral element 1 in case S does not already possess one.

A semigroup S is said to be *A-generated* if a mapping $\iota : A \to S$ is given such that $\iota(A)$ generates S. The mapping ι will sometimes be called the *generating mapping*. We will somewhat abuse notation and usually denote the element $\iota(a)$ of S again by a. Note that the mapping ι extends uniquely to a homomorphism $\hat{\iota} : A^+ \to S$ such that the diagram

commutes, that is A^+ is a semigroup freely generated by A. From an A-generated semigroup S one can easily build an automaton whose transition semigroup is S: the *Cayley automaton*, whose states are the elements of S^1, each $a \in A$ acts on S^1 as the transformation $s \mapsto sa$, and the initial state is 1.

---

[a]Some authors prefer to also suppress states from which no final state is accessible as a requirement for trimness but we prefer to deal here with complete automata and the two requirements are frequently incompatible.

If the language $L \subseteq A^+$ is recognized by an onto homomorphism $\varphi : A^+ \to S$ and $\varphi(L)$ is chosen as the set of final states of the Cayley automaton of the A-generated semigroup S, then the language it recognizes is precisely L. Note that the Cayley automaton is complete and deterministic. Thus a language over a finite alphabet is reconized by a finite automaton if and only if it is recognized by a finite complete deterministic automaton, if and only if it is recognized by a homomorphism into a finite semigroup.

Given a subset L of a semigroup S, a congruence ρ on S is said to *saturate* L if L is a union of ρ-classes. Equality on S is of course such a congruence. Denote by ρ_L the relation on S defined by: for $s, t \in S$, $(s,t) \in \rho_L$ if, for all $u, v \in S^1$, $usv \in L \Leftrightarrow utv \in L$. It is easily verified that ρ_L is a congruence on S which saturates L and that it is the largest such congruence with respect to the order relation given by set inclusion. Since the congruence class of an element is determined by the context in which it appears in L, the congruence ρ_L is called the *syntactic congruence* and the quotient S/ρ_L is called the *syntactic semigroup* of L and is denoted Synt(L).

Binary relations on a set Q may be viewed as transformations of the power set $\mathcal{P}(Q)$. This is the basic idea in the so-called *powerset construction* turning a non-deterministic automaton into a complete deterministic automaton recognizing the same language. Rather than describing it in detail, we illustrate the construction with the example in Figure 2. Briefly, the automaton is constructed by starting with the initial states of the given automaton as singleton states and successively determining how these sets are transformed by the action of the letters.

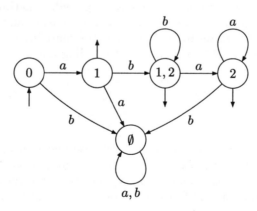

Figure 2. The powerset construction on the automaton of Figure 2

The powerset construction thus provides another means to show that recognizable languages are recognized by finite complete deterministic automata. Note that exchanging final and non-final states, a complete deterministic automaton $\mathcal{A}$ with alphabet A is turned into another automaton which recognizes the language $A^+ \setminus L(\mathcal{A})$ and so the set of recognizable languages of A^+ is closed under complementation.

If $\mathcal{A}$ is a deterministic automaton, if necessary one may change $\mathcal{A}$ to avoid that the initial state be in the image of the transformation associated with a letter, that is, in the graphical representation, to avoid any edge leading to the initial state: just add a new initial state q_0', unmarking the old initial state q_0 as such, and add an edge $q_0' \to q$ labeled a whenever there is an edge $q_0 \to q$ with that label. This procedure does not change the language recognized by the automaton.

If $\mathcal{A}$ and $\mathcal{B}$ are deterministic automata whose initial states have no edges leading to them and the two initial states are identified, we obtain a non-deterministic automaton recognizing $L(\mathcal{A}) \cup L(\mathcal{B})$. On the other hand, if we modify the automaton $\mathcal{A}$ by adding an edge $q_1 \to q_0$ labeled a whenever there is an a-labeled edge $q_1 \to q_2$ into a final state q_2, the resulting automaton recognizes $L(\mathcal{A})^+$. Similarly, if in this construction we use the initial state of $\mathcal{B}$ instead of the initial state of $\mathcal{A}$, the resulting automaton recognizes the concatenation $L(\mathcal{A}) L(\mathcal{B})$. Hence the set of all recognizable languages is also closed under union, concatenation, and the +-operation.

A *rational* subset of a semigroup S is a member of the smallest set of subsets of S which contains the empty set and the singleton subsets, and is closed under union, subset product, and taking the generated subsemigroup. By observing that the empty language and one letter languages are obviously recognizable, the above shows that every rational language over a finite alphabet is recognizable. Kleene [32] showed that the converse is also true, so that the rational and the recognizable subsets of a finitely generated free semigroup are the same.

For a rational language L over a finite alphabet, the above describes an effective procedure to build a finite complete deterministic automaton recognizing L. There are many computer implementations of such procedures (such as AMoRe [36]).

The syntactic congruence ρ_L may be computed from a complete trim deterministic automaton $\mathcal{A}$ recognizing L by a procedure called *minimization*. First, identify any two states from which no final state is accessible. Next, if there are two states q_1 and q_2 such that the edges leading from them into other states have the same labels, then one may identify the states q_1 and q_2 without affecting the language recognized by the automaton but enlarging the transition congruence. Starting with a finite automaton and iterating this procedure, in a number of steps which is bounded by the number of states of the given automaton one obtains an automaton $\mathcal{A}'$ for which no further identifications are possible. By construction, the automaton $\mathcal{A}'$ has the following properties:

- there is at most one state from which no final state is accessible, which is called a *sink state*;

- if two states q_1 and q_2 are such that the labels of paths leading from q_1, respectively q_2, into final states are the same, then $q_1 = q_2$.

Since the transition congruence of any automaton recognizing L saturates L, the transition congruence ρ of $\mathcal{A}'$ is contained in the syntactic congruence ρ_L. If the containment $\rho \subseteq \rho_L$ were strict, then there would be two words w_1 and w_2 such that $(w_1, w_2) \in \rho_L \setminus \rho$. Hence there would be a state q such that the states $q_1 = qw_1$ and $q_2 = qw_2$ do not coincide. Let u be a (perhaps empty) word such that $q_0 u = q$ where q_0 is the initial state. If v is a (perhaps empty) word such that $q_1 v$ is a final state, then $u w_1 v \in L$ and, since w_1 and w_2 are ρ_L-equivalent, we also have $u w_2 v \in L$. Hence the labels of paths leading from q_1, respectively q_2, into final states are the same, which by the above leads to the contradiction $q_1 = q_2$. Hence ρ is the syntactic congruence of L and the correspondence $w/\rho_L \mapsto q_0 w$ defines an isomorphism between the Cayley automaton of the syntactic semigroup of L (with final states the congruence classes of elements of L) and $\mathcal{A}'$ in the sense that the correspondence is a bijection sending initial state to initial state and final states onto final states, and respecting transitions. This justifies calling $\mathcal{A}'$ the *minimal automaton* of L.

Thus one may hope to be able to solve combinatorial problems on rational languages over finite alphabets by studying their syntactic semigroups. Indeed, a number of classical results showed that this is possible such as Schützenberger's characterization of +-free languages (drop the +-operation in the construction of languages but add complementation) as those whose syntactic semigroups are finite and have only trivial subgroups [56] (such semigroups are called *aperiodic*). Eilenberg [22] found a framework to describe just which properties of rational languages can be characterized by properties of their syntactical semigroups. This leads to the notion of a pseudovariety and into the next section.

3   Pseudovarieties

A semigroup S is said to *divide* a semigroup T and we write $S \prec T$ if S is a homomorphic image of a subsemigroup of T. If a congruence ρ saturates a subset L of a semigroup S, then so does every congruence contained in ρ. Hence, if a finite semigroup S recognizes a rational language L, then so does every finite semigroup which S divides. Using the results of Section 2 it follows that a finite semigroup S recognizes a rational language L if and only if the syntactic semigroup of L divides S. Call a semigroup *syntactic* if it is the syntactic semigroup of some rational language over a finite alphabet. Note that a finite semigroup is syntactic if and only if it is has a *disjunctive* subset, that is a subset whose syntactic congruence is equality. In particular, the syntactic semigroup of a subset of a finite semigroup is a syntactic semigroup.

On the other hand, if S is a finite semigroup then one can find syntactic semigroups $S_1, \ldots, S_n$ dividing S such that S divides $S_1 \times \cdots \times S_n$. To prove this, consider for each $s \in S$ the syntactic semigroup $T_s = \text{Synt}(\{s\})$ and the natural quotient homomorphism $\psi_s : S \to T_s$. Then $T_s \prec S$ and the homomorphisms ψ_s ($s \in S$) induce a homomorphism $\psi : S \to \prod_{s \in S} T_s$ which is injective: if s_1, s_2 are identified by ψ, then $\psi_{s_1}(s_1) = \psi_{s_1}(s_2)$ which shows that $s_1 = s_2$ since the syntactic congruence of $\{s_1\}$ saturates this singleton set.

Call a class of finite semigroups a *pseudovariety* if it is closed under taking divisors and finite direct products. The preceding paragraph proves that every pseudovariety is generated by its syntactic semigroups.

Associate with a pseudovariety V the collection $\mathcal{V}$ of all rational languages over finite alphabets whose syntactic semigroups belong to V. More specifically, denote by $\mathcal{V}A^+$ the set of all rational languages L over the finite alphabet A such that $\text{Synt}(L) \in \mathsf{V}$. By the results of Section 2, the collection $\mathcal{V}A^+$ is a *field of subsets* of A^+, that is a subset of $\mathcal{P}(A^+)$ which contains A^+ and is closed under complementation and finitary union. Moreover $\mathcal{V}$ is closed under the operation $L \mapsto \varphi^{-1}(L)$ (*inverse homomorphism*) for homomorphisms $\varphi : A^+ \to B^+$ between finitely generated free semigroups in the sense that $L \in \mathcal{V}B^+$ implies $\varphi^{-1}(L) \in \mathcal{V}A^+$. For languages K and L over an alphabet A, denote by $K^{-1}L$ (resp. LK^{-1}) the set of all words $u \in A^+$ such that there exists $v \in K$ such that $vu \in L$ (resp. $uv \in L$). Since $K^{-1}L$ and LK^{-1} are saturated by every congruence which saturates L, $\mathcal{V}A^+$ is in particular closed under the operations $L \mapsto a^{-1}L$ and $L \mapsto La^{-1}$ (*left* and *right cancellation*) for every letter $a \in A$. A *variety of languages* is a collection $\mathcal{V}$ of rational languages over finite alphabets with the above properties, namely:

1. each $\mathcal{V}A^+$ is a field of subsets of A^+;

2. each $\mathcal{V}A^+$ is closed under left and right cancellation;

3. $\mathcal{V}$ is closed under inverse homomorphism.

To each variety of languages $\mathcal{V}$ one may associate the pseudovariety generated by the syntactic semigroups of members of V. Eilenberg's correspondence theorem [22] states that this correspondence is the inverse of the earlier correspondence $V \mapsto \mathcal{V}$. The proof is quite elementary but would distract us too much from the main aim of these lectures. A complete presentation as well as various extensions may be found in [4, Sections 3.1–3.3].

In the late 1970's and early 1980's it became a research program to exhibit specific instances of Eilenberg's correspondence. Many such examples may be found in the books [22, 33, 42] including classical examples which predate the discovery of Eilenberg's correspondence such as Schützenberger's theorem characterizing $+$-free languages as those whose syntactic semigroups are finite aperiodic. This work was quite useful and led to many interesting problems and results. Nowadays it is not so common to find such results.

In spite of what was stated above, Eilenberg's varieties of languages do not exhaust the framework in which finite semigroups can be useful in studying combinatorial properties of collections of rational languages. Pin [45] has shown that adding further structure, namely order structure, to finite semigroups, finer syntactic properties of rational languages may be classified and most of the theory has been extended to this order context. A further enriching of structure has been considered by Polák [48] leading to a characterization of yet more general syntactic properties.

4 Pro-V algebras

In this section and several of the following we step out to a more general context by considering a (fixed) finitary algebraic signature σ and a pseudovariety V of σ-algebras.

By a *finitary algebraic signature* we mean a set σ of operation symbols together with an arity function with domain σ and image a set of finite sets. The *arity* of an operation symbol is its value under σ. This may appear to be in contrast with the usual definition, according to which arities are natural numbers. In the following it will be convenient to deal with arities as sets and, in any case, as in set theory, natural numbers may be themselves regarded as sets. Usually, the arity will be understood and we will refer simply to the signature σ.

A σ-*algebra* is a nonempty set S together with an *interpretation function* $o \in \sigma \mapsto o_S$ assigning to each operation symbol in σ an operation $o_S : S^A \to S$

where A is the arity of o and S^A is viewed as the set of all mappings $A \to S$. Again, usually the interpretation function will be understood and we will refer to the σ-algebra S. Whenever we refer to algebras without specifying their signature we mean σ-algebras.

A *homomorphism* of σ-algebras is a mapping $\varphi : S \to T$ such that, for every operation symbol $o \in \sigma$, the following diagram commutes, where A is the arity of o:

$$\begin{array}{ccc} S^A & \xrightarrow{o_S} & S \\ {\scriptstyle \varphi o\_} \downarrow & & \downarrow {\scriptstyle \varphi} \\ T^A & \xrightarrow{o_T} & T \end{array}$$

Given two σ-algebras S and T such that $S \subseteq T$ and the inclusion mapping $S \hookrightarrow T$ is a homomorphism, we say that S is a *σ-subalgebra* of T. A *divisor* of a σ-algebra T is a σ-algebra S such that S is a homomorphic image of a subalgebra of T. The Cartesian product $\prod_{i \in I} S_i$ is endowed with a natural structure of σ-algebra by letting, for each operation symbol $o \in \sigma$, of arity A, and $\varphi \in (\prod_{i \in I} S_i)^A$ with components $\varphi_i \in S_i^A$,

$$o_{\prod_{i \in I} S_i}(\varphi)(i) = o_{S_i}(\varphi_i).$$

A *pseudovariety* of σ-algebras is a class of finite σ-algebras which is closed under taking divisors and finite direct products.

A *topological algebra* is an algebra S endowed with a topology such that the interpretations of operations from σ are continuous. If the topology is also compact (for which we require Hausdorff's separation axiom), then we say that S is a *compact algebra*. Finite σ-algebras are viewed as compact algebras by endowing them with the discrete topology. By a *pro-V algebra* we mean a compact algebra S which is *residually in* V in the sense that, for all distinct $s_1, s_2 \in S$, there exists a continuous homomorphism $\varphi : S \to T$ into a member of V such that $\varphi(s_1) \neq \varphi(s_2)$. In case V consists of all finite algebras, a pro-V algebra is said to be a *profinite algebra*. Since homomorphisms preserve identities, if a compact algebra is profinite, then it is pro-V for the pseudovariety V of all finite algebras which satisfy all the σ-identities which are valid in S.

Of course, members of V are pro-V algebras. They are in fact precisely the finite pro-V algebras since a finite pro-V algebra embeds in a finite direct product of members of V. Indeed, for a pro-V algebra S, consider for each two-element subset $\{s_1, s_2\}$ a continuous homomorphism $\varphi_{s_1, s_2} : S \to S_{s_1, s_2}$

into an algebra from V such that $\varphi_{s_1,s_2}(s_1) \neq \varphi_{s_1,s_2}(s_2)$. Then the mapping

$$\varphi : S \to \prod_{\substack{s_1,s_2 \in S \\ s_1 \neq s_2}} S_{s_1,s_2}$$

whose (s_1, s_2)-component is φ_{s_1,s_2} is an injective homomorphism. In particular, if S is finite, then so is the product and, therefore, $S \in \mathsf{V}$.

Say that a topological space is *zero-dimensional* if it has a basis for its topology consisting of clopen (that is both closed and open) subsets and say that it is *totally disconnected* if its connected components are singletons. It is well known that, for compact spaces, zero-dimensionality and total disconnectedness are equivalent properties. Finite algebras are obviously totally disconnected and therefore so are pro-V algebras since continuous homomorphisms from the latter into members of V suffice to separate points.

Say that a topological space X has the *initial topology* for a family of mappings with domain X into certain topological spaces if the topology of X is the least topology on X with respect to which all the mappings in the family are continuous.

Lemma 4.1. *Let S be a pro-V algebra. Then every clopen subset K of S is such that $K = \varphi^{-1}\varphi(K)$ for some continuous homomorphism $\varphi : S \to T$ into a member of V. In particular, S has the initial topology for the continuous homomorphisms into members of V.*

Proof. For a clopen subset K of S, if $K = S$ then the result is obvious. Otherwise, each $s \in K$ may be separated from each $u \in S \setminus K$ by a continuous homomorphism $\varphi_{s,u} : S \to T_{s,u}$ into a member of V. For a fixed $s \in K$, the open sets $\varphi^{-1}\varphi(u)$ with $u \in S \setminus K$ cover the compact set $S \setminus K$ and therefore a finite number of them, say with $u \in \{u_1, \ldots, u_m\}$, suffice to cover $S \setminus K$. Consider the direct product $T_s = T_{s,u_1} \times \cdots \times T_{s,u_m}$ and the continuous homomorphism $\varphi_s : S \to T_s$ whose components are the φ_{s,u_i}. By construction, $\varphi_s^{-1}\varphi_s(s)$ is an open set contained in K. Hence we get this time a cover $\varphi_s^{-1}\varphi_s(s)$ ($s \in K$) of the compact set K consisting of open subsets of K which must contain a finite cover, say for $s \in \{s_1, \ldots, s_n\}$. Then the continuous homomorphism $\varphi : S \to T_{s_1} \times \cdots \times T_{s_n}$ whose components are the φ_{s_i} satisfies the desired condition $K = \varphi^{-1}\varphi(K)$. □

We say that a topological algebra S is *generated* by a set A if a mapping $\iota : A \to S$ is given whose image generates a dense subalgebra of S. As in the discrete semigroup case, the mapping ι will sometimes be called the *generating mapping*.

In some special cases such as semigroups, the separation property in the definition of profinite algebra may be replaced by zero-dimensionality. The following result is due to Numakura [40]. The proof presented here is due to Hunter [29]. See Almeida [2] and the original paper for other situations under which a similar phenomenon occurs. For the proof, we assume the reader is familiar with nets and their significance in topology (see, for instance, [64, Chapter 4]). In particular, recall that a directed set is a partially ordered set I such that, for all $i, j \in I$, there exists $k \in I$ satisfying $k \geq i$ and $k \geq j$.

Proposition 4.2. *A compact semigroup is profinite if and only if it is zero-dimensional.*

Proof. Let S be a compact zero-dimensional semigroup and consider two distinct points of S, s_1 and s_2. Since S is a Hausdorff zero-dimensional space, there is some clopen subset L of S which contains s_1 but not s_2. Consider the syntactic congruence ρ_L of L. If we show that ρ_L has open classes then it must have finite index and, therefore, the natural homomorphism $S \to \mathrm{Synt}(L)$ is a continuous homomorphism onto a finite semigroup which separates s_1 and s_2. Hence it remains to show that each ρ_L-class is open.

Let then $(s_i)_{i \in I}$ be a convergent net in S with limit s, where I is a directed set. We must show that for every $i_0 \in I$ there exists $i \in I$ such that $i \geq i_0$ and s_i lies in the same ρ_L-class as s. Otherwise, there would be a subnet $(s_j)_{j \in J}$ consisting of elements outside the ρ_L-class of s. Hence there are nets $(u_j)_{j \in J}$ and $(v_j)_{j \in J}$ in S^1 such that, for every $j \in J$, one of the products $u_j s_j v_j$ and $u_j s v_j$, but not the other, lies in L. Since S is compact, we may assume that the nets $(u_j)_{j \in J}$ and $(v_j)_{j \in J}$ converge, say to limits $u, v \in S^1$, respectively. Since the semigroup multiplication is continuous, the nets $(u_j s_j v_j)_{j \in J}$ and $(u_j s v_j)_{j \in J}$ both converge to usv. Since both L and its complement are closed, we conclude that usv is simultaneously in L and in its complement, which is absurd. Hence the ρ_L-classes are open. □

Various algebraic-topological constructions preserve the property of being a pro-V algebra.

Proposition 4.3. *The class of pro-V algebras enjoys the following closure properties:*

1. *The direct product of pro-V algebras is again a pro-V algebra.*

2. *A closed subalgebra of a pro-V algebra is also a pro-V algebra under the induced topology.*

3. *A profinite continuous homomorphic image of a pro-V algebra is a pro-V algebra.*

Proof. The first closure property is an immediate consequence of the axiom of choice via Tychonoff's Theorem [64, Theorem 17.8]. The second property follows directly from the definitions.

For (3), consider a continuous homomorphism $\varphi : S \to T$ from a pro-V algebra onto a profinite algebra. We have to show that T is residually in V. Since T is residually in the class of all its finite continuous homomorphic images and a continuous homomorphic image of T is also a continuous homomorphic image of S, it suffices to show that, if T is finite, then T must lie in V.

The inverse images of points $\varphi^{-1}(t)$ with $t \in T$ provide a partition of S into clopen subsets. By Lemma 4.1 for each $t \in T$ there exists some continuous homomorphism $\psi_t : S \to U_t$ onto some member of V such that $\varphi^{-1}(t) = \psi_t^{-1}(\psi_t(\varphi^{-1}(t)))$. Consider the induced continuous homomorphism

$$\psi : S \to \prod_{t \in T} U_t$$
$$s \mapsto (\psi_t(s))_{t \in T}$$

Then $\varphi^{-1}(t) = \psi^{-1}(\psi(\varphi^{-1}(t)))$ holds for every $t \in T$. Hence the homomorphism φ factors through ψ by means of a homomorphism $\theta : \psi(S) \to T$ such that $\theta \circ \psi = \varphi$. Since φ is onto, it follows that T is a homomorphic image of a subalgebra of a finite direct product of members of V and, therefore, T belongs to V. □

The following examples show that it is not sufficient to replace "profinite" by "compact" in Proposition 4.3(3). Recall that the real interval $[0, 1] \subseteq \mathbb{R}$ is a continuous image of the Cantor set $C = \{0, 1\}^{\mathbb{N}}$ but it is not zero-dimensional. This is an elementary exercise in real analysis; more generally, every compact metric space is a continuous image of C [64, Theorem 30.7]. Thus, taking the empty algebraic signature, we see that a continuous homomorphic image of a profinite algebra is not necessarily profinite. One may add more structure say by taking left-zero semigroups, that is by adding the binary operation $xy = x$. Indeed, any compact zero-dimensional space is a profinite semigroup under this operation and any mapping between two left-zero semigroups is a homomorphism.

The situation for groups is simpler, basically because of their high symmetry. Recall that the topology in a topological group is completely characterized by its open subgroups. A compact group is profinite if and only if its open subgroups are all of finite index. Now, if $\varphi : G \to H$ is a continuous homomorphism from a profinite group G onto a topological group H, then certainly H is a compact group. Moreover, if K is an open sibgroup of H, then $\varphi^{-1}K$ is an open subgroup of G and therefore it has finite index in G, so that so is the index of K in H. Hence H is a profinite group. For the rich theory of profinite groups, see [54].

An alternative characterization of pro-V algebras which is quite useful is as projective limits of directed systems of homomorphisms between members of V. Such a system consists of a directed set I together with two families $(S_i)_{i \in I}$ and $(\varphi_{i,j})_{i,j \in I;\ i \leq j}$ where each $S_i \in$ V and each $\varphi_{i,j} : S_i \leftarrow S_j$ is a homomorphism such that $\varphi_{i,i}$ is the identity on S_i and $\varphi_{i,j} \circ \varphi_{j,k} = \varphi_{i,k}$ whenever $i \leq j \leq k$. The corresponding *projective limit* is the subalgebra S of the direct product $\prod_{i \in I} S_i$ consisting of all those $(s_i)_{i \in I}$ such that $s_i \in S_i$ and $\varphi_{i,j}(s_j) = s_i$ whenever $i \leq j$. Note that, since each condition $\varphi_{i,j}(s_j) = s_i$ only involves two components in the product, it defines a closed subset of the product $\prod_{i \in I} S_i$. Moreover, by compactness of $\prod_{i \in I} S_i$, these closed subsets have the finite intersection property if we assume that the $\varphi_{i,j}$ are onto, as we do from hereon, so that the projective limit is nonempty. Hence the projective limit S is a closed subalgebra of the direct product. By earlier observations, we conclude that S is a pro-V algebra. Any topological algebra which is isomorphic, as a topological algebra, to a projective limit, will also be called a projective limit.

Conversely, suppose S is a pro-V algebra. Take a set $\mathcal{S}$ of representatives of isomorphism classes of members of V and consider the set I of all onto continuous homomorphisms $\varphi : S \to T$ with $T \in \mathcal{S}$ and order them by letting $\varphi \leq \psi$ for another continuous homomorphism $\psi : S \to U$ if there is a homomorphism $\theta_{\varphi,\psi} : T \leftarrow U$ such that $\theta_{\varphi,\psi} \circ \psi = \varphi$. Note that I is a directed set: two onto continuous homomorphisms $\varphi : S \to T$ and $\psi : S \to U$ induce a continuous homomorphism $\lambda : S \to T \times U$ in which if we replace the direct product $T \times U$ by a member of $\mathcal{S}$ isomorphic to the image of λ we obtain a member of I which is above both φ and ψ. We thus obtain a directed system of homomorphisms between members of V. Let S' be its projective limit. We claim that S' and S are isomorphic as topological algebras. The homomorphisms $\varphi : S \to T$ in I induce a continuous homomorphism Φ from S into the direct product of all the T's which by construction takes its values in S'. Since S is compact, Φ is a closed mapping. Since S is residually in V, Φ is certainly injective.

It remains to show that Φ is onto. Given $s' = (t_\varphi)_{\varphi \in I}$ in S', for each $\varphi \in I$ the closed set $\varphi^{-1}(t_\varphi)$ is non-empty. The fact that the set I is directed and the given family belongs to S' implies that any finite intersection of such closed subsets is still non-empty. By compactness of S, we deduce that there is some $s \in \bigcap_{\varphi \in I} \varphi^{-1}(t_\varphi)$ and for such s we have $\Phi(s) = s'$. Hence Φ is indeed onto. We have thus established the following result.

Proposition 4.4. *A topological σ-algebra is pro-V if and only if it is a projective limit of algebras of V.*

Given a set A, we say that a pro-V algebra S is a *free pro-V algebra over A* if a mapping $\iota : A \to S$ is given such that the following universal property holds: for every mapping $\varphi : A \to T$ into another pro-V algebra T there exists a unique continuous homomorphism $\hat{\varphi} : S \to T$ such that $\hat{\varphi} \circ \iota = \varphi$, that is such that the following diagram commutes:

By the usual abstract nonsense, up to isomorphism of topological algebras there is no more than one such free pro-V algebra over A. We denote it by $\overline{\Omega}_A \mathsf{V}$.

To prove the existence of $\overline{\Omega}_A \mathsf{V}$, the easiest way is to construct an appropriate projective limit. Suppose for a moment that $\overline{\Omega}_A \mathsf{V}$ exists. Since the closed subalgebra of $\overline{\Omega}_A \mathsf{V}$ generated by the image of ι retains the required universal property, we conclude that $\overline{\Omega}_A \mathsf{V}$ is A-generated and the universal property implies that every A-generated pro-V algebra is a continuous homomorphic image of $\overline{\Omega}_A \mathsf{V}$. This suggests considering the projective limit F_A of all A-generated members of $\mathcal{S}$ under connecting homomorphisms which respect the choice of generators. The pro-V algebra F_A comes naturally endowed with a mapping $\iota : A \to F_A$ which projects on each component as the given choice of generators on that component.

It remains to show that F_A has the above universal property. So, consider a mapping $\varphi : A \to S$ into a pro-V algebra S. We may as well assume that the image of φ generates a dense subalgebra of S. Then S may further be assumed to be a projective limit of A-generated members of $\mathcal{S}$ built from a subset of all those that make up the directed family defining F_A. This implies that there is a natural continuous homomorphism $\hat{\varphi} : F_A \to S$, obtained by forgetting all other components, which also respects the choice of generators, that is such that $\hat{\varphi} \circ \iota = \varphi$, which verifies the desired universal property. This establishes the following result.

Proposition 4.5. *For every set A and every pseudovariety V, there exists a free pro-V algebra $\overline{\Omega}_A V$.*

Let A be a set and let B be a subset of A. Consider the free pro-V algebras respectively on A and B and their generating mappings $\iota : A \to \overline{\Omega}_A V$ and $\kappa : B \to \overline{\Omega}_B V$. Then we may complete the following diagram in a unique way into a commutative square by adding a continuous homomorphism φ:

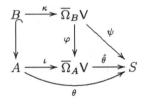

Proposition 4.6. *The mapping φ is injective.*

Proof. Suppose that there are distinct $u, v \in \overline{\Omega}_B V$ such that $\varphi(u) = \varphi(v)$. Since $\overline{\Omega}_B V$ is residually in V, there exists a continuous homomorphism $\psi : \overline{\Omega}_B V \to S$ into some member of V such that $\psi(u) \neq \psi(v)$. Let $\theta : A \to S$ be any mapping extending $\psi \circ \kappa$ and let $\hat{\theta}$ be its unique extension to a continuous homomorphism $\overline{\Omega}_A V \to S$. We then have the following commutative diagram

$$\begin{array}{ccc} B & \xrightarrow{\kappa} & \overline{\Omega}_B V \\ \downarrow & \varphi \searrow & \downarrow \psi \\ A & \xrightarrow{\iota} \overline{\Omega}_A V \xrightarrow{\hat{\theta}} & S \end{array}$$

where the relation $\hat{\theta} \circ \varphi = \psi$ comes from the facts that the composites of both sides with κ are equal to $\theta|_B$, both sides are continuous homomorphisms, and κ is a generating mapping. Since ψ distinguishes u and v, it follows that so must φ. $\square$

5 V-free algebras and their recognizable subsets

Let $\iota : A \to \overline{\Omega}_A V$ be a mapping giving $\overline{\Omega}_A V$ as a free pro-V algebra over the set A. Within $\overline{\Omega}_A V$ lives the subalgebra generated by $\iota(A)$ which is denoted $\Omega_A V$ and which is a dense subalgebra of $\overline{\Omega}_A V$. Given any mapping $\varphi : A \to S$ into a member of V, the universal property of $\overline{\Omega}_A V$ provides a continuous homomorphism $\hat{\varphi} : \overline{\Omega}_A V \to S$. Its restriction ψ to $\Omega_A V$ is therefore a homomorphism $\Omega_A V \to S$ such that $\psi \circ \iota = \varphi$. Moreover, from the uniqueness of

$\hat{\varphi}$ and the fact that $\Omega_A V$ is generated by $\iota(A)$ it follows that ψ is the unique homomorphism such that the following diagram commutes:

For this reason we say that $\Omega_A V$ is a V-*free σ-algebra over A*. For later reference, we state the following result.

Proposition 5.1. *The σ-subalgebra of $\overline{\Omega}_A V$ generated by the image of ι is a V-free σ-algebra over A.*

In the spirit of Section 2, we say that a subset L of a σ-algebra S is V-*recognizable* if there is a homomorphism $\varphi : S \to T$ into a member of V such that $L = \varphi^{-1}\varphi(L)$. Our next aim is to show that the topological structure of $\overline{\Omega}_A V$ is just a structural encoding of the combinatorics of V-recognizable subsets of $\Omega_A V$.

Let L be a V-recognizable subset of $\Omega_A V$. Then there exists a homomorphism $\varphi : \Omega_A V \to S$ into a member of V such that $L = \varphi^{-1}\varphi(L)$. By the universal property of $\overline{\Omega}_A V$ we obtain a continuous homomorphic extension of φ to $\hat{\varphi} : \overline{\Omega}_A V \to S$. By continuity, the set $K = \hat{\varphi}^{-1}\hat{\varphi}(L)$ is clopen. Since $\hat{\varphi}$ is an extension of φ, we have $K \cap \Omega_A V = L$. Moreover, since $\Omega_A V$ is dense in $\overline{\Omega}_A V$ and K is open, L is dense in K.

Conversely, let L be a subset of $\Omega_A V$ and suppose that there is a clopen subset K of $\overline{\Omega}_A V$ such that $L = K \cap \Omega_A V$. By Lemma 4.1 there exists a continuous homomorphism $\varphi : \overline{\Omega}_A V \to S$ into some member of V such that $K = \varphi^{-1}\varphi(K)$. Let ψ be the restriction of φ to $\Omega_A V$. Then we have the following chain of inclusions

$$L \subseteq \psi^{-1}\psi(L) \subseteq \varphi^{-1}\varphi(K) \cap \Omega_A V = K \cap \Omega_A V = L$$

which proves that L is V-recognizable. This establishes the following result.

Proposition 5.2. *The following conditions are equivalent for a subset L of $\Omega_A V$:*

1. *L is V-recognizable;*

2. *$L = K \cap \Omega_A V$ for some clopen subset K of $\Omega_A V$;*

3. *the closure K of L in $\overline{\Omega}_A V$ is open and $L = K \cap \Omega_A V$.*

Recall that, by definition, the rational languages over a given finite alphabet A in any variety of languages $\mathcal{V}$ form a field of subsets of A^+. Often these fields of subsets are described by giving generating sets. This condition has a natural topological formulation in terms of relatively free profinite algebras. Note that an immediate consequence of the definition is that the V-recognizable subsets of $\Omega_A \mathsf{V}$ form a field of subsets since the collection of them is closed under union and complement.

Proposition 5.3. *The following conditions are equivalent for a family $\mathcal{F}$ of V-recognizable subsets of $\Omega_A \mathsf{V}$:*

1. *$\mathcal{F}$ generates the field of all V-recognizable subsets of $\Omega_A \mathsf{V}$;*

2. *the closures in $\overline{\Omega}_A \mathsf{V}$ of the elements of $\mathcal{F}$ generate the topology of $\overline{\Omega}_A \mathsf{V}$;*

3. *the closures in $\overline{\Omega}_A \mathsf{V}$ of the elements of $\mathcal{F}$ suffice to separate any two distinct points.*

Proof. Since $\overline{\Omega}_A \mathsf{V}$ is compact zero-dimensional, the clopen subsets form a field of subsets. By Proposition 5.2, they are precisely the closures of V-recognizable subsets of $\Omega_A \mathsf{V}$. The equivalence between conditions (1) and (2) follows easily. The equivalence between conditions (2) and (3) is true for every pro-V algebra: the argument in the proof of Lemma 4.1 proves the backward direction whereas the forward direction follows from zero-dimensionality of pro-V algebras. □

6 Implicit operations and pseudoidentities

Each element $w \in \overline{\Omega}_A \mathsf{V}$ has a *natural interpretation* as an A-ary operation $w_S : S^A \to S$ on each pro-V algebra S: given $\varphi \in S^A$, which is viewed here as a mapping $A \to S$, the universal property of $\overline{\Omega}_A \mathsf{V}$ provides a unique extension to a continuous homomorphism $\hat\varphi : \overline{\Omega}_A \mathsf{V} \to S$; define $w_S(\varphi)$ to be $\hat\varphi(w)$. An important property of this interpretation is that it commutes with any continuous homomorphism $\theta : S \to T$ between pro-V algebras in the sense that the following diagram commutes:

$$\begin{array}{ccc} S^A & \xrightarrow{w_S} & S \\ {\scriptstyle \theta \circ \_}\downarrow & & \downarrow{\scriptstyle \theta} \\ T^A & \xrightarrow{w_T} & T \end{array} \qquad (6.1)$$

To establish this property, consider $\varphi \in S^A$ and its continuous homomorphic extension $\hat\varphi : \overline{\Omega}_A \mathsf{V} \to S$. Then $\theta \circ \hat\varphi : \overline{\Omega}_A \mathsf{V} \to T$ is a continuous homomorphism whose restriction to A coincides with $\theta \circ \varphi$. Hence $w_T(\theta \circ \varphi) = \theta \circ \hat{\varphi}(w) = \theta \circ w_S(\varphi)$ which establishes commutativity of the diagram (6.1).

An A-ary operation w with an interpretation $w_S : S^A \to S$ on each member of V such that diagram (6.1) commutes for every homomorphism $\theta : S \to T$ is said to be an A-ary implicit operation on V. By the above, every element w of $\overline{\Omega}_A$V determines such an implicit operation. The fact that $\overline{\Omega}_A$V is residually in V implies that this correspondence in injective: if w and v are distinct elements of $\overline{\Omega}_A$V then there exists some homomorphism $\varphi : \overline{\Omega}_A$V $\to S$ into some $S \in$ V which distinguishes them, whence $w_S(\varphi|_A) \neq v_S(\varphi|_A)$.

Proposition 6.1. *The correspondence associating to each $w \in \overline{\Omega}_A$V the A-ary implicit operation on V defined by the natural interpretations $(w_S)_{S \in V}$ is a bijection. In particular, every implicit operation on V has a natural interpretation on every pro-V algebra.*

Proof. Let $o = (o_S)_{S \in V}$ be an A-ary implicit operation on V. By the proof of Proposition 4.5, $\overline{\Omega}_A$V may be represented as a projective limit of a family $(S_i)_{i \in I}$ of A-generated members of V connected by homomorphisms $\varphi_{i,j} : S_i \leftarrow S_j$ such that every A-generated member of V is isomorphic to one of the S_i.

For each $i \in I$, we have a mapping $\iota_i : A \to S_i$ which gives S_i as an A-generated algebra and such that $\varphi_{i,j} \circ \iota_j = \iota_i$ whenever $i \leq j$. Let $s_i = o_{S_i}(\iota_i) \in S_i$. Since o is an implicit operation, we have

$$\varphi_{i,j}(s_j) = \varphi_{i,j}(o_{S_j}(\iota_j)) = o_{S_i}(\varphi_{i,j}(\iota_j)) = o_{S_i}(\iota_i) = s_i.$$

Hence the family $w = (s_i)_{i \in I}$ is a member of the projective limit.

It remains to show that the natural interpretation w_S of w in each $S \in$ V coincides with o_S. Let $\psi \in S^A$. The image of ψ in S is an A-generated member of V and, therefore, it is isomorphic to one of the S_i. Since both w and o are implicit operations, we may as well assume that $S = S_i$ and $\psi = \iota_i$. By construction, the unique continuous homomorphic extension $\hat{\iota}_i : \overline{\Omega}_A$V $\to S_i$ is precisely the component projection for the projective limit. Hence $w_S(\psi) = s_i = o_S(\psi)$, which completes the proof. □

From hereon, we will identify A-ary implicit operations on V with members of $\overline{\Omega}_A$V.

A further property of implicit operations which is worth registering at this time is their continuity.

Proposition 6.2. *The natural interpretation of a finitary implicit operation on a pro-V algebra is continuous.*

Proof. Let w be an A-ary implicit operation on V for some finite set A and let S be a pro-V algebra. By Lemma 4.1, S has the initial topology for continuous homomorphisms into members of V. So, let $\theta : S \to T$ be a continuous homomorphism into a member of V and let $t \in T$. By commutativity of the diagram (6.1), the inverse image X under w_S of the basic open set $\theta^{-1}(t)$ coincides with $\zeta^{-1}(w_T^{-1}(t))$ where $\zeta(\varphi) = \theta \circ \varphi$ is a continuous function since θ is continuous and A is finite. Hence X is open and w_S is continuous. □

The simplest examples of implicit operations are the *component projections*: for $a \in A$, x_a denotes the operation such that $(x_a)_S(\varphi) = \varphi(a)$ whenever $\varphi \in S^A$. As a member of $\overline{\Omega}_A\mathsf{V}$, the operation x_a is precisely the image of a under the generating mapping $A \to \overline{\Omega}_A\mathsf{V}$. The subalgebra of $\overline{\Omega}_A\mathsf{V}$ generated by the x_a ($a \in A$) is therefore the free V-algebra $\Omega_A\mathsf{V}$. Thus the members of $\Omega_A\mathsf{V}$ are constructed from the component projections by applying the operations from σ. For this reason they are said to be *explicit operations*.

More generally, it is worth observing the way implicit operations compose. In this context the arity notation, which has been rather convenient so far, obscures what is going on. So we temporarily change to component notation as we will do elsewhere whenever convenient. For an operation w of finite arity A, if we fix an ordering $a_1, \ldots, a_n$ for the elements of A, we also write $w = w(a_1, \ldots, a_n)$. If w has an interpretation on S, $\varphi \in S^A$, and $s_i = \varphi(a_i)$, then we also denote $w_S(\varphi)$ by $w_S(s_1, \ldots, s_n)$.

Suppose now that $w \in \overline{\Omega}_A\mathsf{V}$, let B be another set, and let $v_1, \ldots, v_n \in \overline{\Omega}_B\mathsf{V}$. Then we have a *composite* implicit operation $u = w_{\overline{\Omega}_B\mathsf{V}}(v_1, \ldots, v_n)$. The interpretation of this operation on a member S of V is the following. For $\psi \in S^B$, consider the following commutative diagram, where ι gives $\overline{\Omega}_B\mathsf{V}$ as a free pro-V algebra over the generating set B:

$$\begin{array}{ccccc} (\overline{\Omega}_B\mathsf{V})^A & \xrightarrow{w_{\overline{\Omega}_B\mathsf{V}}} & \overline{\Omega}_B\mathsf{V} & \xleftarrow{\iota} & B \\ {\scriptstyle \hat\psi \circ -}\downarrow & & \downarrow{\scriptstyle \hat\psi} & \swarrow{\scriptstyle \psi} & \\ S^A & \xrightarrow{w_S} & S & & \end{array}$$

By definition of v_S and the commutativity of the diagram we obtain the following equalities:

$$\begin{aligned} u_S(\psi) &= \hat\psi(u) = \hat\psi(w_{\overline{\Omega}_B\mathsf{V}}(v_1, \ldots, v_n)) \\ &= w_S(\hat\psi(v_1), \ldots, \hat\psi(v_n)) \\ &= w_S((v_1)_S(\psi), \ldots, (v_n)_S(\psi)), \end{aligned}$$

which says that the composite implicit operation is obtained on each member of V by likewise composing the natural interpretations of the intervening operations.

Let W be a pseudovariety contained in V. Then, by restricting implicit operations on V to W we obtain a mapping $p_W : \overline{\Omega}_A V \to \overline{\Omega}_A W$ which we call the *natural projection*. We claim p_W is an onto continuous homomorphism. Consider the two mappings $\iota_V : A \to \overline{\Omega}_A V$ and $\iota_W : A \to \overline{\Omega}_A W$ giving $\overline{\Omega}_A V$ and $\overline{\Omega}_A W$ as free respectively pro-V and pro-W algebras over the set A. Then there is a unique continuous homomorphism $\widehat{\iota_W} : \overline{\Omega}_A V \to \overline{\Omega}_A W$ such that $\widehat{\iota_W} \circ \iota_V = \iota_W$. Since the image of ι_W generates a dense subalgebra of $\overline{\Omega}_A W$, $\widehat{\iota_W}$ is onto. Thus it suffices to show $\widehat{\iota_W}$ is the natural projection. Now, given a mapping $\varphi : A \to S$ into a member of V, let $\hat{\varphi} : \overline{\Omega}_A W \to S$ be its unique extension to a continuous homomorphism such that $\hat{\varphi} \circ \iota_W = \varphi$. The following diagram will help keep track of the notation:

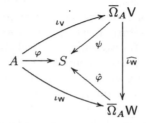

The diagram also makes it clear that the only continuous homomorphism $\psi : \overline{\Omega}_A V \to S$ such that $\psi \circ \iota_V = \varphi$ is the composite $\hat{\varphi} \circ \widehat{\iota_W}$. Hence, for $w \in \overline{\Omega}_A V$, we have

$$(\widehat{\iota_W}(w))_S(\varphi) = \hat{\varphi}(\widehat{\iota_W}(w)) = \psi(w)$$

which shows that indeed the mapping $\widehat{\iota_W}$ is the natural projection.

There is an incorrect claim in the literature concerning the continuity of the homomorphisms $\overline{\Omega}_A V \to S$ into members of V. Since the author is responsible for that claim, which originated in [1], it seems appropriate to show here that it is false. A very simple example is the following.

Example 6.1. Let S denote the pseudovariety of all finite semigroups and let $\psi : \overline{\Omega}_A S \to \{0,1\}$, where $A \neq \emptyset$, denote the characteristic function of $\Omega_A S$ whose value is 1 on an implicit operation if and only if the operation is explicit. Then ψ is a homomorphism into the 2-element multiplicative semilattice which is not continuous.

Proof. The statement that ψ is a homomorphism amounts to saying that the product of explicit operations is explicit, which is obvious by definition, and that the product of any non-explicit operation by any implicit operation is non-explicit. The latter property follows from the easily proved fact that an implicit operation on finite semigroups is explicit if and only if there is some finite commutative nilpotent semigroup in which it assumes a nonzero value. Since $\Omega_A \mathsf{S}$ is dense in $\overline{\Omega}_A \mathsf{S}$ and the topology in $\{0, 1\}$ is discrete, ψ is discontinuous. □

For the argument in the proof of the above counter-example to work it suffices that the pseudovariety contain all finite semilattices and all finite commutative nilpotent semigroups. The least such pseudovariety is described in [4, Exercise 9.1.4] as the pseudovariety of all finite commutative local semilattices.

In the realm of finite semigroups there are another two frequently encountered implicit operations which we now introduce. These are the unary operations x^ω and $x^{\omega-1}$. Their interpretations on a finite semigroup S are as follows: for $s \in S$, s^ω is the unique idempotent power of s and $s^{\omega-1}$ is the inverse of $s^{\omega+1} = s^\omega s$ in the maximal subgroup containing s^ω.

These two examples may be extended beyond profinite semigroups to compact semigroups as follows.

Proposition 6.3. *Let S be a compact semigroup, let $s \in S$, and let C be the closure of the subsemigroup generated by s. Then C possesses a unique idempotent.*

Proof. The intersection of all nonempty closed ideals of a compact semigroup is its minimal such subset. Since C is a compact commutative semigroup, it follows that its minimal closed ideal is a group and therefore contains an idempotent. On the other hand, if e is an idempotent in C, then e is the limit of some sequence s^{k_n} and so, either s itself is idempotent, in which case $C = \{s\}$; or there exists a convergent subsequence of $s^{k_n - 1}$, say to a limit t, in which case $st = e$ by continuity of multiplication in S, so that eC is a closed ideal and a group. Hence e is the unique idempotent in C. □

Thus compact semigroups are rather close to finite semigroups. Recall Green's relations on a semigroup S: for $s, t \in S$, write $s \leq_\mathcal{R} t$ (respectively $s \leq_\mathcal{L} t$, $s \leq_\mathcal{J} t$) if $s \in tS^1$ (respectively $s \in S^1 t$, $s \in S^1 t S^1$); for $\mathcal{K} \in \{\mathcal{J}, \mathcal{L}, \mathcal{R}\}$, write $s \mathcal{K} t$ if $s \leq_\mathcal{K} t$ and $t \leq_\mathcal{K} s$; the equivalence relations $\mathcal{R}$ and $\mathcal{L}$ commute and so their join, denoted $\mathcal{D}$, is their composite $\mathcal{R} \circ \mathcal{L}$; finally, $\mathcal{H} = \mathcal{R} \cap \mathcal{L}$. For instance, using Proposition 6.3 it is a simple exercise to show that compact semigroups are *stable* in the sense that there are no two

25

comparable $\mathcal{R}$-classes under the $\leq_\mathcal{R}$ ordering or comparable $\mathcal{L}$-classes under the $\leq_\mathcal{L}$ ordering in the same $\mathcal{J}$-class. In a stable semigroup, it is easy to show that Green's relations $\mathcal{J}$ and $\mathcal{D}$ coincide (cf. [33, Chapter 2]).

By a V-*pseudoidentity* we mean a formal equality $u = v$ between implicit operations over V of the same arity. The *arity* of a pseudoidentity is the common arity of its sides. When the two sides of the pseudoidentity are actually explicit operations, the pseudoidentity is called an *identity*. The pseudoidentity $u = v$ is said to be *satisfied by* a pro-V algebra S if $u_S = v_S$. In this context, the projections x_a are often called *variables*.

Given a set Σ of pseudoidentities, we denote by $[\![\Sigma]\!]$ the class of all members of V which satisfy all the pseudoidentities from Σ. From the fact that implicit operations commute with homomorphisms between members of V, it follows that $[\![\Sigma]\!]$ is a pseudovariety.

The following result was first proved by Reiterman [49]. There are many alternative proofs, including one by Higgins [28] which uses the Eilenberg-Schützenberger characterization of pseudovarieties by sequences of identities [23].

Theorem 6.4. *Every pseudovariety* W *contained in* V *is of the form* $[\![\Sigma]\!]$ *for some set* Σ *of pseudoidentities.*

Proof. Let Σ be the set of all V-pseudoidentities with finite arities contained in a given countable set X which are satisfied by all members of W. Then obviously W is contained in $[\![\Sigma]\!]$. For the converse, let $S \in [\![\Sigma]\!]$, which we may assume to be A-generated for some finite subset A of X, say through a mapping $\varphi : A \to S$. Let $\hat{\varphi} : \overline{\Omega}_A V \to S$ be its unique extension to a continuous homomorphism.

Consider the natural projection $p_W : \overline{\Omega}_A V \to \overline{\Omega}_A W$ introduced in Section 6. The hypothesis that S satisfies all pseudoidentities from Σ implies that $\hat{\varphi}$ factors through p_W, say $\hat{\varphi} = \psi \circ p_W$ for a homomorphism $\psi : \overline{\Omega}_A W \to S$, as in the following commutative diagram:

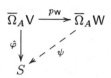

Since the continuous mapping p_W is actually closed, the homomorphism ψ is continuous and it is obviously onto. Hence, by Proposition 4.3(3) the finite algebra S is pro-W and, therefore, it lies in W. □

If $W = [\![\Sigma]\!]$, then we say that Σ is a *basis of pseudoidentities* of W. We also say that Σ is a *basis of pseudoidentities over* X of the pseudovariety W if Σ consists of pseudoidentities of arity X, $W \subseteq [\![\Sigma]\!]$ and every pseudoidentity of arity X which is valid in W is also valid in $[\![\Sigma]\!]$. Here are some examples again from the realm of semigroups or monoids.

1. The pseudovariety Sl of all finite semilattices is defined by the identities $xy = yx$ and $x^2 = x$.

2. The pseudovariety G of all finite groups is defined by the pseudoidentity $x^\omega = 1$ in the world of monoids. For semigroups, we usually write the same formal equality which should then be interpreted as an abbreviation of the pseudoidentities $yx^\omega = x^\omega y = y$ where y is a variable different from x.

3. The pseudovariety N of all finite nil semigroups is defined by the pseudoidentity $x^\omega = 0$. Again, strictly speaking, this is not a pseudoidentity of semigroups but rather of semigroups with zero. We see it as an abbreviation of the pseudoidentities of semigroups $yx^\omega = x^\omega y = x^\omega$.

4. The pseudovariety A of all finite aperiodic semigroups is defined by the pseudoidentity $x^{\omega+1} = x^\omega$.

5. The pseudovariety J of all finite $\mathcal{J}$-trivial semigroups is defined by the pseudoidentities $(xy)^\omega x = (xy)^\omega = y(xy)^\omega$.

For more examples see the end of Section 11.

7 The pro-V topology and the natural metric

Let S be a σ-algebra. We may endow S with the initial topology for the family of all homomorphisms $S \to T$ into members of V. This is called the *pro-V topology* of S. Note that this topology is Hausdorff if and only if S is residually in V as a plain σ-algebra, that is distinct points of S may be separated by homomorphisms into members of V.

Proposition 7.1. *The pro-V topology turns S into a topological σ-algebra.*

Proof. We must show that any basic operation w from σ is continuous on S. Let A be the (finite) arity of w and let $\theta : S \to T$ be a homomorphism into a member of V. We should show that $w_S^{-1}(\theta^{-1}(t))$ is an open subset of S^A. By the definition of homomorphism of σ-algebras, the diagram (6.1) commutes and so $\varphi \in S^A$ belongs to $w_S^{-1}(\theta^{-1}(t))$ if and only if $\theta \circ \varphi \in w_T^{-1}(t)$ and the latter condition defines an open subset of S^A in the pro-V topology. □

Example 6.1 shows that the topology of a free pro-V algebra is not necessarily the pro-V topology. In contrast, we have the following result.

Proposition 7.2. *The topology induced on $\Omega_A \mathsf{V}$ as a subspace of $\overline{\Omega}_A \mathsf{V}$ is the pro-V topology.*

Proof. Since any homomorphism $\varphi : \Omega_A \mathsf{V} \to S$ into a member of V is completely determined by its restriction to the generators x_a ($a \in A$) and this restriction extends uniquely to a continuous homomorphism $\hat{\varphi} : \overline{\Omega}_A \mathsf{V} \to S$, we conclude that φ is continuous for the induced topology of $\Omega_A \mathsf{V}$. Hence the induced topology contains the pro-V topology. For the converse it suffices to observe that the sets of the form $\psi^{-1}(s) \cap \Omega_A \mathsf{V}$ with $\psi : \overline{\Omega}_A \mathsf{V} \to S$ a continuous homomorphism into a member of V and $s \in S$ constitute a basis of open subsets of $\Omega_A \mathsf{V}$ for the induced topology and $\psi^{-1}(s) \cap \Omega_A \mathsf{V} = (\psi|_{\Omega_A \mathsf{V}})^{-1}(s)$. Hence the two topologies coincide. □

Pro-V topologies have been extensively studied over the past decade, particularly for free monoids and free groups with respect to pseudovarieties of groups. See [35, 18, 25, 43, 46, 44, 52, 53, 20, 34, 59, 60, 7] for a trace of results. For pseudovarieties of semigroups V containing the pseudovariety N of all finite nilpotent semigroups, the pro-V topology of the free semigroup A^+ is the discrete topology since the Rees quotient by the ideal consisting of all the words of length at least a given n is a member of N. We next develop a metric structure which is non-trivial even in this case.

Given a topological algebra S and $u, v \in S$, let $r(u, v)$ denote the cardinality of the smallest $T \in \mathsf{V}$ for which there is a continuous homomorphism $\varphi : S \to T$ such that $\varphi(u) \neq \varphi(v)$ if such a φ exists and let $r(u, v) = +\infty$ otherwise. Note that in this definition we may without harm consider only onto continuous homomorphisms. Let

$$d(u, v) = 2^{-r(u,v)}$$

where we take $2^{-\infty} = 0$. This defines a *pseudo-ultrametric* over S that is a function $d : S \times S \to \mathbb{R}$ such that the following properties hold for all $u, v, w \in S$:

1. $d(u, v) \geq 0$;

2. $u = v$ implies $d(u, v) = 0$;

3. $d(u,v) = d(v,u)$;

4. $d(u,w) \leq \max\{d(u,v), d(v,w)\}$.

The only non-trivial verification to show that d_V is indeed a pseudo-ultrametric is that of property (4). In terms of the function r, it states that $r(u,w) \geq \min\{r(u,v), r(v,w)\}$ and indeed if $\varphi : S \to T$ is a continuous homomorphism into a member of V such that $\varphi(u) = \varphi(v)$ and $\varphi(v) = \varphi(w)$, then also $\varphi(u) = \varphi(w)$.

If the converse of property (2) also holds, then d is said to be an *ultrametric*. For the function d defined above, this property is equivalent to the topological algebra S being residually in V.

The following interaction between the metric and algebraic structures should not be surprising.

Lemma 7.3. *The operations from σ are uniformly continuous with respect to the pseudo-ultrametric d of the topological algebra S.*

Proof. Let o be an operation from σ of arity A. For $\varphi, \psi \in S^A$, if the inequality $r(\varphi(a), \psi(a)) \geq n$ holds for every $a \in A$, then also $r(o_S(\varphi), o_S(\psi)) \geq n$. Hence the mapping $o_S : S^A \to S$ is actually a contraction if we endow S^A with the pseudo-metric defined by

$$d(\varphi, \psi) = \max\{d(\varphi(a), \psi(a)) : a \in A\}$$

which proves the lemma. □

Note that the (product) topology of the profinite group $G = (\mathbb{Z}/2\mathbb{Z})^I$ is metrizable if and only if the set I is countable (see [64, Theorem 22.3]). Even in case I is countably infinite, the above ultrametric d does not determine the product topology of G: for any two distinct points $u, v \in G$, the projection into some component $G \to \mathbb{Z}/2\mathbb{Z}$ separates them and so $d(u,v) = 1/4$ so that the metric d is discrete and it therefore induces the discrete topology. However, for the most common pro-V algebras, namely those that are finitely generated, the topological structure is metrizable and it is determined by d as the following result shows.

Proposition 7.4. *For a finitely generated pro-V algebra S, the ultrametric d determines the given topology of S.*

Proof. We first claim that every open ball of the form

$$B_n(v) = \{u \in S : d(u,v) < 2^{-n}\}$$

is open in the given topology. A given $u \in S$ lies in $B_n(v)$ if and only if $r(u,v) > n$, that is, for every onto continuous homomorphism $\varphi : S \to T$ into a member of V with at most n elements, $\varphi(u) = \varphi(v)$. Since S is finitely generated, there are only a finite number of onto continuous homomorphisms $S \to T$ with T a member of V such that $T \subseteq \{1, \ldots, n\}$. Hence there exists some continuous homomorphism $\varphi : S \to T$ into a member of V such that $\varphi(u) = \varphi(v)$ if and only if $r(u,v) > n$, which shows that $B_n(v) = \varphi^{-1}\varphi(v)$ and therefore $B_n(v)$ is open in the topology of S.

On the other hand, by Lemma 4.1 the topology of S is the initial topology for the continuous homomorphisms into members of V. Given a continuous homomorphism $\varphi : S \to T$ into a member of V, $t \in T$ and $v \in \varphi^{-1}(t)$, the ball $B_{|T|}(v)$ is completely contained in $\varphi^{-1}(t)$ and therefore this set is open in the ultrametric topology. Hence the two topologies coincide. □

Taking into account Proposition 7.2, we obtain the following corollary.

Corollary 7.5. *For a finite set A,*

1. *the pro-V topology of $\Omega_A \mathsf{V}$ is that determined by the ultrametric d;*

2. *the completion of $\Omega_A \mathsf{V}$ with respect to the ultrametric d is isomorphic to $\overline{\Omega}_A \mathsf{V}$.*

Proof. Since every homomorphism from $\Omega_A \mathsf{V}$ into a member of V is continuous in the induced topology, the induced ultrametric on $\Omega_A \mathsf{V}$ from that of $\overline{\Omega}_A \mathsf{V}$ is precisely the ultrametric of $\Omega_A \mathsf{V}$. Part (1) follows immediately from Propositions 7.4 and 7.2.

For part (2), just note that the ultrametric induced on $\Omega_A \mathsf{V}$ from that of $\overline{\Omega}_A \mathsf{V}$ is the same as the ultrametric of $\Omega_A \mathsf{V}$. □

More generally, we have the following result.

Proposition 7.6. *The completion of a finitely generated topological σ-algebra S which is residually in V with respect to the ultrametric d is a pro-V algebra.*

Proof. Since S is residually in V, there exists a continuous injective mapping $\iota : S \to T$ into a product of members of V in such a way that composing ι with the component projections we obtain, up to isomorphism, all possible onto continuous homomorphisms from S to members of V. Let U be the closure of the image of ι. Then U is a finitely generated pro-V algebra by Proposition 4.3 and therefore its topology is determined by its ultrametric d by Proposition 7.4. In particular, U is a complete metric space. Moreover, the continuous homomorphisms from S to members of V are the same whether S is taken with its given topology or with the subspace topology induced from U. Hence the induced metric on S as a subspace of U is its own original metric and therefore U is the completion of S. □

8 A compactness theorem

Consider a system of *pseudo-equations*

$$u_i = v_i \quad (i \in I) \tag{8.1}$$

where the u_i and v_i are elements of a fixed $\overline{\Omega}_A \mathsf{V}$ with A finite. We admit some *parameters* in the system which are given by the choice of a subset $P \subseteq A$, where we think of the $p \in P$ as the parameters and we continue to call the elements of $A \setminus P$ the *variables*. For a pro-V algebra S, by an *evaluation in S (of the variables)* of the system we mean a mapping $\varphi : A \setminus P \to S$. We say that the evaluation φ is *consistent with an evaluation* $\gamma : P \to S$ *of the parameters* (or simply *consistent* if there are no parameters) if the two evaluations together provide a solution of the system, that is $(u_i)_S(\xi) = (v_i)_S(\xi)$ for all $i \in I$ where $\xi : A \to S$ coincides with φ on $A \setminus P$ and with γ on P.

A *relational morphism* between topological algebras S and T is a relation $\mu : S \to T$ with domain S such that, as a subset of $S \times T$, μ is a closed subalgebra. Although we are assuming that relational morphisms between topological algebras to deserve that name should be closed, we will sometimes explicitly say so since the notion of relational morphism between algebras makes sense if we drop that requirement and thus some confusion may result. We say that μ is *onto* if its range is T. Note that every continuous homomorphism is a relational morphism as is the inverse of every onto continuous homomorphism. If T is compact, then the composite $\nu \circ \mu : S \to U$ of two relational morphisms $\mu : S \to T$ and $\nu : T \to U$ is again a relational morphism.

The following compactness result is presented here as a side remark since it does not intervene in our development of the theory.

Proposition 8.1. *Let $\mu : S \to T$ be a closed relational morphism from a finite algebra S to a profinite algebra T and let V be a closed subset of T. Then there is a continuous homomorphism $\varphi : T \to F$ onto a finite algebra such that $\mu^{-1} V = (\varphi \circ \mu)^{-1} \varphi V$.*

Proof. Let $U = \mu^{-1} V = \{ s \in S : (\{s\} \times V) \cap \mu \neq \emptyset \}$ so that $S \setminus U$ consists of

all $s \in S$ such that $(\mu s) \cap V = \emptyset$. Consider the set
$$C = \bigcup_{s \in S \setminus U} \mu s.$$
Since μ is closed in $S \times T$, so is $(\{s\} \times V) \cap \mu$ and therefore its projection in T, which is the set μs, is also closed. Since S is finite, it follows that C is a closed subset of T which is disjoint from V. Since T is a compact zero-dimensional space, there is some clopen subset K of T which contains V and which is disjoint from C. By Lemma 4.1 there is some continuous homomorphism onto a finite algebra $\varphi : T \to F$ such that $K = \varphi^{-1}\varphi K$. We claim that φ has the required property, that is $U = (\varphi \circ \mu)^{-1}\varphi V$.

We obviously have
$$U = \mu^{-1}V \subseteq \mu^{-1}\varphi^{-1}\varphi V = (\varphi \circ \mu)^{-1}\varphi V.$$
On the other hand, if $s \in S \setminus U$, then $\mu s \subseteq C$ so that
$$\emptyset = (\mu s) \cap K = (\mu s) \cap \varphi^{-1}\varphi K \supseteq (\mu s) \cap \varphi^{-1}\varphi V$$
which means that $s \notin \mu^{-1}\varphi^{-1}\varphi V = (\varphi \circ \mu)^{-1}\varphi V$. This proves the claim and completes the proof of the proposition. □

For example, in finite semigroup theory one finds a canonical relational morphism $\mu : M \to \overline{\Omega}_A \mathsf{G}$, for a finite A-generated monoid M, which will be defined shortly. The set $\mu^{-1}\{1\}$ is therefore also obtained as $\nu^{-1}\{1\}$ for some relational morphism $\nu : M \to G$ into a finite group.

Let $\mu : S \to T$ be an onto relational morphism between two topological algebras. Two evaluations $\varphi : A \setminus P \to S$ and $\psi : A \setminus P \to T$ of the variables of the system (8.1) are said to be μ-*related* if, for every $a \in A \setminus P$, $(\varphi(a), \psi(a)) \in \mu$. We say that φ is μ-*inevitable* if, for every evaluation $\gamma : P \to T$ of the parameters in T, there exists an evaluation $\psi : A \setminus P \to T$ of the variables which is μ-related with φ and which is consistent with γ. In a sense, we are trying to solve in the topological algebra T a system of pseudo-equations in terms of certain parameters, where the variables are constrained to belong to certain subsets of T described by the images under the relational morphism μ of the elements of S ascribed to those variables. In this context, we may speak of φ as a *constraining evaluation of the variables*.

For a subpseudovariety W of V, the evaluation φ in an algebra $S \in \mathsf{V}$ of the system (8.1) is said to be W-*inevitable* if it is μ-inevitable for all onto relational morphisms $\mu : S \to T$ into members of W. Here, in a sense we are trying to solve the system, with constraints on the variables and perhaps some parameters, in all members of W. Theorem 8.3 below shows that if this is possible then there is a uniform solution.

Consider a B-generated pro-V algebra S, with generating mapping $\psi : B \to S$. Let $\hat{\psi}$ be the unique extension to a continuous homomorphism $\overline{\Omega}_B V \to S$ and let $p_W : \overline{\Omega}_B V \to \overline{\Omega}_B W$ be the natural projection. Then the composite $p_W \circ \hat{\psi}^{-1}$ is an onto relational morphism $S \to \overline{\Omega}_B W$ which is called the *canonical* W-*relational morphism* for S. Note that an evaluation $\varphi : A \setminus P \to S$ of the variables of the system (8.1) is inevitable with respect to the canonical W-relational morphism if and only if, for every evaluation $\gamma : P \to \overline{\Omega}_B W$ of the parameters, there exists an evaluation $\theta : A \setminus P \to \overline{\Omega}_B V$ such that $\hat{\psi} \circ \theta = \varphi$ and the composite $p_W \circ \theta$ is consistent with γ.

Lemma 8.2. *Let φ be an evaluation of the variables of the system (8.1) in a B-generated algebra S which is a member of* V *and let $\hat{\psi}$ be an onto continuous homomorphism $\overline{\Omega}_B V \to S$. Suppose that, for every evaluation $\gamma : P \to \overline{\Omega}_B W$ of the parameters, there exists an evaluation $\theta_\gamma : A \setminus P \to \overline{\Omega}_B V$ satisfying $\hat{\psi} \circ \theta_\gamma = \varphi$ and such that the composite $p_W \circ \theta_\gamma$ is consistent with γ. Then φ is inevitable with respect to every closed relational morphism $S \to T$ onto a pro-W algebra.*

Proof. Denote the restriction of $\hat{\psi}$ to B by ψ. Let $\mu : S \to T$ be an onto relational morphism. Then μ is a closed subalgebra of $S \times T$ whose projection on S is onto. Denote by $p_1 : \mu \to S$ and $p_2 : \mu \to T$ the two component projections. Since p_1 is onto, there exists a mapping $\lambda : B \to \mu$ such that $p_1 \circ \lambda = \psi$. Since μ is itself a pro-V algebra by Proposition 4.3, there exists a unique extension of λ to a continuous homomorphism $\hat{\lambda} : \overline{\Omega}_B V \to \mu$. Since the composite $p_1 \circ \hat{\lambda}$ coincides with ψ on B, it must be equal to $\hat{\psi}$.

Consider next the composite $p_2 \circ \lambda : B \to T$. Since T is a pro-W algebra, that mapping extends uniquely to a continuous homomorphism $\hat{\zeta} : \overline{\Omega}_B W \to T$ which is onto since μ is onto and which satisfies $p_2 \circ \lambda = \hat{\zeta} \circ p_W$. The diagram in Figure 3 should help the reader to keep track of all these mappings.

Let $\delta : P \to T$ be an evaluation of the parameters. Since $\hat{\zeta}$ is onto, there is a lifting to an evaluation $\gamma : P \to \overline{\Omega}_B W$ such that $\hat{\zeta} \circ \gamma = \delta$. By hypothesis there exists an evaluation $\theta_\gamma : A \setminus P \to \overline{\Omega}_B V$ of the variables such that $\hat{\psi} \circ \theta_\gamma = \varphi$ and such that the composite $p_W \circ \theta_\gamma$ is consistent with γ.

Let $\chi = \hat{\zeta} \circ p_W \circ \theta_\gamma = p_2 \circ \hat{\lambda} \circ \theta_\gamma$. By construction, χ is an evaluation of the variables of the system in T which is μ-related with φ. Moreover, since $p_W \circ \theta_\gamma$ is consistent with γ and implicit operations on V commute with continuous homomorphisms between pro-V algebras, χ is consistent with δ. □

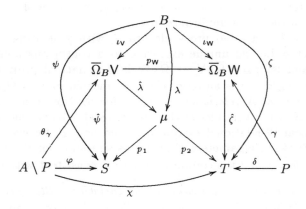

Figure 3. A diagram of mappings for the proof of Lemma 8.2

The converse of Lemma 8.2 is more interesting and is also true. It amounts to a general compactness theorem of which many instances have been encountered (and proved independently). It is meant to put in a general algebraic-topological setting both results involving special types of systems with constraints without parameters (that is $P = \emptyset$, see [5, 47, 58]) as well as unconstrained systems with parameters (corresponding to constraining evaluations of the variables in trivial algebras, see [4, Theorem 5.6.1]). The proof follows closely that of [5, Proposition 3].

Theorem 8.3. *The following conditions are equivalent for a constraining evaluation φ of the variables of the system (8.1) in a member S of V:*

1. *φ is W-inevitable;*

2. *φ is inevitable with respect to every closed relational morphism into every pro-W algebra;*

3. *φ is inevitable with respect to the canonical W-relational morphism corresponding to any specific choice of generators of S.*

Proof. Let $\varphi : A \setminus P \to S$ be a constraining evaluation of the variables of the system (8.1) in a B-generated member of V. Let $\psi : B \to S$ be the generating mapping and let $\hat{\psi}$ be its unique extension to a continuous homomorphism $\overline{\Omega}_B \mathsf{V} \to S$.

Suppose the evaluation φ is W-inevitable but it is not inevitable with respect to the canonical W-relational morphism. Let $\gamma : P \to \overline{\Omega}_B V$ be an evaluation of the parameters for which there is no lifting of φ to an evaluation $\theta : A \setminus P \to \overline{\Omega}_B V$ such that $\hat{\psi} \circ \theta = \varphi$ and $p_W \circ \theta$ is consistent with $p_W \circ \gamma$.

Consider the function
$$\nabla : (\overline{\Omega}_B V)^{A \setminus P} \to (\overline{\Omega}_B W \times \overline{\Omega}_B W)^I$$
$$\theta \mapsto \left(\left(p_W(u_i(\theta')), p_W(v_i(\theta')) \right) \right)_{i \in I}$$

where θ' denotes the unique continuous homomorphism $\overline{\Omega}_A V \to \overline{\Omega}_B V$ whose restriction to $A \setminus P$ is θ and whose restriction to P is γ. Since each u_i and each v_i (or rather their natural interpretations in $\overline{\Omega}_B V$) are continuous and the natural projection $p_W : \overline{\Omega}_B V \to \overline{\Omega}_B W$ is also continuous, ∇ is continuous when the set $(\overline{\Omega}_B W \times \overline{\Omega}_B W)^I$ is endowed with the product topology.

Consider next the set
$$\mathcal{L} = \{ \theta \in (\overline{\Omega}_B V)^{A \setminus P} : \hat{\psi} \circ \theta = \varphi \}$$
of all liftings of the given evaluation φ of the variables of the system in S to evaluations in $\overline{\Omega}_B V$. Since
$$\mathcal{L} = \prod_{a \in A \setminus P} \hat{\psi}^{-1}(\varphi(a)),$$
the set $\mathcal{L}$ is a closed subset of $(\overline{\Omega}_B V)^{A \setminus P}$. Hence $\mathcal{L}$ is compact and therefore $\nabla(\mathcal{L})$ is a closed subset of $(\overline{\Omega}_B W \times \overline{\Omega}_B W)^I$. Note that $\mathcal{L}$ consists of all evaluations of the variables of the system (8.1) in $\overline{\Omega}_B W$ which are related with φ under the canonical W-relational morphism.

By the choice of γ, the set $\nabla(\mathcal{L})$ has empty intersection with the *diagonal*
$$\Delta = \{(u, u) : u \in \overline{\Omega}_B W\}^I.$$
Since Δ is a closed subset of $(\overline{\Omega}_B W \times \overline{\Omega}_B W)^I$, which is a compact zero-dimensional space, there is a clopen neighbourhood K of Δ such that $K \cap \nabla(\mathcal{L}) = \emptyset$. By definition of the product topology, we may take K to be of the form $K = K_1 \cup \ldots \cup K_p$ where each K_j is of the form $\prod_{i \in I} K_{i,j} \times K_{i,j}$ with each $K_{i,j}$ a clopen subset of $\overline{\Omega}_B W$ and only finitely many of them are proper subsets of $\overline{\Omega}_B W$. By Lemma 4.1, there is some continuous homomorphism $\zeta : \overline{\Omega}_B W \to T$ into some member of W such that $K_{i,j} = \zeta^{-1} \zeta(K_{i,j})$ for all i and j.

Let $\ker \zeta$ denote the kernel congruence of the homomorphism ζ consisting of all pairs (u, v) of elements of $\overline{\Omega}_B W$ such that $\zeta(u) = \zeta(v)$. Then one may easily check that $\Delta \subseteq (\ker \zeta)^I \subseteq K$ and so $(\ker \zeta)^I \cap \nabla(\mathcal{L}) = \emptyset$.

Consider the relational morphism $\mu : S \to T$ given by $\mu = \zeta \circ p_W \circ \hat{\psi}^{-1}$ and let $\delta : P \to T$ be the evaluation of the parameters given by the composite $\delta = \zeta \circ p_W \circ \gamma$. Since we are assuming that the evaluation φ of the system (8.1) is W-inevitable, there exists an evaluation $\chi : A \backslash P \to T$ of the variables which is μ-related with φ and which is consistent with δ. By the choice of μ, χ lifts to an evaluation $\theta : A \backslash P \to \overline{\Omega}_B V$ such that $\zeta \circ p_W \circ \theta = \chi$ and $\hat{\psi} \circ \theta = \varphi$. Hence $\theta \in \mathcal{L}$. On the other hand, the consistency of χ with δ means that $\nabla(\theta) \in (\ker \zeta)^I$. This contradicts $\nabla(\mathcal{L}) \cap (\ker \zeta)^I = \emptyset$ and proves that (1) implies (3). That (2) implies (1) is obvious and that (3) implies (2) is given by Lemma 8.2. □

The following examples of application of Theorem 8.3 in the case where there are no constraints are taken from [4, Section 5.6]. Further examples will be given later. For the last part of Corollary 8.5, we require the following well-known combinatorial observation.

Lemma 8.4. *Let S be a finite semigroup of cardinality m and consider m, not necessarily distinct, elements $s_1, s_2, \ldots, s_m$ of S. Then there exist i and j with $1 \leq i < j \leq m$ such that*
$$s_1 s_2 \cdots s_m = s_1 \cdots s_i (s_{i+1} \cdots s_j)^\omega s_{j+1} \cdots s_m.$$

Proof. Consider the products $t_k = s_1 \cdots s_k$ with $k = 1, \ldots, m$. If one of the t_k is an idempotent, then we are done. Otherwise, since S has at least one idempotent, there are i and j with $i < j$ such that $t_i = t_j$. Then we have
$$t_i = t_j = t_i s_{i+1} \cdots s_j = t_i (s_{i+1} \cdots s_j)^\ell = t_i (s_{i+1} \cdots s_j)^\omega$$
from which the desired equality follows. □

Corollary 8.5. *Let V be a pseudovariety of semigroups and let $u, v \in \overline{\Omega}_B V$.*

1. *The element u is regular in $\overline{\Omega}_B V$ if and only if every value of u in every member of V is regular.*

2. *If $\mathcal{K}$ denotes any of Green's relations, then u and v are $\mathcal{K}$-equivalent in $\overline{\Omega}_B V$ if and only if, under every evaluation of B in every member of V, the values of u and v are $\mathcal{K}$-equivalent.*

3. *If u is not explicit, then there exist $u_1, w, u_2 \in \overline{\Omega}_B V$ such that $u = u_1 w^\omega u_2$.*

Proof. (1) Let $A = B \mathbin{\mathring{\cup}} \{x\}$, where x is a new variable and consider the equation $uxu = u$ with variable x and parameters the elements of B. The existence of an unconstrained solution in $\overline{\Omega}_B \mathsf{V}$ means that u is a regular element of $\overline{\Omega}_B \mathsf{V}$. On the other hand, the existence of an unconstrained solution in every member S of V means that u always evaluates to a regular element of S. Note that, since there are no constraints, the existence of a solution for the identity evaluation of the parameters implies that a solution for all other evaluations is obtained by mere substitution.

(2) The proof is similar to that of part of (1) but working here with suitably chosen equations. For instance, for $\mathcal{R}$-equivalence, if $u = v$ then the result is obvious; otherwise, take the system consisting of the equations $ux = v$ and $vy = u$ with new variables x, y and parameters the elements of B.

(3) Use the same type of argument for the equation $u = xy^\omega z$ with new variables x, y, z and parameters the elements of B. The existence of solutions of the equation in finite semigroups follows from Lemma 8.4 taking into account that, since every implicit operation is the limit of a sequence of explicit operations, a non-explicit operation must admit a factorization into n factors for every $n \geq 2$. □

9 Decidability

A pseudovariety of σ-algebras V is said to be *decidable* if its membership problem is decidable, that is there exists an algorithm which, given a finite σ-algebra S, decides whether or not $S \in \mathsf{V}$. In the applications of the theory of finite semigroups via Eilenberg's correspondence, this is a very important property.

For the description of a pseudovariety V in terms of a basis of pseudoidentities Σ to directly provide an algorithm to decide membership in V, the basis should be easy to verify. A first approximation to this question is to require Σ to be finite. Although there are pseudovarieties which are known to be decidable which are not finitely based, inspired by the commutative semigroup case, the author asked [4, Problem 8] whether that property would be verified for pseudovarieties whose membership problem can be efficiently solved (meaning it can be done in polynomial time). Negative solutions have been given by Volkov [63], Sapir (see [31]), and Trahtman [62].

Quite often a given pseudovariety V is *recursively enumerable* in the sense that there is an effective procedure that step by step outputs σ-algebras and the set of its outputs, up to isomorphism, is the pseudovariety V. If we can effectively recognize whether two given finite σ-algebras are isomorphic, then half the task of deciding membership in V will be accomplished. A simple hypothesis which yields that property and which is certainly verified in the applications of finite semigroup theory is to assume that σ is finite, and we will do so from hereon.

To do the other half of the task, we need some means of generating (representatives of the isomorphism classes of) the σ-algebras which do not lie in V. For this purpose, we may consider a basis Σ of pseudoidentities of V and generate systematically, in sequence, all counter-examples for these pseudoidentities. This may be achieved if Σ is recursively enumerable and each of its members can be effectively verified: just generate successively all pairs consisting of a finite σ-algebra S and a member $u = v$ of Σ and test whether S satisfies $u = v$, outputting S in the negative case. As in [11], we say that V is *recursively definable* if it has such a basis of pseudoidentities.

Proposition 9.1. *Every recursively enumerable, recursively definable pseudovariety is decidable.*

Proof. An algorithm deciding membership in such a pseudovariety V is obtained by running in parallel two processes which generate, up to isomorphism, respectively the set of all members of V and the set of all finite σ-algebras which do not lie in V. Once a given finite σ-algebra S is observed as an output of one of the two processes, we may conclude respectively that S is in V or not. □

Of course we do not claim that the algorithm in the above proof is efficient. On the contrary, it is absolutely impractical. For instance, with at most 8 elements there are up to isomorphism and anti-isomorphism approximately 1.8×10^9 semigroups [55]. It is nevertheless of theoretical value in the sense that it provides a sufficient condition for decidability. We do not know whether the condition of recursive definability is also a necessary condition.

Various strong forms of decidability have been considered which tie it with the ideas of Section 8. For instance, one may be interested in a certain class C of systems of pseudo-equations with parameters over a pseudovariety V, and consider the following decision problem:

Determine whether, given an evaluation φ in a given member of V of the variables of a system from C, φ is W-inevitable.

The decidability of such a problem is a decidability property of W which may perhaps be useful if the class C is conveniently chosen. For shortness, let us say that W is C-*decidable* if the above decision problem is decidable. We present below a few examples where this idea has been carried out, in all of them the systems having no parameters. This will also give us an excuse to introduce some important operators on pseudovarieties, some of them only defined here in the semigroup case.

For a σ-algebra S, denote by $\mathcal{P}(S)$ the set of all its subsets. We turn $\mathcal{P}(S)$ easily into a σ-algebra, called the *power algebra of S*, by letting, for each A-ary operation w from σ and each $\varphi \in (\mathcal{P}(S))^A$,

$$w_{\mathcal{P}(S)}(\varphi) = \{w_S(\psi) : \psi \in S^A \text{ such that } \psi(a) \in \varphi(a) \ (\forall a \in A)\}.$$

For a pseudovariety W of σ-algebras, $\mathcal{P}$W denotes the pseudovariety generated by all power algebras $\mathcal{P}(S)$ with $S \in$ W. This is called the *power pseudovariety of* W.

Example 9.1. Let $\mathcal{C}$ consist of the systems of the form

$$x_1 = x_2 = \cdots = x_n.$$

An evaluation of the variables in a member S of V consists in the choice of n, perhaps not distinct, elements $s_1, \ldots, s_n$ of S. The evaluation is W-inevitable if and only if for every relational morphism $\mu : S \to T$ into a member of W, there exists $t \in T$ such that $(s_i, t) \in \mu$ for $i = 1, \ldots, n$. In case V = S, this property of the n-tuple $(s_1, \ldots, s_n)$ has been referred by Henckell [24] by saying that the n-tuple is W-*pointlike*. In the cited paper, Henckell gives a simple algorithm that shows that A is $\mathcal{C}$-decidable. Henckell and Rhodes [26] reduced the computation of the power pseudovariety $\mathcal{P}$G to the $\mathcal{C}$-decidability of G which was shortly after proved by Ash [16]. Steinberg [57] has shown how such a property intervenes in various contexts such as computing joins. The instance of Theorem 8.3 corresponding to such systems and semigroups was first observed in [6] where $\mathcal{C}$-decidability was called *strong decidability* and formulations of this property of W in terms of rational languages were also presented.

The following easy observation is often useful.

Lemma 9.2. *Let S be a member of* V *and let* W *be a subpseudovariety of* V. *Then S belongs to* W *if and only if S has no two-element* W-*pointlike subsets.*

Proof. Let $\psi : \overline{\Omega}_B V \to S$ be an onto continuous homomorphism with B finite. Then $S \in$ W if and only if ψ factors through the natural projection p_W. By Theorem 8.3, the two-element W-pointlike subsets of S are precisely the obstructions to the existence of such a factorization which proves the lemma. □

The following is an immediate corollary.

Corollary 9.3. *Every strongly decidable pseudovariety is decidable.*

The *Mal'cev product* of two pseudovarieties U and W of semigroups is the pseudovariety U ⓜ W generated by the class of all finite semigroups S for which there exists a homomorphism $\theta : S \to T$ into a member T from W such that,

for every idempotent e of T, the subsemigroup $\theta^{-1}(e)$ of S belongs to U.

It is easy to check that U ⓜ W consists of all finite semigroups S for which there exists a relational morphism $\theta : S \to T$ into some $T \in$ W with the above property. (See Section 11 for an extension of this operation to a much more general context.)

Example 9.2. Take once more V = S and consider for $\mathcal{C}$ the systems of the form
$$x_1 = x_2 = \cdots = x_n = x_n^2.$$
An evaluation of such a system in a finite semigroup S consists again in the choice of an n-tuple $(s_1, \ldots, s_n)$. The evaluation is W-inevitable if and only if, for every relational morphism $\mu : S \to T$ into $T \in$ W, there exists an idempotent $e \in T$ such that $(s_i, e) \in \mu$ for $i = 1, \ldots, n$. Steinberg [58] says then that the n-tuple is an *idempotent-pointlike* with respect to W. Pin and Weil [47] proved the corresponding instance of Theorem 8.3 and showed that the Mal'cev product U ⓜ W may be computed if U is recursively enumerable and recursively definable, and W is $\mathcal{C}$-decidable.

Although graphs already appeared in Section 2 as a means of representing automata, we will need for the next couple of examples a more formal treatment.

By a *graph* we mean a set $\Gamma = V \mathbin{\mathring{\cup}} E$ with two parts, where the elements of V are called *vertices* and the elements of E are called *edges*, together with two operations $\alpha, \omega : E \to V$ which give, respectively, the *beginning* and the *end* vertices of each edge. The graph is *finite* if Γ is finite. Normally, the operations α and ω are understood and we refer simply to the graph Γ. Graphs are represented pictorially in the usual way with an edge e represented by

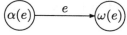

A *path* in the graph Γ is a sequence of edges $e_1, \ldots, e_n$ such that $\omega(e_i) = \alpha(e_{i+1})$ for $i = 1, \ldots, n-1$; the path is *closed* if $\alpha(e_1) = \omega(e_n)$; the path is *simple* if there are no distinct $i, j \in \{1, \ldots, n\}$ such that $\alpha(e_i) = \alpha(e_j)$.

Every graph Γ may be embedded in a graph $\hat{\Gamma}$ by adding a *dual* edge e' for each edge e in Γ in the sense that $\alpha(e) = \omega(e')$ and $\omega(e) = \alpha(e')$. An *undirected path* in Γ is a sequence of edges $p = (e_1, \ldots, e_n)$ such that, dualizing some of the edges e_i in $\hat{\Gamma}$, the resulting sequence p' is a path in the graph $\hat{\Gamma}$; edges which are not dualized in this process are said *to appear in the direction of the path* whereas those that are dualized are said *to appear in the reverse direction* of the path. The undirected path p is said to be *simple*, respectively *closed*, if p' has that property.

Example 9.3. Take here V to be the pseudovariety M of all finite monoids. To a finite graph Γ we associate a system of pseudo-equations by first choosing a variable x_e for each edge e. In the system there is a pseudo-equation

$$x_{e_1}^{\epsilon_1} \cdots x_{e_n}^{\epsilon_n} = 1$$

for each closed simple undirected path $e_1, \ldots, e_n$ where $\epsilon_i = 1$ if e_i appears in the direction of the path and $\epsilon_i = \omega - 1$ otherwise. Let $\mathcal{C}$ denote the class of all systems of this form. Ash [16] proved that G is $\mathcal{C}$-decidable and Steinberg [59] extended this result to other pseudovarieties of groups including the pseudovariety G_p of all finite p-groups for prime p. Connections with model theory can be found in [27, 8, 9].

Given two semigroups S and T and a monoid homomorphism $\theta : T^1 \to \mathrm{End}(S)$ (the *action*), we write simply $^t s$ for $(\theta(t))(s)$. The *semidirect product* $S *_\theta T$ is the Cartesian product $S \times T$ endowed with the following operation:

$$(s_1, t_1)(s_2, t_2) = (s_1 {}^{t_1}\! s_2, t_1 t_2).$$

Usually the action being understood, we write $S * T$ instead of $S *_\theta T$. The *semidirect product* of two pseudovarieties U and W is the pseudovariety $\mathsf{U} * \mathsf{W}$ generated by all semidirect products $S * T$ with $S \in \mathsf{U}$ and $T \in \mathsf{W}$. There is an alternative definition of this operation, in terms of wreath products, which shows that it is associative (cf. [22, 4]).

There is an extension of the theory of pseudovarieties to some special types of partial algebras, namely semigroupoids and categories. A *semigroupoid* is a graph Γ endowed with an associative partial composition of edges $E \times E \to E$ such that the composite ef is defined if and only if $\omega(e) = \alpha(f)$ and then $\alpha(ef) = \alpha(e)$ and $\omega(ef) = \omega(f)$ as depicted in the diagram

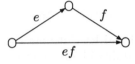

A *category* has additionally the requirement that for every vertex v there is a *local identity* 1_v such that $1_{\alpha(e)}e = e1_{\omega(e)} = e$ for every edge e. A *homomorphism* of such structures is a function between two structures of the same kind which sends vertices to vertices, edges to edges, respects all operations, namely α, ω, composition, and also local identities in the category case. Such a homomorphism $\varphi : S \to T$ is said to be a *quotient* if it is surjective but its restriction to vertices is a bijection between the vertices of S and T; on the other hand, φ is said to be *faithful* if, for all vertices u and v of S, the restriction of φ to the set of all edges e in S which begin in u and end in v is injective.

Following Tilson [61], we say that a semigroupoid (or a category) S *divides* another, say T, if there exists a structure of the same type U and two homomorphisms $U \to S$ and $U \to T$ such that the first is a quotient and the latter is faithful. A *pseudovariety of semigroupoids* is a class of finite semigroupoids containing the one-vertex one-edge semigroupoid which is closed under taking divisors, finite products, and finite coproducts. A *pseudovariety of categories* is a non-empty class of finite categories which is closed under taking divisors and finite products.

The role of generating sets for semigroupoids and categories is played by graphs. Otherwise, basically everything that has been done in the general algebraic-topological setting in Sections 4–8 carries over to this partial algebra context. In particular, for a pseudovariety of semigroupoids W, there exists a free pro-W semigroupoid over a generating graph Γ whose elements may be viewed as implicit operations of arity Γ and pseudovarieties of semigroupoids are defined by pseudoidentities over finite graphs. (See [61, 30, 13] for motivation and details.)

A semigroup may be viewed as a semigroupoid by considering its elements as edges with both ends at a virtual vertex and composing them in the way they multiply in the semigroup. Naturally, if we start with a monoid, we end up with a category. For a pseudovariety W of semigroups (resp. monoids), denote by gW the pseudovariety of semigroupoids (resp. categories) generated by W.

Example 9.4. A variation of the preceding example consists in associating to a finite graph Γ a system of equations by first choosing a variable x_g for each $g \in \Gamma$ and considering an equation

$$x_{\alpha(e)}x_e = x_{\omega(e)}$$

for each edge e. Let $\mathsf{V} = \mathsf{M}$ and let $\mathcal{C}$ denote the class of all systems of this form. The notion of $\mathcal{C}$-decidability of a pseudovariety W of monoids was introduced in [5] under the name of *hyperdecidability*. The corresponding version of Theorem 8.3 was proved in the same paper where, starting from a *basis theorem for semidirect products* due to Weil and the author [13], it was also shown that the semidirect product U∗W is decidable if a "reasonable" pseudoidentity basis is known for gU and W is hyperdecidable. Also in [5], it was shown that a pseudovariety of groups is hyperdecidable if and only if it is $\mathcal{C}$-decidable for the class $\mathcal{C}$ of systems of Example 9.3. The "basis theorem" asserts that a basis of pseudoidentities for U ∗ W is given by a certain set of pseudoidentities depending on a basis of pseudoidentities for gU. As was already observed in the introduction the proof in [13] is faulty which means that this basis for the semidirect product may be incomplete, at least for some pseudovarieties. The proof in [13] works nevertheless if W is locally finite (that is, finitely generated free W-algebras are finite) or gU has a basis of pseudoidentities which involves only finitely many graphs. The applications in [5] using hyperdecidability only use these special cases of the "basis theorem".

An extension of the semidirect product is the so-called two-sided semidirect product. Given two semigroups S and T and two monoid endomorphisms $T^1 \to \text{End}(S)$, respectively the *left* and *right actions* where, for the latter, endomorphisms of S are applied on the right, denote the images of $s \in S$ under the respective endomorphisms associated with $t \in T^1$ by $^t s$ and s^t. The associated *two-sided semidirect product* $S \ast\ast T$ is again the Cartesian product $S \times T$ but now endowed with the operation

$$(s_1, t_1)(s_2, t_2) = (s_1^{t_2} \cdot {}^{t_1}s_2, t_1 t_2).$$

For two pseudovarieties U and W, their *two-sided semidirect product* is the pseudovariety U ∗∗ W generated by all two-sided semidirect products $S \ast\ast T$ with $S \in $ U and $T \in $ W.

Example 9.5. Here, to a finite graph Γ we associate a system of equations by first choosing a variable y_e for each edge e and a pair of variables x_v and z_v for each vertex v and then considering a pair of equations

$$x_{\alpha(e)} y_e = x_{\omega(e)} \quad \text{and} \quad y_e z_{\omega(e)} = z_{\alpha(e)}$$

for each edge e. Denote by $\mathcal{C}$ the class of all such systems of equations. Adapting the arguments of [5] based again on the results of [13], one may show that if U is recursively enumerable, gU is definable by a "reasonable" pseudoidentity basis, and W is $\mathcal{C}$-decidable, then U ∗∗ W is decidable. Unfortunately, the difficulty here is the same as for the semidirect product so that the basis of pseudoidentities for U ∗∗ W given in [13] which underlies the above decidability result is only known to be complete under the additional hypotheses indicated at the end of Example 9.4.

10   Tameness

Although Theorem 8.3 reduces W-inevitability of an evaluation to inevitability with respect to a single relational morphism, this relational morphism takes its values in a free pro-W algebra $\overline{\Omega}_B \mathsf{W}$. Since $\overline{\Omega}_B \mathsf{W}$ is often uncountable, there seems to be no direct way to derive decidability results from it such as $\mathcal{C}$-decidability of W for a given class $\mathcal{C}$ of systems of pseudo-equations. Taking into account various examples, particularly that of G in the work of Ash [16], Steinberg and the author [12, 11] introduced the notion of *tameness* of a pseudovariety.

Let $\mathcal{C}$ be a class of finite systems of pseudo-equations. A basic property of such systems that should be assumed if we are to get any decidability results concerning their solutions is to be able to verify if a candidate for a solution is indeed a solution. So we assume that the implicit operations u_i, v_i which define the sides of the pseudo-equations $u_i = v_i$ in the systems are *computable* in the sense that for each of them there is an algorithm that, given a member S of V, computes the natural interpretation of the implicit operation in S. We will refer to systems of pseudo-equations with this property as *effectively verifiable systems*.

The idea of tameness is to replace the single relational morphism into $\overline{\Omega}_B \mathsf{W}$ that allows testing W-inevitability of an evaluation of the variables of a system from $\mathcal{C}$ by an equally natural relational morphism into a suitable countable subalgebra of $\overline{\Omega}_B \mathsf{W}$. The hope is that after all one may not need all implicit operations from $\overline{\Omega}_B \mathsf{W}$ to find an evaluation in $\overline{\Omega}_B \mathsf{W}$ which is related with the original evaluation and which is consistent with a given evaluation of the parameters. Hopefully, many fewer such implicit operations will suffice.

Exploring this idea, we enlarge the algebraic signature σ to a signature τ by adding finitary implicit operations over V. Such a signature is called an *implicit signature*. Again because we are aiming at establishing decidability properties, we assume τ is accessible to computations in the sense that the following two properties hold:

1. the set τ is recursively enumerable;

2. every member of τ is a computable operation.

We then say that τ is *highly computable*.

Pro-V algebras immediately inherit a structure of τ-algebras since every implicit operation has a natural interpretation as a continuous operation on each of them (cf. Propositions 6.1 and 6.2). Moreover, because continuous homomorphisms between pro-V algebras commute with these natural interpretations, in going to the τ-algebra world subpseudovarieties of V remain pseudovarieties of τ-algebras and we lose neither pro-V algebras nor any continuous homomorphisms between them.

For a subpseudovariety W of V, denote by $\Omega_B^\tau W$ the W-free τ-algebra over the set B. By Proposition 5.1, $\Omega_B^\tau W$ is the τ-subalgebra of $\overline{\Omega}_B W$ generated by the component projections. Let $p_W^\tau : \Omega_B^\tau V \to \Omega_B^\tau W$ denote the natural projection obtained by restriction of the natural projection $p_W : \overline{\Omega}_B V \to \overline{\Omega}_B W$.

The proof of the following result is essentially the same as that of its special case which is Lemma 8.2: just replace throughout relatively free profinite σ-algebras by relatively free τ-algebras and continuous σ-homomorphisms by τ-homomorphisms.

Lemma 10.1. *Let φ be an evaluation of the variables of the system (8.1) in a B-generated algebra S which is a member of V and let $\hat{\psi}$ be an onto τ-homomorphism $\Omega_B^\tau V \to S$. Suppose that, for every evaluation $\gamma : P \to \Omega_B^\tau W$ of the parameters, there exists an evaluation $\theta_\gamma : A \setminus P \to \Omega_B^\tau V$ satisfying $\hat{\psi} \circ \theta_\gamma = \varphi$ and such that the composite $p_W \circ \theta_\gamma$ is consistent with γ. Then φ is W-inevitable.*

By a τ-term of arity B we mean a member of the absolutely free τ-algebra F_B^τ over the set B, defined by the usual universal property: there is a mapping $\kappa : B \to F_B^\tau$ such that, given any τ-algebra S and any mapping $\varphi : B \to S$, there exists a unique extension to a homomorphism $\hat{\varphi} : F_B^\tau \to S$ for which $\hat{\varphi} \circ \kappa = \varphi$. In particular, each $w \in F_B^\tau$ induces a member of $\Omega_B^\tau W$ by taking $\widehat{\iota_W}(w)$.

We will be interested in the *word problem* for $\Omega_B^\tau W$ which consists in determining, for two given τ-terms u and v of arity B, whether they induce the same element of $\Omega_B^\tau W$. We say that the τ-*word problem is decidable* for W if that word problem admits an algorithmic solution.

Given a B-generated member S of V, a generating mapping $\varphi : B \to S$ extends uniquely to a homomorphism of τ-algebras $\hat{\varphi} : \Omega_B^\tau V \to S$; the composite $p_W^\tau \circ \hat{\varphi}^{-1} : S \to \Omega_B^\tau W$ is called the *canonical (τ, W)-relational morphism*. We say that W is τ-*reducible with respect to $\mathcal{C}$* if, for every W-inevitable evaluation φ of the variables of any system from $\mathcal{C}$ in a B-generated member of V and every evaluation γ of the parameters in $\Omega_B^\tau W$, there exists an evaluation of the variables in $\Omega_B^\tau W$ which is consistent with γ and which is related with φ with respect to the canonical (τ, W)-relational morphism.

We further say that W is *τ-tame with respect to* $\mathcal{C}$ for a highly computable signature $τ$ if W is recursively enumerable and $τ$-reducible, and the $τ$-word problem for W is decidable. We say simply that W is *tame with respect to* $\mathcal{C}$ if it is $τ$-tame with respect to $\mathcal{C}$ for some highly computable implicit signature $τ$. We say that W is *absolutely tame* if it is tame with respect to the class of all finite systems of pseudo-equations. Finally, we say that W is *completely $τ$-tame* if is $τ$-tame with respect to the calls of all finite systems of $τ$-equations. Note that tameness with respect to a signature implies tameness with respect to any larger signature.

Say that a subpseudovariety W of V is *$τ$-equational* for an implicit signature $τ$ if W admits a basis of pseudoidentities whose sides are determined by $τ$-terms. The following observation is a straightforward generalization of [11, Proposition 4.2].

Proposition 10.2. *If W is $τ$-reducible with respect to the equation $x = y$ then W is $τ$-equational.*

Proof. Let Σ stand for the set of all pseudoidentities with finite arities contained in a fixed countable set of variables X which are satisfied by W and whose sides are determined by $τ$-terms. Clearly W is contained in $[\![\Sigma]\!]$. We show that every member S of $V \setminus W$ fails some pseudoidentity in Σ which will establish that $W = [\![\Sigma]\!]$.

By Lemma 9.2, there exists a two-element W-pointlike subset $\{s_1, s_2\}$ of S. Since W is $τ$-reducible with respect to the equation $x = y$, where x and y are variables, there exists an onto continuous homomorphism $\varphi : \Omega_B^τ V \to S$ and $u, v \in \Omega_B^τ V$ for some finite set $B \subseteq X$ such that $\varphi(u) = s_1$, $\varphi(v) = s_2$ but $p_W(u) = p_W(v)$, which shows that S fails the pseudoidentity $u = v$ which in turn holds in W. Since $u = v$ lies in Σ, this proves the claim. □

Note that, as a particular case of Proposition 10.2, we obtain Theorem 6.4: just take $τ$ to consist of all implicit operations over V.

For instance, by a result of Baumslag [17], every pseudovariety H of groups containing the pseudovariety G_p of all finite p-groups for some prime p is such that free groups are residually in H and, therefore, H is not $κ$-equational for the signature $κ$ consisting of multiplication together with the unary implicit operation $x \mapsto x^{ω-1}$, unless H = G. By Proposition 10.2, H is not $κ$-reducible with respect to any class of systems which includes the system containing only the equation $x = y$.

The following result extends [11, Theorem 4.7] with a basically identical proof.

Theorem 10.3. *Let σ be a finite finitary signature and let $\mathcal{C}$ be a recursively enumerable class of finite systems of pseudo-equations, without parameters, over a recursively enumerable pseudovariety V. If W is a subpseudovariety which is τ-tame with respect to $\mathcal{C}$, where τ is a highly computable signature such that the sides of the pseudo-equations of systems from $\mathcal{C}$ are determined by τ-terms, then W is $\mathcal{C}$-decidable.*

Proof. Let φ be an evaluation in a member S of V of the variables of a system from $\mathcal{C}$. We wish to decide whether or not φ is W-inevitable. The idea is as in the proof of Proposition 9.1: to start two automatic processes which generate evaluations of the variables of systems from $\mathcal{C}$ in members of V, one producing only W-inevitable such evaluations and eventually producing, up to isomorphism, every W-inevitable evaluation, the other producing only those that are not W-inevitable and eventually producing, up to isomorphism, every evaluation which is not W-inevitable. Then we just watch the output of each process until we recognize that our evaluation φ is produced; since σ is finite and finitary, it poses no difficulties to decide whether the algebra produced in each step is isomorphic with our algebra S. In other words, we should show that, up to isomorphism of σ-algebras, both the set of W-inevitable evaluations and its complement in the set of all evaluations of systems of $\mathcal{C}$ in members of V are recursively enumerable.

Now, if the evaluation φ is not W-inevitable then there exists a relational morphism $\mu : S \to T$ into some $T \in \mathsf{W}$ such that there is no consistent evaluation of the variables ψ in T which is μ-related with φ. Since $\mathcal{C}$, V and W are all recursively enumerable and σ is finite and finitary, we may successively generate all relational morphisms $\mu : S \to T$ from members of V to members of W as well as evaluations φ and ψ of variables from systems of $\mathcal{C}$ respectively in S and T; check whether φ and ψ are μ-related and whether ψ is consistent, which we can do since the sides of the pseudo-equations of $\mathcal{C}$ are determined by τ-terms and τ is highly computable; and output as being not W-inevitable those evaluations φ for which some T is found for which no ψ verifies the above conditions.

On the contrary, if φ is W-inevitable, then we use the tameness hypothesis to generate it. Suppose the algebra S, where the evaluation φ takes its values, is B-generated where B is a finite set. Since W is τ-reducible with respect to $\mathcal{C}$ and φ is W-inevitable, there exists an evaluation ψ of the variables in $\Omega_B^\tau \mathsf{V}$ which projects to φ in S under the unique τ-homomorphism $\hat{\lambda}$ extending the given generating function $\lambda : B \to S$ and such that $p_{\mathsf{W}} \circ \psi$ is consistent. Since τ is recursively enumerable, so is $\Omega_B^\tau \mathsf{V}$. On the other hand, since both $\mathcal{C}$ and V are also recursively enumerable, we may thus successively generate quintuples

$(B, S, \lambda, \Sigma, \psi)$ consisting of a finite set B, a B-generated member S of V, a generating mapping $\lambda : B \to S$, a system Σ from $\mathcal{C}$, and an evaluation ψ of the variables of the system in $\Omega_B^\tau \mathsf{V}$. For each quintuple, since the sides of pseudo-equations from $\mathcal{C}$ are τ-terms, we may use the solution of the τ-word problem for W to check whether the evaluation $p_\mathsf{W} \circ \psi$ is consistent. In the affirmative case, we output the evaluation $\hat{\lambda} \circ \psi$ which will then be W-inevitable by Lemma 10.1. By τ-tameness, this process will eventually output every given W-inevitable evaluation. □

Thus, decidability results such as those given in the examples of Section 9 which depend on some sort of $\mathcal{C}$-decidability may, under some mild decidability hypotheses, be established by proving a suitable abstract τ-reducibility property. It turns out that it is often easier to deal with this abstract property than trying to construct directly an algorithm to prove $\mathcal{C}$-decidability. But, of course, one also needs that the τ-word problem for our pseudovariety be solvable and that τ be highly computable and so the implicit signature τ must be carefully chosen. For pseudovarieties of semigroups, the signature κ introduced after the proof of Proposition 10.2 is often adequate for this purpose. But Proposition 10.2 shows that we should not expect this to be the case for all tame pseudovarieties of semigroups. For G_p, which is not κ-tame with respect to the class of systems of Example 9.4 by Proposition 10.2 but which is hyperdecidable by a result of Steinberg [59], the author has constructed in [7] an infinite signature for which G_p is tame.

11 Generalized Mal'cev and wreath products

This section does not contain any deep results but just a syntactic framework for computing certain kinds of pseudovarieties. The ideas are formulated in a rather abstract setting which may render a first reading somewhat difficult. Nevertheless, it should be clear that the hypotheses are fixed so that the natural arguments work. Perhaps working harder under a more restrictive setting one may find deeper results. The present aim is to set up common tools to deal with operations such as the Mal'cev and similar products and the semidirect product. There is however a fundamental difference between the two types of operations which this section exhibits and which makes the latter harder to compute by our syntactical approach.

Let Σ be a system of pseudo-equations over the pseudovariety V, without parameters, on the variables $x_1, \ldots, x_m$. Write $\Sigma(s_1, \ldots, s_m)$ if $(s_1, \ldots, s_m)$ is a solution of the system Σ in a pro-V algebra S.

The following result is a simple corollary of our general compactness theorem, Theorem 8.3. The apparently heavy notation with the superfluous index n is adopted to facilitate later reference in a context where the index will be required.

Theorem 11.1. *Let Φ_n and Σ_n be two systems of pseudo-equations over the pseudovariety* V, *both without parameters, and both on the same ν_n variables $x_1, \ldots, x_{\nu_n}$. If* W *is a subpseudovariety of* V, *then the following properties are equivalent for every $S \in$* V *and every choice of elements $s_1, \ldots, s_{\nu_n} \in S$:*

1. *There exists a relational morphism $\mu : S \to T$ into a member of* W *such that*

$$\left.\begin{array}{l}(s_i, t_i) \in \mu \ (i = 1, \ldots, \nu_n) \\ \Sigma_n(t_1, \ldots, t_{\nu_n})\end{array}\right\} \quad \Rightarrow \quad \Phi_n(s_1, \ldots, s_{\nu_n}) \qquad (11.1)$$

2. *There exists a closed relational morphism $\mu : S \to T$ into a pro-*W *algebra such that condition (11.1) holds.*

3. *Condition (11.1) holds for the canonical* W*-relational morphism $\mu : S \to \overline{\Omega}_B$*W *with respect to any (or every) finite generating set B.*

Proof. The negation of property (1) is equivalent to saying that the ν_n-tuple $(s_1, \ldots, s_{\nu_n})$ is such that $\Phi_n(s_1, \ldots, s_{\nu_n})$ fails and $(s_1, \ldots, s_{\nu_n})$ defines a constraining evaluation of the variables of the system Σ_n which is W-inevitable. The negation of property (2) may be formulated similarly by replacing W-inevitability by inevitability with respect to every closed relational morphism into every pro-W algebra. Given any finite generating set B for S, by Theorem 8.3 both these negations are equivalent to the ν_n-tuple $(s_1, \ldots, s_{\nu_n})$ failing the condition $\Phi_n(s_1, \ldots, s_{\nu_n})$ and being inevitable for the canonical W-relational morphism associated with the choice of generators. This property in turn is equivalent to the negation of (3). Hence properties (1), (2) and (3) are equivalent. □

Note that property (3) in the theorem is independent of the choice of elements of S. Thus, from the assumption that, for every ν_n-tuple $s_1, \ldots, s_{\nu_n}$ of elements of S, there exists a relational morphism satisfying (11.1), we deduce that there is such a relational morphism which works for all such ν_n-tuples. This could also be proved directly by noting that, S being finite, only finitely many relational morphisms $\mu_i : S \to T_i$ ($i = 1, \ldots, r$) are needed to handle all ν_n-tuples of elements of S. Letting $\mu : S \to T_1 \times \cdots \times T_r$ be the naturally associated relational morphism, (11.1) is satisfied for all ν_n-tuples of elements of S.

For the remainder of this section, let $(\Phi_n, \Sigma_n, \nu_n)_n$ be a triple sequence where $(\nu_n)_n$ is an increasing sequence of positive integers and Φ_n and Σ_n are systems of pseudo-equations over V, without parameters, depending on ν_n variables $x_1, \ldots, x_{\nu_n}$. For a pseudovariety W, consider the following two subclasses of V:

- $\mathcal{M}(\Phi_n, \Sigma_n, \nu_n, \mathsf{W})$ consists of all $S \in \mathsf{V}$ such that, for every $n \geq 1$ there exists $T_n \in \mathsf{W}$ and a relational morphism $\mu_n : S \to T_n$ such that, whenever $(s_i, t_i) \in \mu_n$ $(i = 1, \ldots, \nu_n)$ and $\Sigma_n(t_1, \ldots, t_{\nu_n})$, then $\Phi_n(s_1, \ldots, s_{\nu_n})$;

- $\mathcal{W}(\Phi_n, \Sigma_n, \nu_n, \mathsf{W})$ consists of all $S \in \mathsf{V}$ such that, there exists $T \in \mathsf{W}$ and a relational morphism $\mu : S \to T$ such that, whenever $n \geq 1$, $(s_i, t_i) \in \mu$ $(i = 1, \ldots, \nu_n)$, and $\Sigma_n(t_1, \ldots, t_{\nu_n})$, then $\Phi_n(s_1, \ldots, s_{\nu_n})$.

Thus the two definitions differ basically by the exchange of an existential with a universal quantifier. Obviously we have the inclusion

$$\mathcal{W}(\Phi_n, \Sigma_n, \nu_n, \mathsf{W}) \subseteq \mathcal{M}(\Phi_n, \Sigma_n, \nu_n, \mathsf{W})$$

but there is no *a priori* reason to expect equality.

Proposition 11.2. *The classes $\mathcal{W}(\Phi_n, \Sigma_n, \nu_n, \mathsf{W})$ and $\mathcal{M}(\Phi_n, \Sigma_n, \nu_n, \mathsf{W})$ are subpseudovarieties of V.*

Proof. The proof is immediate from the fact that the validity of a system of pseudo-equations on members of V is preserved by homomorphisms. To be more precise, we consider here the case of closure under taking homomorphic images for the class $\mathcal{M}(\Phi_n, \Sigma_n, \nu_n, \mathsf{W})$. The other closure properties are proved similarly.

Let S be a member of $\mathcal{M}(\Phi_n, \Sigma_n, \nu_n, \mathsf{W})$ and let $\varphi : S \to U$ be an onto homomorphism. Given $n \geq 1$, let $\mu_n : S \to T_n$ be a relational morphism into a member of W verifying (11.1). Then the composite $\theta_n = \mu_n \circ \varphi^{-1}$ is a relational morphism $U \to T_n$. If $(u_i, t_i) \in \theta$ for $i = 1, \ldots, \nu_n$ and $\Sigma_n(t_1, \ldots, t_{\nu_n})$, then by definition of θ there exist $s_1, \ldots, s_{\nu_n} \in S$ such that $\varphi(s_i) = u_i$ and $(s_i, t_i) \in \mu_n$ $(i = 1, \ldots, \nu_n)$. Since μ_n satisfies (11.1), it follows that $\Phi_n(s_1, \ldots, s_{\nu_n})$ and so also $\Phi_n(u_1, \ldots, u_{\nu_n})$ as implicit operations on V commute with homomorphisms between members of V. Hence U belongs to $\mathcal{M}(\Phi_n, \Sigma_n, \nu_n, \mathsf{W})$. □

Usually, it is pseudovarieties of the form $\mathcal{W}(\Phi_n, \Sigma_n, \nu_n, \mathsf{W})$ that one is led to consider. On the other hand, because of Theorem 11.1(3), pseudovarieties of the form $\mathcal{M}(\Phi_n, \Sigma_n, \nu_n, \mathsf{W})$ are more suitable for handling using the tameness approach of Section 10 as described further below. Thus it is important to examine the relationships between the two families of pseudovarieties.

One obvious situation under which the equality

$$\mathcal{W}(\Phi_n, \Sigma_n, \nu_n, \mathsf{W}) = \mathcal{M}(\Phi_n, \Sigma_n, \nu_n, \mathsf{W}) \tag{11.2}$$

holds is when Φ_n is trivial over V, in the sense that all members of V satisfy Φ_n, for all but finitely many values of n. Indeed then, for a relational morphism $\mu : S \to T$ into a member of W to witness that $S \in \mathcal{W}(\Phi_n, \Sigma_n, \nu_n, \mathsf{W})$, it only needs to verify condition (11.1) for finitely many values of n. But, if each $\mu_i : S \to T_i$ satisfies (11.1) for n_i ($i = 1, \ldots, r$), then the naturally associated relational morphism $\mu : S \to T_1 \times \cdots \times T_r$ also satisfies (11.1) for $n_1, \ldots, n_r$. This establishes the equality (11.2).

Another case which is of interest is considered in the following result which generalizes [47, Theorem 3.1] and whose proof amounts to purely combinatorial arguments. The hypotheses are still rather strong and it would be of interest to find weaker hypotheses that yield the equality (11.2).

Theorem 11.3. *Suppose there exists m such that, for all k, ℓ such that $\ell \geq k \geq m$, for every $T \in \mathsf{W}$ and all $t_1, \ldots, t_{\nu_\ell} \in T$, if $\Sigma_\ell(t_1, \ldots, t_{\nu_\ell})$ then, for all $i_1, \ldots, i_{\nu_k}$ such that $1 \leq i_1 < \cdots < i_{\nu_k} \leq \nu_\ell$, we have $\Sigma_k(t_{i_1}, \ldots, t_{i_{\nu_k}})$. Suppose further that either (a) the sequence $(\nu_n)_n$ is bounded or (b) for all k, ℓ such that $\ell \geq k \geq m$, $\Phi_k(s_1, \ldots, s_{\nu_k})$ implies $\Phi_\ell(s'_1, \ldots, s'_{\nu_\ell})$ whenever $\{s'_1, \ldots, s'_{\nu_\ell}\} = \{s_1, \ldots, s_{\nu_k}\} \subseteq S$ for an arbitrary element S of V. Then the equality (11.2) holds.*

Proof. Let S be a member of $\mathcal{M}(\Phi_n, \Sigma_n, \nu_n, \mathsf{W})$. Let m be as in the statement of the theorem. Note that every larger value of m possesses the same property. We also observe that the first hypothesis is inherited by all pro-W algebras.

(a) Consider first the case when the sequence $(\nu_n)_n$ is bounded. Since $(\nu_n)_n$ is assumed to be an increasing sequence, we may assume that it is constant for $n \geq m$. Moreover, since S is finite, there are only finitely ν_m-tuples of elements of S and so there exists $r \geq m$ such that, if $n > r$ and $\Phi_n(s_1, \ldots, s_{\nu_m})$ fails with the $s_i \in S$, then there exists k such that $r \geq k \geq m$ and $\Phi_k(s_1, \ldots, s_{\nu_m})$ also fails. Then the first hypothesis implies that, for $n > r$ and elements $(t_1, \ldots, t_{\nu_m})$ of a pro-W algebra T, for k such that $n \geq k \geq m$,

$$\Sigma_n(t_1, \ldots, t_{\nu_m}) \Rightarrow \Sigma_k(t_1, \ldots, t_{\nu_m}). \tag{11.3}$$

Now, for each $n \leq r$ there exists a relational morphism $\theta_n : S \to F_n$ with $F_n \in \mathsf{W}$ such that condition (11.1) holds. Let $F = F_1 \times \cdots \times F_r$ and let $\theta : S \to F$ be the induced relational morphism. Then θ verifies (11.1) for every $n \leq r$. Given $n > r$ and elements $s_1, \ldots, s_{\nu_m}$ of S, by the choice of r there exists k such that if $\Phi_n(s_1, \ldots, s_{\nu_m})$ fails, then $\Phi_k(s_1, \ldots, s_{\nu_m})$ also fails. Now, if the t_i are elements of F such that $(s_i, t_i) \in \theta$ and $\Sigma_n(t_1, \ldots, t_{\nu_m})$, then

$\Phi_n(s_1, \ldots, s_{\nu_m})$ must hold since, otherwise, $\Phi_k(s_1, \ldots, s_{\nu_m})$ would also have to fail and, since θ satisfies (11.1) for k, $\Sigma_n(t_1, \ldots, t_{\nu_m})$ would have to fail, in contradiction with (11.3). Hence θ also satisfies (11.1) for $n > r$. This shows that $S \in \mathcal{W}(\Phi_n, \Sigma_n, \nu_n, \mathsf{W})$.

(b) Suppose next that the sequence $(\nu_n)_n$ is unbounded and the hypothesis (b) holds. Then, we may take m so that $\nu_m \geq |S|$. Proceeding as above, consider for each $n \leq m$ a relational morphism $\theta_n : S \to F_n$ with $F_n \in \mathsf{W}$ such that condition (11.1) holds. Let $F = F_1 \times \cdots \times F_m$ and let $\theta : S \to F$ be the induced relational morphism. Then θ verifies (11.1) for every $n \leq m$. We claim θ satisfies (11.1) for arbitrary values of n, which will show that $S \in \mathcal{W}(\Phi_n, \Sigma_n, \nu_n, \mathsf{W})$. Indeed, let $n \geq m$ and suppose $(s_i, f_i) \in \theta$ $(i = 1, \ldots, \nu_n)$ and $\Sigma_n(f_1, \ldots, f_{\nu_n})$. Since $\nu_n \geq \nu_m \geq |S|$, we may choose indices i_j such that $1 \leq i_1 < \cdots < i_{\nu_m} \leq \nu_n$ and $\{s_{i_1}, \ldots, s_{i_{\nu_m}}\} = \{s_1, \ldots, s_{\nu_n}\}$. By the first hypothesis, we have $\Sigma_m(f_{i_1}, \ldots, f_{i_{\nu_m}})$. Since condition (11.1) holds for θ and m, it follows that $\Phi_m(s_{i_1}, \ldots, s_{i_{\nu_m}})$. Then, by the hypothesis (b), we obtain $\Phi_n(s_1, \ldots, s_{\nu_n})$. This proves the claim and completes the proof of the theorem. $\square$

Before proceeding with the study of pseudovarieties of the forms $\mathcal{W}(\ldots)$ and $\mathcal{M}(\ldots)$, it is worth presenting some examples mostly from the realm of semigroups, that is with $\mathsf{V} = \mathsf{S}$.

Example 11.1. For two subpseudovarieties $\mathsf{U}, \mathsf{W} \subseteq \mathsf{S}$, we have the equalities

$$\mathsf{U} \,\textcircled{m}\, \mathsf{W} = \mathcal{W}(\Phi_n, \Sigma_n, n, \mathsf{W}) = \mathcal{M}(\Phi_n, \Sigma_n, n, \mathsf{W}) \tag{11.4}$$

where Φ_n consists of pseudoidentities in n variables such that $\bigcup_{n \geq 1} \Phi_n$ is a basis of pseudoidentities for U which is closed under identification of variables and Σ_n consists of the identities $x_1 = \cdots = x_n = x_n^2$. The first equality in (11.4) is immediately verified from the definitions, whereas the second equality follows from Theorem 11.3. The second equality is a reformulation of [47, Theorem 3.1].

Example 11.2. For two subpseudovarieties U and W of V, following Pin [41, 42] we denote by $\mathsf{U}^{-1}\mathsf{W}$ the class of all $S \in \mathsf{V}$ for which there exists a relational morphism $\mu : S \to T$ with $T \in \mathsf{W}$ and, whenever U is a subalgebra of T such that $U \in \mathsf{U}$, then $\mu^{-1}U \in \mathsf{U}$ too. Denoting by Σ_n a basis of the pseudoidentities in n variables of U, we note that both hypotheses of Theorem 11.3 hold with $\Phi_n = \Sigma_n$ and $\nu_n = n$. It follows that

$$\mathsf{U}^{-1}\mathsf{W} = \mathcal{W}(\Sigma_n, \Sigma_n, n, \mathsf{W}) = \mathcal{M}(\Sigma_n, \Sigma_n, n, \mathsf{W}).$$

Example 11.3. Consider two subpseudovarieties U and W of S. Let Φ be a basis of pseudoidentities (over finite graphs) for gU. Let $\Gamma_1, \Gamma_2, \ldots$ be an enumeration of all finite graphs, up to isomorphism. Let ν_n be the size of the graph Γ_n (number of edges plus the number of vertices) and let Σ_n be the system consisting of all equations of the form

$$x_{\alpha e} x_e = x_{\omega e} \quad (e \in E(\Gamma_n)) \tag{11.5}$$

which express consistency of a labeling of Γ_n. Finally, let Φ_n consist of all pseudoidentities from Φ which are written over the graph Γ_n. Then we have the equality

$$\text{U} * \text{W} = \mathcal{W}(\Phi_n, \Sigma_n, \nu_n, \text{W}),$$

whose proof follows from the results of [13, Proposition 3.8]. Due to a mistake which basically amounts to an exchange of universal and existential quantifiers, it is asserted in [13] that the equality

$$\text{U} * \text{W} = \mathcal{M}(\Phi_n, \Sigma_n, \nu_n, \text{W}) \tag{11.6}$$

also holds.[b] The case of the equality (11.6) which is correctly proved in [13, Theorem 5.2] is when only finitely many graphs are used in the basis Φ of gU. This case is also established by the discussion preceding Theorem 11.3 but it should be pointed out that the compactness theorems upon which the present results depend are obtained by generalizing the arguments in [13]. Similar results and remarks apply to the bilateral semidirect product (cf. [13, Section 6] and Example 9.5).

Proceeding with the study of pseudovarieties of the forms $\mathcal{M}(\ldots)$ and $\mathcal{W}(\ldots)$, a question which naturally arises is the dependence of the pseudovarieties $\mathcal{W}(\Phi_n, \Sigma_n, \nu_n, \text{W})$ and $\mathcal{M}(\Phi_n, \Sigma_n, \nu_n, \text{W})$ on the systems of pseudo-equations Φ_n. More specifically, do these pseudovarieties depend on the sequence $(\Phi_n)_n$ or just on the pseudovariety defined by the set of pseudoidentities $\bigcup_n \Phi_n$?

For a set Φ of pseudoidentities over V written on finite subsets of the countable set of variables $\{x_1, x_2, \ldots\}$, let $(\tilde{\Phi}_n)_n$ denote the sequence of systems of pseudo-equations defined by taking $\tilde{\Phi}_n$ to consist of all pseudoidentities from Φ involving only the variables $x_1, \ldots, x_{\nu_n}$.

Proposition 11.4. *Let Φ and Ψ be two sets of pseudoidentities over V written on finite subsets of the countable set of variables $\{x_1, x_2, \ldots\}$ which, up to*

---

[b]The error was found by Rhodes and Steinberg. It is hidden in the first sentence of the proof of [13, Proposition 3.6].

equivalence of pseudoidentities on the same number of variables, are closed under identification of variables. If $[\![\Phi]\!] = [\![\Psi]\!]$ *then the following equalities hold:*

$$\mathcal{M}(\tilde{\Phi}_n, \Sigma_n, \nu_n, \mathsf{W}) = \mathcal{M}(\tilde{\Psi}_n, \Sigma_n, \nu_n, \mathsf{W}) \tag{11.7}$$

$$\mathcal{W}(\tilde{\Phi}_n, \Sigma_n, \nu_n, \mathsf{W}) = \mathcal{W}(\tilde{\Psi}_n, \Sigma_n, \nu_n, \mathsf{W}). \tag{11.8}$$

Proof. The proofs of the two equalities are identical: in the case of (11.7) one should consider a closed relational morphism $S \to T$ into a pro-W algebra T whereas in (11.8) T should be assumed to be also finite. The result follows from the definitions together with the observation that the hypotheses yield that $\tilde{\Phi}_n(s_1, \ldots, s_{\nu_n})$ implies $\tilde{\Psi}_n(s_1, \ldots, s_{\nu_n})$. The details are left for the reader to fill in. □

Proposition 11.4 allows us to define, for two subpseudovarieties U and W of V, and a sequence $\Sigma = (\Sigma_n)_n$ of finite systems of pseudo-equations, the following pseudovarieties:

- the *generalized Mal'cev product* $\mathsf{U} \, \textcircled{m}_\Sigma \, \mathsf{W}$ to be $\mathcal{M}(\tilde{\Phi}_n, \Sigma_n, \nu_n, \mathsf{W})$

- the *generalized wreath product* $\mathsf{U} \circ_\Sigma \mathsf{W}$ to be $\mathcal{W}(\tilde{\Phi}_n, \Sigma_n, \nu_n, \mathsf{W})$

where Φ is any basis of pseudoidentities for U which is closed under the identification of variables. This definition explains the title of this section. Note that:

- by Example 11.1, $\mathsf{U} \, \textcircled{m} \, \mathsf{W} = \mathsf{U} \, \textcircled{m}_\Sigma \, \mathsf{W}$ where $\Sigma_n = \{x_1 = \cdots = x_n = x_n^2\}$;

- by Example 11.2, $\mathsf{U}^{-1}\mathsf{W} = \mathsf{U} \, \textcircled{m}_\Sigma \, \mathsf{W}$ where Σ_n is a basis of the pseudoidentities in n variables of U;

- by Example 11.3, $\mathsf{U} * \mathsf{W} = \mathsf{U} \circ_\Sigma \mathsf{W}$ where Σ_n is the system (11.5) describing consistency of a labeling of the nth finite graph while $\mathsf{U} ** \mathsf{W} = \mathsf{U} \circ_\Lambda \mathsf{W}$ where Λ_n is the system

$$x_{\alpha(e)}y_e = x_{\omega(e)} \quad \text{and} \quad y_e z_{\omega(e)} = z_{\alpha(e)}$$

describing consistency of a "double labeling" of the nth graph.

Under fairly mild hypotheses, $\mathcal{M}(\ldots)$ and $\mathcal{W}(\ldots)$ may be defined in terms of generators by replacing relational morphisms in Theorem 11.1(1) by homomorphisms.

Proposition 11.5. *Suppose that* $\Sigma_n(t_1, \ldots, t_{\nu_n})$ *implies* $\Phi_n(t_1, \ldots, t_{\nu_n})$ *for arbitrary elements* $t_1, \ldots, t_{\nu_n}$ *of members of* W. *Then*

(a) $\mathcal{W}(\Phi_n, \Sigma_n, \nu_n, \mathsf{W})$ *is generated by those* $S \in \mathsf{V}$ *for which there exists a homomorphism* $\mu : S \to T$ *into* $T \in \mathsf{W}$ *such that condition (11.1) holds for every* n;

(b) $\mathcal{M}(\Phi_n, \Sigma_n, \nu_n, \mathsf{W})$ *is generated by those* $S \in \mathsf{V}$ *such that, for every* n *there exists a homomorphism* $\mu_n : S \to T_n$ *into some* $T_n \in \mathsf{W}$ *such that condition (11.1) holds.*

Proof. That such elements S of V belong to $\mathcal{W}(\ldots)$ and $\mathcal{M}(\ldots)$, respectively, is obvious since homomorphisms are relational morphisms. On the other hand, the hypothesis implies that, if a relational morphism $\mu : S \to T$ satisfies condition (11.1), then the second component projection $\mu \to T$ is a homomorphism that also satisfies the same condition. Since the first component projection $\mu \to S$ is an onto homomorphism, the result follows. □

To conclude this section, we show how the tameness approach allows us to prove decidability of pseudovarieties of the form $\mathcal{M}(\ldots)$.

Theorem 11.6. *Let σ be a finite finitary signature and suppose each Σ_n is a finite system of effectively verifiable pseudo-equations over a recursively enumerable pseudovariety V, the sequence $(\nu_n)_n$ is recursive, and each Φ_n is an effectively verifiable system of pseudo-equations. Let W be a recursively enumerable subpseudovariety of V and let $\mathsf{Z} = \mathcal{M}(\Phi_n, \Sigma_n, \nu_n, \mathsf{W})$.*

(a) *Suppose there exists m such that, for all k, ℓ such that $\ell \geq k \geq m$, every $T \in \mathsf{W}$, and all $t_1, \ldots, t_{\nu_\ell} \in T$, if $\Sigma_\ell(t_1, \ldots, t_{\nu_\ell})$ then, for all $i_1, \ldots, i_{\nu_k}$ such that $1 \leq i_1 < \cdots < i_{\nu_k} \leq \nu_\ell$, we have $\Sigma_k(t_{i_1}, \ldots, t_{i_{\nu_k}})$. Suppose further that either the sequence $(\nu_n)_n$ is bounded or for all k, ℓ such that $\ell \geq k \geq m$, $\Phi_k(s_1, \ldots, s_{\nu_k})$ implies $\Phi_\ell(s'_1, \ldots, s'_{\nu_\ell})$ whenever $\{s'_1, \ldots, s'_{\nu_\ell}\} = \{s_1, \ldots, s_{\nu_k}\} \subseteq S$ for an arbitrary element S of V. Then Z is recursively enumerable.*

(b) *If W is τ-tame with respect to each Σ_n, where τ is a highly computable signature such that the sides of the pseudo-equations of systems from each Σ_n are determined by τ-terms, then the complement $\mathsf{V} \setminus \mathsf{Z}$ is recursively enumerable. Moreover, assuming furthermore that each system Φ_n is recursively enumerable and consists of τ'-pseudo-equations, then Z is definable by a recursively enumerable basis of $(\tau \cup \tau')$-identities.*

Proof. (a) We may effectively enumerate, up to isomorphism, all relational morphisms $\mu : S \to T$ with $S \in \mathsf{V}$ and $T \in \mathsf{W}$. Let m be as in the statement of the theorem. Without loss of generality, we may assume that m, r are

55

sufficiently large as in the proof of Theorem 11.3. Since both S_n and Φ_n are effectively verifiable systems of pseudo-equations, we may test, for each $n \leq m$ (respectively $n \leq r$), whether condition (11.1) holds. If the results of all these tests are positive, then output S. The argument in the proof of Theorem 11.3 shows that in this way we effectively enumerate, up to isomorphism, precisely the pseudovariety Z.

(b) Under the hypotheses of (b), we may successively enumerate (up to isomorphism) members S of V, ν_n, τ-terms w_i ($i = 1, \ldots, \nu_n$), use the τ-word problem for W to verify whether $\Sigma_n(w_1, \ldots, w_{\nu_n})$ in the appropriate free pro-W algebra and, if so, check whether S satisfies the pseudoidentities $\Phi_n(w_1, \ldots, w_{\nu_n})$, outputting S in the negative case. By τ-reducibility of W with respect to each of the systems Σ_n, this provides a recursive enumeration, up to isomorphism, of the class Z. Moreover, consider the set Θ of all pseudoidentities of the form

$$\Phi_n(w_1, \ldots, w_{\nu_n})$$

where the w_i are τ-terms such that the system $\Sigma_n(w_1, \ldots, w_{\nu_n})$ is verified in W. Then, by τ-reducibility of W, Θ is a basis of pseudoidentities for Z. Finally, under the additional assumption on the systems Φ_n, Θ is a recursively enumerable basis of $(\tau \cup \tau')$-identities for Z. □

The following application is apparently new.

Corollary 11.7. *Suppose* V *is a recursively enumerable pseudovariety,* τ *is a highly computable signature,* U *is a subpseudovariety which, for each n, has a finite basis Σ_n for pseudoidentities on n variables which consists of τ-identities, and* W *is another subpseudovariety which is τ-tame with respect to each Σ_n. Then* $U^{-1}W$ *is decidable.*

Proof. By Example 11.2, we have $U^{-1}W = \mathcal{M}(\Sigma_n, \Sigma_n, n, W)$. Moreover, the hypotheses of Theorem 11.6 are verified with $\nu_n = n$ and $\Phi_n = \Sigma_n$, from which we conclude that $U^{-1}W$ is decidable. □

The hypothesis on U in Corollary 11.7 may seem very restrictive but it holds for most common pseudovarieties of semigroups with $\tau = \kappa$. So, for example, if W is completely κ-tame (or just κ-tame for the corresponding basis of pseudoidentities for U), then $U^{-1}W$ is decidable whenever U is any of the following pseudovarieties, which are some of the most studied pseudovarieties of semigroups:

- G, A, SI, Com = $[\![xy = yx]\!]$, Ab = G ∩ Com,

- $\mathsf{CR}_n = [\![x^{n+1} = x]\!]$, $\mathsf{CS} = [\![(xy)^\omega x = x]\!]$,
- $\mathsf{R}_n = [\![(xy)^\omega x = (xy)^\omega, x^{n+1} = x^n]\!]$, $\mathsf{L}_n = [\![y(xy)^\omega = (xy)^\omega, x^{n+1} = x^n]\!]$, $\mathsf{J}_n = \mathsf{R}_n \cap \mathsf{L}_n$,
- $\mathsf{K}_n = [\![x^n y = x^n]\!]$, $\mathsf{D}_n = [\![yx^n = x^n]\!]$, $\mathsf{N}_n = \mathsf{K}_n \cap \mathsf{D}_n$,
- $\mathsf{LI}_n = [\![x^n y x^n = x^n]\!]$, $\mathsf{LG}_n = [\![(x^n y)^\omega x^n = x^n]\!]$,
- $\mathsf{DA} = [\![((xy)^\omega x)^2 = (xy)^\omega x]\!]$, $\mathsf{DS} = [\![((xy)^\omega x)^{\omega+1} = (xy)^\omega x]\!]$,
- $\mathsf{DG} = [\![(xy)^\omega = (yx)^\omega]\!]$, $\mathsf{DO} = [\![(xy)^\omega (yx)^\omega (xy)^\omega = (xy)^\omega]\!]$,
- $\mathsf{ZE} = [\![x^\omega y = yx^\omega]\!]$, $\mathsf{ESl} = [\![x^\omega y^\omega = y^\omega x^\omega]\!]$, $\mathsf{BG} = [\![(x^\omega y)^\omega = (yx^\omega)^\omega]\!]$,

where $n \in \{1, 2, \ldots, \omega\}$.

This is by no means a catalog of all decidability results for pseudovarieties of the form $\mathsf{U}^{-1}\mathsf{W}$ which can be obtained by applying Corollary 11.7 but rather just a list of examples based on a partial table for this Mal'cev-type operation which can be found in Pin's thesis [41]. It should also be emphasized that these results are only concerned with decidability. For more precise results leading to specific calculations of pseudovarieties of the form $\mathsf{U}^{-1}\mathsf{W}$, one should explore the present approach to extract bases of pseudoidentities with a recognizable structural meaning, namely by refining the basis given by Theorem 11.6 in the manner exemplified in the calculations of [13]. The table in Pin's thesis aims at these more precise calculations.

12 Examples of completely tame pseudovarieties

Say that a pseudovariety W is *order-computable* if, for every finite set B, the W-free algebra over B, $\Omega_B \mathsf{W}$, is finite and computable. Since $\Omega_B \mathsf{W}$ is dense in $\overline{\Omega}_B \mathsf{W}$, then $\overline{\Omega}_B \mathsf{W} = \Omega_B \mathsf{W}$ and so $\overline{\Omega}_B \mathsf{W}$ is finite and computable.

Let φ be an evaluation of the variables of a finite system of pseudo-equations in a B-generated member S of V which is W-inevitable and consider an evaluation γ in $\overline{\Omega}_B \mathsf{W}$ of the parameters of the system. By W-inevitability, there is some evaluation θ of the variables of the system in $\overline{\Omega}_B \mathsf{V}$ which projects to φ under the unique continuous homomorphism $\hat{\psi} : \overline{\Omega}_B \mathsf{V} \to S$ extending the choice of generators and such that $p_\mathsf{W} \circ \theta$ is consistent with γ. Now, since, for every $v \in \overline{\Omega}_A \mathsf{V}$,

$$U = \hat{\psi}^{-1}(\hat{\psi}(v)) \cap p_\mathsf{W}^{-1}(p_\mathsf{W}(v))$$

is a clopen set containing v and $\Omega_B \mathsf{V}$ is dense in $\overline{\Omega}_B \mathsf{V}$, U contains some element v' of $\Omega_B \mathsf{V}$. Then, replacing each value v of θ by a so-obtained v', we construct an evaluation θ' of the variables of the system in $\Omega_B\mathsf{V}$ which projects in S and in $\Omega_B\mathsf{W}$ precisely in the same way as θ. This proves the following observation.

Proposition 12.1. *An order-computable pseudovariety is completely tame for the implicit signature reduced to σ.*

For other examples of completely tame pseudovarieties, we turn back into semigroup world, that is we take again $\mathsf{V} = \mathsf{S}$. Recall the signature κ introduced in Section 10 which consists of multiplication together with the unary operation $x \mapsto x^{\omega-1}$. For aperiodic pseudovarieties, we may as well work with the operation $x \mapsto x^\omega$ since the latter is expressible in terms of the former and the two coincide over aperiodic semigroups.

The following improves on a result of Zeitoun and the author [15] on the pseudovariety N of all finite nilpotent semigroups, with basically the same proof.

Proposition 12.2. *The pseudovariety N is completely κ-tame.*

Proof. The pseudovariety N is clearly recursively enumerable. By Corollary 8.5(3), any two non-explicit operations of the same arity coincide over N, as both have constant value 0 over any finite nilpotent semigroup. Thus $\overline{\Omega}_A\mathsf{N} = \Omega_A\mathsf{N} \cup \{x_a^\omega\} = \Omega_A^\kappa\mathsf{N}$ for any $a \in A$ and the κ-word problem is decidable: two κ-terms are equal over N if and only if they are equal or they both involve the unary operation. Thus all we need to prove is κ-reducibility.

The proof follows similar lines to the order-computable case. So, let S be a finite semigroup and consider an evaluation φ, in S, of the variables of a finite system of pseudo-equations which is N-inevitable. Let γ be an evaluation of the parameters in $\Omega_B^\kappa \mathsf{N}$ where B is a finite generating set for S. By N-inevitability of φ, there exists an evaluation θ of the variables in $\overline{\Omega}_B S$ such that

(a) $p_\mathsf{N} \circ \theta$ is consistent with γ

(b) $\hat{\psi} \circ \theta = \varphi$

for the unique continuous homomorphism $\hat{\psi} : \overline{\Omega}_B S \to S$ which extends the given generating mapping $\psi : B \to S$.

We modify θ so as to retain the above properties (a) and (b) but so that the resulting evaluation θ' takes its values in $\Omega_B^\kappa S$. If $\theta(a)$ is explicit, then we let $\theta'(a) = \theta(a)$. Otherwise, we use again Corollary 8.5(3) to obtain a factorization $\theta(a) = uv^\omega w$ for some $u, v, w \in \overline{\Omega}_B S$. Since $\Omega_B S$ is dense in $\overline{\Omega}_B \mathsf{S}$, there are $u_0, v_0, w_0 \in \Omega_B \mathsf{S}$ such that $\hat{\psi}(u_0) = \hat{\psi}(u)$, $\hat{\psi}(v_0) = \hat{\psi}(v)$, and $\hat{\psi}(w_0) = \hat{\psi}(w)$. Then taking $\theta'(a) = u_0 v_0^\omega w_0$ does the required job. □

A more complicated case is that of the pseudovariety J of all finite $\mathcal{J}$-trivial semigroups. The κ-word problem for J was solved by the author [3] (see also [4, Section 8.2]). The solution consists in the application of the following reduction rules transforming any κ-term to a canonical form:

- replace a subterm of the form $(w)^\omega$ by $(u)^\omega$ where u is the product in a fixed order of the distinct variables which occur in w;
- if u and v are subterms such that every variable which occurs in u also occurs in v, then replace subterms of the forms $u(v)^\omega$ and $(v)^\omega u$ by $(v)^\omega$.

Moreover, every implicit operation $w \in \overline{\Omega}_A \mathsf{S}$ admits a factorization of the form $w = w_0 w_1 \cdots w_n$ where:

1. each factor w_i is either explicit or $p_\mathsf{J}(w_i)$ is an idempotent;

2. no two consecutive non-explicit factors w_i, w_{i+1} have comparable contents;

3. if w_i is explicit and $i < n$, then w_{i+1} is non-explicit and the last letter of w_i does not belong to $c(w_{i+1})$;

4. if w_i is explicit and $i > 0$, then w_{i-1} is non-explicit and the first letter of w_i does not belong to $c(w_{i-1})$.

Here, the *content* of implicit operations in $\overline{\Omega}_A \mathsf{S}$ is given by the natural projection $c : \overline{\Omega}_A \mathsf{S} \to \overline{\Omega}_A \mathsf{SI}$. In view of the solution of the κ-word problem for J, two factorizations satisfying these four properties give implicit operations which coincide on J if and only if they have the same number of factors, equal explicit factors in corresponding positions, and equal-content non-explicit factors in corresponding positions. Note that, in particular, $\overline{\Omega}_A \mathsf{J} = \Omega_A^\kappa \mathsf{J}$.

A further important observation is that an implicit operation $w \in \overline{\Omega}_A \mathsf{S}$ is an idempotent on J if and only if, for every $n \geq 1$, w admits a factorization in n factors with the same content.

Theorem 12.3. *The pseudovariety J is completely κ-tame.*

Proof. Since J is obviously recursively enumerable, it remains to show that J is κ-reducible. Let S be a finite semigroup and consider an evaluation φ, in S, of the variables of a finite system of pseudo-equations which is J-inevitable. Let γ be an evaluation of the parameters in $\Omega_B^\kappa \mathsf{J}$ where B is a finite generating

59

set for S. Since φ is J-inevitable, there exists an evaluation θ of the variables in $\overline{\Omega}_B S$ such that

(a) $p_\mathsf{J} \circ \theta$ is consistent with γ

(b) $\hat{\psi} \circ \theta = \varphi$

for the unique continuous homomorphism $\hat{\psi} : \overline{\Omega}_B S \to S$ which extends the given generating mapping $\psi : B \to S$.

We again modify θ so as to retain the above properties (a) and (b) but so that the resulting evaluation θ' takes its values in $\Omega_B^\kappa S$. We start by factorizing each $\theta(a)$ in the form $\theta(a) = w_0 w_1 \cdots w_n$ satisfying the above properties (1)–(4). We leave explicit factors in such factorizations as they are. Let m be the cardinality of S. For each non-explicit factor w_i, we take a factorization $w_i = w_{i,1} w_{i,2} \cdots w_{i,m}$ into m factors with the same content. By Lemma 8.4, there exist j and k, with $1 \leq j < k \leq m$, such that

$$\hat{\psi}(w_i) = \hat{\psi}(w_{i,1} \cdots w_{i,j} (w_{i,j+1} \cdots w_{i,k})^\omega w_{i,k+1} \cdots w_{i,m}).$$

Next, for each factor $w_{i,\ell}$, we choose an explicit operation $v_{i,\ell}$ with the same content such that $\hat{\psi}(w_{i,\ell}) = \hat{\psi}(v_{i,\ell})$. Finally, we replace the non-explicit factor w_i by the product

$$v_{i,1} \cdots v_{i,j} (v_{i,j+1} \cdots v_{i,k})^\omega v_{i,k+1} \cdots v_{i,m}.$$

This produces a product in $\Omega_B^\kappa S$ replacing $w_1 w_2 \cdots w_n$ which we denote by $\theta'(a)$. By construction, $\hat{\psi} \circ \theta' = \hat{\psi} \circ \theta$ and $p_\mathsf{J} \circ \theta' = p_\mathsf{J} \circ \theta$ so that the variable evaluation θ' has the required properties. □

Hyperdecidability of J was proved by Zeitoun and the author [14] by exhibiting more reasonable algorithms but the proof is a lot longer than the above proof of a much stronger result. There are several other instances where it turns out that it is apparently easier to solve a word problem and prove reducibility than trying to construct direct algorithms to prove hyperdecidability.

Recently, Delgado and the author [10] have shown that the pseudovariety Ab of all finite Abelian groups is completely κ-tame. In contrast, results of Coulbois and Khélif [19] show that G is not completely κ-tame.

Acknowledgement

The author wishes to express his thanks to the anonymous referees for their comments and suggestions which helped to remove incorrections and to improve readability of these notes.

References

[1] J. Almeida, *The algebra of implicit operations*, Algebra Universalis **26** (1989) 16–32.

[2] ———, *Residually finite congruences and quasi-regular subsets in uniform algebras*, Portugal. Math. **46** (1989) 313–328.

[3] ———, *Implicit operations on finite J-trivial semigroups and a conjecture of I. Simon*, J. Pure and Applied Algebra **69** (1990) 205–218.

[4] ———, *Finite Semigroups and Universal Algebra*, World Scientific, Singapore, 1995. English translation.

[5] ———, *Hyperdecidable pseudovarieties and the calculation of semidirect products*, Int. J. Algebra Comput. **9** (1999) 241–261.

[6] ———, *Some algorithmic problems for pseudovarieties*, Publ. Math. Debrecen **54 Suppl.** (1999) 531–552.

[7] ———, *Dynamics of implicit operations and tameness of pseudovarieties of groups*, Trans. Amer. Math. Soc. **354** (2002) 387–411.

[8] J. Almeida and M. Delgado, *Sur certains systèmes d'équations avec contraintes dans un groupe libre*, Portugal. Math. **56** (1999) 409–417.

[9] ———, *Sur certains systèmes d'équations avec contraintes dans un groupe libre—addenda*, Portugal. Math. **58** (2001) 379–387.

[10] ———, *Tameness of the pseudovariety of Abelian groups*, Tech. Rep. CMUP 2001-24, Univ. Porto, 2001.

[11] J. Almeida and B. Steinberg, *On the decidability of iterated semidirect products and applications to complexity*, Proc. London Math. Soc. **80** (2000) 50–74.

[12] ———, *Syntactic and Global Semigroup Theory, a Synthesis Approach*, in Algorithmic Problems in Groups and Semigroups, J. C. Birget, S. W. Margolis, J. Meakin, and M. V. Sapir, eds., Birkhäuser, 2000, 1–23.

[13] J. Almeida and P. Weil, *Profinite categories and semidirect products*, J. Pure and Appl. Algebra **123** (1998) 1–50.

[14] J. Almeida and M. Zeitoun, *The pseudovariety J is hyperdecidable*, Theoretical Informatics and Applications **31** (1997) 457–482.

[15] ———, *Tameness of some locally trivial pseudovarieties*, Comm. Algebra. To appear.

[16] C. J. Ash, *Inevitable graphs: a proof of the type II conjecture and some related decision procedures*, Int. J. Algebra Comput. **1** (1991) 127–146.

[17] G. Baumslag, *Residual nilpotence and relations in free groups*, J. Algebra **2** (1965) 271–282.

[18] J. Berstel, M. Crochemore, and J.-É. Pin, *Thue-Morse sequence and p-adic topology for the free monoid*, Discrete Math. **76** (1989) 89–94.

[19] T. Coulbois and A. Khélif, *Equations in free groups are not finitely approximable*, Proc. Amer. Math. Soc. **127** (1999) 963–965.

[20] M. Delgado, *Abelian pointlikes of a monoid*, Semigroup Forum **56** (1998) 127–146.

[21] S. Eilenberg, *Automata, Languages and Machines*, vol. A, Academic Press,

New York, 1974.
[22] ———, *Automata, Languages and Machines*, vol. B, Academic Press, New York, 1976.
[23] S. Eilenberg and M. P. Schützenberger, *On pseudovarieties*, Advances in Math. **19** (1976) 413–418.
[24] K. Henckell, *Pointlike sets: the finest aperiodic cover of a finite semigroup*, J. Pure and Appl. Algebra **55** (1988) 85–126.
[25] K. Henckell, S. Margolis, J.-É. Pin, and J. Rhodes, *Ash's type II theorem, profinite topology and Malcev products. Part I*, Int. J. Algebra Comput. **1** (1991) 411–436.
[26] K. Henckell and J. Rhodes, *The theorem of Knast, the PG=BG and Type II Conjectures*, in Monoids and Semigroups with Applications, J. Rhodes, ed., Singapore, 1991, World Scientific, 453–463.
[27] B. Herwig and D. Lascar, *Extending partial automorphisms and the profinite topology on free groups*, Trans. Amer. Math. Soc. **352** (2000) 1985–2021.
[28] P. M. Higgins, *An algebraic proof that pseudovarieties are defined by pseudoidentities*, Algebra Universalis **27** (1990) 597–599.
[29] R. P. Hunter, *Certain finitely generated compact zero-dimensional semigroups*, J. Austral. Math. Soc., Ser. A **44** (1988) 265–270.
[30] P. R. Jones, *Profinite categories, implicit operations and pseudovarieties of categories*, J. Pure and Appl. Algebra **109** (1996) 61–95.
[31] O. G. Kharlampovich and M. Sapir, *Algorithmic problems in varieties*, Int. J. Algebra Comput. **5** (1995) 379–602.
[32] S. C. Kleene, *Representations of events in nerve nets and finite automata*, in Automata Studies, C. E. Shannon, ed., vol. 3-41, Princeton, N.J., 1956, Princeton University Press. reprinted in [39].
[33] G. Lallement, *Semigroups and Combinatorial Applications*, Wiley, New York, 1979.
[34] S. Margolis, M. Sapir, and P. Weil, *Closed subgroups in pro-V topologies and the extension problem for inverse automata*, Int. J. Algebra Comput. **11** (2001) 405–445.
[35] S. W. Margolis and J.-É. Pin, *Varieties of finite monoids and topology for the free monoid*, in Proc. 1984 Marquette Semigroup Conference, Milwaukee, 1984, Marquette University, 113–129.
[36] O. Matz, A. Miller, A. Potthoff, W. Thomas, and E. Valkema, *Report on the program AMoRe*, Tech. Rep. 9507, Christian Albrechts Universität, Kiel, 1995.
[37] J. McCammond, *Normal forms for free aperiodic semigroups*, Int. J. Algebra Comput. **11** (2001) 565–580.
[38] J. McCammond and J. Rhodes, *Geometric semigroup theory*, tech. rep., Texas A&M Univ. and Univ. California at Berkeley, 2000.
[39] E. F. Moore (Ed.), *Sequential Machines: Selected Papers*, Reading, Mass., 1964, Addison-Wesley.
[40] K. Numakura, *Theorems on compact totally disconnetced semigroups and lat-*

tices, Proc. Amer. Math. Soc. **8** (1957) 623–626.
[41] J.-É. Pin, *Variétés de langages et variétés de semigroupes*, Thèse d'état, Univ. Paris 6, 1981.
[42] ———, *Varieties of Formal Languages*, Plenum, London, 1986. English translation.
[43] ———, *Topologies for the free monoid*, J. Algebra **137** (1991) 297–337.
[44] ———, *Topologie p-adique sur les mots*, J. Théor. Nombres Bordeaux **5** (1993) 263–281.
[45] ———, *A variety theorem without complementation*, Russian Math. (Iz. VUZ) **39** (1995) 80–90.
[46] J.-É. Pin and C. Reutenauer, *A conjecture on the Hall topology for the free group*, Bull. London Math. Soc. **23** (1991) 356–362.
[47] J.-É. Pin and P. Weil, *Profinite semigroups, Mal'cev products and identities*, J. Algebra **182** (1996) 604–626.
[48] L. Polák, *A classification of rational languages by semilattice-ordered monoids*, tech. rep., Masaryk Univ., 2001.
[49] J. Reiterman, *The Birkhoff theorem for finite algebras*, Algebra Universalis **14** (1982) 1–10.
[50] J. Rhodes, *Complexity c is decidable for finite automata and semigroups*, tech. rep., Univ. California at Berkeley, 2000.
[51] ———, *Flows on automata*, tech. rep., Univ. California at Berkeley, 2000.
[52] L. Ribes and P. A. Zalesskiĭ, *On the profinite topology on a free group*, Bull. London Math. Soc. **25** (1993) 37–43.
[53] ———, *The pro-p topology of a free group and algorithmic problems in semigroups*, Int. J. Algebra Comput. **4** (1994) 359–374.
[54] ———, *Profinite Groups*, Ergeb. Math. und ihrer Grenzgebiete 3. Folge 40, Springer, Berlin, 2000.
[55] S. Satoh, K. Yama, and M. Tokizawa, *Semigroups of order 8*, Semigroup Forum **49** (1994) 7–30.
[56] M. P. Schützenberger, *On finite monoids having only trivial subgroups*, Inform. and Control **8** (1965) 190–194.
[57] B. Steinberg, *On pointlike sets and joins of pseudovarieties*, Int. J. Algebra Comput. **8** (1998) 203–231.
[58] ———, *On algorithmic problems for joins of pseudovarieties*, Semigroup Forum **62** (2001) 1–40.
[59] ———, *Inevitable graphs and profinite topologies: some solutions to algorithmic problems in monoid and automata theory, stemming from group theory*, Int. J. Algebra Comput. **11** (2001) 25–71.
[60] ———, *Inverse automata and profinite topologies on a free group*, J. Pure and Appl. Algebra **167** (2002) 341–359.
[61] B. Tilson, *Categories as algebra: an essential ingredient in the theory of monoids*, J. Pure and Applied Algebra **48** (1987) 83–198.
[62] A. N. Trahtman, *Algorithms verifying local threshold and piecewise testability*

of semigroup nd solving Almeida problem, RIMS Kokyuroku Proceedings of Kyoto Univ. **1222** (2001) 145–151.

[63] M. V. Volkov, *On a class of semigroup pseudovarieties without finite pseudoidentity basis*, Int. J. Algebra Comput. **5** (1995) 127–135.

[64] S. Willard, *General Topology*, Addison-Wesley, Reading, Mass., 1970.

[65] I. Y. Zhil'tsov, *On identities of finite aperiodic epigroups*, tech. rep., Ural State Univ., 1999.

ON EXISTENCE VARIETIES OF REGULAR SEMIGROUPS

KARL AUINGER*

Institut für Mathematik, Strudlhofgasse 4, A-1090 Wien, Austria
E-mail: karl.auinger@univie.ac.at

This article presents a summary of (the main parts of) the lectures on existence varieties of regular semigroups given by the author at the Thematic Term on Semigroups, Algorithms, Automata and Languages in Coimbra, July 2001. The selection of the presented topics very much reflects the author's personal taste under the constraint of the limited time he was given. So this is by no means an attempt to give a complete overview over the field. The main emphasis is on the development of the "bi-equational" approach to orthodox, locally inverse and E-solid e-varieties. This is then applied to the study and comparison of two kinds of semidirect product operators $*_r$ and $*_{rr}$. For other topics, the reader may consult the survey papers by Jones [19] and Trotter [36].

1 Introduction

An element x of a semigroup S is *regular* (in S) if there is an element $y \in S$ such that $x = xyx$; for $x' = yxy$ then both equalities $x = xx'x$ and $x' = x'xx'$ hold. In this case x' is an *inverse* of x (and x is an inverse of x'). A semigroup is regular if each element of S is regular, or, equivalently, if each element has an inverse. A standard example of a regular semigroup is the semigroup of all mappings on a set, another the semigroup of all linear mappings on a vector space (both under composition of mappings). The class $\mathcal{RS}$ of all regular semigroups as well as many of its much studied subclasses do not form varieties. This is essentially because of the existential quantifier in the definition of regularity: subsemigroups of regular semigroups are not necessarily regular. Here the central definition of this survey comes into play:

Definition 1.1. A class $\mathcal{V}$ of regular semigroups closed under taking direct products, regular subsemigroups and morphic images is an *existence variety*

---

*THE AUTHOR WOULD LIKE TO ACKNOWLEDGE THE FINANCIAL SUPPORT OF FUNDAÇÃO CALOUSTE GULBENKIAN (FCG), FUNDAÇÃO PARA A CIÊNCIA E A TECNOLOGIA (FCT), FACULDADE DE CIÊNCIAS DA UNIVERSIDADE DE LISBOA (FCUL) AND REITORIA DA UNIVERSIDADE DO PORTO.

or *e-variety*.

This definition has been given independently by Hall [15] and Kaďourek and Szendrei [23] in the late '80s (the latter authors first considered only orthodox semigroups and used the term *bivariety* instead).

The purpose of this definition is to provide a tool for the hierarchic classification of regular semigroups as is common in universal algebra. Implicitly, the concept occurred already earlier. Many previously studied classes of regular semigroups such as: completely regular, inverse, orthodox, E-solid semigroups, etc., have turned out to be e-varieties. What do we mean by "hierarchic classification"? Consider a variety of universal algebras, $\mathcal{V}$, say. Then the collection of all subvarieties of $\mathcal{V}$ forms a complete lattice under inclusion. "Classification" now can be viewed in this context as "describing this lattice" or at least "providing information on this lattice."

The collection of all e-varieties of regular semigroups also forms a complete lattice under inclusion. This has been observed by Hall [15] and can be seen as follows. Call a unary operation ′ on a (regular) semigroup S an *inverse unary operation* if it satisfies the identities $x \simeq xx'x$ and $x' \simeq x'xx'$ and an algebra $(S, \cdot, ')$ a *regular unary semigroup* if $(S, \cdot)$ is a (regular) semigroup and ′ is an inverse unary operation on S. For an e-variety $\mathcal{V}$ set

$$\mathcal{V}' = \{(S, \cdot, ') \mid {}' \text{ is an inverse unary operation and } (S, \cdot) \in \mathcal{V}\}$$

— then $\mathcal{V}'$ is a variety of regular unary semigroups and the assignment $\mathcal{V} \mapsto \mathcal{V}'$ is injective. In particular, the collection of all e-varieties is a set[a] and therefore a complete lattice under inclusion. Moreover, the mapping $\mathcal{V} \mapsto \mathcal{V}'$ from the lattice of e-varieties to the lattice of varieties of regular unary semigroups respects complete meets, but in general does not respect joins. Thus the lattice of all e-varieties of regular semigroups is isomorphic with a complete meet-subsemilattice of the lattice of all regular unary semigroup varieties. As far as we know this has not yet been used to obtain interesting information about the lattice of all e-varieties.

Let us list some prominent members of the family of all e-varieties:

$\mathcal{RS}$ all regular semigroups
$\mathcal{G}$ groups
$\mathcal{I}$ inverse semigroups: each element has a unique inverse
$\mathcal{CR}$ completely regular semigroups: unions of groups
$\mathcal{B}$ bands: idempotent semigroups
$\mathcal{O}$ orthodox semigroups: the idempotents form a subsemigroup

---

[a] The cardinality of this set is that of the continuum.

$\mathcal{ES}$ E-solid semigroups: the idempotents generate a completely regular subsemigroup
$\mathcal{S}l$ semilattices
$\mathcal{CS}$ completely simple semigroups
$\mathcal{RB}$ rectangular bands
$\mathcal{RZ}$ right zero semigroups

Moreover, there are two important operators E and L to obtain new e-varieties from old ones; denote by $E(S)$ the set of all idempotents of the semigroup S. For an e-variety $\mathcal{V}$ set

$$L\mathcal{V} = \{S \in \mathcal{RS} \mid \forall e \in E(S) : eSe \in \mathcal{V}\}$$

and

$$E\mathcal{V} = \{S \in \mathcal{RS} \mid \langle E(S) \rangle \in \mathcal{V}\}.$$

The subsemigroups eSe of S, which are monoids with identities e, are usually referred to as the *local submonoids* of S, and the idempotent generated subsemigroup $\langle E(S) \rangle$ is the *core* of S; note that the local submonoids and the core of a regular semigroup are always regular. Three particular e-varieties will play a central role in this paper:

$\mathcal{O} = E\mathcal{B}$ orthodox semigroups
$\mathcal{ES} = E\mathcal{CR}$ E-solid semigroups
$L\mathcal{I}$ locally inverse semigroups.

Two other e-varieties also play an exceptional role, namely inverse semigroups $\mathcal{I}$ and completely regular semigroups $\mathcal{CR}$. Both can be viewed as varieties of regular unary semigroups. On an inverse semigroup each element has a unique inverse which gives rise to a naturally defined inverse unary operation on any inverse semigroup. All semigroup morphisms among inverse semigroups respect this operation; moreover, the regular subsemigroups of an inverse semigroup are precisely the *inverse* subsemigroups, that is, the substructures with respect to this enriched signature. From this it follows that there is a one-to-one correspondence between varieties of inverse semigroups (considered as unary semigroups) and e-varieties of inverse semigroups (considered as regular semigroups). The same is true for completely regular semigroups: on such a semigroup one chooses as inverse unary operation the operation which assigns to each element s the unique inverse s^{-1} inside the maximal subgroup H_s of the element s. Then again, $^{-1}$ is respected by all semigroup morphisms and the regular subsemigroups of a completely regular semigroup are precisely the *completely regular* subsemigroups. Again, varieties of completely regular semigroups are essentially the same as e-varieties. In particular this means that for the study of the lattices $\mathcal{L}(\mathcal{I})$ and $\mathcal{L}(\mathcal{CR})$ of e-varieties of inverse and completely regular semigroups, respectively, we can apply all well established tools from universal algebra such as free objects, identities, etc. This was done in both cases already before the notion of e-variety was invented, see Petrich [28] for the inverse case and Polák [30, 31, 32] and Petrich and Reilly [29] for the completely regular case.

So the question arises: Can we do the same outside $\mathcal{I}$ and $\mathcal{CR}$? The answer is: No! Indeed, take any e-variety $\mathcal{V}$ and suppose we can define on each member of $\mathcal{V}$ additional operations so that the class $\mathcal{V}$ becomes a variety in this enriched signature and such that the subvarieties of this variety are in one-to-one correspondence to the sub-e-varieties of $\mathcal{V}$. Then, for any $S \in \mathcal{V}$ and any non-empty subset $X \subseteq S$ the semigroup reduct of the subalgebra $\langle X \rangle$ of S generated by X is a regular subsemigroup of S and, moreover, this regular semigroup contains each regular subsemigroup T of S which contains the set X. In other words, for each member S of $\mathcal{V}$ and each subset $X \subseteq S$ the intersection of all regular subsemigroups of S which contain X must again be a regular subsemigroup of S. However, this is not the case as soon as we go outside inverse or completely regular semigroups. Indeed, if $\mathcal{V}$ is an e-variety not contained in $\mathcal{I}$ nor in $\mathcal{CR}$ then $\mathcal{V}$ contains a non-trivial right zero or left zero semigroup and also the Brandt semigroup B_2. From this it follows easily that $\mathcal{V}$ also contains the combinatorial completely 0-simple semigroup with non-zero $\mathcal{D}$-class

or its dual. Here the stars * indicate the positions of the idempotents. It is clear that there is no least regular subsemigroup containing the element a.

However, it turns out that inside the larger e-varieties $\mathcal{LI}$ and $\mathcal{ECR}$ we do have a behaviour which is very close to that of varieties. This has been discovered first by Kaďourek and Szendrei [23] for the class of orthodox semigroups.

2 Characterization of E-varieties in Terms of Bi-identities

2.1 The orthodox case

In the above example, if we choose an inverse a' of a then we do have a unique smallest regular subsemigroup containing the set $\{a, a'\}$. In the following, the essential idea will be to replace the set X of variables (which in the variety-context represent arbitrary elements of an algebra in the variety) by the doubled set $\overline{X} := X \cup X'$ where $X' = \{x' \mid x \in X\}$ is a disjoint copy of X with $x \mapsto x'$ being a bijection, and build up an equational logic on this doubled alphabet. Here the elements $x \in X$ are supposed to represent arbitrary elements of a regular semigroup while the element $x' \in X'$ represents an inverse element of the element represented by x. In order that such an approach works — for the moment only in the context of orthodox semigroups — the following lemma is crucial.

Lemma 2.1. *Let S be orthodox and $A \subseteq S$ be such that each element of A has an inverse in A. Then the semigroup $\langle A \rangle$ generated by A is a regular subsemigroup of S.*

In the following we present the "bi-equational" approach to e-varieties of orthodox semigroups developed by Kaďourek and Szendrei [23]. Let X be a countably infinite set (of variables) and put $\overline{X} := X \cup X'$ as above. Set

$$T(X) := \overline{X}^+, \text{ the free semigroup on } \overline{X}.$$

$T(X)$ shall be the semigroup of *terms*; as in the context of varieties of universal algebras, the terms are aimed at representing elements in orthodox semigroups, given that the variables are substituted with elements. However, we have to be careful as far as the substitutions are concerned and have to take into account that x' stands for an inverse element of x:

Definition 2.1. *A mapping $\varphi : \overline{X} \to S$ is matched or a substitution if $x'\varphi$ is an inverse of $x\varphi$ for each $x \in X$.*

We note that this definition includes the possibility that for $x \neq y$ we may have $x\varphi = y\varphi$ but $x'\varphi \neq y'\varphi$. It is important to understand that $'$ here is not a unary operation but is just a *marker to distinguish different sorts of variables*. Each substitution $\varphi : \overline{X} \to S$ extends uniquely to a morphism $\Phi : T(X) \to S$; then for a term $w \in T(X)$, $w\Phi$ is the *value of w under the substitution φ*.

A *bi-identity* is a formal equality $u \simeq w$ among terms u, w. A bi-identity $u \simeq w$ *holds* in S [or S *satisfies* the bi-identity] ($S \models u \simeq w$, for short) if u and w have the same value under each substitution $\varphi : \overline{X} \to S$. Likewise, a class $\mathcal{C}$ satisfies $u \simeq w$ if each member of $\mathcal{C}$ satisfies $u \simeq w$. Note that a bi-identity is a special form of a quasi-identity (i.e. an implication). In the following we shall not distinguish between the bi-identity $u \simeq w$ and the pair (u, w). In particular, a set of bi-identities (over the variables X) is the same as a binary relation on $T(X)$. Let ρ be a set of bi-identities, and $[\rho]$ be the class of all orthodox semigroups satisfying all bi-identities from ρ. As a first result we may state:

Proposition 2.2. *The class $[\rho]$ is an e-variety.*

It is clear that $[\rho]$ is closed under taking direct products and orthodox subsemigroups because the bi-identities are in fact quasi-identities. In order that $[\rho]$ be also closed under taking morphic images we need that if T is a morphic image of S via the surjective morphism f then each substitution $\varphi : \overline{X} \to T$ can be lifted to a substitution $\widetilde{\varphi} : \overline{X} \to S$ (that is, there is a substitution $\widetilde{\varphi} : \overline{X} :\to S$ such that $\widetilde{\varphi} f = \varphi$). The problem here is that the lift must satisfy the premises $x \simeq xx'x, x' \simeq x'xx'$ occurring in the quasi-identities. Fortunately this is solved by an old observation of Hall [14]:

Proposition 2.3. *Let S and T be regular semigroups and $f : S \to T$ be a surjective morphism. Then each pair of mutually inverse elements of T can be lifted through f to a pair of mutually inverse elements of S.*

Proof. For t, t' mutually inverse in T take $r \in tf^{-1}$, $r^* \in t'f^{-1}$; then a pair of mutually inverse lifts is given by $s = rr^*((rr^*)^2)'r$ and $s' = r^*rr^*((rr^*)^2)'rr^*$ where $((rr^*)^2)'$ denotes any inverse of $(rr^*)^2$. □

From now on everything goes through in the same way as it does for varieties of universal algebras (keeping always in mind the necessary modifications concerning variables and substitutions). Let $\mathcal{C}$ be a class of orthodox semigroups and $\Sigma(\mathcal{C})$ [$\Sigma(\mathcal{C}, X)$] be the set of all bi-identities [over $\overline{X}$] which hold in $\mathcal{C}$. As a binary relation on $T(X)$, $\Sigma(\mathcal{C})$ is a congruence, and it satisfies

1. $x \simeq xx'x, x' \simeq x'xx' \in \Sigma(\mathcal{C})$ for all $x \in X$,

2. if, for $w, v, p, q \in T(X)$, $w \simeq v, p \simeq pqp, q \simeq qpq \in \Sigma(\mathcal{C})$ then also $w(x \to p, x' \to q) \simeq v(x \to p, x' \to q) \in \Sigma(\mathcal{C})$

where $w(x \to p, x' \to q)$ is the term obtained from w by substituting all occurrences of x by p and of x' by q. A congruence on $T(X)$ satisfying these two conditions shall be called *bi-invariant*. Such congruences play the role of fully invariant congruences on free algebras in the context of varieties. So we need the analogue of free objects:

Definition 2.2. Let X be a non-empty set, $\mathcal{V}$ be a class of orthodox semigroups. A member $F \in \mathcal{V}$ together with a substitution $\iota : \overline{X} \to F$ is a *bifree object on X in $\mathcal{V}$* if, for each $S \in \mathcal{V}$ and each substitution $\theta : \overline{X} \to S$ there exists a unique morphism $\Theta : F \to S$ extending θ, that is, $\theta = \iota \Theta$.

If a bifree object exists then it is unique up to isomorphism and up to the cardinality $|X|$, and is denoted by $BF\mathcal{V}(X)$. We arrive at the first central result:

Theorem 2.4. *In any class $\mathcal{V}$ of orthodox semigroups which is closed under taking direct products and regular subsemigroups and for any (non-empty) set X there is a unique bifree object $BF\mathcal{V}(X)$ and it is isomorphic with $T(X)/\Sigma(\mathcal{V})$.*

The proof to this is essentially the same as in the classical case: informally, the bifree object on X is the $\overline{X}$-generated subdirect product of all $\overline{X}$-generated members of $\mathcal{V}$. More precisely one does the following.

Proof. If $\mathcal{V}$ consists only of the trivial (one-element) semigroup then nothing is to prove. So assume that $\mathcal{V}$ is non-trivial. Let $\{u_i \simeq v_i \mid i \in I\}$ be the set of all bi-identities over X which do not hold in $\mathcal{V}$. For each $i \in I$ choose $S_i \in \mathcal{V}$ and a substitution $\theta_i : \overline{X} \to S_i$ such that $u_i\Theta_i \neq v_i\Theta_i$ where Θ_i is the unique extension of θ_i to $T(X)$. Let $S = \prod S_i$ (which is a member of $\mathcal{V}$) and consider the substitution $\theta : \overline{X} \to S$, $x\theta = (x\theta_i)_{i \in I}$ for each $x \in \overline{X}$. Let F be the subsemigroup of S generated by $\overline{X}\theta$. Then: (i) F is orthodox because of Lemma 2.1 and since $x\theta$ and $x'\theta$ are mutually inverse elements; (ii) F is in $\mathcal{V}$ because S is in $\mathcal{V}$ and $\mathcal{V}$ is closed under regular (i.e. orthodox) subsemigroups and (iii) the kernel of the morphism $\Theta : T(X) \to F$ induced by $x \mapsto x\theta$ is precisely $\Sigma(\mathcal{V})$ whence $F \cong T(X)/\Sigma(\mathcal{V})$. We show that F is a bifree object in $\mathcal{V}$. Take any $S \in \mathcal{V}$ and any substitution $\varphi : \overline{X} \to S$ and consider the associated morphism $\Phi : T(X) \to S$. By definition, $\Sigma(\mathcal{V}) \subseteq \ker\Phi$. It follows that the morphism $\Phi : T(X) \to S$ factors through $T(X)/\Sigma(\mathcal{V})$ and this gives rise to a unique morphism $\widetilde{\varphi} : T(X)/\Sigma(\mathcal{V}) \to S$ which extends φ. Since $T(X)/\Sigma(\mathcal{V})$ is generated by the $\Sigma(\mathcal{V})$-classes of the elements of $\overline{X}$, any extension of φ to a morphism on $T(X)/\Sigma(\mathcal{V})$ must be unique. The claim is completely proved. □

We are thus ready to prove the following Birkhoff type theorem for e-varieties of orthodox semigroups.

Theorem 2.5. *A class $\mathcal{V}$ of orthodox semigroups is an e-variety if and only if $\mathcal{V} = [\rho]$ for some set ρ of bi-identities. In this case, $\mathcal{V} = [\Sigma(\mathcal{V})]$.*

Proof. We have already seen that $[\rho]$ is an e-variety (Proposition 2.2). For the converse, we show that $\mathcal{V} = [\Sigma(\mathcal{V})]$ and the non-trivial inclusion here is $[\Sigma(\mathcal{V})] \subseteq \mathcal{V}$. Take any $S \in [\Sigma(\mathcal{V})]$ and choose a set $Y = \{y_s \mid s \in S\}$ of variables in bijective correspondence with the set S. As in the proof of

Theorem 2.4, $T(Y)/\Sigma(\mathcal{V},Y) = BF\mathcal{V}(Y)$ is in $\mathcal{V}$ and is a bifree object in $\mathcal{V}$ on Y. For each $s \in S$ choose an inverse s'; then the substitution $y_s \mapsto s$, $y_s' \mapsto s'$ extends to a surjective morphism $T(Y) \to S$ and the kernel ρ of this morphism contains $\Sigma(\mathcal{V}, Y)$. Consequently, the morphism factors through $BF\mathcal{V}(Y)$. In particular, S is a morphic image of $BF\mathcal{V}(Y) \in \mathcal{V}$. Since $\mathcal{V}$ is closed under morphic images, S is in $\mathcal{V}$. □

The lattice of all subvarieties of a variety of universal algebras is anti-isomorphic with the lattice of all fully invariant congruences on the free algebra (on an infinite set) in this variety, or, equivalently, with the lattice of equational theories in infinitely many variables. We shall obtain an analogue to this for e-varieties of orthodox semigroups. Fully invariant congruences correspond to bi-invariant congruences. Let X be a fixed (countably) infinite set of variables and $\Sigma(\mathcal{O})$ be the bi-invariant congruence on $T(X)$ consisting of all bi-identities (over X) which hold in all orthodox semigroups. First we show that essentially all bi-invariant congruences are of the form $\Sigma(\mathcal{V})$.

Lemma 2.6. *Let ρ be a bi-invariant congruence on $T(X)$ containing $\Sigma(\mathcal{O})$. Then $\rho = \Sigma([\rho])$.*

Proof. Let ρ be a bi-invariant congruence containing $\Sigma(\mathcal{O})$; then $[\rho]$ is the class of all orthodox semigroups satisfying all bi-identities of ρ and $\Sigma([\rho])$ is the set of all bi-identities satisfied by each member of $[\rho]$; since all members of $[\rho]$ satisfy ρ we have the trivial inclusion $\rho \subseteq \Sigma([\rho])$. So we need to show the reverse inclusion $\Sigma([\rho]) \subseteq \rho$. Now

$$T(X)/\rho = (T(X)/\Sigma(\mathcal{O})) / (\rho/\Sigma(\mathcal{O}))$$

is an orthodox semigroup. Since ρ is bi-invariant, $T(X)/\rho$ satisfies each bi-identity in ρ, that is $T(X)/\rho \in [\rho]$. By construction, $T(X)/\rho$ does not satisfy any other bi-identities than those of ρ. Consequently, $\Sigma(T(X)/\rho) \subseteq \rho$ and therefore

$$\Sigma([\rho]) \subseteq \Sigma(T(X)/\rho) \subseteq \rho$$

and the claim is proved. □

We arrive at the main theorem in this context.

Theorem 2.7. *The mappings $\mathcal{V} \mapsto \Sigma(\mathcal{V})$ and $\rho \mapsto [\rho]$ are mutually inverse, order inverting bijections between the lattice of all orthodox e-varieties and the lattice of bi-invariant congruences on $T(X)$ containing $\Sigma(\mathcal{O})$ (where X is a fixed infinite set).*

In the theory of varieties the concept of an *identity basis* is fundamental. Because of the Completeness Theorem of Equational Logic one can give two equivalent definitions for this notion. We shall develop an analogue for bi-identities of orthodox semigroups. First we have to define two concepts of "consequence" of a set of bi-identities.

Definition 2.3. Let Σ be a set of bi-identities; the bi-identity $u \simeq v$ is a *syntactic consequence* of Σ, or $\Sigma \vdash u \simeq v$, if $u \simeq v$ can be obtained from the bi-identities in Σ in finitely many steps by application of any of the following rules:

1. $\emptyset \vdash u \simeq u$
2. $u \simeq v \vdash v \simeq u$
3. $u \simeq v, v \simeq w \vdash u \simeq w$
4. $u_1 \simeq u_2, v_1 \simeq v_2 \vdash u_1 v_1 \simeq u_2 v_2$
5. $\emptyset \vdash x \simeq xx'x, x' \simeq x'xx'$
6. $u \simeq v, p \simeq pqp, q \simeq qpq \vdash u(x \to p, x' \to q) \simeq v(x \to p, x' \to q)$

where u, v, p, q, u_i, v_i represent arbitrary terms and x an arbitrary variable.

In this definition we have formalized in terms of equational logic what it means for a bi-identity $u \simeq v$ to be contained in the bi-invariant congruence generated by Σ. The other, the semantic concept of consequence, is defined as follows.

Definition 2.4. Let Σ be a set of bi-identities; a bi-identity $u \simeq v$ is a *semantic consequence*, or $\Sigma \models u \simeq v$, if, for each orthodox semigroup S with $S \models \Sigma$ then also $S \models u \simeq v$.

From Theorem 2.7 we immediately get the Completeness Theorem of Bi-equational Logic.

Theorem 2.8. *For any set of bi-identities Σ and any bi-identity $u \simeq v$,*

$$\Sigma \models u \simeq v \text{ if and only if } \Sigma \cup \Sigma(\mathcal{O}) \vdash u \simeq v.$$

It is clear that in this theorem $\Sigma(\mathcal{O})$ may be replaced with any subset ρ which generates $\Sigma(\mathcal{O})$ *as a bi-invariant congruence*. For example, $\Sigma(\mathcal{O})$ may be replaced with the single bi-identity $xx'yy' \simeq (xx'yy')^2$ (see Auinger and Szendrei [5]). In this sense, the single bi-identity $xx'yy' \simeq (xx'yy')^2$ is a basis for the bi-identities of $\mathcal{O}$. If we have any e-variety $\mathcal{V}$ of orthodox semigroups, and any set of bi-identities B which defines $\mathcal{V}$ within the class of all orthodox semigroups then $B \cup \{xx'yy' \simeq (xx'yy')^2\}$ is a basis for the bi-identities in $\mathcal{V}$ in the sense that this set of bi-identities generates $\Sigma(\mathcal{V})$ as a bi-invariant congruence.

2.2 The locally inverse and E-solid cases

At this point the question arises if we can do something similar outside orthodox semigroups. The most important result in this context is due to Yeh [37]. His main result asserts that in an e-variety $\mathcal{V}$ all bifree objects exist if and only if $\mathcal{V}$ consists of locally inverse or E-solid semigroups. We state the negative part of his result first.

Theorem 2.9. *If $|X| \geq 2$ and $\mathcal{V}$ is an e-variety which is neither locally inverse nor E-solid then the bifree object $BF\mathcal{V}(X)$ does not exist.*

From private communication with Tom Hall the author is aware that monogenic bifree objects exist in other e-varieties, as well. The proof of Theorem 2.9 comes about as follows: take a member S of $\mathcal{V}$ and $A \subseteq S$ such that each element of A has an inverse in A. If the bifree object $BF\mathcal{V}(A)$ exists then inside S there exists a (unique) least regular subsemigroup containing A. Indeed, we choose a set $X = \{x_a \mid a \in A\}$ of variables which are in bijective correspondence to the elements of A. Then the mapping $x_a \mapsto a$ can be extended to a substitution $\overline{X} \to A$ and hence there is a canonical morphism $BF\mathcal{V}(A) = BF\mathcal{V}(X) \to S$ and the image of $BF\mathcal{V}(A)$ in S is this least regular subsemigroup. So, in order to show that a bifree object $BF\mathcal{V}(X)$ does not exist it is sufficient to find a semigroup $S \in \mathcal{V}$ and a subset A of cardinality $|X|$ for which each element of A has an inverse in A and such that there is no least regular subsemigroup of S containing A. Let K be the combinatorial completely 0-simple semigroup whose non-zero $\mathcal{D}$-class is

a	b	
*	*	c

where a, b, c and the stars * indicate idempotents (thus the only non-idempotent element is the element in the empty box). Now consider an ideal extension L of K by two further idempotents e, f where the action of e, f on K is given by saying that $a, b \leq e$ and $c \leq f$ (here $\leq$ denotes to the natural order on the set of idempotents). This completely defines the semigroup L. It is obvious that for the set $\{e, f\}$ there is no smallest regular subsemigroup of L containing $\{e, f\}$. Moreover, from a result of Hall [16] one can deduce that any e-variety $\mathcal{V}$ not consisting of locally inverse or E-solid semigroups must contain the semigroup L or its dual.

On the positive side, Yeh showed that in each locally inverse and each E-solid e-variety $\mathcal{V}$ all bifree objects exist. His approach was different from the one developed by Kaďourek and Szendrei for orthodox e-varieties. What is more, he did not provide an appropriate concept of bi-identity to get a Birkhoff type theorem. However, all the necessary prerequisites to develop such a theory are implicitly contained in his work. The crucial step in his paper was to prove that if S is locally inverse or E-solid then for each subset $A \subseteq S$, which contains an inverse a' for each element a, there exists a (unique) least regular subsemigroup of S containing A.[b] In order to prove this, one has to consider the two cases (locally inverse and E-solid) separately.

Let us treat the locally inverse case first. For an element a of a semigroup S denote by $V(a)$ the set of all inverses of a in S. Define the *sandwich set* $S(a, b)$ of two elements a, b of S by setting $S(a, b) := bV(ab)a$. If S is regular then the sandwich set is never empty. Notice that each element of a sandwich set is idempotent and $V(b)S(a,b)V(a) \subseteq V(ab)$; for more on sandwich sets the reader may consult Nambooripad [26] and Trotter [35]. Among other things, Nambooripad showed that a (regular) semigroup is locally inverse if and only if each sandwich set $S(a, b)$ is a singleton. Therefore, on any locally inverse semigroup S we can define an additional binary operation $\wedge$ by setting

$$a \wedge b := \text{ the unique element of } S(b, a).$$

Yeh then proved:

Theorem 2.10. *Let S be locally inverse and $A \subseteq S$ be such that each element of A has an inverse in A. Then the closure of A under multiplication and sandwich operation $\wedge$ is the (unique) least regular subsemigroup of S containing A.*

As a consequence Yeh showed that in each class $\mathcal{V}$ of locally inverse semigroups which is closed under taking direct products and regular subsemigroups the bifree object $BF\mathcal{V}(X)$ exists for any non-empty set X. His main idea was to consider the unary semigroup variety $(\mathcal{LI})'$ of all locally inverse unary semigroups and to show that $BF\mathcal{LI}(X)$ is the least regular subsemigroup of

---

[b]We note that an equivalent result has been obtained by Johnston [17]: she proved that an e-variety $\mathcal{V}$ has the property that for each member $S \in \mathcal{V}$ the partially ordered set of all regular subsemigroups of S is a lattice if and only if $\mathcal{V} \subseteq \mathcal{LI}$ or $\mathcal{V} \subseteq \mathcal{ES}$.

the free locally inverse unary semigroup containing $X \cup X^*$ (where $*$ denotes the unary operation). However, with the use of the sandwich operation $\wedge$ and by observing that (i) semigroup morphisms among locally inverse semigroups respect this operation and (ii) the regular subsemigroups of a locally inverse semigroup are precisely those substructures with respect to multiplication and $\wedge$ which are generated by sets of the form $\overline{X}\varphi$ for a substitution φ, we can proceed completely analogously to the orthodox case.

Again, fix a (countably) infinite set X and a disjoint copy X' of X and set $\overline{X} = X \cup X'$. Call a semigroup which is equipped with an additional binary operation $\wedge$ a *binary* semigroup. Then the class of all binary semigroups is a variety of $\langle 2, 2 \rangle$-algebras. As term algebra $T_{\mathcal{LI}}(X) = T(X)$ now serves the *free binary semigroup* on $\overline{X}$. Note that a model of $T(X)$ is given by the smallest subset T of the set of all finite strings in the alphabet $X \cup X' \cup \{(, \wedge,)\}$ such that

1. $\overline{X} \subseteq T$,

2. $u, v \in T \Rightarrow uv \in T$,

3. $u, v \in T \Rightarrow (u \wedge v) \in T$.

Given this, all definitions of the previous sub-section can be made analogously for locally inverse e-varieties, the results may be restated verbatim and proved analogously. This includes: the existence of bifree objects $BF\mathcal{V}(X)$; that bifree objects are given as the term algebra modulo the bi-identities which hold in the respective class: $BF\mathcal{V}(X) = T(X)/\Sigma(\mathcal{V})$; the anti-isomorphism between the lattice of sub-e-varieties and the lattice of bi-invariant congruences; the Completeness Theorem of Bi-equational Logic. For a detailed treatment the reader may consult [1]. Moreover, from the main result of [2] one may deduce that an independent basis for the bi-identities of $\mathcal{LI}$ is given by

1. $(x \wedge y) \simeq xx'(x \wedge y)(x \wedge y)y'y$,

2. $yx \simeq y(x \wedge y)x$,

3. $(z \wedge x)(z \wedge y)zz' \simeq (z \wedge y)(z \wedge x)zz'$.

Now let us turn to the E-solid case. We first introduce the concept of the *self conjugate core* of a regular semigroup S (see Trotter [35]). Set $C_1(S) := \langle E(S) \rangle$ and for $n \geq 2$

$$C_n(S) = \langle sts' \mid t \in C_{n-1}(S), s \in S^1, s' \in V(s) \rangle$$

and $C_\infty(S) = \bigcup_{n \geq 1} C_n(S)$. Trotter [35] proved that a regular semigroup is E-solid if and only if $C_\infty(S)$ is completely regular. Therefore, on each E-solid semigroup we may define a partial unary operation $^{-1}$ as follows. For $s \in S$ set

$$s^{-1} := \begin{cases} \text{the unique group inverse of } s \text{ if } s \in C_\infty(S) \\ \text{undefined otherwise.} \end{cases}$$

Yeh [37] essentially showed that if S is E-solid and $A \subseteq S$ is such that each element of A has an inverse in A then the (unique) least regular subsemigroup of S containing A is the closure of A under multiplication and the partial operation $^{-1}$. Yeh then proceeded as in the locally inverse case: he considered the free unary E-solid semigroup on X and showed that the least regular subsemigroup containing $X \cup X^*$ is the bifree E-solid semigroup. More generally, any class $\mathcal{V}$ of E-solid semigroups closed under direct products and regular subsemigroups has bifree objects on any non-empty set X.

However, to get the stronger results, including the Birkhoff type theorem we can again proceed as in the previous subsection. We have to find the appropriate term algebra first. Since the additional operation is now partial this term set becomes more complicated. Let us start with the free unary semigroup $U(\overline{X})$ on $\overline{X}$ and consider the unique morphism (of unary semigroups) $\varphi : U(\overline{X}) \to FG(X)$ onto the free group on X which is determined by $x \mapsto x$ and $x' \mapsto x^{-1}$; here x' is the variable from X' — the unary operation in $U(\overline{X})$ is denoted by $^{-1}$. Let $K(X) := 1\varphi^{-1}$ be the kernel of φ. Now let $T_{\mathcal{ES}}(X) = T(X)$ be the smallest subsemigroup T of $U(\overline{X})$ which contains $\overline{X}$ and is closed under the partial unary operation $^{-1}|K(X)$. Then $T(X)$ is a semigroup endowed with a partial unary operation $^{-1}$ which is defined on $T(X) \cap K(X)$. It is easy to see that each substitution $\varphi : \overline{X} \to S$ with S E-solid can be extended in a unique way to a morphism $\Phi : T(X) \to S$. Now one can restate all definitions and results from the previous subsection almost verbally and prove the assertions completely analogously — this has been done in detail by Kaďourek and Szendrei [24]. At one point we need a modification: the substitution rule (6) in the definition of syntactic consequence has to be restricted appropriately. That is, the rule (6) now has to be formulated as

$$u \simeq v, p \simeq pqp, q \simeq qpq, pq \in K(X) \vdash u(x \to p, x' \to q) \simeq v(x \to p, x' \to q).$$

Likewise, since $^{-1}$ is partial only, the correct rule for application of $^{-1}$ to a bi-identity is

$$u \simeq v, u, v \in K(X) \vdash (u)^{-1} \simeq (v)^{-1}.$$

One major difference to the locally inverse and orthodox cases is that in the E-solid case we no longer have finite bases for the bi-equational theories. Indeed it has been shown in [5] that no non-orthodox E-solid e-variety $\mathcal{V}$ has a finite basis for its bi-identities. More precisely the following holds.

Theorem 2.11. *For each E-solid e-variety $\mathcal{V}$ which is not an e-variety of orthodox semigroups and for each finite set ρ of bi-identities which hold in $\mathcal{V}$ there exists a positive integer M such that*

$$\rho \not\vdash x_1 x_1' \ldots x_M x_M' \simeq ((x_1 x_1' \ldots x_M x_M')^{-1})^{-1}.$$

Informally this is because substitution here is of limited use. We note that from the Completeness Theorem of First Order Logic it follows that an e-variety $\mathcal{V}$ of orthodox or locally inverse semigroups has a finite basis for its bi-identities if and only if $\mathcal{V}$ is finitely axiomatizable as a class of first order structures. This is no longer true for E-solid e-varieties with respect to the signature presented above.

2.3 Refinements and further developments

Here we mention only a few. For some e-varieties of E-solid semigroups one can use the same concept of bi-identity as for orthodox semigroups. Indeed let, for a positive integer n, the class $\mathcal{ES}_n$ be defined by

$$\mathcal{ES}_n = \{S \in \mathcal{RS} \mid \forall\, e, f \in E(S) : (ef)^{n+1} = ef\}.$$

Then each $\mathcal{ES}_n$ is an e-variety of E-solid semigroups and the classes form an infinite ascending chain

$$\mathcal{O} = \mathcal{ES}_1 \subseteq \mathcal{ES}_2 \subseteq \cdots \subseteq \mathcal{ES}_n \subseteq \mathcal{ES}_{n+1} \subseteq \cdots$$

of e-varieties. For each of these e-varieties one can develop a theory of bi-identities, a Birkhoff theorem, a Completeness Theorem, etc., using the same concept of bi-identity as in the orthodox case; for more details the reader is referred to Broeksteeg [12].

The concepts of bi-identity and bifree object have been extended to *tri-identity* and *trifree object* by Kaďourek in [22]. The principal idea is to once more enrich the set of variables. Instead of the alphabet $\overline{X} = X \cup X'$ one considers here the bigger alphabet $\overline{X} \cup (\overline{X} \wedge \overline{X})$ where $(\overline{X} \wedge \overline{X})$ is a set of new variables $(y \wedge z)$ which are in bijective correspondence with the set of all pairs (y, z) with $y, z \in \overline{X}$ but $(y, z) \neq (x, x'), (x', x)$. As the reader will expect already, these new variables are aimed at representing sandwich elements. The notions of substitution and trifree object then are defined accordingly. Kaďourek [22] shows that in each e-variety of locally orthodox semigroups all trifree objects exist. Trifree objects behave similarly to free and bifree objects: for example, each member of a locally orthodox e-variety is the morphic image of a trifree object on a set of the appropriate size. In the same paper the existence of such objects is heavily used to prove several deep decomposition results for locally orthodox e-varieties. The corresponding theory of tri-identitites no longer yields so nice results as in the orthodox, locally inverse and E-solid cases: there is no Birkhoff-type theorem. In fact, the class defined by a set of tri-identities need no longer be closed under morphic images. This is caused by the fact that for a surjective morphism $\varphi : S \to T$, a substitution (appropriate for this tri-partite alphabet) $\alpha : \overline{X} \cup (\overline{X} \wedge \overline{X}) \to T$ in general cannot be lifted to a substitution $\tilde{\alpha} : \overline{X} \cup (\overline{X} \wedge \overline{X}) \to S$. Concerning existence of trifree objects, Churchill and Trotter [13] have shown that in each e-variety of locally E-solid semigroups all trifree objects exist while in e-varieties which are not locally E-solid there are no trifree objects on sets of at least three variables, a result which was already asserted without proof by Kaďourek [22].

Finally it should be mentioned that in the same paper [13], Churchill and Trotter have developed a unified approach to bi-identities for locally inverse and E-solid e-varieties by using the same term algebra for both types of semigroups. Their idea is the following: on each locally inverse semigroup consider the *partial* binary operation $\wedge | E(S) \times E(S)$: that is, restrict the sandwich operation to the set of idempotents. Likewise on an E-solid semigroup consider the partial binary operation $(e, f) \mapsto e(fe)^{-1}f$ which selects for each pair of idempotents a canonical sandwich element. Consider each locally inverse and each E-solid semigroup endowed with this partial binary operation. The regular subsemigroups of a locally inverse or an E-solid semigroup now are precisely those subsemigroups which are closed under this partial binary operation and are generated (with respect to both operations) by sets of the form $\overline{X}\varphi$ for a substitution φ. The term algebra $T(X)$, which can be used in both cases, now is a certain subsemigroup of the free binary semigroup on $\overline{X}$ the definition of which is slightly more complicated than the definition of term algebra for the E-solid case presented above. However, using this one can proceed along the lines presented above and can derive the analogous (appropriately reformulated) theorems. The price for this unification is to give up the totality of the sandwich operation in the locally inverse case which in turn restricts the use and applicability of substitutions in the corresponding bi-equational logic.

3 The regular and the restricted regular semidirect products

This section is concerned with semidirect product decompositions of various locally inverse e-varieties and the constructions of bifree objects. Most of the proofs depend heavily on the use of bi-equational logic. So, the presented results can be viewed as applications of the theory developed in the previous section.

3.1 Preliminaries

In regular semigroup theory, the use of semidirect products has played a less important role than in finite semigroup theory. This is caused by the fact that a semidirect product of two regular semigroups in general is no longer regular. Still it has been used for various purposes, in particular in the context of inverse and orthodox semigroups (see the survey of Szendrei [33]). A crucial step forward in the applicability of semidirect products for regular semigroups was the following result by Jones and Trotter [18]:

Theorem 3.1. *Let S and T be regular semigroups, one of them being completely simple; let T act on S by endomorphisms on the left. Then within the usual semidirect product $S * T$ the set $\mathrm{Reg}(S * T)$ of all regular elements of $S * T$ forms a (regular) subsemigroup.*

On the level of e-varieties this leads to a partial binary operation $*_r$ by setting
$$\mathcal{U} *_r \mathcal{V} := \langle \mathrm{Reg}(S * T) \mid S \in \mathcal{U}, T \in \mathcal{V} \rangle$$
which is defined whenever $\mathcal{U}$ or $\mathcal{V}$ is contained in $\mathcal{CS}$, and this latter assumption cannot be removed [18, Theorem 3.1]. Throughout, for a class $\mathcal{C}$ of regular semigroups, $\langle \mathcal{C} \rangle$ denotes the smallest e-variety containing $\mathcal{C}$. We call $\mathcal{U} *_r \mathcal{V}$ the *regular semidirect product* of the e-varieties $\mathcal{U}$ and $\mathcal{V}$. In their paper [18] Jones and Trotter gave a lot of examples of $*_r$-decompositions of e-varieties and also models of bifree objects in various e-varieties. A subsequent work by Jones [20] undertakes a careful study of the operator $\_ *_r \mathcal{RZ}$ and characterizes the e-varieties $\mathcal{V}$ satisfying the equation $L\mathcal{V} = \mathcal{V} *_r \mathcal{RZ}$ (thereby giving a regular version of Tilson's Delay Theorem [34]). Concerning associativity of the partial product $*_r$, the most advanced result is by Billhardt and Szendrei [10]: Let $\mathcal{U}, \mathcal{V}, \mathcal{W}$ be e-varieties (at least) two of them being completely simple then
$$(\mathcal{U} *_r \mathcal{V}) *_r \mathcal{W} = \mathcal{U} *_r (\mathcal{V} *_r \mathcal{W})$$
holds provided that
$$\mathcal{V} *_r \mathcal{W} = \bigcup_{n \geq 1} (HS)^n P\{\mathrm{Reg}(V * W) \mid V \in \mathcal{V}, W \in \mathcal{W}\}.$$

Here H, S, P denote the operators of taking all morphic images, regular subsemigroups and direct products, respectively, of all members of a class of regular semigroups. We note that an analogue to the usual HSP-theorem for varieties of universal algebras does not hold for e-varieties in general (see Kaďourek [22]), yet it does hold for locally inverse or E-solid e-varieties (see Yeh [37]). In particular, associativity holds if $\mathcal{V} *_r \mathcal{W}$ is contained in $\mathcal{ES}$ or $\mathcal{LI}$. It is not known if associativity ever fails.

Now we turn to another kind of semidirect product. Let S be a regular semigroup and T be a locally inverse semigroups which acts on S by endomorphisms on the left. Set

$$S *_{rr} T = \{(a,x) \in S \times T \mid \exists x' \in V(x) : {}^{xx'}a = a\}$$

and endow this set with the multiplication

$$(a,x)(b,y) = ({}^{x(y \wedge x)x'}a \cdot {}^{x}b, xy).$$

Then $S *_{rr} T$ becomes a regular semigroup. The present definition comes from [4] but it should be mentioned that for T inverse the definition goes back to Billhardt [8]. Moreover, in the recent paper [9], Billhardt has extended this definition to the case when the second factor T is locally $\mathcal{R}$-unipotent, that is, T is locally orthodox with the idempotents in the local submonoids eTe forming a left regular band. Again we get a partial binary operator $*_{rr}$ on the lattice of e-varieties by setting

$$\mathcal{U} *_{rr} \mathcal{V} := \langle U *_{rr} V \mid U \in \mathcal{U}, V \in \mathcal{V}\rangle$$

which is defined as soon as the second factor $\mathcal{V}$ is locally $\mathcal{R}$-unipotent. We call $\mathcal{U} *_{rr} \mathcal{V}$ the *restricted regular semidirect product* of $\mathcal{U}$ and $\mathcal{V}$. Concerning associativity, it was shown in [4] that associativity holds provided all three factors are locally inverse (the proof there was syntactic, that is, by use of bi-identities) while the most general result in this context is by Billhardt [9] showing that associatvity holds under the same condition as for the regular semidirect product (by use of semantic arguments).

We note that if T happens to be a completely simple semigroup then

$$S *_{rr} T \leq \operatorname{Reg}(S * T)$$

and therefore

$$\mathcal{U} *_{rr} \mathcal{V} \subseteq \mathcal{U} *_r \mathcal{V}$$

whenever $\mathcal{V} \subseteq \mathcal{CS}$. At this point we mention that the inclusion $\subseteq$ can be strict but can be equality, as well. For example $\mathcal{Sl} *_{rr} \mathcal{RB}$ is the e-variety of all normal bands while $\mathcal{Sl} *_r \mathcal{RB}$ is the e-variety $\mathcal{LSl}$ of all combinatorial strict

81

regular semigroups. On the other hand $\mathcal{S}l *_{rr} \mathcal{CS} = \mathcal{S}l *_r \mathcal{CS} = \mathcal{LI}$ (see [18]). The study of the inclusion $\mathcal{U} *_{rr} \mathcal{V} \subseteq \mathcal{U} *_r \mathcal{V}$ will be one of the main topics of the subsequent subsections.

Next we remind the reader of the very familiar construction of a *Rees matrix semigroup*. Let S be a semigroup, I, Λ index sets and $P = (p_{\lambda i})$ a $\Lambda \times I$-matrix with entries in S^1; note that we explicitly allow entries being the identity element even in case S does not have an identity. Then the $I \times \Lambda$-matrix semigroup over S with matrix P, denoted $\mathcal{M}(I, S, \Lambda; P)$ consists of the set $I \times S \times \Lambda$ endowed with multiplication $(i, s, \lambda)(j, t, \mu) = (i, sp_{\lambda j}t, \mu)$. Note that $\mathcal{M}(I, S, \Lambda; P)$ need not be regular even in case S is regular. However, if S is regular then the set $\text{Reg}\mathcal{M}(I, S, \Lambda; P)$ of all regular elements always forms a (regular) subsemigroup of $\mathcal{M}(I, S, \Lambda; P)$ which we denote by $\mathcal{RM}(I, S, \Lambda; P)$ (see McAlister [25]).

The next construction is in a sense a mixture between a Rees matrix semigroup and a semidirect product. Let S be a semigroup and $T = \mathcal{M}(I, G, \Lambda; P)$ be a completely simple semigroup represented as a Rees matrix semigroup. Suppose that the group G acts on S by automorphism on the left. The *Pastijn product* $S \odot T$ of S by T, is defined by

$$S \odot T = I \times S \times G \times \Lambda$$

endowed with the multiplication

$$(i, s, g, \lambda)(j, t, h, \mu) = (i, s \cdot {}^{gp_{\lambda j}}t, gp_{\lambda j}h, \mu).$$

This construction occurred first in a paper by Pastijn [27] but was then systematically studied by Kaďourek [21, 22]. He pointed out that the product $S \odot T$ depends on T but not on the Rees matrix representation of T. Also, it is shown that $S \odot T$ is regular if S is regular, and locally inverse [locally orthodox, locally E-solid] if S is inverse [orthodox, E-solid]. The Pastijn product $S \odot T$ for $T \cong \mathcal{M}(I, G, \Lambda; P)$ can be interpreted as a Rees matrix semigroup, as well. Indeed, consider the usual semidirect product $S * G$ and let $1 \times P$ be the $\Lambda \times I$-matrix over $S^1 * G$ with entries $(1, p_{\lambda i})$; then

$$S \odot T \cong \mathcal{M}(I, S * G, I; 1 \times P).$$

Moreover, the Pastijn product again gives rise to a partial binary operation on the lattice of e-varieties of regular semigroups: for e-varieties $\mathcal{U}, \mathcal{V}$ with $\mathcal{V} \subseteq \mathcal{CS}$ let

$$\mathcal{U} \odot \mathcal{V} = \langle U \odot V \mid U \in \mathcal{U}, V \in \mathcal{V} \rangle.$$

Finally the reader is reminded of the definition of the *Mal'cev product* as it is understood here: for classes $\mathcal{U}, \mathcal{V}$ of regular semigroups the Mal'cev product

82

is defined as

$$\mathcal{U} \circ \mathcal{V} = \{S \in \mathcal{RS} \mid \exists \text{ congruence } \rho : S/\rho \in \mathcal{V} \text{ and } e\rho \in \mathcal{U} \ \forall e \in E(S)\}.$$

The Mal'cev product of two e-varieties is closed under taking direct products, but in general not under taking regular subsemigroups or morphic images.

3.2 E-varieties of the form $\mathcal{Sl} *_{rr} \mathcal{Q}$

Throughout the following, $\mathcal{Q}$ stands for any e-variety of completely simple semigroups containing all rectangular bands. First we shall be interested in the structure of the bifree objects in e-varieties of the form $\langle \mathcal{Sl} \circ \mathcal{Q} \rangle$. For this purpose we need a model of the bifree object in $\mathcal{Q}$ first. We start with the case $\mathcal{Q} = \mathcal{CS}$. Let X be a non-empty set, $\overline{X} = X \cup X'$ as in section 2.2. Choose and fix an element $z \in X$. For each pair (x, y) with $x, y \in \overline{X} \setminus \{z\}$ choose a new symbol p_{xy} and let G be the free group on the set $X \cup \{p_{xy} \mid x, y \in \overline{X} \setminus \{z\}\}$. Moreover, set $p_{zx} = p_{xz} = 1$ (the identity of G) for all x and let $P = (p_{xy})$ be the so obtained $\overline{X} \times \overline{X}$-matrix. We get the Rees matrix semigroup $\mathcal{M}(\overline{X}, G, \overline{X}; P)$ and this is already a model of the bifree completely simple semigroup $BFCS(X)$, as proved by Kaďourek [21]. More precisely, the following holds

Theorem 3.2. *The Rees matrix semigroup $\mathcal{M}(\overline{X}, G, \overline{X}; P)$ together with the substitution*

$$\iota : \overline{X} \to \mathcal{M}(\overline{X}, G, \overline{X}; P), \quad x \mapsto (x, x, x), \quad x' \mapsto (x', p_{xx'}^{-1} x^{-1} p_{x'x}^{-1}, x')$$

is a model of the bifree completely simple semigroup.

Moreover, for each $\mathcal{Q} \in [\mathcal{RB}, \mathcal{CS}]$ there is a unique normal subgroup $N_\mathcal{Q}$ of G such that $BF\mathcal{Q}(X) \cong \mathcal{M}(\overline{X}, G/N_\mathcal{Q}, \overline{X}, P_\mathcal{Q})$ where $P_\mathcal{Q} = (p_{xy} N_\mathcal{Q})$. The details can be found in [21]. Next let $Y_\mathcal{Q}$ be the set of all finite subsets of $(G/N_\mathcal{Q}) \times X$; then $Y_\mathcal{Q}$ is a semilattice under forming the union. Moreover, $G/N_\mathcal{Q}$ naturally acts on $Y_\mathcal{Q}$ by automorphisms on the left. (Note that $Y_\mathcal{Q}$ is a semilattice of finite subgraphs of the Cayley graph of $G/N_\mathcal{Q}$, but this graph is considered with respect to the alphabet X only, so this graph is not connected.) In particular we may form the Pastijn product $Y_\mathcal{Q} \odot \mathcal{M}(\overline{X}, G/N_\mathcal{Q}, \overline{X}; P_\mathcal{Q})$. Kaďourek [21] has proved the following

Theorem 3.3. *The bifree object in $\langle \mathcal{Sl} \circ \mathcal{Q} \rangle$ can be embedded into the Pastijn product $Y_\mathcal{Q} \odot \mathcal{M}(\overline{X}, G/N_\mathcal{Q}, \overline{X}; P_\mathcal{Q})$. The corresponding canonical substitution $\overline{X} \to BF \langle \mathcal{Sl} \circ \mathcal{Q} \rangle (X)$ is given by*

$$x \mapsto (x, \{(\overline{1}, x)\}, x, x), \quad x' \mapsto (x', \{(\overline{p_{xx'}^{-1} x^{-1}}, x)\}, \overline{p_{xx'}^{-1} x^{-1} p_{x'x}^{-1}}, x').$$

83

Here $\bar{z}$ stands for the coset $zN_\mathcal{Q}$. The proof of this theorem is entirely syntactic: it is shown that the Pastijn product $Y_\mathcal{Q} \odot \mathcal{M}(\overline{X}, G/N_\mathcal{Q}, \overline{X}, P_\mathcal{Q})$ does not satisfy more bi-identities over X than the class $\langle \mathcal{Sl} \circ \mathcal{Q} \rangle$ whence the result follows. In particular, the bifree object in the e-variety $\langle \mathcal{Sl} \circ \mathcal{Q} \rangle$ is the same as the bifree object in the quasi-e-variety $\mathcal{Sl} \circ \mathcal{Q}$. An important consequence of Theorem 3.3 is the e-variety equation

Corollary 3.4. $\mathcal{Sl} \odot \mathcal{Q} = \langle \mathcal{Sl} \circ \mathcal{Q} \rangle$

The relation beween the e-varieties $\mathcal{Sl} \odot \mathcal{Q}$ and $\mathcal{Sl} *_{rr} \mathcal{Q}$ has been fully clarified by Billhardt and Szendrei [11]:

Theorem 3.5. *Let S be regular and T be completely simple. Then the following hold. (a) Each restricted semidirect product $S *_{rr} T$ is isomorphic with a Pastijn product $S_1 \odot T$ where S_1 is a regular subsemigroup of S. (b) Each Pastijn product $S \odot T$ is isomorphic with a regular subsemigroup of a restricted regular semidirect product $S^R *_{rr} T$ of a direct power S^R of S by T.*

As an immediate consequence we have

Corollary 3.6. $\mathcal{Sl} *_{rr} \mathcal{Q} = \mathcal{Sl} \odot \mathcal{Q} = \langle \mathcal{Sl} \circ \mathcal{Q} \rangle$

In particular, Theorem 3.3 provides us with a model of the bifree object in $\mathcal{Sl} *_{rr} \mathcal{Q}$.

Suppose that the group G acts on the semigroup Y by automorphisms on the left and denote by $Y * G$ the corresponding semidirect product. Then each Rees matrix semigroup $\mathcal{M}(I, Y * G, \Lambda; P)$ admits a natural projection $\pi_{\mathcal{CS}}$: $(i, (y, g), \lambda) \mapsto (i, g, \lambda)$ onto the completely simple semigroup $\mathcal{M}(I, G, \Lambda, P|_G)$ where the entries of the matrix $P|_G$ are the group components of the entries of P. In view of the interpretation of a Pastijn product as a Rees matrix semigroup we may restate the equality $\mathcal{Sl} *_{rr} \mathcal{Q} = \mathcal{Sl} \odot \mathcal{Q}$ as follows.

Corollary 3.7. *The e-variety $\mathcal{Sl} *_{rr} \mathcal{Q}$ is generated by all Rees matrix semigroups $S = \mathcal{M}(I, Y * G, \Lambda; P)$ where Y is a semilattice, $S\pi_{\mathcal{CS}}$ is in $\mathcal{Q}$ and the entries of P are in the group of units of $Y^1 * G$.*

As a comparison with the product $\mathcal{Sl} *_r \mathcal{Q}$ we note that a model of the bifree object in $\mathcal{Sl} *_r \mathcal{Q}$ in terms of a subsemigroup of a Rees matrix semigroup has been obtained in [6]. We will not give the details of this construction here, but mention that an analogue of Corollary 3.7 can be obtained as follows

Corollary 3.8. *The e-variety $\mathcal{Sl} *_r \mathcal{Q}$ is generated by all regular Rees matrix semigroups $S = \mathcal{RM}(I, Y * G, \Lambda; P)$ where Y is a semilattice and $S\pi_{\mathcal{CS}}$ is in*

Q.

Finally it should be mentioned that (less transparent) models of the bifree objects in $Sl *_r Q$ and $Sl *_{rr} Q$ can be obtained by the general method of [18].

3.3 Comparing $Sl *_{rr} Q$ and $Sl *_r Q$

Here we present the two main results of [7] and try to give a rough idea of the proofs. As already mentioned, for $Q \in [\mathcal{RB}, \mathcal{CS}]$ in the inclusion $Sl *_{rr} Q \subseteq Sl *_r Q$ both — strict inclusion as well as equality — can happen. The first result states that equality only occurs when $Q = \mathcal{CS}$.

Theorem 3.9. *If $Q \neq \mathcal{CS}$ then $Sl *_{rr} Q$ is properly contained in $Sl *_r Q$.*

The method of proving this is to find a bi-identity which holds in $Sl *_{rr} Q$ but fails in $Sl *_r Q$. First, since $Sl *_{rr} Q$ coincides with the Mal'cev product $\langle Sl \circ Q \rangle$, whenever $Q \models u \simeq u^2$ then also $Sl *_{rr} Q \models u \simeq u^2$. It suffices therefore to find a term $u \in T(X)$ such that $u \simeq u^2$ holds in Q but fails in $Sl *_r Q$. Such a term u can be found as follows. Within $T(X)$ lies the semigroup $L(X)$ generated by all elements of $\overline{X} \cup (\overline{X} \wedge \overline{X})$ where $(\overline{X} \wedge \overline{X}) = \{(x \wedge y) \mid x, y \in \overline{X}\}$. Call the elements of $L(X)$ *linear terms*. A linear term is *reduced* if it does not contain segments of the form $a(b \wedge a), (b \wedge a)b, (a \wedge b)(a \wedge c), (b \wedge a)(c \wedge a)$ with $a, b, c \in \overline{X}$ nor segments of the form $xx'x, x'xx', (x \wedge x'), (x' \wedge x)$ with $x \in X$. It is shown in [3] that each linear term admits a unique reduced form obtained by reductions of the form $a(b \wedge a) \to a, (x \wedge x') \to xx'$, etc. The set of all reduced forms provides a model of the bifree completely simple semigroups on X [3]. A linear term u is *cyclically reduced* if uu is reduced. Denote by λu respectively $u\rho$ the first letter, respectively, last letter appearing in u. In [7] the following has been proved.

Proposition 3.10. *For each e-variety Q of completely simple semigroups properly contained in $\mathcal{CS}$ and containing all rectangular bands there exists a cyclically reduced linear term u such that*

1. $Q \models u \simeq u^2$,

2. $\{\lambda u, u\rho\} \neq \{x, x'\}$ *for each $x \in X$.*

For each term u which satisfies the conditions of this proposition the bi-identity $u \simeq u^2$ then holds in $Sl *_{rr} Q$ but fails in $Sl *_r Q$. To verify the latter one can use the model of the bifree object in $Sl *_r Q$ constructed in [6].

The second major result of [7] asserts that if we replace, in $\mathcal{S}l *_{rr} \mathcal{Q}$, $\mathcal{Q}$ by a slightly bigger e-variety then the reverse inclusion relation holds. We are going to make this more precise. For a prime p denote by $\mathcal{A}_p$ the e-variety of all abelian groups of exponent p. Note that for the completely simple e-variety $\mathcal{Q}$ the Mal'cev product $\mathcal{A}_p \circ \mathcal{Q}$ is also an e-variety of completely simple semigroups (see [29]). The precise meaning of the above vague statement is

Theorem 3.11. *For each e-variety $\mathcal{Q}$ of completely simple semigroups containing all rectangular bands, the inclusion*

$$\mathcal{S}l *_r \mathcal{Q} \subseteq \bigcap_p \mathcal{S}l *_{rr} (\mathcal{A}_p \circ \mathcal{Q})$$

holds where the intersection is over all primes p.

Here one has to show that for any prime p the inclusion

$$\mathcal{S}l *_r \mathcal{Q} \subseteq \mathcal{S}l *_{rr} (\mathcal{A}_p \circ \mathcal{Q})$$

holds. For this purpose we prove that each bi-identity which holds in $\mathcal{S}l *_{rr} (\mathcal{A}_p \circ \mathcal{Q})$ also holds in $\mathcal{S}l *_r \mathcal{Q}$. This is in a sense a converse to the above procedure. Of course, in that task we may restrict ourselves to a basis of bi-identitites of $\mathcal{S}l *_{rr} (\mathcal{A}_p \circ \mathcal{Q})$ (within $L\mathcal{I}$). Such a basis is given by the infinite set

$$\{u \simeq u^2 \mid \mathcal{A}_p \circ \mathcal{Q} \models u \simeq u^2, u \text{ a linear term}\}.$$

It suffices to consider bi-identities in linear terms since each term of $T(X)$ is equivalent in $L\mathcal{I}$ to some linear term (see [3, 21]). The verification that $\mathcal{S}l *_r \mathcal{Q}$ satsifies indeed all such bi-identities involves quite heavy calculations the details of which we omit here.

Finally we mention that, although the proofs of the above results are syntactic and use the theory of bi-identities one can use this approach to obtain also results for e-pseudovarieties, that is, for classes of finite regular semigroups closed under finite direct products, regular subsemigroups and morphic images. For example, if **Q** is an e-pseudovariety of completely simple semigroups containing all (finite) rectangular bands, $\mathbf{A}_p$ the e-pseudovariety of all finite abelian groups of exponent p then one can show (by approximation from below by locally finite e-pseudovarieties) that

$$\mathbf{Sl} *_r \mathbf{Q} \subseteq \mathbf{Sl} *_{rr} (\mathbf{A}_p \circ \mathbf{Q})$$

for any prime p. In particular, if **Q** is closed under co-extension by elementary abelian p-groups, that is, if $\mathbf{A}_p \circ \mathbf{Q} = \mathbf{Q}$ then we get $\mathbf{Sl} *_r \mathbf{Q} = \mathbf{Sl} *_{rr} \mathbf{Q}$. In contrast to the (infinitary) e-variety case, this can happen even if **Q** does not coincide with the e-pseudovariety of all completely simple semigroups. Moreover, the above inclusion, in combination with associativity of $*_r$ on the lattice of e-pseudovarieties (Billhardt and Szendrei [10]) and the results of Jones [20], can be used [7] to verify the equation

$$\mathbf{Sl} \odot \mathbf{CS} = \langle \mathbf{Sl} \circ \mathbf{CS} \rangle = L\mathbf{I}$$

in the context of e-pseudovarieties. Its infinitary analogue is an old result by Pastijn [27] later reproved by Kaďourek [21] and the author [3].

Acknowledgement

The author is very grateful to Gracinda Gomes, Jean-Éric Pin and Pedro Silva, the organizers of the Thematic Term on Semigroups, Algorithms, Automata and Languages, held at the Centro Internacional de Matemática at Coimbra, Portugal, May to July 2001, for being invited to this meeting.

References

[1] K. Auinger, *The bifree locally inverse semigroup on a set*, J. Algebra **166** (1994), 630–650.

[2] K. Auinger, *A system of bi-identities for locally inverse semigroups*, Proc. Amer. Math. Soc. **123** (1995), 979–988.

[3] K. Auinger, *On the bifree locally inverse semigroup*, J. Algebra **178** (1995), 581–613.

[4] K. Auinger and L. Polák, *A semidirect product for locally inverse semigroups*, Acta Sci. Math. (Szeged) **63** (1997), 405–435.

[5] K. Auinger and M. B. Szendrei, *E-solid e-varieties are not finitely based*, Isr. J. Math. **108** (1998), 327–343.

[6] K. Auinger and M. B. Szendrei, *Rees matrix semigroups and the regular semidirect product*, in preparation.

[7] K. Auinger and M. B. Szendrei, *Comparing the regular and the restricted regular semidirect products*, in preparation.

[8] B. Billhardt, *A wreath product embedding and idempotent pure congruences on inverse semigroups*, Semigroup Forum **45** (1992), 45–54.

[9] B. Billhardt, *On λ-semidirect products by locally $\mathcal{R}$-unipotent semigroups*, Acta Sci. Math. (Szeged) **67** (2001), 161–176.

[10] B. Billhardt and M. B. Szendrei, *Associativity of the regular semidirect product of existence varieties*, J. Austral. Math. Soc. **69** (2000), 85–115.

[11] B. Billhardt and M. B. Szendrei, *Weakly E-unitary locally inverse semigroups*, preprint.

[12] R. Broeksteeg, *E-solid e-varieties of regular semigroups which are bi-equational*, Internat. J. Algebra Comput. **6** (1996), 277-290.

[13] G. A. Churchill and P. G. Trotter, *A unified approach to bi-identities for e-varieties*, Semigroup Forum **60** (2000), 208-230.

[14] T. E. Hall, *Congruences and Green' relations in regular semigroups*, Glasgow Math. J. **13** (1972), 167-175.

[15] T. E. Hall, *Identities for existence varieties of regular semigroups*, Bull. Austral. Math. Soc. **40** (1989), 59-77.

[16] T. E. Hall, *Regular semigroups: amalgamation and the lattice of existence varieties*, Algebra Universalis **28** (1991), 79-102.

[17] K. G. Johnston, *Existence varieties with lattices of regular subsemigroups*, Algebra Universalis **30** (1993), 463-468.

[18] P. R. Jones and P. G. Trotter, *Semidirect products of regular semigroups*, Trans. Amer. Math. Soc. **349** (1997), 4265-4310.

[19] P. R. Jones, *An introduction to existence varieties of regular semigroups*, Southeast Asian Bull. Math. **19** (1995), 107-118.

[20] P. R. Jones, *Rees matrix covers and semidirect products of regular semigroups*, J. Algebra **218** (1999), 287-306.

[21] J. Kaďourek, *On some existence varieties of locally inverse semigroups*, Internat. J. Algebra Comput. **6** (1996), 761-788.

[22] J. Kaďourek, *On some existence varieties of locally orthodox semigroups*, Internat. J. Algebra Comput. **7** (1997), 93-131.

[23] J. Kaďourek and M. B. Szendrei, *A new approach in the theory of orthodox semigroups*, Semigroup Forum **40** (1990), 257-296.

[24] J. Kaďourek and M. B. Szendrei, *On existence varieties of E-solid semigroups*, Semigroup Forum **59** (1999), 470-521.

[25] D. B. McAlister, *Rees matrix covers for locally inverse semigroups*, Trans. Amer. Math. Soc. **277** (1983), 727-738.

[26] K. S. S. Nambooripad, *Structure of Regular Semigroups I*, Mem. Amer. Math. Soc. **224** (1979).

[27] F. Pastijn, *The structure of pseudo-inverse semigroups*, Trans. Amer. Math. Soc. **273** (1982), 631-655.

[28] M. Petrich, *Inverse Semigroups*, Wiley, New York, 1984.

[29] M. Petrich and N. R. Reilly, *Completely Regular Semigroups*, Canadian Mathematical Society Series of Monographs and Advanced Texts, Wiley, New York, 1999.

[30] L. Polák, *On varieties of completely regular semigroups I*, Semigroup Forum **32** (1985), 97-123.

[31] L. Polák, *On varieties of completely regular semigroups II*, Semigroup Forum **36** (1987), 253-284.

[32] L. Polák, *On varieties of completely regular semigroups III*, Semigroup Forum **37** (1988), 1–30.

[33] M. B. Szendrei, *Regular semigroups and semidirect products*, pp 233–246 in: Semigroups, Automata and Languages, J. Almeida, G. M. S. Gomes, P. V. Silva eds., World Scientific, Singapore, 1996.

[34] B. Tilson, *Categories as algebra: an essential ingredient in the theory of monoids*, J. Pure Appl. Algebra **48** (1987), 83–198.

[35] P. G. Trotter, *Congruence extensions in regular semigroups*, J. Algebra **137** (1991), 166–179.

[36] P. G. Trotter, *E-varieties of regular semigroups*, pp 247–262 in: Semigroups, Automata and Languages, J. Almeida, G. M. S. Gomes, P. V. Silva eds., World Scientific, Singapore, 1996.

[37] Y. T. Yeh, *The existence of e-free objects in e-varieties of regular semigroups*, Internat. J. Algebra Comput. **2** (1992), 471–484.

VARIETIES OF LANGUAGES

MÁRIO J. J. BRANCO*
Centro de Álgebra da Universidade de Lisboa
Av. Prof. Gama Pinto, 2, 1649-003 Lisboa, PORTUGAL
and
Departamento de Matemática
Faculdade de Ciências da Universidade de Lisboa
Campo Grande, Edifício C1, 1749-016 Lisboa, PORTUGAL
E-mail: mbranco@lmc.fc.ul.pt

In this article we give an overview of the algebraic theory of recognizable languages, namely Eilenberg's variety theory and its analogue for ordered monoids due to Pin. It also includes the descriptions of varieties of finite (ordered) monoids in terms of identities and some results on the power operator and Schützenberger and Malcev products, including their interpretation in terms of languages.

1 Introduction

The aim of this paper is to present the basis of the theory of varieties of languages as well as some recent research work in this area. The positive varieties of languages are also considered. The material presented here does not cover all existing literature on this field. However, this paper is essentially self-contained although the proofs are omitted.

Varieties of languages concern recognizable languages, that is languages recognized by finite automata. It was in the fifties that there was an upswing in the study of formal languages, which was due to the work of linguists, mainly Chomsky, in the attempt to formalize and study natural languages from a mathematical point of vue. Recognizable languages form the lower level of a Chomsky's hierarchy, which is based on his classification of formal grammars. Also in the fifties, Kleene [42] showed that these languages are precisely the rational languages (i. e. languages that can be expressed by finite sets using the symbols of union, product and the closure by product).

---

*THIS WORK WAS SUPPORTED BY THE PROJECT POCTI/32440/MAT/2000 OF FEDER, CENTRO DE ÁLGEBRA DA UNIVERSIDADE DE LISBOA, FUNDAÇÃO CALOUSTE GULBENKIAN, FUNDAÇÃO PARA A CIÊNCIA E A TECNOLOGIA, FACULDADE DE CIÊNCIAS DA UNIVERSIDADE DE LISBOA AND REITORIA DA UNIVERSIDADE DO PORTO.

This Kleene's theorem is usually considered as the foundation of the algebraic theory of recognizable languages. In the same decade, Rabin and Scott [75] presented the notion of syntactic monoid of a language, which allows us to establish the equivalence between finite automata and finite monoids. In particular, recognizable languages are exactly the languages whose syntactic monoids are finite. This approach permits, in certain cases, to describe classes of languages in terms of properties of monoids. An important example was shown by Schützenberger [79] in the mid sixties concerning the subclass of rational languages formed by the star-free languages (i. e. languages that can be described by expressions where only finite sets and the boolean and product symbols are used). He proved that a language is star-free if and only if its syntactic monoid is finite and aperiodic (i. e. its subgroups are trivial). Two other important results of this type were proved in the early seventies: Simon's theorem [81], which characterizes the piecewise testable languages, and a theorem, proved by Brzozowski and Simon [26] and, independently, by McNaughton [52], which characterizes the locally testable languages. In 1976, Eilenberg [32] introduced the appropriate tools to formulate results of this type. He gave the notions of variety of finite monoids and variety of languages and established a one to one correspondence between them. For instance, Schützenberger's theorem can now be stated as follows: the star-free languages correspond to the variety of all finite aperiodic monoids.

Recently, Pin [61] introduced the notion of syntactic ordered monoid of a language and proved a result similar to Eilenberg's variety theorem. He showed that there exists a one to one correspondence between certain classes of recognizable languages, the positive varieties of languages, and certain classes of ordered monoids, the varieties of finite ordered monoids. With this new tools we can also characterize in an algebraic form certain classes of recognizable languages that are not necessarily closed under complement.

A similar theory holds for semigroups and recognizable languages without the empty word.

With this machinery the theory of recognizable languages has been growing very fast. One of main problems is to translate certain properties or constructions on recognizable languages into properties or constructions on finite monoids and vice versa.

The paper is organized as follows. Section 2 contains the basic notions on semigroups, languages and automata that are used in the next sections. Section 3 is devoted to Eilenberg's variety theory. In Section 4 we introduce the notion of recognizability of a language by an ordered monoid as well as the notion of syntactic ordered monoid of a language. The analogue of Eilenberg's variety theory to ordered monoids, which is due to Pin, is presented in Section 5. The descriptions of varieties of monoids, varieties of ordered monoids, varieties of finite monoids and varieties of finite ordered monoids due to Birkhoff, Bloom, Reiterman and Pin and Weil, respectively, are stated in Section 6. Finally, Section 7 is devoted to some operators on varieties of finite monoids and varieties of finite semigroups, namely the power operator and Schützenberger and Malcev products, and their relations with languages.

For an introduction to the theory of recognizable languages and semigroups the reader is referred to Pin's book [58]. Eilenberg's book [32] is also a good reference although its presentation is not simple. See also the survey articles [59, 64], whose orientation is followed in this paper.

2 Semigroups, recognizable languages and automata

We start this section by defining a few notions on semigroups that are used in this paper. For a course on semigroups the reader is referred to [41, 45]. Subsection 2.2 contains some basic notions about formal languages, including the basic operation on languages. The terminology comes from linguistics. Next, we will defined the concepts of recognizability of a language by an automaton and by a semigroup and we will show they are equivalent. We will end this section by stating Kleene's theorem.

2.1 Semigroups

A *semigroup* is a pair $(S, \cdot)$ formed by a set S and an associative binary operation on S. A *monoid* is a triple $(S, \cdot, 1)$, where $(S, \cdot)$ is a semigroup and 1 is an identity. A semigroup or a monoid is usually represented by its underlying set S.

Given a semigroup S, a monoid S^1 can be constructed as follows: if S is a monoid, then S^1 is S; otherwise, $S^1 = S \cup \{1\}$, where 1 is an element that is not in S, and the multiplication on S^1 extends the multiplication on S by setting $s1 = 1s = s$, for all $s \in S^1$.

Given sets X and Y, a relation $\theta : X \to Y$ and an element x of X, we denote by $x\theta$ the set $\{y \in Y \mid (x, y) \in \theta\}$. In particular, if θ is a function and $x\theta \neq \emptyset$, then $x\theta$ is the image of x by θ. The composition of a relation $\theta : X \to Y$ with a relation $\rho : Y \to Z$ is the relation $\theta \circ \rho : X \to Z$ defined by $\theta \circ \rho = \{(x, z) \mid \exists y \in Y : (x, y) \in \theta \text{ and } (y, z) \in \rho\}$.

Concerning semigroups, the notion of subsemigroup, direct product, quotient and morphism are defined as usual. The same happens for monoids. In this case, for a morphism of monoids it is required that the image of the identity is the identity.

Given two monoids M and N, we say that M *divides* N, and we write $M \prec N$, if M is a homomorphic image of a submonoid of N.

A *relational morphism* between semigroups S and T is a relation $\sigma : S \to T$ satisfying

(1) $s\sigma \neq \emptyset$ for all $s \in S$;

(2) $(s\sigma)(t\sigma) \subseteq (st)\sigma$ for all $s, t \in S$.

If S and T are monoids and σ also satisfies

(3) $1 \in 1\sigma$

we say that σ is a monoid relational morphism. In what follows every relational morphism between monoids will always be a monoid relational morphism.

Let S be a semigroup. Given a subset A of S, we denote by $\langle A \rangle$ the subsemigroup of S generated by A.

If two elements a and b of S satisfy $a = aba$ and $b = bab$, we say that b is an *inverse* of a. If each element of S has a unique inverse, the semigroup is said to be *inverse*. If each element of S has at most one inverse, S is called a *block group*.

An element e of S is said to be an *idempotent* if $e^2 = e$. We denote by $E(S)$ the set of all idempotents of S. The semigroup S is *idempotent* if all its elements are idempotent. If e is an idempotent of S, then eSe is a subsemigroup of S, which is a monoid with e as identity. It is called a *local submonoid* of S.

In a finite semigroup each element has a unique idempotent power. So, there exists the smallest $n \in \mathbb{N}$ such that s^n is idempotent, for all $s \in S$, which is denoted by ω. In particular, s^ω is the unique power of s that is idempotent.

Green's relations on a semigroup S are five equivalence relations on S, denoted by $\mathcal{R}, \mathcal{L}, \mathcal{J}, \mathcal{D}$ and $\mathcal{H}$, which can be defined as follows:

$$a \, \mathcal{R} \, b \Leftrightarrow aS^1 = bS^1,$$
$$a \, \mathcal{L} \, b \Leftrightarrow S^1 a = S^1 b,$$
$$a \, \mathcal{J} \, b \Leftrightarrow S^1 a S^1 = S^1 b S^1,$$

for all $a, b \in S$; $\mathcal{H} = \mathcal{R} \cap \mathcal{L}$ and $\mathcal{D} = \mathcal{R} \circ \mathcal{L} \, (= \mathcal{L} \circ \mathcal{R})$.

One has $\mathcal{R}, \mathcal{L} \subseteq \mathcal{J}$, and so one also has $\mathcal{D} \subseteq \mathcal{J}$. The next result is essential to the study of finite semigroups.

Proposition 2.1. *In a finite semigroup, $\mathcal{D} = \mathcal{J}$.*

A semigroup S is said to be *aperiodic* if there exists a natural number n such that $x^n = x^{n+1}$, for every element x of S. The next proposition gives some characterizations of finite aperiodic semigroups.

Proposition 2.2. *Let S be a finite semigroup. The following conditions are equivalent:*

(1) *S is aperiodic.*
(2) *S is $\mathcal{H}$-trivial.*
(3) *Every subsemigroup of S that is a group is trivial.*

2.2 Languages

An *alphabet* is simply a finite set. Its elements are called *letters*. For instance, two important alphabets are the binary alphabet, $\{0, 1\}$, and the latin alphabet, $\{a, b, \ldots, z\}$. In this paper, alphabets and letters will be represented by letters of the beginning of the latin alphabet: capital letters for alphabets: $A, B, \ldots$; small letters for *letters*: $a, b, \ldots$.

A *word* on the alphabet A is a finite sequence (including the empty sequence) $a_1 a_2 \ldots a_n$ of letters of A. We say that a such word has *length* n. The length of a word u is represented by $|u|$. The empty sequence, which has length 0, is denoted by 1. For instance, the sequences 1, a, b, aa, ab, ba, bb, aab, aba, baa are words on the binary alphabet $\{a, b\}$. Given a word u and a letter a, we represent by $|u|_a$ the number of occurrences of a in u. We denote the set of all words on the alphabet A by A^* and the set of all non-empty words on A by A^+.

One can define a binary operation in A^*, called *concatenation*, as follows:

$$(a_1 a_2 \ldots a_m) \cdot (b_1 b_2 \ldots b_n) = a_1 a_2 \ldots a_m b_1 b_2 \ldots b_n,$$
$$1 \cdot u = u \cdot 1 = u,$$

for every $m, n \in \mathbb{N}$, $a_1, a_2, \ldots, a_m, b_1, b_2, \ldots, b_n \in A$ and $u \in A^*$. With this product, A^* is a monoid and A^+ is a semigroup. They are called the *free monoid* and *free semigroup* on A, respectively. This terminology comes from the following fact: A^* (resp. A^+) is a monoid (resp. semigroup) generated by A and, for every map $\alpha : A \to S$, where S is a monoid (resp. semigroup), there exists a unique monoid morphism $\varphi : A^* \to S$ (resp. morphism $\varphi : A^+ \to S$) extending α.

If $u = xyz$, where $x, y, z \in A^*$, we say that x is a *prefix* of u, z is a *suffix* of u and y is a *factor* of u.

Let $u = a_1 a_2 \ldots a_n \in A^*$, where $a_1, a_2, \ldots, a_n \in A$. We say that u is a *subword* of a word $v \in A^*$ if $v = v_0 a_1 v_1 a_2 v_2 \cdots a_n v_n$, for some $v_0, v_1, \ldots, v_n \in A^*$.

A *language* on the alphabet A is a set of words on A, that is a subset of A^*.

Let us define some basic operations on languages.

95

The *product* of two languages $K, L \subseteq A^*$ is defined as usual:
$$KL = \{uv \mid u \in K, v \in L\}.$$

Now, we define the *plus* and *star* operations. For $L \subseteq A^*$, let L^+ be the set of all non-empty finite products of elements of L:
$$L^+ = \{u_1 u_2 \cdots u_n \mid n \in \mathbb{N}, u_1, u_2, \ldots, u_n \in L\}.$$
We also define $L^* = L^+ \cup \{1\}$. Clearly, L^+ (resp. L^*) is the subsemigroup (resp. submonoid) of A^* generated by L.

Let $K, L \subseteq A^*$. The *left quotient* of L by K is the language
$$K^{-1}L = \{v \in A^* \mid \exists u \in K : uv \in L\}.$$
The elements of $K^{-1}L$ are the words of A^* obtained from the words of L by deleting their prefixes that are in K. Dually, the *right quotient* of L by K is the language:
$$LK^{-1} = \{v \in A^* \mid \exists u \in K : vu \in L\}.$$
Left and right quotients are defined similarly in the context of languages of A^+. In this case one has to replace A^* by A^+ in the definition.

Together with the boolean operations, plus, star, product and quotients are the basic operations considered when studying varieties of recognizable languages.

Given a word $u \in A^*$, we frequently denote the language $\{u\}$ simply by u. Sometimes we write $+$ for union, as usual. With this conventions, for instance the language $\{a\}^* \cup \{a\}^*\{a, b\}\{c, ab\}^*\{a\}$ on the alphabet $\{a, b, c\}$ can be represented by the expression $a^* + a^*(a+b)(c+ab)^*a$.

2.3 Automata and recognizable languages

An *automaton* is a quintuple $\mathcal{A} = (Q, A, E, I, F)$, where

- Q is a set, whose elements are called *states*;
- A is an alphabet;
- $E \subseteq Q \times A \times Q$; the elements of E are called *transitions*;
- $I \subseteq Q$; the elements of I are called *initial states*;
- $F \subseteq Q$; the elements of F are called *final states*.

An automaton is usually represented by a graph: the vertices are the states; the edges represent the transitions; initial states are indicated by $\rightarrow\!\bigcirc$ and final states are indicated by $\bigcirc\!\rightarrow$.

96

Example 2.1. Let $\mathcal{A}$ be the automaton

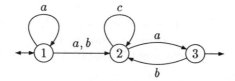

According to the notation of the definition, this means that
$$Q = \{1, 2, 3\}, \quad A = \{a, b, c\},$$
$$E = \{(1, a, 1), (1, a, 2), (1, b, 2), (2, c, 2), (2, a, 3), (3, b, 2)\},$$
$$I = \{1\}, \quad F = \{1, 3\}.$$

We represent a transition (p, a, q) by an arrow whose origin is p, the end is q and the label is a, that is $p \xrightarrow{a} q$.

A word $w = a_1 a_2 \ldots a_n$, with $a_1, a_2, \ldots, a_n \in A$, is *recognized* by the automaton $\mathcal{A}$ if there exists a sequence of transitions of the form
$$q_0 \xrightarrow{a_1} q_1 \xrightarrow{a_2} q_2 \cdots \xrightarrow{a_n} q_n,$$
where $q_0 \in I$ and $q_n \in F$. In a such case, we say that w is the label of a *successful path* in $\mathcal{A}$.

Define
$$L(\mathcal{A}) = \{w \in A^* \mid w \text{ is recognized by } \mathcal{A}\},$$
which is said to be the *language recognized* by $\mathcal{A}$.

Example 2.2. In the previous example, $L(\mathcal{A}) = a^* + a^*(a+b)(c+ab)^*a$.

We say that an automaton is finite if its set of states is finite. A language $L \subseteq A^*$ is *recognizable* if it is the language recognized by some finite automaton.

Let $\mathcal{A} = (Q, A, E, I, F)$ be an automaton. We can associate to each word of A^* a binary relation on Q as follows: if $w = a_1 a_2 \ldots a_n \in A^*$, with $a_1, a_2, \ldots, a_n \in A$, define
$$\overline{w} = \left\{(p, q) \in Q \times Q \mid \text{there exists a path } p \xrightarrow{a_1} q_1 \xrightarrow{a_2} q_2 \cdots \xrightarrow{a_n} q\right\}.$$

It is clear that, for evey $u, v \in A^*$, one has $\overline{uv} = \overline{u} \circ \overline{v}$ in the monoid $\mathcal{B}(Q)$ of the binary relations on Q with composition. Let
$$M(\mathcal{A}) = \{\overline{w} \mid w \in A^*\},$$

which is a submonoid of $\mathcal{B}(Q)$, called the *transition monoid* of $\mathcal{A}$. It is obvious that $M(\mathcal{A}) = \langle \overline{a} \mid a \in A \rangle$.

Remark. Suppose that Q is a finite set with cardinal n, let $Q = \{q_1, q_2, \ldots, q_n\}$. Consider the boolean semiring $(\{0,1\}, +, \cdot)$ and the monoid of square matrices on this semiring $(\mathcal{M}_{n \times n}(\{0,1\}), \cdot)$. Then the map

$$(\mathcal{B}(Q), \circ) \to (\mathcal{M}_{n \times n}(\{0,1\}), \cdot)$$
$$\theta \mapsto [\theta], \quad [\theta]_{ij} = \begin{cases} 1 \text{ if } (q_i, q_j) \in \theta \\ 0 \text{ otherwise} \end{cases}$$

is an isomorphism. So, every transition monoid is a monoid of boolean matrices.

Example 2.3. For the automaton we gave before as example, we have

$$\overline{a} \mapsto \begin{bmatrix} 1 & 1 & 0 \\ 0 & 0 & 1 \\ 0 & 0 & 0 \end{bmatrix} = M_1, \quad \overline{b} \mapsto \begin{bmatrix} 0 & 1 & 0 \\ 0 & 0 & 0 \\ 0 & 1 & 0 \end{bmatrix} = M_2, \quad \overline{c} \mapsto \begin{bmatrix} 0 & 0 & 0 \\ 0 & 1 & 0 \\ 0 & 0 & 0 \end{bmatrix} = M_3.$$

So, $M(\mathcal{A}) \simeq \langle M_1, M_2, M_3 \rangle$.

2.4 Semigroups and recognizable languages

Now we give an algebraic notion of recognizability of a language, which is equivalent to the one of recognizability by an automaton.

We say that a monoid M *recognizes* a language $L \subseteq A^*$ if there exist a morphism $\varphi : A^* \to M$ and a set $P \subseteq M$ such that $L = P\varphi^{-1}$. So, given φ and P, in order to see if a word belongs to L we can calculate its image by φ and check whether it belongs to P.

If a language L is the language recognized by some automaton $\mathcal{A} = (Q, A, E, I, F)$, then the map

$$\varphi : A^* \to M(\mathcal{A})$$
$$u \mapsto \overline{u}$$

is a surjective morphism and $L = (L\varphi)\varphi^{-1}$. So, L is recognized by the monoid $M(\mathcal{A})$.

Conversely, suppose that $L \subseteq A^*$ is recognized by a monoid M. There exist a morphism $\varphi : A^* \to M$ and a set $P \subseteq M$ such that $L = P\varphi^{-1}$. Then L is recognized by the automaton on the alphabet A described as follows:

- States: elements of M;
- Transitions: $(m, a, m(a\varphi))$, where $m \in M$ and $a \in A$;

- Initial states: 1;
- Final states: elements of P.

Therefore, a language is recognized if and only if it is recognized by a finite monoid.

2.5 Minimal automaton and syntactic monoid

Given a finite automaton $\mathcal{A}$, we can effectively construct a finite automaton $\mathcal{B} = (Q, A, E, \{i\}, F)$ with a unique initial state and such that each $\bar{a}$ is a total map ($a \in A$) (such an automaton is called *deterministic*) and $L(\mathcal{A}) = L(\mathcal{B})$. It is clear that, given a word $u \in A^*$, the automaton $\mathcal{B}_u = (Q, A, E, \{i\bar{u}\}, F)$ recognizes the language $u^{-1}L(\mathcal{A})$. So, if $L \subseteq A^*$ is a recognizable language, then the set $\{u^{-1}L \mid u \in A^*\}$ is finite.

Let $L \subseteq A^*$. Define an automaton $\mathcal{A}(L)$ on the alphabet A as follows:

- States: $u^{-1}L$, with $u \in A^*$;
- Transitions: $(u^{-1}L, a, (ua)^{-1}L)$, with $u \in A^*$ and $a \in A$;
- Initial states: L $(= 1^{-1}L)$;
- Final states: $u^{-1}L$, with $u \in L$.

Let $w = a_1 a_2 \ldots a_n \in A^*$, where $a_1, a_2, \ldots, a_n \in A$. In the automaton $\mathcal{A}(L)$ one has the path

$$L \xrightarrow{a_1} a_1^{-1}L \xrightarrow{a_2} (a_1 a_2)^{-1}L \cdots \xrightarrow{a_n} w^{-1}L,$$

which is the unique path in $\mathcal{A}(L)$ beginning at L and labelled by w. Thus,

$$w \in \mathcal{A}(L) \Leftrightarrow w^{-1}L \text{ is a final state}$$
$$\Leftrightarrow w \in L.$$

Then, $\mathcal{A}(L)$ recognizes L.

Therefore, L is recognizable if and only if $\mathcal{A}(L)$ is finite. The automaton $\mathcal{A}(L)$ is called the *minimal automaton* of L. This terminology is due to the fact that $\mathcal{A}(L)$ is, in a certain sense, a minimal automaton among the automata recognizing L with a certain property. In the case that L is recognizable, $\mathcal{A}(L)$ can be effectively constructed from any finite automaton recognizing L.

Now, consider the transition monoid of the minimal automaton of L, $M(\mathcal{A}(L))$, and the morphism

$$\varphi : A^* \to M(\mathcal{A}(L)).$$
$$u \mapsto \bar{u}$$

For any $u, v \in A^*$, one has
$$\overline{u} = \overline{v} \Leftrightarrow \forall x \in A^*, \, (x^{-1}L)\,\overline{u} = (x^{-1}L)\,\overline{v}$$
$$\Leftrightarrow \forall x, y \in A^*, \, (xuy \in L \Leftrightarrow xvy \in L).$$
Thus, $M(\mathcal{A}(L)) \simeq A^*/\sim_L$, where $\sim_L$ is the congruence on A^* defined by
$$u \sim_L v \Leftrightarrow \forall x, y \in A^*, \, (xuy \in L \Leftrightarrow xvy \in L).$$
This congruence is called the *syntactic congruence* of L. The quotient $A^*/\sim_L$ is called the *syntactic monoid* of L and is denoted by $M(L)$. The *syntactic morphism* of L is the canonical morphism $A^* \to M(L)$. So, we saw that $M(L)$ recognizes L. Moreover, L is recognizable if and only if $M(L)$ is finite.

The next proposition shows that $M(L)$ is, in a certain sense, a minimal monoid among the finite monoids recognizing L.

Proposition 2.3. *If $L \subseteq A^*$ and M is a finite monoid, M recognizes L if and only if $M(L)$ divides M.*

Let A be an alphabet. Define $Rec(A^*)$ as being the set of recognizable languages of A^* and define $Rat(A^*)$ as being the set of rational languages of A^*, that is the smallest set of languages of A^* containing $\emptyset$ and $\{a\}$, for all $a \in A$, and closed under finite union, product and star. So, the rational languages of A^* are the languages of A^* that can be described as a finite expression using only the sets $\emptyset$, $\{a\}$ ($a \in A$) and the symbols of union, product and star.

It is easy to see that $Rat(A^*) \subseteq Rec(A^*)$. In the fifties Kleene proved the following result.

Theorem 2.4 (Kleene). *For every alphabet A, one has $Rec(A^*) = Rat(A^*)$.*

Recognizable languages are also the languages generated by right (or left) linear grammars.

We note that there exists an algorithm that allows us to construct an automaton recognizing a rational language given any rational expression that describes it. From this automaton we can then construct the minimal automaton of the language. So, there exists an algorithm to construct the syntactic monoid of a recognizable language given by a rational expression or given any automaton recognizing it. See, for instance, [35, 39, 40, 58]. On the other hand, there exists an algorithm to obtain a rational expression of a recognizable language given any automaton that recognizes it. See, for instance, [19].

Similarly, the previous notions and results can be given for languages of A^+ and semigroups. In this case, the *syntactic semigroup* of $L \subseteq A^+$, $A^+/\sim_L$, is denoted by $S(L)$.

3 Varieties of semigroups and varieties of languages

The notion of recognizability of a language by a monoid, as well as the concept of syntactic monoid, allows us to classify classes of recognizable languages according to the algebraic properties of their syntactic monoids. This is the aim of this section. Following the historic appearance of the results, we start by the algebraic characterizations of star-free languages, piecewise testable languages and locally testable languages. Then, we present the framework due to Eilenberg [32] that makes possible to state that type of results: the notions of variety of finite monoids and variety of languages.

It is easy to see that the recognizable languages on an alphabet A form a boolean algebra. So, by Kleene's theorem, the set of rational languages of A^* can be defined as being the smallest set of languages of A^* containing $\emptyset$ and $\{a\}$, for all $a \in A$, and closed under finite boolean operations, product and star. If we omit the star operation from this definition we obtain a very important set. Its members are called the *star-free languages* of A^*.

Examples 3.1. For any alphabet A, the language A^* is star-free, since $A^* = A^* \setminus \emptyset$. Then the languages A^+ and $\{1\}$ are also star-free, since $A^+ = A^*A$ and $\{1\} = A^* \setminus A^+$.

If $A = \{a, b\}$, then a^* and $(ab)^*$ are star-free languages, since

$$a^* = A^* \setminus A^*bA^*$$

and

$$(ab)^* = A^* \setminus (bA^* + A^*a + A^*aaA^* + A^*bbA^*).$$

In the sixties Schützenberger used the syntactic monoid of a language to give the following algebraic description of the star-free languages.

Theorem 3.1 (Schützenberger). *A language $L \subseteq A^*$ is star-free if and only if the monoid $M(L)$ is finite and aperiodic.*

Thus, there exists an algorithm to decide whether a given recognizable language is star-free: given a recognizable language described by a rational expression or by a finite automaton, we construct its minimal automaton and, from it, the syntactic monoid; then, it suffices to analyse its multiplication table in order to see if it is aperiodic. However, in certain cases we can do this directly, as shown in the next example.

Example 3.2. Let $A = \{a, b\}$ be an alphabet and let $L = (aa)^*$. For each $n \in \mathbb{N}$, we have $a^n \in L$ if and only if $a^{n+1} \notin L$. So, for each $n \in \mathbb{N}$, we have $a^n \not\sim_L a^{n+1}$. Then the monoid $M(L) = A^*/\sim_L$ is not aperiodic and, by Schützenberger's theorem, L is not star-free.

Now let us see another important class of languages. A language $L \subseteq A^*$ is *piecewise testable* if it is a finite boolean combination of languages of the form $A^* a_1 A^* a_2 A^* \cdots a_n A^*$, where $n \geq 0$ and $a_1, \ldots, a_n \in A$. This means that we can check whether a given word belongs to L by looking at its subwords of bounded length.

In the seventies Simon gave the following characterization of these languages.

Theorem 3.2 (Simon). *A language $L \subseteq A^*$ is piecewise testable if and only if the monoid $M(L)$ is finite and $\mathcal{J}$-trivial.*

Thus, also in this case there is an algorithm to decide whether a given recognizable language is piecewise testable.

Let us see a third example, now about languages of A^+. A language $L \subseteq A^+$ is *locally testable* if it is a finite boolean combination of languages of the form uA^*, A^*v or A^*wA^*, with $u, v, w \in A^+$. This means that we can check whether a given word belongs to L by looking at its factors of bounded length.

Example. Take the alphabet $A = \{a, b\}$. The language $L = a(b + ba)^*b$ is locally testable because $L = (aA^* \cap A^*b) \setminus A^*aaA^*$.

The next result describes algebraicly these languages. To state it we need the following definition: a semigroup is *locally idempotent and commutative* if all its local submonoids are idempotent and commutative.

Theorem 3.3 (Brzozowski and Simon; McNaughton). *A language $L \subseteq A^+$ is locally testable if and only if the semigroup $S(L)$ is finite and locally idempotent and commutative.*

Once more, this theorem gives an algorithm to decide whether a given recognizable language is locally testable.

At the beginning of the seventies, Eilenberg [32] presented the theoretical tools to formulate these type of results. The main tool of his approach is the notion of variety. See [32, 58].

An **M**-*variety*, a *variety of finite monoids* or a *pseudovariety of monoids*, is a class of finite monoids closed under submonoids, homomorphic images and finite direct products. Similarly, an **S**-*variety*, a *variety of finite semigroups* or a *pseudovariety of semigroups*, is a class of finite semigroups closed under subsemigroups, homomorphic images and finite direct products. We usually represent **M**-varieties and **S**-varieties by bold capital letters.

Examples 3.3. The following classes of finite monoids are **M**-varieties:
  I – trivial monoids. This is called the *trivial* **M**-*variety*.
  M – monoids.
  G – groups.
  Com – commutative monoids.
  $\mathbf{J_1}$ – idempotent and commutative monoids.
  R – $\mathcal{R}$-trivial monoids.
  L – $\mathcal{L}$-trivial monoids.
  A – aperiodic monoids.
  J – $\mathcal{J}$-trivial monoids.
  BG – block groups.

We also represent by **I** the *trivial* **S**-*variety*, that is the class formed by the empty semigroup and by the semigroups of cardinal one. The **S**-variety of all finite semigroups is denoted by **S**. We also use the other notations of the previous list to represent **S**-varieties described similarly. The context will make clear whether they represent **M**-varieties or **S**-varieties, otherwise it will be mentioned.

If **V** is an **M**-variety, then the class of finite monoids

$$\mathbf{EV} = \{M \mid \langle E(M) \rangle \in \mathbf{V}\}$$

is an **M**-variety and the class of finite semigroups

$$\mathbf{LV} = \{S \mid eSe \in \mathbf{V}, \forall e \in E(S)\}$$

is an **S**-variety. Obviously, **EI** = **G**. Margolis and Pin [50] proved that **EJ** = **BG**.

Given an **S**-variety **V**, the class of finite monoids **DV** formed by the finite monoids M such that its regular $\mathcal{J}$-classes are subsemigroups of M in **V** is an **M**-variety. For instance, if we take the trivial **S**-variety **I**, then **DI** = **J**.

Now, we define the corresponding notion for languages. A $*$-*variety* is a correspondence $\mathcal{V} : A \mapsto A^*\mathcal{V} \subseteq Rec(A^*)$ that associates to each alphabet A a set of recognizable languages $A^*\mathcal{V}$ of A^* such that, for every alphabets A and B, the following conditions are satisfied:

  – $A^*\mathcal{V}$ is a boolean algebra (finite union, finite intersection and complementation);

  – If $L \in A^*\mathcal{V}$ and $\varphi : B^* \to A^*$ is a morphism, then $L\varphi^{-1} \in B^*\mathcal{V}$;

  – If $L \in A^*\mathcal{V}$ and $a \in A$, then $a^{-1}L, La^{-1} \in A^*\mathcal{V}$.

Similarly, we define a +-*variety* by replacing in the previous definition ∗ by +.

Given an **M**-variety **V**, let $\mathcal{V}$ be the correspondence $\mathcal{V} : A \mapsto A^*\mathcal{V}$, where

$$A^*\mathcal{V} = \{L \subseteq A^* \mid L \text{ is recognized by a monoid of } \mathbf{V}\}$$
$$= \{L \subseteq A^* \mid M(L) \in \mathbf{V}\}.$$

Similarly, given an **S**-variety **V**, we also define $\mathcal{V}$ as being the correspondence $\mathcal{V} : A \mapsto A^+\mathcal{V}$, where

$$A^+\mathcal{V} = \{L \subseteq A^+ \mid L \text{ is recognized by a semigroup of } \mathbf{V}\}$$
$$= \{L \subseteq A^+ \mid S(L) \in \mathbf{V}\}.$$

Now, we state Eilenberg's variety theorem.

Theorem 3.4 (Eilenberg). *The correspondence* $\mathbf{V} \mapsto \mathcal{V}$ *defines a bijection between the* **M**-*varieties and the* ∗-*varieties.*

The corresponding version for **S**-varieties and +-varieties also holds.

Thus, Kleene's theorem says that $A^*\mathcal{M} = Rat(A^*)$, for every alphabet A. Similarly, Theorems 3.1, 3.2 and 3.3 can now be stated as follows: for every alphabet A,

$A^*\mathcal{A} = \{\text{star-free languages of } A^*\}$;

$A^*\mathcal{J} = \{\text{piecewise testable languages of } A^*\}$;

$A^+\mathcal{LJ}_1 = \{\text{locally testable languages of } A^+\}$.

Given a class **C** of finite monoids, the intersection of all **M**-varieties containing **C** is the smallest **M**-variety containing **C**, which is called the **M**-*variety generated* by **C**. We represent it by (**C**). One has

$$(\mathbf{C}) = \{M \mid \exists M_1, \ldots, M_n \in \mathbf{C} : M \prec M_1 \times \cdots \times M_n\}.$$

Given two **M**-varieties **V** and **W**, the *join* of **V** and **W** is the **M**-variety $\mathbf{V} \vee \mathbf{W} = (\mathbf{V} \cup \mathbf{W})$.

If $\mathcal{V}$ is a ∗-variety, then the class

$$\mathbf{V} = \{M \mid \text{there exist an alphabet } A \text{ and languages } L_1, \ldots, L_n \in A^*\mathcal{V} :$$
$$M \prec M(L_1) \times \cdots \times M(L_n)\}$$

is an **M**-variety (the **M**-variety generated by the syntactic monoids of languages of $\mathcal{V}$) and $\mathbf{V} \mapsto \mathcal{V}$.

It is easy to prove the following proposition.

Proposition 3.5. *Let* **C** *be a class of finite monoids and let* $\mathbf{V} = (\mathbf{C})$. *Then, for each alphabet* A, $A^*\mathcal{V}$ *is the boolean algebra generated by the set of languages of* A^* *recognized by some monoid of* **C**.

Let us see some examples. See [32, 58] for the details.
It is well-known that $\mathbf{J}_1 = (U_1)$, where $U_1 = \{0,1\}$ and $0 \cdot 0 = 0 \cdot 1 = 1 \cdot 0 = 0$ and $1 \cdot 1 = 1$. If $\varphi : A^* \to U_1$ is a morphism, then $\emptyset\varphi^{-1} = \emptyset$, $U_1\varphi^{-1} = A^*$, $0\varphi^{-1} = \bigcup_{a \in A \cap 0\varphi^{-1}} A^*aA^*$ and $1\varphi^{-1} = B^*$, where $B = 1\varphi^{-1} \cap A$. Since $1\varphi^{-1} = A^* \setminus 0\varphi^{-1}$, we have the following result.

Proposition 3.6. *For each alphabet* A,
– $A^*\mathcal{J}_1$ *is the boolean algebra generated by the languages of the form* A^*aA^*, *where* $a \in A$; *and*
– $A^*\mathcal{J}_1$ *is the boolean algebra generated by the languages of the form* B^*, *where* $B \subseteq A$.

Given an alphabet A, a letter $a \in A$, $n \in \mathbb{N}$ and $k \in \mathbb{N}_0$, define
$$F(a,k,n) = \{u \in A^* \mid |u|_a \equiv k(\mathrm{mod}\ n)\}.$$
The same technique lead us to the following result.

Proposition 3.7. *Fix* $n \in \mathbb{N}$ *and let* $\mathbf{V} = (\mathbb{Z}_n)$. *Then, for each alphabet* A, $A^*\mathcal{V}$ *is the boolean algebra generated by the languages of the form* $F(a,k,n)$, *where* $a \in A$ *and* $0 \leq k < n$.

Let **Gcom** be the **M**-variety of finite commutative groups. It is well-known that $\mathbf{Gcom} = (\{\mathbb{Z}_n \mid n \in \mathbb{N}\})$. This proves the following result.

Proposition 3.8. *For each alphabet* A, $A^*\mathcal{G}com$ *is the boolean algebra generated by the languages of the form* $F(a,k,n)$, *where* $a \in A$, $n \in \mathbb{N}$ *and* $0 \leq k < n$.

Now, let **Acom** be the **M**-variety of finite aperiodic and commutative groups.

Given an alphabet A, a letter $a \in A$ and $k \in \mathbb{N}_0$, define
$$F(a,k) = \{u \in A^* \mid |u|_a \geq k\}.$$

Proposition 3.9. *For each alphabet* A, $A^*\mathcal{A}com$ *is the boolean algebra generated by the languages of the form* $F(a,k)$, *where* $a \in A$ *and* $k \in \mathbb{N}_0$.

Since **Com** = **Acom** $\vee$ **Gcom**, we have the following result.

Proposition 3.10. *For each alphabet* A, $A^*\mathcal{C}om$ *is the boolean algebra gene-*

rated by the languages of the form $F(a, k, n)$ or $F(a, k)$, where $a \in A$, $n \in \mathbb{N}$ and $0 \le k < n$.

Now, consider the following **S**-varieties:

$$\mathbf{K} = \{S \mid \forall e \in E(S), s \in S,\ es = e\},$$
$$\mathbf{K^r} = \{S \mid \forall e \in E(S), s \in S,\ se = e\} \text{ and}$$
$$\mathbf{Nil} = \{S \mid \forall e \in E(S), s \in S,\ es = e = se\}\ (= \mathbf{K} \cap \mathbf{K^r}).$$

A semigroup of **Nil** is called *nilpotent*. The next proposition gives a characterization of the languages recognized by semigroups of these **S**-varieties.

Proposition 3.11. *For each alphabet A,*
$A^+\mathcal{N}il = \{L \subseteq A^+ \mid L \text{ is finite or cofinite}\}$,
$A^+\mathcal{K} = \{XA^* \cup Y \mid X, Y \subseteq A^+ \text{ are finite}\}$,
$A^+\mathcal{K}^r = \{A^*X \cup Y \mid X, Y \subseteq A^+ \text{ are finite}\}$ *and*
$A^+\mathcal{LI} = \{XA^*Y \cup Z \mid X, Y, Z \subseteq A^+ \text{ are finite}\}$.

Schützenberger's and Simon's theorems describe the ∗-varieties corresponding to the **M**-varieties **A** and **J**, respectively. The following two statements regard the other two Green's relations.

Proposition 3.12. *For each alphabet A, $A^*\mathcal{R}$ consists of all finite disjoint unions of languages of the form $A_0^* a_1 A_1^* a_2 A_2^* \cdots a_n A_n^*$, where $n \in \mathbb{N}_0$, $A_0, \ldots, A_n \subseteq A$ and $a_i \in A \setminus A_{i-1}$ for all $i \in \{1, \ldots, n\}$.*

Proposition 3.13. *For each alphabet A, $A^*\mathcal{L}$ consists of all finite disjoint unions of languages of the form $A_0^* a_1 A_1^* a_2 A_2^* \cdots a_n A_n^*$, where $n \in \mathbb{N}_0$, $A_0, \ldots, A_n \subseteq A$ and $a_i \in A \setminus A_i$ for all $i \in \{1, \ldots, n\}$.*

4 Ordered semigroups and recognizable languages

There are important classes of languages that are not closed under complement. In [61] Pin has proved that some of them can also be algebraically studied. The algebraic tool for such an approach is the ordered semigroup. In this section we present some notions concerning ordered semigroups as well as the notion of recognizability of a language by an ordered semigroup.

An *ordered monoid* (resp. *ordered semigroup*) is a pair $(M, \le)$, where M is a monoid (resp. semigroup) and $\le$ is a partial order relation on M compatible with the product on M, that is, for every $x, y, z \in M$, $x \le y$ implies $xz \le yz$ and $zx \le zy$.

Examples 4.1. 1) For every semigroup S, the pair $(S, =)$ is an ordered semigroup.

2) In a finite group there is a unique partial order compatible with the product. In fact, if $\leq$ is a partial order relation on a finite group G of order n, one has

$$x \leq y \Rightarrow 1 \leq x^{-1}y$$
$$\Rightarrow 1 \leq x^{-1}y \leq (x^{-1}y)^2 \leq \ldots \leq (x^{-1}y)^n = 1$$
$$\Rightarrow x^{-1}y = 1$$
$$\Rightarrow x = y.$$

3) Let S be an inverse semigroup. Then, the *natural order* on S, which is defined by

$$x \leq y \Leftrightarrow \exists e \in E(S) : x = ye$$
$$(\Leftrightarrow \exists e \in E(S) : x = ey),$$

is compatible with the product. So, $(S, \leq)$ is an order semigroup. In the case that S is an idempotent and commutative semigroup one has

$$x \leq y \Leftrightarrow x = xy.$$

We notice that, given an ordered idempotent and commutative monoid $(M, \leq)$, its order is the natural order if and only if $x \leq 1$, for all $x \in M$.

4) Let S be a semigroup. Consider the *power semigroup* of S, $\mathcal{P}(S)$, with multiplication $XY = \{xy \mid x \in X, y \in Y\}$. Then $(\mathcal{P}(S), \subseteq)$ is an ordered semigroup.

A map $\varphi : (S, \leq) \to (T, \leq)$ between ordered semigroups is a *morphism of ordered semigroups* if, for all $x, y \in S$, one has

$$(xy)\varphi = x\varphi y\varphi$$

and

$$x \leq y \Rightarrow x\varphi \leq y\varphi.$$

If, in addition, φ is surjective and, for every $x, y \in S$, $x\varphi \leq y\varphi$ implies $x \leq y$, we say that φ is an *isomorphism of ordered semigroups*. For a *morphism of ordered monoids* we also require that the image of the identity is the identity. If $\varphi : M \to N$ is a monoid morphism and $(N, \leq)$ is an ordered monoid, then $\varphi : (M, =) \to (N, \leq)$ is a morphism of ordered monoids. So, for each alphabet A, the ordered monoid $(A^*, =)$ is the free ordered monoid on A.

The *direct product* $\prod_{i \in I}(M_i, \leq)$ of a family of ordered monoids is the direct product $\prod_{i \in I} M_i$ of the monoids equipped with the product order.

If $(M, \leq)$ and $(N, \leq)$ are ordered monoids, we say that $(M, \leq)$ is an *ordered submonoid* of $(N, \leq)$ if M is a submonoid of N and the order on M is the restriction to M of the order on N.

We say that an ordered monoid $(M, \leq)$ *divides* an ordered monoid $(N, \leq)$, and we write $(M, \leq) \prec (N, \leq)$, if $(M, \leq)$ is a homomorphic image of an ordered submonoid of $(N, \leq)$.

An *order ideal* of a partially ordered set $(P, \leq)$ is a subset I of P such that, for all $x, y \in P$,

$$x \leq y, \ y \in I \Rightarrow x \in I.$$

Now, let us see how ordered monoids can be applied to the study of recognizable languages. An ordered monoid $(M, \leq)$ *recognizes* a language $L \subseteq A^*$ if there exist a morphism $\varphi : A^* \to M$ and an order ideal I of $(M, \leq)$ such that $L = I\varphi^{-1}$. This definition generalizes the definition of recognizability of a language by a monoid when we take the ordered monoid $(M, =)$.

Let $L \subseteq A^*$. Define a quasi-order relation $\preceq_L$ on A^* by

$$u \preceq_L v \Leftrightarrow \forall x, y \in A^*, \ (xvy \in L \Rightarrow xuy \in L).$$

This relation is compatible with the product and we have

$$u \sim_L v \Leftrightarrow u \preceq_L v \text{ and } v \preceq_L u.$$

So, in the monoid $M(L) = A^*/\sim_L$ we can define a partial order $\leq_L$, called the *syntactic order* of L, by

$$[u]_{\sim_L} \leq_L [v]_{\sim_L} \Leftrightarrow u \preceq_L v.$$

Thus, $(M(L), \leq_L)$ is an ordered monoid, called the *syntactic ordered monoid* of L. If $\varphi : A^* \to M(L)$ is the syntactic morphism, then $L\varphi$ is an order ideal of $(M(L), \leq_L)$ and $L = (L\varphi)\varphi^{-1}$. Therefore $(M(L), \leq_L)$ recognizes L.

We notice that $M(L) = M(A^* \setminus L)$ and the syntactic ordered monoid of $A^* \setminus L$ is $(M(L), \geq_L)$.

We also have the following result.

Proposition 4.1. *If $L \subseteq A^*$ and $(M, \leq)$ is a finite ordered monoid, $(M, \leq)$ recognizes L if and only if $(M(L), \leq_L)$ divides $(M, \leq)$.*

All these definitions and results can be given similarly for ordered semigroups and languages of A^+.

5 Varieties of ordered semigroups and positive varieties of languages

In this section we see how the definitions of varieties of Section 3 can be modified in order to obtain a theorem analogous to Eilenberg's theorem for certain classes of recognizable languages that are not necessarily closed under complement, the positive varieties of languages, and certain classes of finite ordered monoids, the **OM**-varieties.

We say that a class of finite ordered monoids is an **OM**-*variety*, a *variety of finite ordered monoids* or a *pseudovariety of ordered monoids*, if it is closed under ordered submonoids, homomorphic images and finite direct products. Similarly, an **OS**-*variety*, a *variety of finite ordered semigroups* or a *pseudovariety of ordered semigroups*, is a class of finite ordered semigroups closed under ordered subsemigroups, homomorphic images and finite direct products. We also represent **OM**-varieties and **OS**-varieties by bold capital letters.

If **V** is an **M**-variety, the class of all ordered monoids $(M, \leq)$, with $M \in \mathbf{V}$, is an **OM**-variety, called the **OM**-*variety generated* by **V**, and is also denoted by **V**. The context will make clear whether **V** is considered as an **M**-variety or as an **OM**-variety, otherwise it will be mentioned.

Examples 5.1. The following classes of finite ordered monoids are **OM**-varieties:

  I – trivial ordered monoids. This is called the *trivial* **OM**-*variety*.
  M – ordered monoids.
  G – groups, that is ordered groups of the form $(G, =)$ (see example 2 in Section 4).
  $\mathbf{J}_1^+$ – ordered idempotent and commutative monoids with the natural order (see example 3 in Section 4).
  $\mathbf{J}^+$ – ordered monoids $(M, \leq)$ such that $x \leq 1$, for all $x \in M$.

If **V** is an **OM**-variety, then the class of finite ordered monoids

$$\mathbf{EV} = \{(M, \leq) \mid \langle E(M) \rangle \in \mathbf{V}\}$$

is an **OM**-variety and the class of finite ordered semigroups

$$\mathbf{LV} = \{(S, \leq) \mid eSe \in \mathbf{V}, \forall e \in E(S)\}$$

is an **OS**-variety.

A *positive* ∗-*variety* is a correspondence $\mathcal{V} : A \mapsto A^*\mathcal{V} \subseteq Rec(A^*)$ that associates to each alphabet A a set of recognizable languages $A^*\mathcal{V}$ of A^* such that, for every alphabets A and B, the following conditions are satisfied:

- $A^*\mathcal{V}$ is a positive boolean algebra (finite union and finite intersection);
- If $L \in A^*\mathcal{V}$ and $\varphi : B^* \to A^*$ is a morphism, then $L\varphi^{-1} \in B^*\mathcal{V}$;
- If $L \in A^*\mathcal{V}$ and $a \in A$, then $a^{-1}L, La^{-1} \in A^*\mathcal{V}$.

Similarly, we define *positive +-variety* by replacing in the previous definition $*$ by $+$.

If **V** is an **OM**-variety, let $\mathcal{V}$ be the correspondence $\mathcal{V} : A \mapsto A^*\mathcal{V}$ defined by

$$A^*\mathcal{V} = \{L \subseteq A^* \mid L \text{ is recognized by an ordered monoid of } \mathbf{V}\}$$
$$= \{L \subseteq A^* \mid (M(L), \leq_L) \in \mathbf{V}\}.$$

Similarly, if **V** is an **OS**-variety, we also denote by $\mathcal{V}$ the correspondence $\mathcal{V} : A \mapsto A^+\mathcal{V}$, where

$$A^+\mathcal{V} = \{L \subseteq A^+ \mid L \text{ is recognized by an ordered semigroup of } \mathbf{V}\}$$
$$= \{L \subseteq A^+ \mid (S(L), \leq_L) \in \mathbf{V}\}.$$

The next result is similar to Eilenberg's variety theorem.

Theorem 5.1 (Pin [61]). *The correspondence* $\mathbf{V} \mapsto \mathcal{V}$ *defines a bijection between the* **OM**-*varieties and the positive* $*$-*varieties.*

The version of this theorem for **OS**-varieties and positive +-varieties also holds.

Given an **OM**-variety **V**, let $\mathbf{V}^c$ be the class of ordered monoids $(M, \leq)$ such that $(M, \geq) \in \mathbf{V}$. This class is an **OM**-variety, which is called the *dual of* **V**.

A set I is an order ideal of a partially ordered set $(P, \leq)$ if and only if $P \setminus I$ is an order ideal of $(P, \geq)$. So, for every alphabet A, $A^*\mathcal{V}^c$ is the set of languages of the form $A^* \setminus L$, with $L \in A^*\mathcal{V}$.

If **C** is a class of finite ordered monoids, we also represent by (**C**) the **OM**-*variety generated* by **C**, that is the smallest **OM**-variety containing **C**. The *join* of two **OM**-varieties, **V** and **W**, is the **OM**-variety $\mathbf{V} \vee \mathbf{W} = (\mathbf{V} \cup \mathbf{W})$.

The proposition corresponding to Proposition 3.5 for ordered monoids is the following.

Proposition 5.2. *Let* **C** *be a class of finite ordered monoids and let* $\mathbf{V} = (\mathbf{C})$. *Then, for each alphabet* A, $A^*\mathcal{V}$ *is the positive boolean algebra generated by the set of languages of* A^* *recognized by some ordered monoid of* **C**.

Let us see a few examples of algebraic characterizations of positive $*$-varieties and +-varieties.

From the fact $\mathbf{J}_1 = (U_1)$ it follows that $\mathbf{J}_1^+ = (U_1^+)$, where U_1^+ is the ordered monoid $(U_1, \leq)$, with $0 \leq 1$. The order ideals of U_1^+ are $\emptyset$, $\{0\}$ and U_1. Then, we have the following result.

Proposition 5.3. *For each alphabet A, $A^* \mathcal{J}_1^+$ is the positive boolean algebra generated by the languages of the form $A^* a A^*$, where $a \in A$.*

Let $\mathbf{J}_1^- = (\mathbf{J}_1^+)^c$. We also have $\mathbf{J}_1^- = (U_1^-)$, where U_1^- is the ordered monoid $(U_1, \leq)$, with $1 \leq 0$. Then, one has the following proposition.

Proposition 5.4. *For each alphabet A, $A^* \mathcal{J}_1^-$ is the positive boolean algebra generated by the languages of the form B^*, where $B \subseteq A$.*

The next proposition describes the positive $*$-variety corresponding to $\mathbf{J}^+$. Compare it with Theorem 3.2.

Proposition 5.5. [61] *For each alphabet A, $A^* \mathcal{J}^+$ is formed by the finite unions of languages of the form $A^* a_1 A^* a_2 A^* \cdots a_n A^*$, where $n \geq 0$ and $a_1, \ldots, a_n \in A$.*

The following example concerns an **OS**-variety. Compare it with Proposition 3.11.

Proposition 5.6. [61] *Let $\mathbf{V}$ be the **OS**-variety of all finite ordered nilpotent semigroups $(S, \leq)$ such that $s \leq 0$, for all $s \in S$. Then, for each alphabet A, $A^+ \mathcal{V}$ is formed by the finite languages of A^+ and by A^+.*

We have presented the definition of **OM**-variety generated by an **M**-variety. The following proposition allows us to define the **M**-variety generated by an **OM**-variety.

Proposition 5.7. [69] *Let $\mathbf{V}$ be an **OM**-variety. For every finite monoid M, the following conditions are equivalent:*
  (1) $(M, \leq) \in \mathbf{V} \vee \mathbf{V}^c$, *for some order $\leq$ on M.*
  (2) $(M, =) \in \mathbf{V} \vee \mathbf{V}^c$.
  (3) *There exists an ordered monoid $(N, \leq) \in \mathbf{V}$ such that M is a homomorphic image of N.*

Now, given an **OM**-variety $\mathbf{V}$, we define the **M**-*variety generated by* $\mathbf{V}$, denoted by $\mathbf{V} \vee \mathbf{V}^c$, as follows: $\mathbf{V} \vee \mathbf{V}^c$ is the **M**-variety of all monoids M such that $(M, \leq) \in \mathbf{V} \vee \mathbf{V}^c$, for some order $\leq$ on M.

We notice that if $\mathbf{V}$ is an **M**-variety and $\mathbf{W}$ is the **OM**-variety generated by $\mathbf{V}$, then $\mathbf{W} = \mathbf{W}^c$, and so the **M**-variety generated by $\mathbf{W}$ is precisely $\mathbf{V}$.

Corollary 5.8. *Let* **V** *be an* **OM**-*variety and let* **W** *be the* **M**-*variety generated by* **V**. *Let* $\mathcal{V}$ *be the positive* *-*variety corresponding to* **V** *and let* $\mathcal{W}$ *be the* *-*variety corresponding to* **W**.

Then, for every alphabet A, $A^*\mathcal{W}$ *is the boolean algebra generated by* $A^*\mathcal{V}$.

Let $\mathbf{J}^- = (\mathbf{J}^+)^c$.

From Propositions 3.6, 5.3, 5.5, Theorem 3.2 and previous corollary we deduce the following result.

Corollary 5.9. *One has* $\mathbf{J}_1 = \mathbf{J}_1^+ \vee \mathbf{J}_1^-$ *and* $\mathbf{J} = \mathbf{J}^+ \vee \mathbf{J}^-$.

We notice that the equality $\mathbf{J} = \mathbf{J}^+ \vee \mathbf{J}^-$ can also be deduced from Proposition 5.7 and a theorem of Straubing and Thérien [92].

6 Identities

Birkhoff's variety theorem for algebras was extended by Reiterman to finite algebras in terms of implicit operations. A result similar to Birkhoff's variety theorem was proved by Bloom for ordered algebras. More recently Pin and Weil also extended Bloom's theorem to finite ordered algebras.

In this section we present Reiterman's and Pin and Weil's theorems for finite monoids and ordered finite monoids. However, we will not follow the original presentation of Reiterman's theorem. Instead we will use profinite completions. See Almeida's book [1] and Almeida and Weil's article [14].

An *equality* (resp. *inequality*) on an alphabet A is a formal equality (resp. inequality) of the form $u = v$ (resp. $u \leq v$), where $u, v \in A^*$. An *identity* is an equality or an inequality.

A monoid M (resp. An ordered monoid $(M, \leq)$) *satisfies* the identity $u = v$ (resp. $u \leq v$) on the alphabet A if, for every morphism $\varphi : A^* \to M$, we have $u\varphi = v\varphi$ (resp. $u\varphi \leq v\varphi$). This means that, when we replace the letters in u and v by elements of M, we obtain the same element (resp. related elements), since every morphism from A^* to M only depends on the images of the letters.

If Σ is a set of equalities (not necessarily involving a finite number of letters), let $[\Sigma]$ be the class of monoids (not necessarily finite) satisfying all equalities of Σ and let $[\![\Sigma]\!]$ be the class of finite monoids satisfying all equalities of Σ. Similarly, if Σ is a set of identities, $[\Sigma]$ represents the class of ordered monoids satisfying all identities of Σ, and $[\![\Sigma]\!]$ represents the class of finite ordered monoids satisfying all identities of Σ. We say that the classes $[\Sigma]$ and $[\![\Sigma]\!]$ are defined by the set Σ. When we explicit the identities, the symbols represent letters if nothing is said in contrary (see the next examples). If Σ is a set of equalities, the notation $[\Sigma]$ and $[\![\Sigma]\!]$ can be interpreted both as classes of monoids or as classes of ordered monoids. However, the context will make their meaning clear. Moreover, if, for instance, $[\![\Sigma]\!]$ is interpreted as a class of finite ordered monoids, then it is the class of all finite ordered monoids $(M, \leq)$ such that the monoid M satisfies all equalities of Σ.

Examples 6.1. The class of monoids $[xy = yx]$ is formed by all commutative monoids, since a monoid M is commutative if and only if $st = ts$ for all $s, t \in M$. The trivial **M**-variety **I** is $[\![x = y]\!]$ and the **M**-variety **M** is $[\![x = x]\!]$. One also has $\mathbf{J}_1 = [\![x^2 = x,\ xy = yx]\!]$ and $\mathbf{J}_1^+ = [\![x^2 = x,\ xy = yx,\ x \leq 1]\!]$.

A *variety* of monoids is a class of monoids (not necessarily finite) closed under submonoids, homomorphic images and direct products.

The following result is a particular case of a general theorem for algebras due to Birkhoff. See, for instance, [30, 94].

Theorem 6.1 (Birkhoff). *The varieties of monoids are precisely the classes of monoids of the form $[\Sigma]$, where Σ is a set of equalities.*

A *variety* of ordered monoids is a class of ordered monoids (not necessarily finite) closed under ordered submonoids, homomorphic images and direct products.

Bloom has proved a general result for ordered algebras similar to Birkhoff's variety theorem that we also state for ordered monoids. See [25, 94].

Theorem 6.2 (Bloom). *The varieties of ordered monoids are precisely the classes of ordered monoids of the form $[\Sigma]$, where Σ is a set of identities.*

The previous definitions and theorems can also be given for semigroups. As usual, we only have to replace A^* by A^+.

With this notion of identity the analogues of these statements for finite monoids and finite ordered monoids do not hold in general. For instance, the **M**-variety **G** is not a class of the form $[\![\Sigma]\!]$ in the previous sense. However, we have the following result.

Proposition 6.3. *Every **M**-variety (resp. **OM**-variety) generated by a monoid (resp. an ordered monoid) is defined by a set of equalities (resp. identities).*

In order to obtain the analogues of Theorems 6.1 and 6.2 for **M**-varieties and **OM**-varieties, we have to extend the notion of identity.

Let A be an alphabet and let $u, v \in A^*$. We say that a monoid M

separates u and v if there exists a morphism $\varphi : A^* \to M$ such that $u\varphi \neq v\varphi$. Let

$$r(u,v) = \min\{|M| \mid M \text{ is a finite monoid separating } u \text{ and } v\}$$

and

$$d(u,v) = 2^{-r(u,v)},$$

with the conventions $\min \emptyset = +\infty$ and $2^{-\infty} = 0$.

Given words u and v of A^* such that $u \neq v$, there exists a finite monoid separating u and v. In fact, the quotient monoid A^*/θ of A^* by the congruence $\theta = \{(x,y) \in A^* \times A^* \mid x = y \text{ or } |x|,|y| > |u|\}$ separates u and v. Let $u,v,w,x,y \in A^*$. Clearly, if a monoid separates u and v, then it separates u and w or it separates w and v. Also, if a monoid separates ux and vy, then it separates u and v or it separates x and y. These remarks prove the following properties:

- $d(u,v) = 0$ if and only if $u = v$;
- $d(u,v) = d(v,u)$;
- $d(u,v) \leq \max\{d(u,w), d(w,v)\}$;
- $d(ux,vy) \leq \max\{d(u,v), d(x,y)\}$.

In particular, d is a metric on A^*. In this metric, two words are close if a large monoid is required to separate them. The last property implies that the product on A^* is uniformly continuous, so that A^* is a topological monoid. The completion of the metric space (A^*, d) is also a topological monoid, called the *free profinite monoid* on A and denoted by $\widehat{A^*}$. So, every element of $\widehat{A^*}$ is a limit of a Cauchy sequence of words of A^*.

Given a finite monoid M, every morphism from A^* to M is uniformly continuous, so it can be extended in a unique way into a continuous morphism from $\widehat{A^*}$ to M. Therefore, every continuous morphism from $\widehat{A^*}$ to M only depends on the letters of A.

Thus, we can extend the notion of identity as follows. An *equality* (resp. *inequality*) is a formal equality (resp. inequality) of the form $u = v$ (resp. $u \leq v$), where $u, v \in \widehat{A^*}$. An *identity* is an equality or an inequality.

A finite monoid M (resp. ordered monoid $(M, \leq)$) *satisfies* the identity $u = v$ (resp. $u \leq v$), with $u, v \in \widehat{A^*}$ if, for every continuous morphism $\varphi : \widehat{A^*} \to M$, one has $u\varphi = v\varphi$ (resp. $u\varphi \leq v\varphi$).

Every element of $\widehat{A^*}$ is a limit of a Cauchy sequence of words of A^*. The following proposition gives an example of such an important limit.

Proposition 6.4. *For each $x \in \widehat{A^*}$, the sequence $\left(x^{n!}\right)_{n \geq 0}$ converges.*

The limit of the sequence $\left(x^{n!}\right)_{n \geq 0}$ is denoted by x^ω. Thus, given a finite monoid M and a continuous morphism $\varphi : \widehat{A^*} \to M$, the element $x^\omega \varphi$ is the unique idempotent power of $x\varphi$, which we have represented by $(x\varphi)^\omega$. So, we can forget the topological aspects of the definition and interpret x^ω as an idempotent.

Example. A finite monoid M satisfies the identity $(xy)^\omega = (xy)^\omega x$ if and only if $(st)^\omega = (st)^\omega s$, for all $s, t \in M$.

The definition of $[\![\Sigma]\!]$ given previously can also be extended as follows. If Σ is a set of equalities (not necessarily involving a finite number of letters), we define $[\![\Sigma]\!]$ as being the class of finite monoids satisfying all identities of Σ. When Σ is set of identities, we also denote by $[\![\Sigma]\!]$ the class of finite ordered monoids satisfying all identities of Σ.

The previous notions can also be given for semigroups with the natural adaptations.

Examples 6.2. The M-varieties corresponding to Green's relations can be described as follows:

$$\mathbf{A} = [\![x^\omega = x^{\omega+1}]\!] \quad (x^{\omega+1} \text{ represents } x^\omega x),$$
$$\mathbf{R} = [\![(xy)^\omega x = (xy)^\omega]\!],$$
$$\mathbf{L} = [\![y(xy)^\omega = (xy)^\omega]\!],$$
$$\mathbf{J} = [\![(xy)^\omega = (yx)^\omega, x^\omega = x^{\omega+1}]\!].$$

One also has $\mathbf{BG} = [\![(x^\omega y^\omega)^\omega = (y^\omega x^\omega)^\omega]\!]$ and $\mathbf{EJ^+} = [\![x^\omega \leq 1]\!]$.

For the S-varieties **Nil** and **LI** we have

$$\mathbf{Nil} = [\![x^\omega y = x^\omega = yx^\omega]\!] \quad \text{and} \quad \mathbf{LI} = [\![x^\omega y x^\omega = x^\omega]\!].$$

It is also clear that $\mathbf{LJ^+} = [\![x^\omega y x^\omega \leq x^\omega]\!]$.

Now, we have the tools to state the analogues of Birkhoff and Bloom's theorems.

Theorem 6.5 (Reiterman [76]). *The M-varieties are precisely the classes of finite monoids of the form $[\![\Sigma]\!]$, where Σ is a set of equalities.*

For ordered monoids one has the following.

Theorem 6.6 (Pin and Weil [70]). *The OM-varieties are precisely the classes of finite ordered monoids of the form $[\![\Sigma]\!]$, where Σ is a set of identities.*

These theorems also hold for semigroups.

7 Some operators

We saw in the previous sections how recognizable languages can be algebraicly studied by using finite monoids. So, it is natural to use this bridge to interpret operations or properties on finite monoids in recognizable languages and vice versa. In this section we use the variety approach to broach this problem. Concerning varieties of finite (ordered) monoids (or semigroups) we only present some operators, namely the power operator and Schützenberger and Malcev products. Important operators, such as the semidirect product, will not be mentioned in this article.

7.1 The power operator

Let **V** be an **M**-variety (resp. **S**-variety). We denote by **PV** the **M**-variety (resp. **S**-variety) generated by the class of monoids (resp. semigroups) of the form $(\mathcal{P}(M), \cdot)$, where $M \in \mathbf{V}$. Similarly, we denote by **P'V** the **M**-variety (resp. **S**-variety) generated by the class of monoids (resp. semigroups) of the form $(\mathcal{P}(M) \setminus \{\emptyset\}, \cdot)$, where $M \in \mathbf{V}$. We define $\mathbf{P}^0\mathbf{V} = \mathbf{V}$, $\mathbf{P'}^0\mathbf{V} = \mathbf{V}$, $\mathbf{P}^{n+1}\mathbf{V} = \mathbf{P}(\mathbf{P}^n\mathbf{V})$ and $\mathbf{P'}^{n+1}\mathbf{V} = \mathbf{P'}(\mathbf{P'}^n\mathbf{V})$, for any nonnegative integer n.

From the fact that $\mathcal{P}(\{1\}) = \{\emptyset, \{1\}\} \simeq U_1$, one has $\mathbf{PI} = \mathbf{J}_1$. Also, given a semigroup S, the semigroup $\mathcal{P}(S) \setminus \{\emptyset\}$ is a subsemigroup of $\mathcal{P}(S)$ and it is easy to see that $\mathcal{P}(S)$ is a quotient of $(\mathcal{P}(S) \setminus \{\emptyset\}) \times U_1$. Therefore, $\mathbf{PV} = \mathbf{P'V} \vee \mathbf{J}_1$, for any **M**-variety or **S**-variety **V**. If M is a non-trivial monoid, the submonoid $\{\{1\}, M\}$ of $\mathcal{P}(M) \setminus \{\emptyset\}$ is isomorphic to U_1. Thus, one has $\mathbf{PV} = \mathbf{P'V}$, for any non-trivial **M**-variety **V**. However, this is not true for non-trivial **S**-varieties. For this reason, in the context of semigroups, numerous results on **PV** have required the study of **P'V**.

Working with these two operators is difficult and it needs some powerful theorems and machinery. Here, we present just a few results, but many others can be found in the references.

The power operator on finite monoids was interpreted on recognizable languages by Reutenauer and Straubing in the following way.

Theorem 7.1. [77, 88] *Let **V** be an **M**-variety and let $\mathcal{V}$ and $\mathcal{W}$ be the *-varieties corresponding to **V** and **PV**, respectively. Then, for each alphabet A, $A^*\mathcal{W}$ is the boolean algebra generated by the languages of the form $L\varphi$, where $L \in B^*\mathcal{V}$ for some alphabet B and $\varphi : B^* \to A^*$ is a morphism such that $B\varphi \subseteq A$.*

Another description of the ∗-variety corresponding to **PV** can be found in [55]. Also, see [49] for a variant of the operators **P** and **P′** on **M**-varieties and its interpretation in recognizable languages.

A result similar to Theorem 7.1 also holds for **S**-varieties by replacing ∗ by +. As Straubing noted in [88], one obtains a description of the +-variety corresponding to **P′V** by considering surjective morphisms.

Examples 7.1. We saw that $\mathbf{PI} = \mathbf{J}_1$. Independently, Perrot [53] and Straubing [88] have proved that $\mathbf{PJ}_1 = \mathbf{Acom}$ and $\mathbf{P}(\mathbf{Acom}) = \mathbf{Acom}$.

Margolis and Pin [48] proved that, in the monoid case, $\mathbf{P}^2\mathbf{J} = \mathbf{M}$, and Almeida [6] proved that, in the semigroup case, $\mathbf{P}^2\mathbf{J} = \mathbf{S}$. However, it remains an open problem to give an explicit characterization of **PJ**, although a description of the corresponding languages is known (see [1]).

One has $\mathbf{PG} = \mathbf{BG}$. The inclusion $\mathbf{PG} \subseteq \mathbf{BG}$ is due to Margolis and Pin [50, 51] and the another one is due to Henckell [36] (see also [37, 38, 47, 60]). Their proofs are difficult, they make use of the solution of the type-II conjecture [18], the notions of semidirect, Malcev and Schützenberger products (see Subsections 7.3 and 7.4) and arguments on languages.

The following three statements summarize some important results on the power operator on **M**-varieties. They were obtained by Margolis, Perrot, Pin and Straubing [48, 53, 55, 88].

Theorem 7.2. *If* **V** *is a non-trivial* **M**-*variety contained in* **Com**, *then* **PV** *is the* **M**-*variety of all finite commutative monoids such that all their groups are in* **V** *and* $\mathbf{P}^2\mathbf{V} = \mathbf{PV}$.

Theorem 7.3. *Let* **V** *be an* **M**-*variety.*
If **V** *contains a non-commutative group, then* $\mathbf{P}^2\mathbf{V} = \mathbf{M}$.
If **V** *contains a non-commutative monoid, then* $\mathbf{P}^3\mathbf{V} = \mathbf{M}$.

In general $\mathbf{P}^2\mathbf{V} \neq \mathbf{P}^3\mathbf{V}$ (see [55]). However, one has the following statement, which is an obvious consequence of the previous examples and these theorems.

Corollary 7.4. *One has* $\mathbf{P}^3\mathbf{V} = \mathbf{P}^4\mathbf{V}$, *for any* **M**-*variety* **V**.

Almeida studied the iteration of the power operator on **S**-varieties. His methods are different from those applied in the monoid case, since extra difficulties were encountered. The role played by commutative monoids in the context of **M**-varieties is played in the context of **S**-varieties by permutative semigroups, a natural generalization of commutative semigroups. A *permu-*

tative semigroup is a semigroup that satisfies a non-trivial *permutation identity*, that is an identity of the form $x_1 x_2 \ldots x_n = x_{1\sigma} x_{2\sigma} \ldots x_{n\sigma}$, where σ is a non-identical permutation of $\{1, \ldots, n\}$ and $x_1, x_2, \ldots, x_n$ are distinct variables. For a finite semigroup, this is equivalent to satisfying the identity $x^\omega yzt^\omega = x^\omega zyt^\omega$ (see [2]). Note that the permutative monoids are precisely the commutative ones. A class of semigroups is *permutative* if all its members are permutative. A straightforward calculation shows that if a semigroup S satisfies a non-trivial permutation identity, then $\mathcal{P}(S)$ also satisfies the same identity. Hence, $\mathbf{P}(\mathbf{Perm}) = \mathbf{Perm} = \mathbf{P'}(\mathbf{Perm})$, where **Perm** is the S-variety of all permutative semigroups.

For a permutative S-variety **V**, Almeida [1] described $\mathbf{P'V}$ and $\mathbf{P^2V}$ in terms of "simpler" operators on **V** and intersection. For non-permutative S-varieties one has the following result, also due to Almeida [6].

Theorem 7.5. [6] *If* **V** *is a non-permutative* S*-variety, then* $\mathbf{P^3V} = \mathbf{S}$.

The following theorem gives crucial relations on the operators **P** and **P'** on S-varieties.

Theorem 7.6. [1, 6] (a) *The semigroup of operators on permutative* S*-varieties generated by* $\{\mathbf{P'}, \mathbf{P}\}$ *has three elements,* $\mathbf{P'}$, $\mathbf{P}$ *and* $\mathbf{P^2}$, *and is defined by the relations*

$$\mathbf{P'^2} = \mathbf{P'}, \quad \mathbf{PP'} = \mathbf{P}, \quad \mathbf{P'P} = \mathbf{P^2}.$$

(b) *The semigroup of operators on* S*-varieties generated by* $\{\mathbf{P'}, \mathbf{P}\}$ *has eight elements,* $\mathbf{P'}$, $\mathbf{P'^2}$, $\mathbf{P'^3}$, $\mathbf{P}$, $\mathbf{P^2}$, $\mathbf{P^3}$, $\mathbf{PP'}$ *and* $\mathbf{PP'^2}$, *and is defined by the relations*

$$\mathbf{P'P} = \mathbf{P^2}, \quad \mathbf{P^2P'} = \mathbf{P^3}, \quad \mathbf{P'^4} = \mathbf{P'^3}, \quad \mathbf{P'^3P} = \mathbf{P'^2P}, \quad \mathbf{PP'^3} = \mathbf{PP'^2}.$$

Now, as for M-varieties, we conclude that the iteration of the operator **P** on S-varieties stabilizes at the end of three steps.

Corollary 7.7. *One has* $\mathbf{P^3V} = \mathbf{P^4V}$, *for any* S*-variety* **V**.

As we saw, the stabilization of the iteration of the operator **P** on a permutative M-variety (resp. S-variety) occurs at an M-variety (resp. S-variety) contained in **Com** (resp. **Perm**). For non-permutative M-varieties (resp. S-varieties) the stabilization occurs at **M** (resp. **S**). Given a non-permutative M-variety (resp. S-variety), the least nonnegative integer n such that $\mathbf{P^n V} = \mathbf{M}$ (resp. $\mathbf{P^n V} = \mathbf{S}$) is called the *exponent* of **V**. So, the exponent of **V** is at most three, by Corollaries 7.4 and 7.7.

Margolis [46] determined the **M**-varieties and the **S**-varieties of exponent one and Escada [33, 34] determined the **M**-varieties and the **S**-varieties of exponent two and three (see also [12]). Here, we only state Margolis' result.

Let
$$B_2 = \left\{ \begin{bmatrix} 0 & 0 \\ 0 & 0 \end{bmatrix}, \begin{bmatrix} 1 & 0 \\ 0 & 0 \end{bmatrix}, \begin{bmatrix} 0 & 1 \\ 0 & 0 \end{bmatrix}, \begin{bmatrix} 0 & 0 \\ 1 & 0 \end{bmatrix}, \begin{bmatrix} 0 & 0 \\ 0 & 1 \end{bmatrix} \right\}$$
be the subsemigroup of the monoid of square matrices $(\mathcal{M}_{2\times 2}(\{0,1\}), \cdot)$ over the boolean semiring $(\{0,1\}, +, \cdot)$.

Theorem 7.8. [46] *For an* **S**-*variety* **V**, *one has* $\mathbf{PV} = \mathbf{S}$ *if and only if* $B_2 \in \mathbf{V}$.
For an **M**-*variety* **V**, *one has* $\mathbf{PV} = \mathbf{M}$ *if and only if* $(B_2)^1 \in \mathbf{V}$.

As a consequence of this theorem, we have $\mathbf{PA} = \mathbf{M}$ in the context of **M**-varieties and $\mathbf{PA} = \mathbf{S}$ in the context of **S**-varieties. This last equality was originally proved by Straubing [88].

Given an **M**-variety **V**, we denote by $\mathbf{PV}^+$ the **OM**-variety generated by the class of ordered monoids of the form $(\mathcal{P}(M), \supseteq)$, where $M \in \mathbf{V}$.

Theorem 7.1 can be easily adapted to the **OM**-variety $\mathbf{PV}^+$.

Theorem 7.9. *Let* **V** *be an* **M**-*variety and let* $\mathcal{V}$ *be the* $*$-*variety corresponding to* **V**. *Let* $\mathcal{W}$ *be the positive* $*$-*variety corresponding to* $\mathbf{PV}^+$. *Then, for each alphabet* A, $A^*\mathcal{W}$ *is the positive boolean algebra generated by the languages of the form* $L\varphi$, *where* $L \in B^*\mathcal{V}$ *for some alphabet* B *and* $\varphi : B^* \to A^*$ *is a morphism such that* $B\varphi \subseteq A$.

7.2 Polynomial closure

Consider a correspondence $\mathcal{C} : A \mapsto A^*\mathcal{C} \subseteq Rec(A^*)$. The *polynomial closure* of the set $A^*\mathcal{C}$, denoted $A^*\operatorname{Pol}\mathcal{C}$, is the set of languages that are finite unions of languages of the form
$$L_0 a_1 L_1 a_2 L_2 \cdots a_n L_n,$$
where $n \geq 0$, $a_1, \ldots, a_n \in A$ and $L_0, \ldots, L_n \in A^*\mathcal{C}$. Let $A^*\operatorname{BPol}\mathcal{C}$ be the boolean algebra generated by $A^*\operatorname{Pol}\mathcal{C}$.

In the context of languages of A^+, the polynomial closure is defined in a different manner as follows. Let $\mathcal{C} : A \mapsto A^+\mathcal{C} \subseteq Rec(A^+)$ be a correspondence. The *polynomial closure* of the set $A^+\mathcal{C}$, denoted $A^+\operatorname{Pol}\mathcal{C}$, is the set of languages that are finite unions of languages of the form
$$u_0 L_1 u_1 L_2 u_2 \cdots L_n u_n,$$

119

where $n \geq 0$, $u_0, \ldots, u_n \in A^*$ and $L_1, \ldots, L_n \in A^+\mathcal{C}$ (if $n = 0$, then $u_0 \neq 1$). The boolean algebra generated by $A^+\text{Pol}\,\mathcal{C}$ is also denoted by $A^+\text{BPol}\,\mathcal{C}$.

In the following subsections we will see that if $\mathcal{V}$ is a $*$-variety (resp. $+$-variety), then $\text{Pol}\,\mathcal{V}$ is a positive $*$-variety (resp. positive $+$-variety) and $\text{BPol}\,\mathcal{V}$ is a $*$-variety (resp. $+$-variety).

7.3 Schützenberger product

The Schützenberger product was originally defined by Schützenberger [79] for two monoids and extended by Straubing [89] for any number of monoids. Pin [66] generalized it to ordered semigroups. Here we only present the non-ordered case for monoids.

If M is a monoid, then $(\mathcal{P}(M), +, \cdot)$ is a semiring, where $+$ represents the union. We denote by $(\mathcal{M}_{n \times n}(\mathcal{P}(M)), +, \cdot)$ the semiring of the square matrices of size n with entries in $\mathcal{P}(M)$.

Let $M_1, \ldots, M_n$ be monoids and let $M = M_1 \times \cdots \times M_n$. The *Schützenberger product* of $M_1, \ldots, M_n$, denoted by $\Diamond_n(M_1, \ldots, M_n)$, is the submonoid of the multiplicative monoid $\mathcal{M}_{n \times n}(\mathcal{P}(M))$ formed by all the matrices $P = [P_{ij}]$ satisfying the three following conditions:

(1) for all $i, j \in \{1, \ldots, n\}$ such that $i > j$, $P_{ij} = 0$;
(2) for all $i \in \{1, \ldots, n\}$, $P_{ii} = \{(1, \ldots, 1, s_i, 1, \ldots, 1)\}$ for some $s_i \in M_i$;
(3) for all $i, j \in \{1, \ldots, n\}$ such that $i < j$, $P_{ij} \subseteq \{1\} \times \cdots \times \{1\} \times M_i \times \cdots \times M_j \times \{1\} \times \cdots \times \{1\}$.

Condition (1) means that the matrix is upper triangular. Condition (2) means that the diagonal coefficient P_{ii} can be identified with an element of M_i. Condition (3) means that if $1 \leq i < j \leq n$, then the coefficient P_{ij} can be identified with a subset of $M_i \times \cdots \times M_j$. With this identification, for instance the elements of $\Diamond_3(M_1, M_2, M_3)$ have the form

$$\begin{bmatrix} s_1 & P_{12} & P_{13} \\ 0 & s_2 & P_{23} \\ 0 & 0 & s_3 \end{bmatrix},$$

where $s_i \in M_i$ ($i = 1, 2, 3$), $P_{12} \subseteq M_1 \times M_2$, $P_{13} \subseteq M_1 \times M_2 \times M_3$ and $P_{23} \subseteq M_2 \times M_3$. In the case $n = 2$, Schützenberger product $\Diamond_2(M_1, M_2)$ can be defined as being the set of all matrices of the form

$$\begin{bmatrix} s_1 & P \\ 0 & s_2 \end{bmatrix},$$

where $s_1 \in M_1$, $s_2 \in M_2$ and $P \subseteq M_1 \times M_2$, with the product

$$\begin{bmatrix} s_1 & P \\ 0 & s_2 \end{bmatrix} \begin{bmatrix} t_1 & Q \\ 0 & t_2 \end{bmatrix} = \begin{bmatrix} s_1 t_1 & s_1 Q + P t_2 \\ 0 & s_2 t_2 \end{bmatrix},$$

where $s_1 Q = \{(s_1 x_1, x_2) : (x_1, x_2) \in Q\}$ and $P t_2 = \{(x_1, x_2 t_2) : (x_1, x_2) \in P\}$.
One can define an order in $\Diamond_n(M_1, \ldots, M_n)$ by

$$[P_{ij}] \leq [Q_{ij}] \text{ if and only if, for all } i, j \in \{1, \ldots, n\}, \ Q_{ij} \subseteq P_{ij}.$$

In this way $(\Diamond_n(M_1, \ldots, M_n), \leq)$ is an order monoid.

The following results relate the Schützenberger product with the product of languages of the form $L_0 a_1 L_1 \cdots a_n L_n$. The next one is the version for ordered monoids of a Straubing's result [89].

Proposition 7.10. [69] *Let $L_0, \ldots, L_n \subseteq A^*$ be languages recognized by monoids $M_0, \ldots, M_n$, respectively, and let $a_1, \ldots, a_n \in A$. Then the language $L_0 a_1 L_1 \cdots a_n L_n$ is recognized by the ordered monoid $(\Diamond_n(M_1, \ldots, M_n), \leq)$.*

The following statement is, in a certain sense, the converse of Proposition 7.10. It was proved by Reutenauer [77] for $n = 1$ and by Pin [57] in the general case (see also [82]).

Proposition 7.11. [57] *If a language $L \subseteq A^*$ is recognized by $\Diamond_n(M_1, \ldots, M_n)$, then L belongs to the boolean algebra generated by the set of the languages of the form $L_{i_0} a_1 L_{i_1} \cdots a_k L_{i_k}$, where $0 \leq i_0 < i_1 \ldots < i_k \leq n$, $L_{i_0}, \ldots, L_{i_k}$ are recognized by $M_{i_0}, \ldots, M_{i_k}$, respectively, and $a_1, \ldots, a_k \in A$.*

Given an **M**-variety **V**, denote by

- $\Diamond \mathbf{V}$ the **M**-variety generated by all monoids of the form $\Diamond_n(M_1, \ldots, M_n)$, with $n \in \mathbb{N}$ and $M_1, \ldots, M_n \in \mathbf{V}$;

- $\Diamond^+ \mathbf{V}$ the **OM**-variety generated by all ordered monoids of the form $(\Diamond_n(M_1, \ldots, M_n), \leq)$, with $n \in \mathbb{N}$ and $M_1, \ldots, M_n \in \mathbf{V}$.

Combining Propositions 7.10 and 7.11, one obtains the following theorem.

Theorem 7.12. *For any **M**-variety **V**, the $*$-variety associated to $\Diamond \mathbf{V}$ is $\mathrm{BPol}\,\mathcal{V}$.*

Using arguments on languages Margolis and Pin [50] proved that $\Diamond \mathbf{G} = \mathbf{PG}$.

7.4 Malcev products

Given an **S**-variety **V** and an **M**-variety (resp. **S**-variety) **W**, the *Malcev product* of **V** and **W**, denoted **V**Ⓜ**W**, is defined as being the class of all finite monoids (resp. semigroups) M such that there exists a relational morphism $\sigma : M \to N$, with $N \in \mathbf{W}$, satisfying $e\sigma^{-1} \in \mathbf{V}$, for every $e \in E(N)$. Such class is an **M**-variety (resp. **S**-variety). If, in this definition, we take morphisms instead of relational morphisms, we obtain a generator class of the Malcev product.

Similarly, given an **OS**-variety **V** and an **M**-variety (resp. **S**-variety) or an **OM**-variety (resp. **OS**-variety) **W**, the *Malcev product* of **V** and **W**, also denoted **V**Ⓜ**W**, is defined as being the class of all finite ordered monoids (resp. ordered semigroups) $(M, \leq)$ such that there exists a relational morphism $\sigma : M \to N$, with $N \in \mathbf{W}$, satisfying $e\sigma^{-1} \in \mathbf{V}$, for every $e \in E(N)$. Such a class is also an **OM**-variety (resp. **OS**-variety).

Before present some examples of Malcev products, let us give a few definitions.

Let **Inv** be the **M**-variety generated by the class of all finite inverse monoids. A deep result of Ash [17] states that **Inv** = **ECom**.

Let **Inv**$^+$ be the **OM**-variety generated by the class of all finite ordered inverse monoids ordered by their natural orders. Let $\mathbf{Com^+} = [\![xy = yx,\, x \leq 1]\!]$. Then $\mathbf{ECom^+} = [\![x^\omega y^\omega = y^\omega x^\omega,\, x^\omega \leq 1]\!]$.

Examples 7.2. In [50] Margolis and Pin proved that **J**Ⓜ**G** = **EJ**. Based on the solution of the type-II conjecture [18], Henckell, Margolis, Pin and Rhodes [37] proved that $\mathbf{J_1}$Ⓜ**G** = **ECom** (= **Inv**) and, more generally, **V**Ⓜ**G** = **EV**, for any **M**-variety **V** of finite idempotent monoids.

Pin and Weil [69, 72] extended these results to ordered monoids as follows: **LJ**$^+$Ⓜ**G** = **EJ**$^+$ and $\mathbf{J_1^+}$Ⓜ**G** = **ECom**$^+$ = **Inv**$^+$. The proofs of these equalities use the description of Malcev products in terms of identities, which is due to Pin and Weil [71], Ash's theorem on the type-II conjecture [18] and the previous equalities.

Given an **S**-variety **V**, a relational morphism $\sigma : M \to N$ between monoids or semigroups is said to be a **V**-*relational morphism* if, for every subsemigroup S of N in **V**, the subsemigroup $S\sigma^{-1}$ of M is also in **V**. Similarly, if **V** is an **OS**-variety, a relational morphism $\sigma : M \to N$ between ordered monoids or ordered semigroups is a **V**-*relational morphism* if, for every ordered subsemigroup S of N in **V**, the ordered subsemigroup $S\sigma^{-1}$ of M is also in **V**.

Now the notion of Malcev product can be slightly modified as follows.

Given an **S**-variety **V** and an **M**-variety (resp. **S**-variety) **W**, let $\mathbf{V}^{-1}\mathbf{W}$ be the class of all finite monoids (resp. semigroups) M such that there exists a **V**-relational morphism $\sigma : M \to N$, with $N \in \mathbf{W}$. If **V** is an **OS**-variety and **W** is an **M**-variety (resp. **S**-variety) or an **OM**-variety (resp. **OS**-variety), $\mathbf{V}^{-1}\mathbf{W}$ denotes the class of all finite ordered monoids (resp. ordered semigroups) $(M, \leq)$ such that there exists a **V**-relational morphism $\sigma : M \to N$, with $N \in \mathbf{W}$. Such classes are varieties.

Given an ordered semigroup $(S, \leq)$ and an element $s \in S$, define $\downarrow s = \{t \in S : t \leq s\}$. If e is an idempotent of S, then the subsemigroup $e(\downarrow e)e$ is a monoid satisfying the identity $x \leq 1$ and so it belongs to the **OS**-variety $[\![x^\omega y x^\omega \leq x^\omega]\!] = \mathbf{LJ^+}$. Based in this fact, Pin and Weil [66] proved the following proposition.

Proposition 7.13. [66] *A relational morphism $\sigma : S \to T$ between finite ordered semigroups is a $\mathbf{LJ^+}$-relational morphism if and only if $(e(\downarrow e)e)\sigma^{-1} \in \mathbf{LJ^+}$, for any $e \in E(T)$.*

If we take an ordered semigroup of the form $(T, =)$, then $e(\downarrow e)e = \{e\}$, for any $e \in E(T)$. Thus, we have the following result.

Corollary 7.14. [66] *For any **S**-variety or **M**-variety **V**, one has $(\mathbf{LJ^+})^{-1}\mathbf{V} = \mathbf{LJ^+} \ⓜ \mathbf{V}$.*

This corollary does not extend to **OS**-varieties nor to **OM**-varieties. Let us see an example. Take the **OM**-variety $\mathbf{J_1^+}$. The identity map from $(U_1, =)$ to U_1^+ is a relational morphism and so the ordered monoid $(U_1, =)$ belongs to $\mathbf{LJ^+} \ⓜ \mathbf{J_1^+}$. However, $(U_1, =)$ does not belong to $(\mathbf{LJ^+})^{-1}\mathbf{J_1^+}$, since $\mathbf{J_1^+} \subseteq \mathbf{LJ^+}$ and for any relational morphism $\sigma : (U_1, =) \to (M, \leq)$, where $(M, \leq) \in \mathbf{J_1^+}$, the order monoid $M\sigma^{-1} = (U_1, =)$ is not in $\mathbf{LJ^+}$.

Now, consider the product

$$L = L_0 a_1 L_1 a_2 L_2 \cdots a_n L_n,$$

where $a_1, \ldots, a_n \in A$ and $L_0, \ldots, L_n \subseteq A^*$. The aim is to relate the syntactic monoid $M(L)$ with the syntactic monoids $M(L_0), \ldots, M(L_n)$. Let $\mu : A^* \to (M(L), \leq_L)$ and $\eta_i : A^* \to (M(L_i), \leq_{L_i})$ be the syntactic morphisms, $i \in \{0, \ldots, n\}$. Take the morphism

$$\eta : A^* \to (M(L_0) \times \cdots \times M(L_n), \leq).$$
$$x \mapsto (x\eta_0, \ldots, x\eta_n)$$

Consider the relational morphism $\pi = \mu^{-1}\eta$. The situation is as follows:

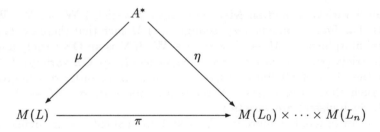

Pin proved the following property concerning this relational morphism π.

Proposition 7.15. [66] π is a $\mathbf{LJ^+}$-relational morphism.

Hence, given an **M**-variety or an **OM**-variety **V**, if $L_0, \ldots, L_n \in A^*\mathcal{V}$, then $M(L) \in (\mathbf{LJ^+})^{-1}\mathbf{V}$. So, every language of $A^*\mathrm{Pol}\,\mathcal{V}$ is recognized by an ordered monoid in $(\mathbf{LJ^+})^{-1}\mathbf{V}$. The following proposition tells us more. The part concerning **M**-varieties was proved by Pin and Weil in [69] and the part concerning **OM**-varieties was proved by Pin in [66].

Theorem 7.16. [66, 69] Let **V** be an **M**-variety or an **OM**-variety. Then Pol $\mathcal{V}$ is a positive $*$-variety of languages, which is associated to the **OM**-variety $(\mathbf{LJ^+})^{-1}\mathbf{V}$. In the case that **V** is an **M**-variety one has $(\mathbf{LJ^+})^{-1}\mathbf{V} = \diamondsuit^+\mathbf{V}$.

The version of this theorem for semigroups is the following.

Theorem 7.17. [66, 69] Let **V** be an **S**-variety or an **OS**-variety. Then Pol $\mathcal{V}$ is a positive $+$-variety of languages, which is associated to the **OS**-variety $(\mathbf{LJ^+})^{-1}\mathbf{V}$.

It is obvious that $\left((\mathbf{LJ^+})^{-1}\mathbf{V}\right)^c = (\mathbf{LJ^-})^{-1}\mathbf{V}$. So, by Corollary 5.8, we obtain the following corollaries.

Corollary 7.18. Let **V** be an **M**-variety or an **OM**-variety. Then BPol $\mathcal{V}$ is a $*$-variety of languages, which is associated to the **M**-variety $(\mathbf{LJ^+})^{-1}\mathbf{V} \vee (\mathbf{LJ^-})^{-1}\mathbf{V}$.

Corollary 7.19. Let **V** be an **S**-variety or an **OS**-variety. Then BPol $\mathcal{V}$ is a $+$-variety of languages, which is associated to the **S**-variety $(\mathbf{LJ^+})^{-1}\mathbf{V} \vee (\mathbf{LJ^-})^{-1}\mathbf{V}$.

Interesting descriptions of the $*$-varieties corresponding to Malcev products of the form $\mathbf{A} \malcev \mathbf{V}$, $\mathbf{K} \malcev \mathbf{V}$, $\mathbf{K}^r \malcev \mathbf{V}$ and $\mathbf{LI} \malcev \mathbf{V}$ in terms of products of languages are also known. See [29, 54, 56, 67, 87].

Now, let us take the trivial **S**-variety **I**. For any alphabet A, one has $A^+\mathcal{I} = \{\emptyset, A^+\}$. So, the languages of $A^+\mathrm{Pol}\,\mathcal{I}$ are the finite unions of languages of the form

$$u_0 A^+ u_1 A^+ u_2 \cdots A^+ u_n,$$

where $n \geq 0$ and $u_0, \ldots, u_n \in A^*$. The following characterization of these languages comes from the fact that $(\mathbf{LJ^+})^{-1}\mathbf{I} = \mathbf{LJ^+}$.

Corollary 7.20. *The* **OS***-variety associated to* $\mathrm{Pol}\,\mathcal{I}$ *is* **LJ$^+$**.

We recall that $\mathbf{LJ^+} = [\![x^\omega y x^\omega \leq x^\omega]\!]$. We notice that, considering the **M**-variety **I**, Proposition 5.5 shows that the **OM**-variety associated to $\mathrm{Pol}\,\mathcal{I}$ is $\mathbf{J^+} = [\![x \leq 1]\!] = [\![x^\omega y x^\omega \leq x^\omega]\!]$, and Simon's theorem (Theorem 3.2) says that the **OM**-variety associated to $\mathrm{BPol}\,\mathcal{I}$ is **J**.

A characterization of the **S**-variety corresponding to $\mathrm{BPol}\,\mathcal{I}$ is given by the following theorem and is due to Knast.

Theorem 7.21. [43, 44] *The* **S***-variety associated to* $\mathrm{BPol}\,\mathcal{I}$ *is*

$$\mathbf{B}_1 = [\![(x^\omega p y^\omega q x^\omega)^\omega\, x^\omega p y^\omega s x^\omega\, (x^\omega r y^\omega s x^\omega)^\omega = (x^\omega p y^\omega q x^\omega)^\omega\, (x^\omega r y^\omega s x^\omega)^\omega]\!].$$

From Corollaries 7.19, 7.20 and Theorem 7.21 it follows that the **S**-variety $\mathbf{B}_1$ is generated by the **OS**-variety **LJ$^+$**, as stated in the following corollary.

Corollary 7.22. *One has* $\mathbf{B}_1 = \mathbf{LJ^+} \vee \mathbf{LJ^-}$.

We notice that Corollaries 5.9 and 7.22 state that, for monoids,

$$\mathbf{J} = [\![x^\omega y x^\omega \leq x^\omega]\!] \vee [\![x^\omega \leq x^\omega y x^\omega]\!]$$

and, for semigroups,

$$\mathbf{B}_1 = [\![x^\omega y x^\omega \leq x^\omega]\!] \vee [\![x^\omega \leq x^\omega y x^\omega]\!].$$

By Proposition 5.7, this first equality means that a finite monoid M is $\mathcal{J}$-trivial if and only if there exists an ordered monoid $(N, \leq)$ satisfying $x^\omega y x^\omega \leq x^\omega$ such that M is a homomorphic image of N. Similarly, by the analogue of Proposition 5.7 for semigroups, the second equality means that a finite semigroup S is in $\mathbf{B}_1$ if and only if there exists an ordered semigroup $(T, \leq)$ satisfying $x^\omega y x^\omega \leq x^\omega$ such that S is a homomorphic image of T.

Now, let us see the case of the polynomial closure of group languages (i.e. languages recognized by finite groups). For a characterization of the group languages see [50, 63]. We have already mentioned that $\mathbf{LJ^+} \text{\textcircled{M}} \mathbf{G} =$

$\mathbf{EJ}^+$. Thus, by applying Theorem 7.16, we obtain the following proposition. However, we notice that this result has been proved before, see [62, 63].

Proposition 7.23. *The* **OM**-*variety associated to* $\mathrm{Pol}\,\mathcal{G}$ *is* $\mathbf{EJ}^+$.

For the following proposition see [50, 51, 60]. Its proof is based on a description of the languages recognized by a Schützenberger product of monoids (see Subsection 7.3) and on the equality $\mathbf{PG} = \mathbf{BG}$.

Proposition 7.24. *The* **M**-*variety associated to* $\mathrm{BPol}\,\mathcal{G}$ *is* **BG** *(= **EJ**)*.

So, we have the following result.

Corollary 7.25. *One has* $\mathbf{BG} = \mathbf{EJ}^+ \vee \mathbf{EJ}^-$.

We saw that if $\mathbf{V}$ is an **M**-variety (resp. **S**-variety), then the **OM**-variety (resp. **OS**-variety) corresponding to $\mathrm{Pol}\,\mathcal{V}$ is $\mathbf{LJ}^+ \text{\textcircled{M}} \mathbf{V}$ and the **M**-variety (resp. **S**-variety) corresponding to $\mathrm{BPol}\,\mathcal{V}$ is $(\mathbf{LJ}^+ \text{\textcircled{M}} \mathbf{V}) \vee (\mathbf{LJ}^- \text{\textcircled{M}} \mathbf{V})$. Thus, we can make a natural question: Is $(\mathbf{LJ}^+ \text{\textcircled{M}} \mathbf{V}) \vee (\mathbf{LJ}^- \text{\textcircled{M}} \mathbf{V})$ a Malcev product of the form $\mathbf{W} \text{\textcircled{M}} \mathbf{V}$, where $\mathbf{W}$ is an **S**-variety that does not depend on $\mathbf{V}$? If the answer is yes, then, in particular, by taking the trivial **S**-variety $\mathbf{I}$ we obtain

$$\begin{aligned} \mathbf{W} &= \mathbf{W} \text{\textcircled{M}} \mathbf{I} \\ &= (\mathbf{LJ}^+ \text{\textcircled{M}} \mathbf{I}) \vee (\mathbf{LJ}^- \text{\textcircled{M}} \mathbf{I}) \\ &= \mathbf{LJ}^+ \vee \mathbf{LJ}^- \\ &= \mathbf{B}_1. \end{aligned}$$

So, the previous question can be formulate as follows: Given an **M**-variety (resp. **S**-variety) $\mathbf{V}$, is $\mathbf{B}_1 \text{\textcircled{M}} \mathbf{V}$ the **M**-variety (resp. **S**-variety) corresponding to $\mathrm{BPol}\,\mathcal{V}$? Pin and Weil, in [69], conjectured that the answer is affirmative. In fact, $(\mathbf{LJ}^+ \text{\textcircled{M}} \mathbf{V}) \vee (\mathbf{LJ}^- \text{\textcircled{M}} \mathbf{V}) \subseteq \mathbf{B}_1 \text{\textcircled{M}} \mathbf{V}$, since $\mathbf{B}_1 = \mathbf{LJ}^+ \vee \mathbf{LJ}^-$. Moreover, the answer is affirmative in the following cases:

- $\mathbf{V} = \mathbf{I}$, the trivial **S**-variety. One has $\mathbf{B}_1 \text{\textcircled{M}} \mathbf{I} = \mathbf{B}_1$, which corresponds to $\mathrm{BPol}\,\mathcal{I}$ by Theorem 7.21.

- $\mathbf{V} = \mathbf{I}$, the trivial **M**-variety. Obviously, $\mathbf{B}_1 \text{\textcircled{M}} \mathbf{I} = \mathbf{J}$, which corresponds to $\mathrm{BPol}\,\mathcal{I}$ by Simon's theorem.

- $\mathbf{V} = \mathbf{G}$. It is clear that $\mathbf{B}_1 \text{\textcircled{M}} \mathbf{G} = \mathbf{J} \text{\textcircled{M}} \mathbf{G}$. So, $\mathbf{B}_1 \text{\textcircled{M}} \mathbf{G} = \mathbf{EJ} = \mathbf{BG}$, which correponds to $\mathrm{BPol}\,\mathcal{G}$ by Proposition 7.24.

However, in a recent paper [73] Pin and Weil showed that the answer is negative by giving a counterexample. Independently, Steinberg gave another counterexample. In [83] he showed that, for any **M**-variety **H** of groups, the **M**-variety corresponding to BPol $\mathcal{H}$ is the semidirect product of **J** and **H**, and in [85] he showed that this semidirect product is not **J**Ⓜ**H** (= **B**$_1$Ⓜ**H**) if **H** is a non-trivial **M**-variety contained in **Gcom** or in an **M**-variety of the form $[\![x^n = 1]\!]$, with $n > 1$. Another conjecture was formulated in [73], but its resolution remains open.

References

[1] J. Almeida, "Finite Semigroups and Universal Algebra", World Scientific, Singapore (1994).
[2] J. Almeida, *Minimal non-permutative pseudovarieties of semigroups I*, Pacific J. Math. **121**, No. 2, (1986) 257–270.
[3] J. Almeida, *Minimal non-permutative pseudovarieties of semigroups II*, Pacific J. Math. **121**, No. 2, (1986) 271–279.
[4] J. Almeida, *Minimal non-permutative pseudovarieties of semigroups III*, Algebra Universalis **21** (1985) 256–279.
[5] J. Almeida, *Power pseudovarieties of semigroups I*, Semigroup Forum **33** (1986) 357–373.
[6] J. Almeida, *Power pseudovarieties of semigroups II*, Semigroup Forum **33** (1986) 375–390.
[7] J. Almeida, *On power varieties of semigroups*, J. Algebra **120** (1989) 1–17.
[8] J. Almeida, *The equation PX = PJ*, in "Proceedings of the International Symposium on the Semigroup Theory and its Related Fields", Kyoto, Matsue, Japan (1990) 1–11.
[9] J. Almeida, *On the power semigroup of a finite semigroup*, Portugaliæ Math. **49**, No. 3, (1992) 295–331.
[10] J. Almeida, *A unified syntactical approach to theorems of Putcha, Margolis, and Straubing on finite power semigroups*, Semigroup Forum **46** (1993) 90–97.
[11] J. Almeida, *A classification of aperiodic power monoids*, J. Algebra **170** (1994) 355–387.
[12] J. Almeida, *Power exponents of aperiodic pseudovarieties*, Semigroup Forum **59** (1999) 18–32.
[13] J. Almeida, *Power semigroups: results and problems*, in "Proceedings of Kyoto Workshop on Formal Languages and Computer Systems", Kyoto, Japan (1997).

[14] J. Almeida and P. Weil, *Relatively free profinite monoids: an introduction and examples*, in "Semigroups, Formal Languages and Groups", J. Fountain (ed.), Kluwer Academic Publishers (1995) 73–117.

[15] M. Arfi, *Polynomial operations and rational languages*, 4th STACS, Lect. Notes in Comput. Sci. **247**, Springer-Verlag, (1987) 198–206.

[16] M. Arfi, *Opérations polynomiales et hiérarchies de concaténation*, Theoret. Comput. Sci. **91** (1991) 71–84.

[17] C. J. Ash, *Finite semigroups with commuting idempotents*, J. Austral. Math. Soc. (Series A) **43** (1987) 81–90.

[18] C. J. Ash, *Inevitable graphs: a proof of the type II conjecture and some related decision procedures*, Int. J. Algebra and Comput. **1**, No. 1, (1991) 127–146.

[19] J.-M. Autebert, J. Berstel and L. Boasson, *Context-free languages and pushdown automata*, in "Handbook of Formal Languages", G. Rozenberg and A. Salomaa (eds.), vol. 1, Springer-Verlag (1997) 111–174.

[20] F. Blanchet-Sadri, *Games, equations and the dot-depth hierarchy*, Comput. Math. Appl. **18** (1989) 809–822.

[21] F. Blanchet-Sadri, *On dot-depth two*, Inform. Théor. Appl. **24** (1990) 521–529.

[22] F. Blanchet-Sadri, *Equations and dot-depth one*, Semigroup Forum **47** (1993) 305–317.

[23] F. Blanchet-Sadri, *On a complete set of generators for dot-depth two*, Discrete Appl. Math. **50** (1994) 1–25.

[24] F. Blanchet-Sadri, *Games, equations and the dot-depth two monoids*, Discrete Appl. Math. **39** (1992) 99–111.

[25] S. L. Bloom, *Varieties of ordered algebras*, J. Comput. System Sci. **13** (1976) 200–212.

[26] J. A. Brzozowski and I. Simon, *Characterizations of locally testable languages*, Discrete Math. **4** (1973) 243–271.

[27] M. J. J. Branco, *On the Pin-Thérien expansion of idempotent monoids*, Semigroup Forum **49** (1994), 329–334.

[28] M. J. J. Branco, "Le produit de concaténation et ses variantes", Thèse de l' Université de Paris VI (1997).

[29] M. J. J. Branco, *The kernel category and variants of the concatenation product*, Int. J. Algebra and Comput. **7**, No. 4, (1997) 487–509.

[30] S. Burris and H. P. Sankappanavar, "A Course in Universal Algebra", Springer-Verlag, New York (1981).

[31] G. L. Cain, "Introduction to General Topology", Addison-Wesley Publ. Co., Inc. (1994).

[32] S. Eilenberg, "Automata, Languages and Machines", vol. B, Academic

Press, New York (1976).
[33] A. Escada, "Contribuições para o estudo de operadores potência sobre pseudovariedades de semigrupos", Ph. D. Thesis, Universidade do Porto (1999).
[34] A. Escada, *The power exponent of a pseudovariety*, Semigroup Forum **64** (2002) 101–129.
[35] M. A. Harrison, "Introduction to Formal Language Theory", Addison-Wesley Publ. Co., Inc. (1978).
[36] K. Henckell, *Blockgroups = powergroups: A consequence of Ash's proof of the Rhodes type II conjecture*, in "Monash Conference on Semigroup Theory", T. E. Hall et al. (eds.), World Scientific, Singapore (1991) 117-134.
[37] K. Henckell, S. Margolis, J.-É. Pin and J. Rhodes, *Ash's type II theorem, profinite topology and Malcev products*, Int. J. Algebra and Comput. **1**, No. 4, (1993) 535–555.
[38] K. Henckell and J. Rhodes, *The theorem of Knast, the $PG = BG$ and type-II conjectures*, in "Monoids and semigroups with applications", J. Rhodes (ed.), World Scientific, Singapore (1991) 453-463.
[39] J. E. Hopcroft and J. D. Ullman, "Introduction to Automata Theory, Languages and Computation", Addison-Wesley Publ. Co. (1979).
[40] J. M. Howie, "Automata and Languages", Clarendon Press, Oxford (1991).
[41] J. M. Howie, "Fundamentals of semigroup theory", Clarendon Press, Oxford (1995).
[42] S. C. Kleene, *Representation of events in nerve nets and finite automata*, in "Automata Studies", C. E. Shannon and J. McCarthy (eds.), Princeton University Press, Princeton, New Jersey (1956) 3–42.
[43] R. Knast, *A semigroup characterization of dot-depth one languages*, RAIRO Inform. Théor. **17** (1983) 321–330.
[44] R. Knast, *Some theorems on graphs congruences*, RAIRO Inform. Théor. **17** (1983) 331–342.
[45] G. Lallement, "Semigroups and combinatorial applications", John Wiley & Sons, New York (1979).
[46] S. Margolis, *On M-varieties generated by power monoids*, Semigroup Forum **22** (1981) 339–353.
[47] S. Margolis, *Consequences of Ash's proof of the Rhodes type II conjecture*, in "Monash Conference on Semigroup Theory", T. E. Hall et al. (eds.), World Scientific, Singapore (1991) 180-205.
[48] S. Margolis and J.-É. Pin, *Minimal noncommutative varieties and power varieties*, Pacific J. Math. **111**, No. 1, (1984) 125–135.

[49] S. Margolis and J.-É. Pin, *Power monoids and finite $\mathcal{J}$-trivial monoids*, Semigroup Forum **29** (1984) 99–108.

[50] S. Margolis and J.-É. Pin, *Varieties of finite monoids and topology for the free monoid*, in "Proceedings of the 1984 Marquette Conference on Semigroups", K. Byleen et al. (eds.), Marquette University (1985) 113–130.

[51] S. Margolis and J.-É. Pin, *Product of group languages*, FCT Conference, Lect. Notes in Comput. Sci. **199**, Springer-Verlag, (1985) 285–299.

[52] R. McNaughton, *Algebraic decision procedures for local testability*, Math. Syst. Theor. **8** (1974) 60–76.

[53] J.-F. Perrot, *Variétés de langages et opérations*, Theoret. Comput. Sci. **7** (1978) 197–210.

[54] J.-É. Pin, *Propriétés syntactiques du produit non ambigu*, 7th ICALP, Lect. Notes in Comput. Sci. **85**, Springer-Verlag, (1980) 483–499.

[55] J.-É. Pin, *Variétés de langages et monoïde des parties*, Semigroup Forum **20** (1980) 11–47.

[56] J.-É. Pin, "Variétés de langages et variétés de semigroupes", Thèse d'Etat, Université de Paris VI (1981).

[57] J.-É. Pin, *Hiérarchies de concaténation*, RAIRO Inform. Théor. **18** (1984) 23–46.

[58] J.-É. Pin, "Variétés de langages formels", Masson, Paris (1984). English translation: "Varieties of Formal Languages", North Oxford, London, and Plenum, New York (1986).

[59] J.-É. Pin, *Finite semigroups and recognizable languages*, in "Semigroups, Formal Languages and Groups", J. Fountain (ed.), Kluwer Academic Publishers (1995) 1–32.

[60] J.-É. Pin, $BG = PG$: *A success story*, in "Semigroups, Formal Languages and Groups", J. Fountain (ed.), Kluwer Academic Publishers (1995) 33–47.

[61] J.-É. Pin, *A variety theorem without complementation*, Izvestiya VUZ. Matematika **39** (1995) 80–90. English version, Russian Mathem. (Iz. VUZ) **39** (1995) 74–83.

[62] J.-É. Pin, *Polynomial closure of group languages and open sets of the Hall topology*, in 21th ICALP, Berlin, Lect. Notes in Comput. Sci. **820**, Springer-Verlag, (1994) 424–435.

[63] J.-É. Pin, *Polynomial closure of group languages and open sets of the Hall topology*, Theoret. Comput. Sci. **169**, No. 2, (1996) 185–200.

[64] J.-É. Pin, *Syntactic semigroups*, in "Handbook of Formal Languages", G. Rozenberg and A. Salomaa (eds.), vol. 1, Springer-Verlag (1997) 679–746.

[65] J.-É. Pin, *Bridges for the concatenation hierarchies*, 25th ICALP, Berlin, Lect. Notes in Comput. Sci. **1443**, Springer-Verlag, (1998) 431–442.
[66] J.-É. Pin, *Algebraic tools for the concatenation product*, Theoret. Comput. Sci., to appear.
[67] J.-É. Pin, H. Straubing and D. Thérien, *Locally trivial categories and unambiguous concatenation*, J. Pure and Appl. Algebra **52** (1988) 297–311.
[68] J.-É. Pin and D. Thérien, *The bideterministic concatenation product*, Int. J. Algebra and Comput. **3**, No. 4, (1993) 535–555.
[69] J.-É. Pin and P. Weil, *Polynomial closure and unambiguous product*, Theory Comput. Systems **30** (1997) 1–39.
[70] J.-É. Pin and P. Weil, *A Reiterman theorem for pseudovarieties of finite first-order structures*, Algebra Universalis **35** (1996) 577–595.
[71] J.-É. Pin and P. Weil, *Profinite semigroups, Mal'cev products and identities*, J. of Algebra **182** (1996) 604–626.
[72] J.-É. Pin and P. Weil, *Semidirect products of ordered semigroups*, Communications in Algebra, **30** (2002) 149–169.
[73] J.-É. Pin and P. Weil, *A conjecture on the concatenation product*, Theoret. Informatics Appl. **35**, (2001), 597–618.
[74] M. Putcha, *Subgroups of the power semigroup of a finite semigroup*, Canad. J. Math. **31** (1979) 1077–1083.
[75] M. O. Rabin and D. Scott, *Finite automata and their decision problems*, IBM J. Research and Development **3** (1959) 114–125. Reprinted in Sequencial Machines, E. F. Moore (ed.), Addison-Wesley, Reading, Massachussetts (1964) 63–91.
[76] J. Reiterman, *The Birkhoff theorem for finite algebras*, Algebra Universalis **14** (1982) 1–10.
[77] C. Reutenauer, *Sur les variétés de langages et de monoïdes*, Lect. Notes in Comput. Sci. **67**, Springer-Verlag, (1979) 260–265.
[78] L. Ribes and P. A. Zalesskii, *On the profinite topology on a free group*, Bull. London Math. Soc. **25** (1993) 37–43.
[79] M. P. Schützenberger, *On finite monoids having only trivial subgroups*, Information and Control **8** (1965) 190–194.
[80] M. P. Schützenberger, *Sur le produit de concaténation non ambigu*, Semigroup Forum **13** (1976) 47–75.
[81] I. Simon, *Piecewise testable events*, Proc. 2nd GI Conf., Lect. Notes in Comput. Sci. **33**, Springer-Verlag, (1975) 214–222.
[82] I. Simon, *The product of rational languages*, Proc. of ICALP 1993, Lect. Notes in Comput. Sci. **700**, Springer-Verlag, (1993) 430–444.

[83] B. Steinberg, *Polynomial closure and topology*, Int. J. Algebra and Comput. **10**, No. 5, (2000) 603-624.
[84] B. Steinberg, *BG = PG: Redux*, in "Proceedings of the International Conference on Semigroups", P. Smith et al. (eds.), World Scientific, Singapore (2000) 181-190.
[85] B. Steinberg, *Finite state automata: a geometric approach*, Trans. Amer. Math. Soc. **353** (2001) 3409-3464.
[86] B. Steinberg, *A note on the equation $PH = J*H$*, Semigroup Forum **63** (2001) 469-474.
[87] H. Straubing, *Aperiodic homomorphisms and the concatenation product of recognizable sets*, J. Pure Appl. Algebra **15** (1979) 319-327.
[88] H. Straubing, *Recognizable sets and power sets of finite semigroups*, Semigroup Forum **18** (1979) 331-340.
[89] H. Straubing, *A generalization of the Schützenberger product of finite monoids*, Theoret. Comput. Sci. **13** (1981) 137-150.
[90] H. Straubing, *Relational morphisms and operations on recognizable sets*, RAIRO Inform. Théor. **15** (1981) 149-159.
[91] H. Straubing, *Semigroups and languages of dot-depth two*, Theoret. Comput. Sci. **58** (1988) 361-378.
[92] H. Straubing and D. Thérien, *Partially ordered finite monoids and a theorem of I. Simon*, J. of Algebra **119** (1985) 393-399.
[93] H. Straubing and P. Weil, *On a conjecture concerning dot-depth two languages*, Theoret. Comput. Sci. **104** (1992) 161-183.
[94] W. Wechler, "Universal Algebra for Computer Scientists", Springer-Verlag, Berlin (1992).
[95] P. Weil, *Inverse monoids of dot-depth two*, Theoret. Comput. Sci. **66** (1989) 233-245.
[96] P. Weil, *Products of languages with counter*, Theoret. Comput. Sci. **76** (1990) 251-260.
[97] P. Weil, *Closure of varieties of languages under products with counter*, J. of Comput. System Sci. **45** (1992) 316-339.
[98] P. Weil, *Some results on the dot-depth hierarchy*, Semigroup Forum **46** (1993) 352-370.

A SHORT INTRODUCTION TO AUTOMATIC GROUP THEORY

CHRISTIAN CHOFFRUT[*]
L.I.A.F.A, Université Paris 7
E-mail: cc@liafa.jussieu.fr

This is a brief presentation of the rich topic introduced in the late eighties of discrete groups for which the product can be performed by finite automata. We focus on the simplest family of such groups, the so-called automatic groups and mention briefly the further extensions. We report at the end a few still open issues.

1 Introduction

This contribution is part of an introductory course on automatic groups. It claims no originality or novelty in the treatment of an already well-established theory. In fact it is widely inspired by the historic papers listed in the references, [2, 6]. A good introduction and tutorial concerning a more comprehensive approach of language theory as a tool for group theoreticians can be found in [8]. Our purpose is to leave a faithful image of the Thematic Term held in Coimbra in spring 2001 during which the lectures were given and to hopefully raise the interest of some theoretical computer scientists for the topic. It is indeed a privileged field where mathematicians and computer scientists may collaborate fruitfully.

The basic idea behind automatic groups is to extend the notion of multiplication table from finite to discrete infinite groups. In order to actually manipulate elements of an infinite group, an encoding is needed. This is achieved by ways of normal forms which are specific strings over a suitable set of symbols, the "generators", and whose set is recognized by a finite state machine. Then a computing device is needed to effectively calculate the product of two elements. This second task is done by again using a finite state machine provided with two tapes, one for the input element and the second for the result of the multiplication of that element by some generator.

---

[*]THE AUTHOR WOULD LIKE TO ACKNOWLEDGE THE FINANCIAL SUPPORT OF FUNDAÇÃO CALOUSTE GULBENKIAN (FCG), FUNDAÇÃO PARA A CIÊNCIA E A TECNOLOGIA (FCT), FACULDADE DE CIÊNCIAS DA UNIVERSIDADE DE LISBOA (FCUL) AND REITORIA DA UNIVERSIDADE DO PORTO.

The main properties of automatic group are: 1) the notion is intrinsic and does not depend on a particular set of generators 2) the family of automatic groups is closed under several operations such as finite direct product, free products, subgroups of finite index 3) the word problem is solvable in quadratic time and 4) many natural groups are automatic and the notion seems to be relevant in the attempt of classifying 3-dimensional manifolds by their fundamental groups, see e.g., [3] for a background on the topic. However there exist such groups that are not automatic and not even asynchronously automatic (the non-deterministic extension). In [3, Theorem B] it is shown that by allowing the language of normal forms to be generated by index-grammars (much more powerful devices than finite automata), one is able to capture the subclass of fundamental groups of "compact geometrizable 3-manifolds".

This presentation focuses on synchronous automatic groups and is self-contained except for the mild prerequisite of a certain familiarity with finite automata theory. After all, finite automata theory is taught at an undergraduate level in the computer science curriculum all over the world and dealt with in widely used textbooks (e.g., [1, 12]). It has also become well-known to non-scientists with the use languages like Perl for example. The background on automata theory necessary for asynchronous automatic groups which is the natural continuation of automatic groups is a bit more advanced, so in spite of its great importance, I only mention this extension in the last section.

Several papers apply the notion of automatic structures to semigroups and monoids. We refer to the contribution of Hoffman and al., [10], in these proceedings for more detail and a wider account on the literature concerning this aspect. Interestingly, the so-called geometrical characterization of automatic structures in terms of Cayley graphs can be extended but it seems to be relevant essentially in the case of cancellative monoids, [16].

2 Finite automata and relations in free monoids

Given a finite set Σ, called an *alphabet*, whose elements are *letters*, the free monoid it generates is denoted by Σ^*. Its elements are *words* or *strings*, and its identity is the *empty word* denoted by ϵ. The product of two strings u and v is simply written uv where the operation symbol is understood. The *length* of a string $w \in \Sigma^*$ is denoted by $|w|$. A subset of Σ^* is also called a *language* and a subset of $\Sigma^* \times \Sigma^*$ a *relation*. We recall that for any two subsets $X, Y \subseteq \Sigma^*$, their *concatenation product* or simply *product* is the set $\{xy \mid x \in X, y \in Y\}$.

We recall that a finite automaton is a quadruple $\mathcal{A} = (Q, q_-, Q_+, E)$

where Q is the set of states, $q_- \in Q$ the unique *initial* state, $Q_+ \subseteq Q$ the subset of *final* states and $E \subseteq Q \times \Sigma \times Q$ the set of *transitions*. A sequence of the form

$$(q_0, a_1, q_1), (q_1, a_2, q_2), \ldots, (q_{n-1}, a_n, q_n) \in E$$

is called a *path* taking state q_0 to state q_n and its *label* $a_1 a_2 \ldots a_n$ is the product of the labels of the different transitions. If, in the above sequence, we assume further that $q_0 = q_-$ and $q_n \in Q_+$, then we say that the path is *successful* and that the string $a_1 \ldots a_n$ is *recognized* by the automaton $\mathcal{A}$. It will not matter, unless otherwise stated, whether the finite automata are deterministic or not. However, we will always assume that all states q are *accessible* from the initial state and *coaccessible* to some final state, in other words that they lie on a successful path. It is also well-known that it is no loss of generality to assume that the automaton is *normalized* in the sense that the subset Q_+ possesses exactly one state. A language in Σ^* is *regular* (we also say *recognizable*) if it is the subset of strings which are recognized by some finite automaton.

Theorem 2.1. *The family of recognizable languages is closed under the Boolean operations, the concatenation product, the image under direct and inverse morphism.*

The closure under product is a direct consequence of Kleene Theorem asserting recognizable languages are exactly the *rational* languages which are those that have a well-formed expression using the symbols of the alphabet and the operations of set union, product and Kleene star (which consists of taking the submonoid generated). Closure under direct and inverse morphisms can be easily established by structural induction of rational languages.

The shortest way to reduce a relation over the free monoid to a subset of this free monoid is to pad the shortest component of each pair of words with the minimal number of occurrences of a dummy symbol $\#$ so that both components have the same length, e.g., starting with $(abaa, bb)$ or with $(cab, bbac)$ one gets $(abaa, bb\#\#)$ and $(cab\#, bbac)$ respectively. Formally, consider the *extended alphabet* $\Sigma^\# = \Sigma \cup \{\#\}$ where $\#$ is a new symbol. With every pair $(u, v) \in \Sigma^* \times \Sigma^*$ associate the pair

$$(u, v)^\# = \begin{cases} (u\#^n, v) & \text{if } n = |v| - |u| \geq 0 \\ (u, v\#^n) & \text{if } n = |u| - |v| \geq 0 \end{cases} \quad (2.1)$$

The pair $(z, t) \in \Sigma^*$ is a *synchronous prefix* of the pair (u, v) if the pair (z, t) is componentwise a prefix of the pair (u, v) and if furthermore $|z| = |t| \leq \min\{|u|, |v|\}$ or if $\min\{|z|, |t|\} = \min\{|u|, |v|\}$. E.g., the synchronous prefixes of $(abaa, bb)$ are (ϵ, ϵ), (a, b), (ab, bb), (aba, bb) and $(abaa, bb)$. Those of $(cab, bbac)$ are (ϵ, ϵ), (c, b), (ca, bb), (cab, bba) and $(cab, bbac)$. Now we extend the notation (2.1) to subsets $R \subseteq \Sigma^* \times \Sigma^*$ by setting

$$R^\# = \{(u, v)^\# \mid (u, v) \in R\}$$

By converting R to the relation $R^\#$ we can view it as a subset of the free monoid $(\Sigma^\# \times \Sigma^\#)^*$ and therefore it becomes eligible for automaton recognition. We say the relation $R \subseteq \Sigma^* \times \Sigma^*$ is *synchronous* if the relation $R^\#$ can be recognized by a finite automaton over the alphabet $\Sigma^\# \times \Sigma^\#$. The class of synchronous relations satisfies many closure properties. Here are those needed for a good understanding of our exposition.

Theorem 2.2. *The family of synchronous relations is closed under the Boolean operations and the composition of relations. The projections over either component are regular.*

If X is regular then the relation $\{(u, u) \mid u \in L\}$ is synchronous.

Furthermore, if X and Y are regular languages then $X \times Y$ is a synchronous relation.

Proof. (Sketch). Let R and S be two synchronous relations. Then $(R \cup S)^\# = R^\# \cup S^\#$ which shows the closure under union by Kleene theorem. Concerning the complement $\bar{R}$, define

$$\Sigma_0 = \{(a, b) \mid a, b \in \Sigma\},\ \Sigma_1 = \{(a, \#) \mid a \in \Sigma\} \text{ and } \Sigma_2 = \{(\#, a) \mid a \in \Sigma\}$$

Then we obtain

$$\bar{R}^\# = \bar{R}^\# \bigcap (\Sigma_0^* \Sigma_1^* \cup \Sigma_0^* \Sigma_2^*)$$

which again in virtue of Kleene theorem shows the closure under complementation.

Given two automata recognizing X and Y it is an easy exercise left to the reader to construct an automaton which recognizes $X \times Y$.

Consider a finite automaton recognizing the synchronous relation R. Its projection over the first (resp. second) component is recognized by the same automaton were the second (resp. first) component of the labels of each transition is ignored.

The statement on the "diagonal" of a regular subset is trivial.

Finally, we prove the closure of synchronous relations under composition. We need to make an excursion into ternary relations, i.e., subsets $T \subseteq (\Sigma^*)^3$. Resort to the same trick as for binary relations and complete, if necessary, each triple of strings with occurrences of the symbol $\#$ so that the three components have the same length. Then T is synchronous if $T^\#$ is a regular language over the alphabet $(\Sigma^\#)^3$. This is now applied to R and S in the following way. Consider the ternary relations $R \times \Sigma^*$ and $\Sigma^* \times S$. As for the last assertion of the theorem, it is easy to verify that these two relations are synchronous. Now observe that the composition $R \circ S$ is the image by the projection π_{13} of $(\Sigma^*)^3$ over the first and third components of the following relation

$$R \circ S = \pi_{13}((R \times \Sigma^*) \cap (\Sigma^* \times S))$$

The closures under intersection and projections easily extend to ternary synchronous relations and this allows us to complete the proof. □

3 Automatic groups

3.1 Basic definitions

A *rational structure* for a group G is a pair (Σ, L) consisting of a finite alphabet Σ and a regular subset $L \subseteq \Sigma^*$ such that there exists a morphism φ of Σ^* onto G with $\varphi(L) = G$. It is therefore possible to view Σ as a set of generators of the group as a monoid and the set L as a set of representatives of the equivalence classes defined by the congruence φ^{-1} though it is not a priori required that there exist only one representative per class. Furthermore, it is also assumed, unless otherwise stated, that for all symbols in $a \in \Sigma$ there exists a symbol $b \in \Sigma$ which maps onto the inverse of the image of a, in other words there exists an involution ι of Σ so that

$$\varphi(\iota(a)) = \varphi(a)^{-1} \tag{3.1}$$

holds for all $a \in \Sigma$. It is also convenient to represent $\varphi(u)$ for all $u \in \Sigma^*$ as $\bar{u}$ and by abuse of notation to write a^{-1} for $\iota(a)$ (but a^{-1} belongs to Σ).

Given a rational structure (Σ, L) for all $a \in \Sigma$ we set

$$R_a = \{(u, v) \in \Sigma^* \times \Sigma^* \mid \bar{u} = \overline{va}, u, v \in L\}$$

and furthermore

$$R_\epsilon = \{(u, v) \in \Sigma^* \times \Sigma^* \mid \bar{u} = \bar{v}, u, v \in L\}$$

The structure (Σ, L), is *automatic* if the relations R_a, for all $a \in \Sigma \cup \{\epsilon\}$ are synchronous. We call L the *language of representatives*, R_ϵ the *equality relation* and the different R_a with $a \in \Sigma$, the *multiplier relations*. More generally, we may define multiplier relations R_w for arbitrary strings $w = a_1 a_2 \ldots a_p \in \Sigma^*$ consisting of all the pairs of strings $(u, v) \in L \times L$ such that $\bar{u} = \overline{vw}$ holds. Such a relation is equal to the composition of relations $R_{a_1} \circ R_{a_2} \ldots \circ R_{a_p}$. In [6], instead of languages and relations composing an automatic structure it is spoken of automata defining them but this leads to unnecessary technicalities.

Convention: the morphism φ is usually understood. Observe that whenever the language L is specified, the equality and the multiplier relations are completely determined. Therefore an automatic structure is unambiguously specified by the pair (Σ, L). As a consequence, each time we deal with automatic structures, we shall specify a pair (Σ, L) and verify that it is indeed automatic.

3.2 Examples

Many natural groups are automatic such as the Braid groups, [6, Theorem 9.3.1] and the Coxeter groups, [4, 9]. We content ourselves with the simple example of free groups. In order to keep notation simple, consider the two-generator free group $F(a, b)$. Put $\Sigma = \{a, b, \alpha, \beta\}$ where a and b map onto the generators of F and α and β onto their inverses. The language L of representatives is composed of all the reduced strings in Σ^*, i.e., having no factor of the form $a\alpha$, αa, $b\beta$ or βb and is therefore regular. Furthermore we have

$$R_\epsilon = \{(u, u) \in \Sigma^* \times \Sigma^* \mid u \in L\}$$

and

$$R_a = \{(ua, u) \in \Sigma^* \times \Sigma^* \mid ua \in L\} \cup \{(u, ua) \in \Sigma^* \times \Sigma^* \mid u\alpha \in L\}$$

The relations R_b, R_α and R_β are defined similarly and are clearly synchronous as is R_a as well.

Observe that the language of representatives has two nice properties which will be studied more extensively in section 5. First there exists exactly one representative per class and second the set is prefix-closed because whenever a string is reduced, so are all of its prefixes.

Examples would not be satisfactory if they were not accompanied by counter-examples. In [6, p. 154], the first example of non-automatic group is referred to Thurston. It consists of the so-called Baumslag-Solitar groups whose presentation is

$$< \{a, b\}; ab^p a^{-1} b^{-q} >$$

with $p \neq q$ (for $p = q$, the groups are automatic). The technique for proving that this is indeed a counter-example uses the notion of isoperimetric inequality, see Theorem 6.2.

4 Geometric properties

4.1 Cayley Graphs

Let G be a group generated, as a group, by Σ. Its Cayley graph $\Gamma(G)$ is the graph whose vertices are the elements of G and whose labelled edges are the triples (x, a, xa) where $x \in G$ and $a \in \Sigma \cup \Sigma^{-1}$. Given $x, y \in G$, their distance, denoted by $d(x, y)$, is the minimum length, in the number of edges, of a path from x to y in $\Gamma(G)$ (this clearly defines a distance). Equivalently, it is the minimum length of a path from the identity $1 \in G$ to the element xy^{-1} (or to the element yx^{-1}). The following properties are more or less trivial (where $|x|$ denotes the length of the group element as the length of the minimum number of generators in Σ for representing it)

$$d(x,y) = |y^{-1}x| = |x^{-1}y| \tag{4.1}$$

$$d(zx, zy) = d(x,y) = d(1, y^{-1}x) = d(1, x^{-1}y) \tag{4.2}$$

In case of possible confusion, we shall write $d_G(x, y)$ to emphasize that the distance is relative to the group G (it will always be clear how the group is generated).

Cayley graphs allow us to define automatic groups in geometric terms via the crucial notion of "fellow traveller property". This can be explained as follows. Consider two cyclists riding in the Cayley graph of a group at the same speed and starting at the origin which is the vertex representing the identity of the group. Both will eventually stop. The property captures the idea that at any time of the ride, the distance which separate the two cyclists measured by the minimum number of edges is bounded by some integer. Formally, we consider a pair $(u, v) \in \Sigma^* \times \Sigma^*$. For each integer $0 < i \leq |u|$ we denote by u_i the prefix of u of length i and we set $u_i = u$ whenever $i \geq |u|$ and likewise for v. Now we say that (u, v) has the k-fellow traveller property if $d(\bar{u}_i, \bar{v}_i) \leq k$ for all integers $i > 0$. This property is the crux for characterizing the automatic structures among the rational structures. More generally, a subset $R \subseteq \Sigma^* \times \Sigma^*$ has the k-fellow traveller property if all its elements (u, v) have it.

Theorem 4.1. (Characterization of automatic structures) *Given a rational structure (Σ, L) for G, it is automatic if and only if there exists an integer $k > 0$ such that for all $\Sigma \cup \{\epsilon\}$, all pairs $(u, v) \in R_a$ have the k-fellow traveller property.*

Proof. We first verify that the condition is necessary. Let K_a be the cardinality of an automaton recognizing R_a, which we assume normalized, and consider a path of $(u,v) \in R_a$ taking the initial state q_- to a final state q_+. Let (u_1, v_1) be a synchronous prefix of (u,v) and assume it takes q_- to some state, say q. There exists a path of length less than or equal to $K_a - 1$, labelled by (u_2, v_2) taking q to q_+ which implies $\overline{u_1 u_2} = \overline{v_1 v_2 a}$. Because of properties (4.1) and (4.2) we have

$$d(\bar{u}_1, \bar{v}_1) = d(1, \bar{v}_1^{-1}\bar{u}_1) = d(1, \overline{v_2 a}\ \bar{u}_2^{-1}) \leq |v_2 a u_2^{-1}| \leq 2K_a - 1$$

It suffices to set $k = \max\{2K_a \mid a \in \Sigma \cup \epsilon\}$.

Now we prove that the condition is sufficient. Let B be the sphere of $\Gamma(G)$ centered at the origin 1 and of radius k and consider an automaton $\mathcal{A} = (Q, q_-, q_+, E)$ recognizing L. We assume without loss of generality that there is no outgoing transition from the state q_+. The idea is that the B controls the distance between the two components. For all $a \in \Sigma \cup \{\epsilon\}$ we define the automaton $(Q \times Q \times B, (q_-, q_-, 1), (q_+, q_+, \bar{a}), E_a)$ where the set of transitions E_a is obtained as follows: given two transitions $(q, b, q'), (p, c, p') \in E$ and an element $x \in B$, the quadruple $((q, p, x), b, c, (q', p', \bar{b}^{-1} x \bar{c}))$ belongs to E_a whenever $x, \bar{b}^{-1} x \bar{c} \in B$. We also add the transitions $((q, p, x), b, \#, (q', p, \bar{b}^{-1} x))$ and $((q, p, x), \#, c, (q, p', x\bar{c}))$ for $b, c \in \Sigma$. The conclusion is met by verifying by induction on the length of $\max\{|u|, |v|\}$ that the pair (u, v) labels a path starting in $(q_-, q_-, 1)$ and ending in (q, p, x) for some states $q, p \in Q$ if and only if $\bar{u}^{-1}\bar{v} = x$. $\square$

Corollary 4.2. *Given an automatic structure for a group and a string $w \in \Sigma^*$ the multiplier relation R_w has the fellow traveller property for some integer K.*

Proof. Indeed, let k be the maximum over the relations R_a where $a \in \Sigma \cup \{\epsilon\}$ of the constants as guaranteed by Theorem 4.1. If the string w equals $a_1 a_2 \ldots a_p$ for $a_i \in \Sigma$, $i = 1, \ldots, p$, then the multiplier relation R_w is the composition of the multiplier relations $R_{a_1} \circ R_{a_2} \ldots \circ R_{a_p}$ and has K-fellow traveller property where $K = kp$. $\square$

4.2 Automaticity is intrinsic

This paragraph shows that the notion of automaticity is a property of the group itself, not of a specific generating system for the group.

Theorem 4.3. *Let G be a group and let Σ and Δ be two sets of (semigroup) generators. Then G is automatic relative to Σ if and only if it is automatic relative to Δ.*

Proof. We prove the result under the assumption that there exists an involution as in (3.1) for both generating sets. It makes the proof somewhat easier to read while not modifying it substantially. We assume G has an automatic structure of the form (Σ, L) and we will prove that it is automatic relative to the generating set Δ. We denote by R_a^Σ the multipliers associated with (Σ, L).

Consider first the case where Δ contains an element e which maps onto the identity of the group. Then we can directly define an automatic strucure (Δ, L'). Indeed, with every $a \in \Sigma$ associate a string $u_a \in \Delta$ which maps to the same element of G. Since Δ contains a generator which maps to the identity we may assume the strings u_a for $a \in \Sigma$ have the same length, say ℓ, after possible padding with sufficiently many occurrences of e. Consider the morphism $\theta : \Sigma^* \to \Delta^*$ defined by $\theta(a) = u_a$. Observe that the length of all strings in $\theta(L) = L'$ is a multiple of ℓ. In particular if $d_\Sigma(x, y)$ is the distance of $x, y \in G$ in the Cayley graph defined by the set Σ of generators, then $d_\Delta(x, y) \leq \ell d_\Sigma(x, y)$ where $d_\Delta(x, y)$ is the distance relative to the set of generators Δ. Now we claim that the pair (Δ, L') is an automatic structure for G.

By Theorem 2.1 the language L' is regular. There exists an integer λ such that for all $b \in \Delta$ there exists a string of Σ^* of length less than or equal to λ which maps to the same element of G. Denote by K an upper bound on the constants associated with the fellow traveller property for all multiplier relations R_w^Σ with $|w| \leq \lambda$. For each $b \in \Delta$ denote by R_b^Δ its multiplier in the rational structure (Δ, L') and let $(\theta(u), \theta(v)) \in R_b^\Delta$. By definition of the morphism θ, for each synchronous prefix (z, t) of $(\theta(u), \theta(v))$ there exist two strings u' and v' in Σ^* of the same length and two strings $z_1, t_1 \in \Delta^*$ of length less than ℓ such that $\theta(u')z_1 = z$ and $\theta(v')t_1 = t$ hold. Now $(u, v) \in R_w^\Sigma$ where w is a string of Σ^* which has the same image as b in G and is of length less than or equal to λ. This yields

$$d_\Delta(\bar{z}, \bar{t}) = d_\Delta(\overline{\theta(u')z_1}, \overline{\theta(v')t_1}) \leq \ell d_\Sigma(\bar{u}', \bar{v}') + |z_1| + |t_1| \leq (K+2)\ell$$

It thus suffices to deal with the special case where Σ contains a (unique) element e which maps to the identity $1 \in G$ and where $\Delta = \Sigma - \{e\}$. It is not possible to simply suppress all occurrences of the letter e in the strings of L since this could violate the fellow traveller property (imagine there is no upper bound on the difference of occurrences of e in u and v). Let K be an upper bound of the constants associated with the multipliers R_a^Σ, $a \in \Sigma$ of the automatic structures (Σ, L) guaranteed by the fellow traveller property. Because of our simplification hypothesis there exist two letters $a, b \in \Sigma$ such that $\overline{ab} = 1$. Consider the mapping $\theta : \Sigma^* \to \Delta^*$ which suppresses every second occurrence of e say at odd positions and substitute one occurrence of ab for each occurrence of e at even positions and observe that inequality $0 \leq |v| - |\theta(v)| \leq 1$ holds for all prefixes of $u \in \Sigma^*$. It is easy to construct a synchronous automaton realizing the function. Setting $T = \{(u, \theta(u)) \in \Sigma^* \times \Delta^* \mid u \in \Sigma^*\}$, we see that the subset $L' = \pi_2(T \cap (L \times \Delta^*)$ is rational because of Theorem 2.2 and maps onto the group. We check that (Δ, L') is automatic. Indeed, consider a pair $(u', v') \in R'_a$ for some $a \in \Delta$ and one of its synchronous prefixes $(u'_1, v'_1) \in L' \times L'$. Let $(u, v) \in R_a$ satisfy $\theta(u) = u'$ and $\theta(v) = v'$. Then there exist a pair $u_1, v_1 \in \Sigma^*$ which are prefixes of u and v respectively and such that $\theta(u_1) = u'_1$ and $\theta(v_1) = v'_1$ holds. This yields

$$d(\bar{u}'_1, \bar{v}'_1) \leq d(\bar{u}'_1, \bar{u}_1) + d(\bar{u}_1, \bar{v}_1) + d(\bar{v}_1, \bar{v}'_1) = d(\bar{u}_1, \bar{v}_1) + 2$$

and shows that R'_a satisfies the fellow traveller property. □

5 Simplifying automatic structures

The definition of automatic structures is purely existential. When working with a group that one knows to be automatic one needs to effectively find an automatic structure and hopefully an efficient one for that group. This aspect is probably the most important combinatorial problem that one may have to face and it is intimately connected with the rewrite theory in term algebras as introduced by Knuth and Bendix, [11]. However it is more complex in that it solves at the same time two different problems: defining rewrite rules that reduce words to a finite regular set of representatives and guaranteeing the fellow traveller property for representatives which are at distance one apart in the Cayley graph. We refer the reader to [6, Chapter 6] for techniques and heuristic resolving this problem. Here with deal with existential results only and show in which ways one may simplify automatic structures.

The notion of automatic structure as defined in section 3 does not require the subset L to be a cross-section. The main result of this paragraph is to show that it is always possible to assume that L is indeed a cross-section. The idea is to single out an element in each congruence class. This element is the short-lexicographically minimum element in the class. The second simplification consists of showing that it is always possible to assume the set of representatives is closed under taking prefixes. Whether both conditions can be achieved simultanously seems to be still unsettled.

We recall that given a linear ordering $<$ on the alphabet Σ, it is possible to extend it to a total well ordering on Σ^* by setting $(u,v) \in$ ShortLex if $|u| < |v|$ or if $|u| = |v|$ and there exist $z, u', v' \in \Sigma^*$ $a, b \in \Sigma$ and $a < b$ such that $u = zau'$ and $v = zbv'$. The relation ShortLex is synchronous since it is recognized by the following finite automaton pictured in Figure 1 where x and y stand for arbitrary symbols of the alphabet Σ.

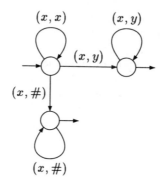

Figure 1. Automaton for ShortLex.

Theorem 5.1. (Unique representative property) *If (Σ, L) is an automatic structure for a group G, then there exists a cross-section $L' \subseteq L$ such that (Σ, L') is also an automatic structure for G.*

Proof. The subset
$$L' = \Sigma - \pi_2(R_\epsilon \cap \text{ShortLex})$$
is clearly a cross-section of the morphism $\varphi : \Sigma^* \longrightarrow G$ since it eliminates all elements in a congruence class which are not minimal. It is also synchronous because of Theorem 2.2. Let us now verify that the pair (Σ, L') defines an automatic structure for G. Is is clear that the relation $R'_\epsilon = \{(u,u) \in \Sigma^* \times \Sigma^* \mid u \in L'\}$ is synchronous. Concerning the relations R'_a with $a \in \Sigma$, we observe that
$$R'_a = R_a \cap (L' \times L') \tag{5.1}$$
holds which proves that they also are synchronous since synchronous relations are closed under the Boolean operations. □

A second natural simplification consists of showing that there always exists a language of representatives which is prefix-closed for all automatic groups.

Theorem 5.2. (Prefix-closed representatives) *If (Σ, L) is an automatic structure for a group G then there exists an automatic structure (Σ, L') for G where the language L' is prefix closed.*

Proof. It suffices to define L' as the prefix closure of the language L and the relations R'_a as the intersections $R_a \cap (L' \times L')$. Indeed, L' is clearly regular: take an automaton recognizing L and make all states final. Furthermore, consider now a pair $(u, v) \in R'_a$. If n is the number of states of the above automaton, then there exist z and t of length less than or equal to n such that $uz, vt \in L$. Let K be an upper bound of the constants of the fellow traveller property associated with the different relations R_w with $|w| \leq 2n + 1$ of the automatic structure (Σ, L). We have $(uz, vt) \in R_{z^{-1}aw}$. Then a synchronous prefix (u', v') of (u, v) is a synchronous prefix of (uz, vt) and therefore one has $d_G(\bar{u}', \bar{v}') \leq K$. □

6 The word problem

A *group presentation* is a pair written $< \Sigma; R >$ composed of a finite alphabet Σ (the set of *generators*) and a subset R of the free group $F(\Sigma)$ generated by Σ (the set of *relators*). The group thus presented is the quotient of the free group $F(\Sigma)$ by the normal subgroup generated by R. The presentation is finite if Σ and R are finite.

Given a presentation $< \Sigma; R >$, the word problem consists of the following

Instance: $w \in F(\Sigma)$

Question: do there exist an integer $m \geq 0$, relators $r_1, r_2, \ldots, r_m \in R$ and elements $y_1, y_2, \ldots, y_m \in F(\Sigma)$ such that

$$w = (y_1^{-1} r_1 y_1)(y_2^{-1} r_2 y_2) \ldots (y_m^{-1} r_m y_m) \tag{6.1}$$

holds.

Novikov in 1953 and Boone in 1956 proved that the problem is undecidable in general, cf. for example [14, Chapter 12]. With every element $w \in F(\Sigma)$ mapping to the identity in the group presented, associate its *area* which is the minimum integer m as defined in (6.1). The *isoperimetric function* of a presentation is the function $\varphi : \mathbb{N} \to \mathbb{N}$ which assigns to every integer n the maximum area of the reduced words of length less than or equal to n that map onto the identity of the group presented. Actually, it can be shown that this function, up to an affine change of variable [7], depends on the group, not on a specific presentation. As a consequence it makes sense to say that the function is, say polynomial of degree three or that it is exponential.

We first start with a technical lemma which asserts that given an arbitrary string $u \in \Sigma^*$, one can find a string in L which is congruent to it in quadratic time relative to the length of u. This is done by incrementally associating a representative to each prefix of a given string.

Lemma 6.1. *Given an automatic structure (Σ, L) there exists an integer N with the following property:*

(1) *Let $u \in L$ and $a \in \Sigma$. Then there exists $v \in L$ satisfying $\overline{ua} = \bar{v}$ and $|v| < |u| + N$.*

(2) *Such a representative v can be computed in linear time relative to $|u|$.*

Proof. Consider the relation R_a recognized by an automaton $\mathcal{A}$ and view it as a (not necessarily deterministic) automaton $\mathcal{B}$ by ignoring the second component. Let N be the maximum size of the automata for all R_b's where $b \in \Sigma$. A path labelled by u can be computed in time $O(|u|)$. Indeed, put $u = a_1 a_2 \ldots a_n$. We paraphrase the famous subset construction for determinizing finite automata by defining the sequence of subsets of Q: $Q_0 = \{q_-\}$, $Q_1, \ldots,$ Q_n where for each $i = 1, \ldots, n$, Q_i is the set of states reached from some state $q \in Q_{i-1}$ by a transition labelled by the letter a_i in the automaton $\mathcal{B}$ (by a transition labelled by (a_i, b) for some $b \in \Sigma^\#$ in the automaton $\mathcal{A}$). If Q_n contains a final state, then this means that there exists a string v of length less than or equal to $|u|$ such that $(u, v) \in R_a$. Otherwise, the shortest v for which $(u, v) \in R_a$ holds has length greater than that of u. Thus, there exists a path starting at some state of Q_n and ending in a final state and labelled by a pair of the form $(\#^p, w)$ for some integer p. By the pigeon hole principle, there exists such a path of length less than N, which completes the proof. $\square$

Theorem 6.2. *Let G be an automatic group. Then the following holds.*

(1) *it has a finite presentation*

(2) *it has quadratic isoperimetric function*

(3) *it has word problem solvable in quadratic time*

Proof. 1) We start with an automatic structure $(\Sigma \cup \Sigma^{-1}, L)$ with unique representative for G and prove that the group has a finite presentation as a monoid which will imply that it has a finite presentation as a group. For all strings $u \in (\Sigma \cup \Sigma^{-1})^*$ we write $u \sim v$ whenever two strings have the same image in the group.

Consider a word $w = a_1 a_2 \ldots a_n$ which maps onto the identity of the

145

group. We devise a double factorization of w. Denoting by $u_i \in L$ the representative of the class of the prefix w_i of length i of w for $i = 0, \ldots, n$ the first factorization is

$$w \sim (u_0 a_1 u_1^{-1})(u_1 a_2 u_2^{-1}) \ldots (u_{n-1} a_n u_n^{-1}) \qquad (6.2)$$

Now we shall verify that each element $u_i a_{i+1} u_{i+1}^{-1}$ in turn can be factored into a product of conjugates of strings of bounded length.

Assume without loss of generality that the length p of u_i is less than or equal to the length m of u_{i+1}. Set $u_i = b_1 b_2 \ldots b_m$ with $b_t \in \Sigma \cup \Sigma^{-1}$ if $0 \leq t \leq p$ and $b_t = \epsilon$ if $p < t \leq m$ and $u_{i+1} = c_1 c_2 \ldots c_m$, where $c_s \in \Sigma \cup \Sigma^{-1}$ for all $s = 1, \ldots, m$. For all $0 \leq t \leq p$ set $u_i(t) = b_1 b_2 \ldots b_t$ and denote by $u_{i+1}(t)$ the prefix of u_{i+1} of length t. Because of $(u_i, u_{i+1}) \in R_{a_{i+1}}$, for all $0 \leq t \leq m$ we have $d(u_i(t), u_{i+1}(t)) \leq k$ where k is the constant of the fellow traveller property. We thus have the equality $u_{i+1}(t) \sim u_i(t) v(t)$ for some string $v(t)$ of length bounded by k. This yields the following (which is actually an equality in the free group)

$$u_i a_{i+1} u_{i+1}^{-1} \sim u_i(1) v(1) u_{i+1}(1)^{-1} \prod_{1 \leq t \leq m} u_{i+1}(t) v(t)^{-1} b_{t+1} v(t+1) u_{i+1}(t+1)^{-1} \qquad (6.3)$$

Because of $u_{i+1}(t+1) = u_{i+1}(t) c_{t+1}$ the t-th factor in the product is a conjugate by $u_{i+1}(t)$ of the element

$$v(t)^{-1} b_{t+1} v(t+1) c_{t+1}^{-1} \qquad (6.4)$$

whose length is bounded by $2k + 2$. The set of relators of the presentation can thus be taken as a subset of strings of length less than or equal to $2k + 2$.

2) The assertion concerning the isoperimetric inequality follows from the previous considerations. Indeed, by Lemma 6.1 the length of each u_i is linearly bounded by the length of w. Now each element in (6.2) consists of a linearly bounded number of relators as in (6.4). This proves the assertion.

3) Given $w = a_1 a_2 \ldots a_n$, by Lemma 6.1 there exists $u_1, u_2, \ldots, u_n \in L$ such that

$$\bar{u}_0 = 1, \bar{u}_1 = \bar{a}_1, \ldots, \bar{u}_i = \overline{u_{i-1} a_i}, \ldots, \bar{u}_n = \overline{u_{n-1} a_n} = \bar{w}$$

where $|u_i| \leq |u_{i-1}| + N$ for $i = 1, \ldots, n$. If u_0 is known and its length is ℓ, the total cost of computing w is bounded by

$$\sum_{i=1}^{n} (\ell + iN) = O(n\ell) + O(n^2)$$

It thus suffices to prove that ℓ is bounded by a constant. It is possible to identify an element of L, say $v = b_1 b_2 \ldots b_r$, in constant time by considering an arbitrary succesful path in the automaton recognizing L. Define the sequence $v_i \in L$ for $1 \leq i \leq r$ by setting $v_0 = v$ and more generally

$$\bar{v}_1 = \bar{v}_0 \bar{b}_1^{-1}, \bar{v}_2 = \bar{v}_1 \bar{b}_2^{-1}, \ldots, \bar{v}_r = \bar{v}_{r-1} \bar{b}_r^{-1}$$

Then we have $\bar{v}_r = 1$ as desired and w maps to the identity if and only if $(v_r, u_n) \in R_\epsilon$ holds, which can be achieved in time proportional to $|u_n|$. □

7 Closure properties

The numerous closure properties enjoyed by the class of automatic groups prove the robustness of the concept. We deal here with two main properties, to wit the closure under direct product and under taking a subgroup of finite index. The closure under free product also holds but uses the same type of techniques (see [10] where is it proven for monoids).

Theorem 7.1. *If G_1 and G_2 automatic groups, then so is their direct product $G_1 \times G_2$.*

Proof. Let (Σ_i, L_i) be automatic structures for G_i with $i = 1, 2$. Without loss of generality we may assume that Σ_1 and Σ_2 are disjoint and by Theorem 5.1 that L_1 and L_2 have the unique representative property. Put $\Sigma = \Sigma_1 \cup \Sigma_2$ and $L = L_1 L_2$ and consider the morphism from Σ^* to $G_1 \times G_2$ which extends the natural morphisms from Σ_1^* and Σ_2^* to G_1 and G_2 respectively. In particular

$$u_1 u_2 \in L_1 L_2 \longrightarrow (\bar{u}_1, \bar{u}_2) \in G_1 \times G_2$$

Then the pair (Σ, L) is a rational structure because the family of rational subsets of a free monoid is closed under product. It now suffices to verify that it satisfies Theorem 4.1. To that purpose, let k_1 and k_2 be two constants for the two structures (Σ_1, L_1) and (Σ_2, L_2) as claimed by Theorem 4.1. Because the two structures have the unique representative property, Lemma 6.1 asserts that there exists an integer N such that for all $a \in \Sigma_1 \cup \Sigma_2$ and for all pairs (u, v) belonging to the multiplier relation R_a of either structure, inequality $||u| - |v|| \leq N$ holds. Set $k = \max\{k_1, k_2\} + N$.

Case 1: $a = \epsilon$, then $R_\epsilon = \{(u, u) \in \Sigma^* \times \Sigma^* \mid u \in L\}$ is clearly synchronous.

Case 2: $a \in \Sigma_2$.

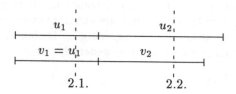

2.1.   2.2.

For all $(u_1 u_2, v_1 v_2) \in R_a$ we have $\overline{u_1 u_2} = \overline{v_1 v_2 a}$. Since the group G is a direct product, this implies $\bar{u}_1 = \bar{v}_1$ and $\bar{u}_2 = \overline{v_2 a}$ and because of the unique representative property for (Σ_1, L_1) we get $u_1 = v_1$. If (u_3, v_3) is a synchronized prefix of (u_1, v_1), we have $u_3 = v_3$ and therefore $d_{G_1 \times G_2}(\bar{u}_3, \bar{v}_3) = 0$. Now let $(u_1 u_3, v_1 v_3)$ be a synchronized prefix of $(u_1 u_2, v_1 v_2)$. We have $d_{G_1 \times G_2}((\bar{u}_1, \bar{u}_3), (\bar{v}_1, \bar{v}_3)) = d_{G_2}(\bar{u}_3, \bar{v}_3) \leq k_2 \leq k$.

Case 3: $a \in \Sigma_1$. This is in turn divided into 3 subcases. We assume without loss of generality that $|u_1| \geq |v_1|$.

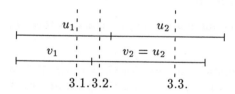

3.1. 3.2.   3.3.

Case 3.1.: (u_3, v_3) is a synchronized prefix of (u_1, v_1). Then we obtain

$$d_{G_1 \times G_2}((\bar{u}_3, 1), (\bar{v}_3, 1)) = d_{G_1}(\bar{u}_3, \bar{v}_3) \leq k_1 \leq k$$

Case 3.2.: $(u_3, v_1 v_3)$ is a synchronized prefix of $(u_1, v_1 v_2)$. Then equality $\overline{v_1 v_3} = (\bar{v}_1, \bar{v}_3)$ implies

$$d_{G_1 \times G_2}(\bar{u}_3, \overline{v_1 v_3}) = d_{G_1}(\bar{u}_3, \bar{v}_1) + d_{G_2}(1, \bar{v}_3) = d_{G_1}(\bar{u}_3, \bar{v}_1) + |v_3| \leq k_1 + N \leq k$$

since inequality $|v_3| \leq N$ is guaranteed by Lemma 6.1 and the unique representative property.

Case 3.3.: $(u_1 u_3, v_1 v_3)$ is a synchronized prefix of $(u_1 u_2, v_1 v_2)$. The string u_3 is a prefix of v_3 and $|v_3| - |u_3| = |u_1| - |v_1| \leq N$. Because of $(\bar{u}_1 \bar{u}_3, \overline{v_1 v_3}) = (\bar{u}_1, \bar{u}_3)(\bar{v}_1, \bar{v}_3)$ we obtain

$$d_{G_1 \times G_2}(\bar{u}_1 \bar{u}_3, \overline{v_1 v_3}) = d_{G_1}(\bar{u}_1, \bar{v}_1) + d_{G_2}(\bar{u}_3, \bar{v}_3) \leq k_1 + N \leq k$$

□

As an immediate corollary we have

Theorem 7.2. *Any finitely generated commutative group is automatic.*

We now turn to the second closure property.

Theorem 7.3. *Let H be a subgroup of finite index of a group G. Then H is automatic if and only if G is.*

Proof. Set $G = \bigcup_{0 \leq i \leq n} Hx_i$ where the x_i's are representatives of the cosets of H in G and where $X = \{x_0, \ldots, x_n\}$ and x_0 is the identity 1 of the group.

Assume first G has an automatic structure (Σ, L) with the unique representative property. Associate with each pair $(x, a) \in X \times \Sigma$ the unique pair $(w, x') \in \Sigma^* \times X$ such that $\overline{xa} = \overline{wx'}$ holds, defining thus an action

$$X \times \Sigma \longrightarrow \Sigma^* \times X \qquad (7.1)$$

We claim that the w's generate the subgroup H. Indeed, for all $z = a_0 a_1 \ldots a_m$ we have $\overline{a_0 a_1 \ldots a_m} = \overline{x_0 a_0 a_1 \ldots a_m}$ and more generally

$$\overline{x_0 a_0 a_1 \ldots a_m} = \overline{w_0 x_{i_1} a_1 \ldots a_m} = \overline{w_0 w_1 x_{i_2} \ldots a_m} = \\ \ldots = \overline{w_0 w_1 w_2 \ldots w_m x_{i_{m+1}}} \qquad (7.2)$$

Thus z belongs to H if and only if $x_{i_{m+1}} = x_0 = 1$ as claimed. Now we construct an automatic structure (Δ, L') for H where Δ is in one-to-one correspondence with the w's as found in (7.1). This yields an action

$$X \times \Sigma \longrightarrow \Delta \times X \qquad (7.3)$$

which extends to $X \times \Sigma^*$ as in (7.2) and associates with every pair $(x, u) \in X \times \Sigma^*$ a pair $(v, x') \in \Delta^* \times X$ where u and v have the same length. We view (7.3) as defining the set of transitions of an automaton whose set of states is X and whose alphabet is $\Sigma \times \Delta$. Take x_0 as initial state and unique final state and define L' as the image of L by the relation recognized by the automaton. This shows that L' is regular and because of the procedure (7.2) that $\bar{L}' = H$. Let us now verify that the pair (Δ, L') defines an automatic structure. Let $b \in \Delta$ and consider a pair $(v_1, v_2) \in \Delta^*$ in the multiplier relation R'_b of the structure (Δ, L'). The reader should follow the proof on the next Figure. By construction there exists a pair $(u_1, u_2) \in L^2$ such that u_1 and u_2 map onto v_1 and v_2 by (7.3). Consider a synchronous prefix (z_1, z_2) of (v_1, v_2). By the same action there exists a synchronous prefix (t_1, t_2) of (u_1, u_2) satisfying for some $y_1, y_2 \in X$ the equalities $\overline{x_0 t_1} = \overline{z_1 y_1}$ and $\overline{x_0 t_2} = \overline{z_2 y_2}$. This yields

$$d_H(\bar{z}_1, \bar{z}_2) = d_H(\overline{x_0 t_1 \bar{y}_1}^{-1}, \overline{x_0 t_2 \bar{y}_2}^{-1}) \leq \lambda d_G(\bar{t}_1, \bar{t}_2) + |y_1| + |y_2|$$

where λ is the maximum of the lengths of the w's obtained in (7.1). Now (u_1, u_2) belongs to the multiplier relation R_w of the structure (Σ, L) where $w \in \Sigma^*$ is the element associated with to the symbol $b \in \Delta$ in the above one-to-one correspondence. Let k be the constant of the fellow traveller property for R_w and let K be an upper bound of the lengths of the elements of X. Then we obtain $d_H(\bar{z}_1, \bar{z}_2) \leq \lambda k + 2K$.

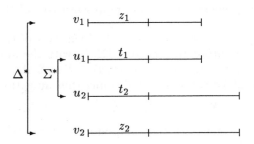

Conversely, assume (Δ, L') is an automatic structure for H. Let Σ be in one-to-one correspondence with X: $a_i \in \Sigma \longleftrightarrow x_i \in X$ and assume $\Sigma \cap \Delta = \emptyset$. Then $(\Sigma \cup \Delta, L'\Sigma)$ is an automatic structure for G. Indeed, $L'\Sigma$ is regular and the mapping $L'\Sigma \longrightarrow G$ is clearly onto. Let now $(ub, vc) \in R_a$ where $u, v \in \Delta^*$, $b, c \in \Sigma$ and $a \in \Sigma \cup \{\epsilon\}$. Then we have $\overline{ub} = \overline{vca}$, i.e., $(u, v) \in R_{ba^{-1}c^{-1}}$. There are only finitely many different values for $ba^{-1}c^{-1}$, implying thus by Corollary 4.2, that the pair ub and vc have the fellow traveller property. □

8 Asynchronous automatic groups and beyond

A wider family of groups, known as asynchronous automatic, is defined in terms of finite 2-tape automata as introduced by Rabin and Scott, [13] and further developed by Elgot and Mezei [5]. They can be viewed as two-tape Turing machines with read-only heads. The program of the machine is a set of instructions consisting of moving one of the heads to the right and changing state depending on the current state and the contents of the cells scanned by the heads. There is a deterministic version where the computation is unique and a non-deterministic one where the machine may choose among a finite set of instructions at each step. The properties enjoyed by asynchronously automatic groups are similar to those for automatic groups: they are closed under finite direct and free products and under taking subgroups of finite index, they have a solvable word problem (whose time complexity has an exponential upper bound though), they have the unique representative property and the property of being asynchronously automatic does not depend on the generating set for the group. These results can be found in [6] where ad hoc methods based on automata constructions are used. I would strongly recommend the version in [2] where the proofs are much better worked out and polished and the tedious constructions on automata avoided as much as possible.

Formally, a two-tape automaton is a construct (Q, q_-, Q_+, E) where Q is a finite set of *states*, q_- is the *initial state*, $Q_+ \subseteq Q$ is the set of *final states* and
$$E \subseteq (Q \times \Sigma \times \{\epsilon\} \times Q) \cup (Q \times \{\epsilon\} \times \Sigma \times Q)$$
is the set of *transitions*. A *path* is a sequence of transitions
$$(q_0, u_1, v_1, q_1), (q_1, u_2, v_2, q_2), \ldots, (q_{n-1}, u_n, v_n, q_n)$$
whose *label* is the pair $(u_1 u_2 \ldots u_n, v_1 v_2 \ldots v_n)$. It is *successful* if $q_0 = q_-$ and $q_n \in Q_+$. The relation *recognized* by the two-tape automaton is the set of labels of the successful computations. The automaton is *deterministic* whenever there exists a partition $Q = Q_1 \cup Q_2$ and $Q_1 \cap Q_2 = \emptyset$ such that the set of transitions may be written as
$$E \subseteq (Q_1 \times \Sigma \times \{\epsilon\} \times Q) \cup (Q_2 \times \{\epsilon\} \times \Sigma \times Q)$$
with the extra condition that $(q_1, a, \epsilon, q), (q_1, a, \epsilon, p) \in E$ implies $q = p$ (resp. $(q_2, \epsilon, a, q), (q_2, \epsilon, a, p) \in E$ implies $q = p$).

The relationship between asynchronously automatic and non-deterministic asynchronously automatic is settled in [15, Theorem 1, p. 301] where it is shown that the two concepts lead to the same class of groups. More precisely, if (Σ, L) is non-deterministic asynchronous automatic then there exists a subset $L' \subseteq L$ such that (Σ, L') is asynchronous automatic. The author argues that non-determinism is more natural. Indeed, characterizing non-deterministic asynchronously automatic structures can be done very much in the same way as automatic structures. More precisely, consider a computation of a pair $(u, v) \in \Sigma^*$ of strings on a two-tape automata, i.e., a successful path labelled by (u, v). Each prefix of this path is labelled by some pair (u', v') where u' and v' are prefixes of u and v respectively. Similarly to synchronous automata, it is clear that the distance $d(\bar{u}', \bar{v}')$ is bounded by some integer k (there is always a path of length less than the number of states which leads to a final state) but unfortunately there is no a priori relationship between the lengths of u' and v'. In other words, given the label (u, v) of a successful computation, one is unable to infer the pairs of strings labelling the prefixes of its computation because the two tapes move separately at different speeds. In the literature this is considered as a parametrization issue. For non-deterministic asynchronous automatic structures, this does not matter because it suffices to consider all possible pairs (u', v') and control their distance in the Cayley graph by verifying that $\bar{u}'^{-1}\bar{v}'$ belongs to the ball centered at the origine and with radius k. The deterministic case is more involved since it is not enough to prevent $\bar{u}'$ and $\bar{v}'$ to be too far apart. One must keep the computation deterministic. This is achieved in two steps. First, it is shown that given a (deterministic) asynchronous automatic structure (Σ, L) for the group G, in is possible to identitfy a subset $L' \subseteq L$ such that (Σ, L') is again an asynchronous automatic structure for G such that every equivalence class has finitely many elements only. It is shown that this guarantees an upper bound of the number of letters that can be read from one tape in a row. Secondly, a deterministic two-tape automaton is devised by using a ball of appropriate radius to keep the images of the prefixes close, but determinism is ensured by switching from one tape to the other "at the last moment" when one tape has exhausted the allowed time of computation.

9 Open problems

The following problems were already presented as open in [6]. We suggest the reader to visit the following site

http://zebra.sci.ccny.cuny.edu/web/problems/probhyp.html

for more references on hyperbolic and automatic groups.

The first question is posed in [6] where it is conjectured to be wrong but I don't have any further reference to the problem.

REPRESENTATIVES. Is it always possible to find an automatic structure for an automatic group such that the language L has the unique representative property and is prefix closed at the same time?

A biautomatic structure for a group G is an automatic structure (Σ, L) such that (Σ, L^{-1}) is also an automatic structure for G where

$$L^{-1} = \{a_1^{-1} a_2^{-1} \ldots a_n^{-1} \mid a_n a_{n-1} \ldots a_1 \in L\}$$

with the convention $\epsilon^{-1} = \epsilon$ for the empty word ($n = 0$).

BIAUTOMATIC. Are automatic groups also biautomatic?

The previous question is interesting for its own sake as a possible closure property but if the answer were positive, then the following question could also be answered positively.

CONJUGACY. Is is possible to decide, given an automatic structure (Σ, L) and two strings $u, v \in \Sigma^*$ whether or not there exists a string $w \in \Sigma^*$ such that $\bar{u} = \bar{w}^{-1}\bar{v}\bar{w}$ holds?

DIRECT PRODUCT. If the direct product $G \times H$ is automatic, is it true that G is automatic?

RETRACT. Let G be automatic and assume there exists a subgroup $H < G$ and a morphism $\varphi : G \to H$ such that φ is the identity on H. Is it true that H is automatic?

The word problem for asynchronous groups is solvable in exponential time, however it is not ruled out that it could be of lower complexity.

WORD PROBLEM. Can the word problem for asynchronous groups be solvable in polynomial time or is exponential time a lower bound?

References

[1] A. Aho and J. Ullman. *Foundations of Computer Science.* W. H. Freeman Company, 1992.
[2] G. Baumslag, S. M. Gersten, M. Shapiro, and H. Short. Automatic groups and amalgams. *J. Pure Appl. Algebra,* 76:229–316, 1991.
[3] M. R. Bridson and R. H. Gilman. Formal languages theory and the geometry of 3-manifolds. *Commentarii Math. Helv.,* 71:525–555, 1996.
[4] B. Brink and R. Howlett. A finiteness property and an automatic structures for Coxeter groups. *Math. Ann.,* 296:179–190, 1993.
[5] C. C. Elgot and J. E. Mezei. On Relations Defined by Finite Automata. *IBM Journal,* 10:47–68, 1965.
[6] D. Epstein, J. Cannon, D. F. Holt, S. V. F. Levy, M. S. Paterson, and W. P. Thurston. *Word processing in groups.* Jones and Bartlett, 1992.
[7] S. M. Gersten. Isoperimetric and isodiametric functions. volume 182 of *LMS Lecture Notes,* pages 79–96. Camb. University Press, 1993.
[8] R. Gilman. Formal languages and infinite groups. *Geometric and Computer Perspectives on Infinite Groups,* DIMACS Ser. Discrete Math. Theoret. Comput. Sci.:27–51, 1996.
[9] Susan M. Hermiller. Rewriting systems for Coxeter groups. *J. Pure Appl. Algebra,* 92:137–148, 1994.
[10] M. Hoffman, D. Kuske, and R. M. Thomas. Some relatives of automatic and hyperbolic groups. these proceedings.

[11] Daniel D. Knuth and P. B. Bendix. Simple word problem in universal algebras. In J. Leech, editor, *Computational Problems in Abstract Algebras*, pages 263–297. Pergamon Press, 1970.
[12] H. R. Lewis and C. H. Papadimitriou. *Elements of the Theory of Computation*. Allyn and Bacon, 1978.
[13] M. O. Rabin and D. Scott. Finite automata and their decision problems. *I.B.M. J. Res. and Develop.*, 3:114–125, 1959.
[14] J. J. Rotman. *The Theory of Groups*. Prentice-Hall, 1981.
[15] M. Shapiro. Deterministic and non-deterministic asynchronous automatic structures. *J. Algebra Comput.*, 2:297–305, 1992.
[16] P. V. Silva and B. Steinberg. A geometric characterization of automatic monoids, 2000. preprint.

AN INTRODUCTION TO COVERS FOR SEMIGROUPS

JOHN FOUNTAIN*

*Department of Mathematics, University of York,
Heslington, York YO10 5DD, UK
E-mail: jbf1@york.ac.uk*

Let T be a subsemigroup of a semigroup S. A T-cover of S is a semigroup $\widehat{S}$ with a subsemigroup $\widehat{T}$ such that $\widehat{S}$ is an extension of $\widehat{T}$ by a group, and there is a surjective homomorphism θ from $\widehat{S}$ onto S which restricts to an isomorphism from $\widehat{T}$ onto T. Of particular interest is the case when both $\widehat{S}$ and S have a least group congruence, and T and $\widehat{T}$ can be taken to be the kernels of these congruences. This is the case when S is E-dense.

In this expository account, we consider under what conditions covers exist, and describe several constructions for covers. The special cases of E-dense, inverse and orthodox semigroups are described in some detail. The use of covers is illustrated in a proof of Reilly's structure theorem for bisimple inverse ω-semigroups, and by giving an account of McAlister's criterion for a finite orthodox semigroup to be a member of the join of the pseudovariety of aperiodic semigroups and the pseudovariety of all finite groups.

This paper is based on a short lecture course for graduate students given at Coimbra in May 2001. I have tried to retain the same level of presentation for the article, with the intention that it will be accessible to anyone with a modest knowledge of semigroup theory and inverse semigroup theory, as provided, for example, by the first two chapters of [21], and the first two sections of Chapter 5 of the same book. For the most part, proofs are given in full detail, and where a proof is omitted, a reference is given. The aim of the paper is to introduce the main concepts associated with covers of semigroups, to give some of the key results, and to illustrate some ways in which some of the theory can be applied.

Before we give details of the content of the paper, we discuss the main concept to be considered. Let T be a subsemigroup of a semigroup S. A

---

*THE AUTHOR WOULD LIKE TO ACKNOWLEDGE THE FINANCIAL SUPPORT OF FUNDAÇÃO CALOUSTE GULBENKIAN (FCG), FUNDAÇÃO PARA A CIÊNCIA E A TECNOLOGIA (FCT), FACULDADE DE CIÊNCIAS DA UNIVERSIDADE DE LISBOA (FCUL), REITORIA DA UNIVERSIDADE DO PORTO, CENTRO DE ÁLGEBRA DA UNIVERSIDADE DE LISBOA (CAUL) AND CENTRO INTERNACIONAL DE MÁTEMATICA (CIM) IN COIMBRA.

T-*cover of* S over a group G is a semigroup $\widehat{S}$ with subsemigroup $\widehat{T}$ such that there are surjective morphisms α, β such that, in the diagram

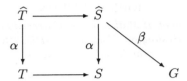

where the horizontal arrows are inclusion maps, the restriction of α to $\widehat{T}$ is an isomorphism and $1\beta^{-1} = \widehat{T}$. The homomorphism α is called the covering homomorphism.

There are a number of questions which immediately spring to mind. First, what conditions must S and T satisfy in order for appropriate $\widehat{S}$ and G to exist? Secondly, if they do exist, what can we say about the structure of $\widehat{S}$ in terms of G and $\widehat{T}$? Finally, and perhaps most importantly, what is the point of studying covers as described above. We consider these questions in the following pages, but first we give a brief explanation of why there are no covering theorems in group theory.

When S is a group and T is a subgroup, it is natural to want $\widehat{S}$ to be a group. Then $\widehat{T}$ has to be a normal subgroup of $\widehat{S}$ and consequently, T is a normal subgroup of S. Thus we may take $\widehat{S}$ to be S, $\widehat{T}$ to be T and G to be S/T, and we are left with the problem of describing S in terms of G and T, that is, the synthesis problem in the theory of group extensions.

In general, however, $\widehat{S}$ will be different from S. One of the early illustrations of this occurs in the work of McAlister [26], [27] on inverse semigroups in the mid 1970s. Groups and semilattices are the natural pieces into which to split an inverse semigroup and this leads to considering the above situation with S inverse and T the commutative subsemigroup of idempotents of S, denoted by $E(S)$ in the sequel. McAlister obtained a covering theorem in which he showed the existence of an inverse semigroup $\widehat{S}$ where we can take $\widehat{T}$ to be $E(\widehat{S})$ and G to be the maximum group homomorphic image of $\widehat{S}$. The finite case of this result was also obtained by Tilson. The semigroup $\widehat{S}$ is said to be E-*unitary* because $E(\widehat{S})$ is a unitary subset of $\widehat{S}$, and we say that $\widehat{S}$ is an E-*unitary cover* of S over G. In the cited papers, McAlister gave a description of E-unitary inverse semigroups in terms of semilattices and groups, in what is now known as the P-theorem.

McAlister's work has been extended in various ways by many authors, including Szendrei, Takizawa, Trotter, Fountain, Almeida, Pin and Weil. Some of these extensions will be discussed in the ensuing pages. In the following section, we consider covers in the contexts of relational morphisms, subdirect products and semidirect products of various kinds. In Section 2, we find necessary and sufficient conditions on a subsemigroup T of a semigroup S for a T-cover of S to exist, and show how such a cover can be constructed. Section 3 is devoted to E-dense semigroups, and starts with some generalities about such semigroups. We introduce the least full weakly self-conjugate subsemigroup $D(S)$ of an E-dense semigroup S, and apply the construction of Section 2 to obtain an E-dense D-unitary cover of S. We concentrate on inverse semigroups in Section 4 starting with a discussion of factorisable monoids which are used to give a short alternative proof of the existence of an E-unitary inverse cover of an inverse semigroup S. We follow this with a brief discussion of McAlister's P-theorem. Next we describe characterisations, due to McAlister and Reilly, of E-unitary inverse covers of an inverse semigroup over a specific group in terms of prehomomorphisms, and dual prehomomorphisms. We conclude the section with an application of covers and the P-theorem to the structure theory of inverse semigroups by giving a new proof of Reilly's theorem on bisimple inverse ω-semigroups. Section 5 also offers an application of covers, this time to orthodox semigroups. Using an appropriate covering theorem, McAlister gave a simple criterion for a finite orthodox semigroup to be a member $\mathbf{A} \vee \mathbf{G}$ where $\mathbf{A}$ is the pseudovariety of all finite aperiodic semigroups, and $\mathbf{G}$ is the pseudovariety of all finite groups. In the final section, we briefly describe some other aspects of covers. First, we mention covers for left ample and weakly left ample semigroups. Secondly, we discuss some work of Auinger and Trotter [5] which extends the results of Section 3 by considering covers of E-dense semigroups over groups in a specific variety of groups. Finally, we say a little about finite covers of finite semigroups, describing just enough to relate the topic to results of Ash [3], Ribes and Zalesskiĭ [36] and Herwig and Lascar [19] which are described in other articles in this volume.

1   Generalities

1.1   Subdirect products and relational morphisms

Let A, B be semigroups. A *subdirect product* of A and B is (a semigroup isomorphic to) a subsemigroup S of $A \times B$ such that S projects onto both A and B. We have the following semigroup version of a standard universal algebra result.

Proposition 1.1. *A semigroup S is a subdirect product of semigroups A and B if and only if there are congruences δ_A, δ_B on S such that $S/\delta_A \cong A$,*

$S/\delta_B \cong B$ and $\delta_A \cap \delta_B = \iota$.

We now recall the associated notion of relational morphism, introduced by Tilson in [8]. Let A and B be semigroups. A *relational morphism* τ from A to B is a mapping $\tau : A \to 2^B$ such that

(1) $a\tau \neq \emptyset$ for all $a \in A$,

(2) $(a_1\tau)(a_2\tau) \subseteq (a_1a_2)\tau$ for all $a_1, a_2 \in A$.

The notation $\tau : A \multimap B$, introduced by Manuel Delgado, is used to indicate that τ is a relational morphism from A to B. The *graph* of τ, that is,

$$\mathrm{gr}(\tau) = \{(a,b) \in A \times B : b \in a\tau\}$$

is a subsemigroup of $A \times B$ which projects onto A. For $b \in B$, we put $b\tau^{-1} = \{a \in A : b \in a\tau\}$, and say that τ is *surjective* if $b\tau^{-1} \neq \emptyset$ for all $b \in B$, that is, $B = \bigcup_{a \in A} a\tau$. In this case, $\mathrm{gr}(\tau)$ is a subdirect product of A and B. Moreover, $\tau^{-1} : B \multimap A$ is also a surjective relational morphism, and $(\tau^{-1})^{-1} = \tau$.

In the case of *inverse* semigroups A and B, we also require

(3) $(a\tau)^{-1} = a^{-1}\tau$ for all $a \in A$

(where $X^{-1} = \{x^{-1} : x \in X\}$ for $X \subseteq A$). In this case, $\mathrm{gr}(\tau)$ is an *inverse* subsemigroup of $A \times B$, and we call τ an *inverse relational morphism*.

For monoids A and B, we impose the condition

(4) $1 \in 1\tau$

so that $\mathrm{gr}(\tau)$ is a submonoid of $A \times B$.

Examples of relational morphisms between semigroups are provided by homomorphisms, and inverses of surjective homomorphisms. It is easy to verify that composing relational morphisms gives a relational morphism. Thus, given semigroups A, B, C and homomorphisms $\alpha : C \to A$, $\beta : C \to B$ with α surjective, the composite $\alpha^{-1}\beta$ is a relational morphism from A to B. In fact, all relational morphisms arise in this way, for, given $\tau : A \multimap B$, we may take C to be $\mathrm{gr}(\tau)$, and α and β to be the projections to A and B respectively; then $\tau = \alpha^{-1}\beta$. Moreover, if α and β are both surjective, then both $\alpha^{-1}\beta$ and $\beta^{-1}\alpha$ are surjective relational morphisms.

If T is a subsemigroup of a semigroup S, we say that a relational morphism $\tau : S \multimap G$ from S to a group G is *T-pure* if $T = 1\tau^{-1}$. Thus if $\widehat{S}$ is a T-cover of S over G, so that we have surjective homomorphisms α and β with

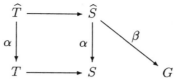

then there are surjective relational morphisms $\tau : S \multimap G$ and $\tau^{-1} : G \multimap S$ ($\tau = \alpha^{-1}\beta$ and $\tau^{-1} = \beta^{-1}\alpha$), and τ is T-pure since $T = \widehat{T}\alpha = (1\beta^{-1})\alpha = 1\tau^{-1}$.

Conversely, if $\tau : S \multimap G$ is a T-pure surjective relational morphism, then $\mathrm{gr}(\tau)$ is a T-cover of S over G. For, we may take α and β to be the projections of $\mathrm{gr}(\tau)$ onto S and G respectively, so that $\tau = \alpha^{-1}\beta$. Then $T = 1\tau^{-1} = 1\beta^{-1}\alpha$ and so α maps $\widehat{T} = 1\beta^{-1}$ onto T. If $x, y \in \widehat{T}$ and $x\alpha = y\alpha$, then $x = (s_1, 1), y = (s_2, 1)$ for some $s_1, s_2 \in S$ so that $s_1 = x\alpha = y\alpha = s_2$ and hence $x = y$.

Thus we have the following.

Proposition 1.2. *Let T be a subsemigroup of a semigroup S, and G be a group. Any T-cover of S over G gives rise to a T-pure surjective relational morphism $\tau : S \multimap G$.*

Conversely, if $\tau : S \multimap G$ is a T-pure surjective relational morphism, then $\mathrm{gr}(\tau)$ is a T-cover of S over G.

We emphasise that, in general, not every T-cover of S over a group arises as the graph of a T-pure surjective relational morphism, as the following example from [5] shows.

Example 1.1. Let $\widehat{S}$ be the monogenic semigroup with generator a and $a^{2n} = a^n$, and let $S = G$ be the maximal subgroup $\{a^n, \ldots, a^{2n}\}$. Put $T = \{a^n\}$, and define α and β by $a^i\alpha = a^i\beta = a^{n+i}$. Then $\widehat{S}$ is a T-cover of S, but, as it is not a group, it is not a subdirect product of S with itself.

In contrast, we have the following result of Auinger and Trotter [5] for regular semigroups which generalises an earlier result of McAlister and Reilly [30] for inverse semigroups.

Proposition 1.3. *Let S be a regular semigroup with subsemigroup T. If $\widehat{S}$ is regular and is a T-cover of S over a group G, then*

(1) T is regular, and

(2) $\widehat{S}$ is a subdirect product of S and G.

Proof. We have the following diagram

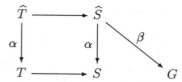

where α and β are surjective homomorphisms, and $\alpha|_{\widehat{T}}$ is an isomorphism.

To see that $\widehat{T}$ (and hence T) is regular, let $a \in \widehat{T}$ and a' be an inverse of a in $\widehat{S}$. Then

$$a'\beta = 1(a'\beta)1 = (a\beta)(a'\beta)(a\beta) = (aa'a)\beta = a\beta = 1,$$

so that $a' \in \widehat{T}$.

Put $\rho = \beta\beta^{-1} \cap \alpha\alpha^{-1}$. We show that $\rho = \iota$, so that $\widehat{S}$ is a subdirect product of S and G.

First, we note that if $a \in \widehat{T}$ and $a \rho b$, then $a = b$. For $1 = a\beta = b\beta$ so that $b \in \widehat{T}$, and so, since $a\alpha = b\alpha$ and α is one-one on $\widehat{T}$, we have $a = b$.

Now let $a, b \in \widehat{S}$ with $a \rho b$, and let a', b' be inverses of a and b respectively. Then $aa' \rho ba'$ and $b'a \rho b'b$. Hence, since $\widehat{T}$ contains the idempotents of $\widehat{S}$, we have $aa' = ba'$ and $b'a = b'b$.

Now $b'ba'a \rho b'aa'a = b'a = b'b$, so that $b'ba'a = b'b$ since $b'b \in \widehat{T}$. Thus

$$a = aa'a = ba'a = bb'ba'a = bb'b = b,$$

and $\rho = \iota$ as required. □

1.2 Semidirect products

We begin by describing semidirect products of semigroups by groups, and giving examples of semigroups which have covers which are such semidirect products. We then extend the notion of semidirect product to semidirect products of semigroupoids by groups and show that every T-cover arising from a T-pure surjective relational morphism can be described using this concept.

A group G *acts by automorphisms* on a semigroup T if, for all elements g, h of G and t, u of T, there is a unique element $g \cdot t$ in T such that

(1) $(gh) \cdot t = g \cdot (h \cdot t)$,

(2) $1 \cdot t = t$,

(3) $g \cdot (tu) = (g \cdot t)(g \cdot u)$.

The *semidirect product* $T \rtimes G$ of T by G is the set $T \times G$ with multiplication
$$(t, g)(u, h) = (t(g \cdot u), gh).$$
It is easily verified that $T \rtimes G$ is a semigroup (a monoid if T is a monoid), and that the projection $\beta : T \rtimes G \to G$ onto G is a surjective homomorphism with
$$1\beta^{-1} = \{(t, 1) : t \in T\} \cong T.$$
Sometimes, a semigroup has a cover which is a semidirect product, as we see in the next example.

Example 1.2. Let $S = M_n(F)$ be the multiplicative monoid of all $n \times n$ matrices over a field F, and let T be the submonoid of all singular matrices together with the identity. Now let G be the general linear group $GL_n(F)$, and note that G acts on T by conjugation. This is an action by automorphisms, and we have

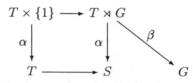

where $(t, g)\alpha = tg$ for $(t, g) \in T \rtimes G$. Thus $T \rtimes G$ is a T-cover of S over G. The associated relational morphism $\tau = \alpha^{-1}\beta$ is given by
$$a\tau = \{g \in G : a = tg \text{ for some } t \in T\}.$$
Hence the association $(t, g) \leftrightarrow (tg, g)$ gives an isomorphism between $T \rtimes G$ and the subdirect product $\{(a, g) : g \in a\tau\}$ of S and G.

In a similar way, we have that, for a finite set X, the full transformation semigroup on X, the monoid of all partial transformations of X, and the symmetric inverse monoid all have covers which are semidirect products of a semigroup by a group. However, not all T-covers are semidirect products in this sense. But, if we use the notion of a semidirect product of a semigroupoid by a group, then we can describe T-covers arising from T-pure surjective relational morphisms as semidirect products.

First, recall that a *semigroupoid* C consists of a set of *objects* denoted by $\text{Obj}\, C$ and a disjoint collection of sets $\text{Mor}(u, v)$ (or $\text{Mor}_C(u, v)$), one for each pair of objects u, v. The elements of the sets $\text{Mor}(u, v)$ are called *morphisms* and the set of all morphisms of C is denoted by $\text{Mor}\, C$. Finally, there is a partial operation on $\text{Mor}\, C$, called *composition* and written $+$ which satisfies the following conditions:

161

(1) if p, q are morphisms of C, the *composite* $p + q$ of p and q is defined if and only if there exist objects u, v, w of C such that $p \in \text{Mor}(u, v)$ and $q \in \text{Mor}(v, w)$; in this case, $p + q \in \text{Mor}(u, w)$;

(2) for all u, v, w, x in $\text{Obj } C$ and all morphisms $p \in \text{Mor}(u, v)$, $q \in \text{Mor}(v, w)$ and $r \in \text{Mor}(w, x)$,
$$(p + q) + r = p + (q + r).$$

A semigroupoid C is a *category* if, for each object u of C, there is a distinguished element 0_u of $\text{Mor}(u, u)$ (called the *identity morphism at* u) such that

(3) for all u, v, w in $\text{Obj } C$ and all morphisms $p \in \text{Mor}(v, u)$, $q \in \text{Mor}(u, w)$, we have
$$p + 0_u = p \text{ and } 0_u + q = q.$$

We use $+$ for composition in semigroupoids and categories, rather than the more conventional multiplicative notation, for increased clarity when we consider actions of groups on semigroupoids. It should be emphasised that there is no implication of commutativity.

A group G *acts* on a semigroupoid C if

(1) G acts on the two sets $\text{Obj } C$ and $\text{Mor } C$ in such a way that, for all objects u, v, if $p \in \text{Mor}(u, v)$, then $g \cdot p \in \text{Mor}(g \cdot u, g \cdot v)$ for all $g \in G$, and

(2) for all $g \in G$, and all $p, q \in \text{Mor } C$ such that $p + q$ is defined,
$$g \cdot (p + q) = g \cdot p + g \cdot q.$$

If C is a category, we also require

(3) $g 0_u = 0_{gu}$ for all $g \in G$ and $u \in \text{Obj } C$.

Let C be a semigroupoid acted upon by a group G. The *semidirect product* $C \rtimes G$ of C by G is a semigroupoid defined as follows:

- $\text{Obj}(C \rtimes G) = \text{Obj } C$,
- $\text{Mor}_{C \rtimes G}(u, v) = \{(f, g) : g \in G \text{ and } f \in \text{Mor}_C(u, gv)\}$

and composition is given by the rule:
$$(f, g)(f', g') = (f + g \cdot f', gg').$$

It is straightforward to check that $C \rtimes G$ is a semigroupoid. Our definition is a very special case of a construction which dates back to the late 1950s, and is often known as the Grothendieck construction. A detailed account of the general construction can be found in [39].

'Choosing a basepoint' means choosing an object u of $C \rtimes G$ and taking the full subsemigroupoid on this one object, that is, the 'local semigroup' at u,

$$L_u(C \rtimes G) = \mathrm{Mor}_{C \rtimes G}(u,u) = \{(f,g) : g \in G \text{ and } f \in \mathrm{Mor}_C(u, gu)\}.$$

This is a special case of a semidirect product with basepoints as described, for example, in [38]. Our special case was also considered in [25] where it was observed that, if the action of G is free and transitive, the local semigroup at any object u can be realised as the collection of orbits of the action on the set of morphisms of the semigroupoid. This approach is generalised in [12].

We now show that for a subsemigroup T of a semigroup S, and a T-pure surjective relational morphism $\tau : S \multimap G$ from S to a group G, the T-cover of S determined by τ, that is, $\mathrm{gr}(\tau)$ can be described as a semidirect product. We use the *weak (or unfactored) derived semigroupoid* W_τ of τ which is defined as follows:

- $\mathrm{Obj}\, W_\tau = G$,

- for $g, h \in G$,

$$\mathrm{Mor}_{W_\tau}(g, h) = \{(g, s, h) \in G \times S \times G : g^{-1}h \in s\tau\}$$

and composition is given by the rule:

$$(g, s_1, h) + (h, s_2, k) = (g, s_1 s_2, h).$$

It is readily verified that W_τ is a semigroupoid, and we note that, for any $g \in G$,

$$\mathrm{Mor}(g,g) = \{(g, s, g) : 1 \in s\tau\} = \{(g, s, g) : s \in T\}$$

which is clearly isomorphic to T.

The group G acts on W_τ as follows. First, it acts on the set of objects, that is, G, by multiplication on the left. The action on $\mathrm{Mor}\, W_\tau$ is given by

$$a \cdot (g, s, h) = (ag, s, ah)$$

where $a \in G$ and $(g, s, h) \in \operatorname{Mor} W_\tau$. It is easy to verify that this is an action. Using this action, we can form $W_\tau \rtimes G$, and we obtain the diagram

$$\begin{array}{ccccc} \{((1,t,1),1) : t \in T\} & \longrightarrow & L_1(W_\tau \rtimes G) & & \\ & & & \searrow^\beta & \\ \alpha \downarrow & & \alpha \downarrow & & \\ & & & & \\ T & \longrightarrow & S & & G \end{array}$$

where $((1, s, g), g)\beta = g$ and $((1, s, g), g)\alpha = s$. We see that $L_1(W_\tau \rtimes G)$ is a T-cover of S over G; an isomorphism θ with a subdirect product of S and G is given by $((1, s, g), g)\theta = (s, g)$.

2 Existence

Let T be a subsemigroup of a semigroup S. We find a necessary and sufficient condition on T for S to have a T-cover over some group. To demonstrate sufficiency we give an explicit construction of a T-cover $\hat{S}$ over the free group on the set S.

2.1 A necessary condition

If a T-cover over a group G exists, we know that there is a T-pure surjective relational morphism $\tau : S \multimap G$ for some group G. Let x be an element of S, and $g \in x\tau$. Since τ is surjective, $g^{-1} \in y\tau$ for some $y \in S$. Hence, for any $a \in T$, we have

$$1 = g 1 g^{-1} \in (x\tau)(a\tau)(y\tau) \subseteq (xay)\tau,$$

so that $xay \in T$. Similarly, $yax \in T$, and also $xy, yx \in T$. Thus T is strongly dense in S, in the sense of the following definition.

A subsemigroup T of a semigroup S is *strongly dense* in S if, for all $x \in S$, there exists $y \in S$ such that $xay, yax \in T$ for all $a \in T^1$.

We say that T is *dense* in S if, for all $x \in S$, there exists $y \in S$ such that $xy, yx \in T$.

We introduce the following useful notation. For $x \in S$, put

$$W_T(x) = \{y \in S : xay, yax \in T \text{ for all } a \in T^1\},$$

so that T is strongly dense in S if and only if $W_T(x) \neq \emptyset$ for all $x \in S$.

Example 2.1. Let S be an inverse semigroup and $E(S)$ be its semilattice of idempotents. For every $x \in S$ and idempotent e, the elements xex^{-1} and

$x^{-1}ex$ are idempotent, so that $x^{-1} \in W_{E(S)}(x)$ and $E(S)$ is strongly dense in S.

Example 2.2. Let T be a strongly dense subgroup of a group K. For $x \in K$, let $y \in W_T(x)$. Then $xy \in T$ so that $y^{-1}x^{-1} \in T$, and hence $xax^{-1} = (xay)y^{-1}x^{-1} \in T$ for all $a \in T$, that is, T is a normal subgroup of K. Conversely, certainly every normal subgroup of K is strongly dense in K.

On the other hand, every subgroup of a group is dense in the group.

It follows from the fact that strongly dense subgroups of a group are normal that if a group K has a T-cover over a group, then it is a T-cover of itself over some group. This is not the case for semigroups in general: if S is a semigroup and $\beta : S \to G$ is a surjective homomorphism onto a group G, then certainly $T = 1\beta^{-1}$ is strongly dense in S, but it has additional properties.

A subset U of a semigroup S is *unitary* in S if for all elements s of S and u of U,

$$su \in U \text{ implies } s \in U, \text{ and } us \in U \text{ implies } s \in U.$$

A subset U of a semigroup S is *reflexive* if for all $x, y \in S$,

$$xy \in U \text{ implies } yx \in U.$$

Lemma 2.1. *If $\beta : S \to G$ is a surjective homomorphism from a semigroup S onto a group G, then $T = 1\beta^{-1}$ is unitary and reflexive in S.*

Proof. If $s \in S$ and $a, sa \in T$, then $s\beta = (s\beta)1 = (s\beta)(a\beta) = (sa)\beta = 1$. Thus $s \in T$, and similarly, if $as \in T$, then $s \in T$ so that T is unitary.

If $x, y \in S$ and $xy \in T$, then $(x\beta)(y\beta) = (xy)\beta = 1$ so that $y\beta = (x\beta)^{-1}$. Hence $(yx)\beta = (y\beta)(x\beta) = 1$, and $yx \in T$. □

In general, if T is a strongly dense subsemigroup of a semigroup S, it will not be unitary and reflexive, for example, if S is inverse with a zero and $S \neq E(S)$, then clearly, $E(S)$ is not unitary in S. But there is the potential for S to have an $E(S)$-cover over a group.

If $\widehat{S}$ is a T-cover of S over a group, we say that $\widehat{S}$ is a T-*unitary* semigroup because $\widehat{T}$ is a unitary subsemigroup of $\widehat{S}$ and $\widehat{T} \cong T$.

2.2 A construction

We have seen that if $\beta : S \to G$ is a surjective homomorphism from a semigroup S onto a group G, then $T = 1\beta^{-1}$ is a unitary, reflexive and dense subsemigroup of S. Conversely, if T is such a subsemigroup of G, then there

165

is a group G and a surjective homomorphism $\beta : S \to G$ with $T = 1\beta^{-1}$. See, for example, [7], [23], [24] or [13]. From [13] we have the following result.

Theorem 2.2. *Let T be a unitary, reflexive, dense subsemigroup of a semigroup S. Then*

$$\rho_T = \{(a,b) \in S \times S : au = vb \text{ for some } u, v \in T\}$$

is a group congruence on S, and $T = 1\beta^{-1}$ where $\beta : S \to S/\rho_T$ is the natural homomorphism.

We have also seen that if T is a subsemigroup of a semigroup S and there is a T-pure surjective relational morphism from S to a group, then T must be strongly dense in S. We now describe a construction which shows that the converse is true. A special case of the construction was given in [10], and the general version is from [12]. Of course, in view of Proposition 1.2, the existence of a T-pure surjective relational morphism from S to a group ensures the existence of a T-cover of S over the group.

Proposition 2.3. *Let S be a semigroup with a strongly dense subsemigroup T. Then S has a T-cover over a group.*

Proof. Let G be the free group on the set S. We construct a surjective relational morphism $\tau : G \multimap S$ with $1\tau = T$. Then τ^{-1} is the required T-pure relational morphism.

The elements of G are equivalences classes of words over X where $X = S \cup \overline{S}$ with $S \cap \overline{S} = \emptyset$ and such that $s \leftrightarrow \overline{s}$ is a bijection between S and $\overline{S}$. We let $\theta : X^* \to G$ be the homomorphism onto G given by $w\theta = [w]$. Then $\theta^{-1} : G \multimap X^*$ is a surjective relational morphism. We find a surjective relational morphism $\varphi : X^* \multimap S$, and put $\tau = \theta^{-1}\varphi$.

Let $x \in S$. Then, since T is strongly dense in S,

$$W_T(x) = \{y \in S : xay, yax \in T \text{ for all } a \in T^1\}$$

is not empty. Choose a non-empty subset $\gamma_T(x)$ of $W_T(x)$ for each x.

We now define φ inductively as follows:

(1) $\epsilon\varphi = T$,

(2) $s\varphi = T^1 s T^1$ for $s \in S$,

(3) $\overline{s}\varphi = T^1 \gamma_T(s) T^1$ for $s \in S$,

and for $v = x_1 \ldots x_n$ where $x_i \in X$, put $v\varphi = (x_1\varphi) \ldots (x_n\varphi)$.

166

Clearly, $v\varphi \neq \emptyset$ for all $v \in X^*$, and $(v\varphi)(w\varphi) = (vw)\varphi$ for nonempty words v, w. Also, since $T^2 \subseteq T$ and $TT^1 \subseteq T^1 \supseteq T^1T$, we have

$$(\epsilon\varphi)(v\varphi) \subseteq v\varphi \supseteq (v\varphi)(\epsilon\varphi)$$

for all $v \in X^*$. Hence φ is a relational morphism, and it is clearly surjective.

Finally, if $w\theta = 1$, we claim that $w\varphi \subseteq \epsilon\varphi = T$ so that

$$1\tau = \bigcup\{w\varphi : w\theta = 1\} = T.$$

We prove the claim by induction on $|w|$. There is nothing to prove when $|w| = 0$. If $|w| > 0$, then it must be the case that $w = u x \overline{x} v$ for some $u, v \in X^*$ and $x \in X$ (where we make the convention that $\overline{\overline{s}} = s$ for $s \in S$).

Now, for $s \in S$ and $a \in s\varphi$, $b \in \overline{s}\varphi$, we have $a = t_1 s t_2$, $b = t_3 y t_4$ for some $t_1, t_2, t_3, t_4 \in T^1$ and $y \in \gamma_T(s)$. Hence $ab = t_1 s t_2 t_3 y t_4$ is in T because $st_2 t_3 y \in T$. Similarly, $ba \in T$, so that $(x\overline{x})\varphi = (x\varphi)(\overline{x}\varphi) \subseteq T$. It follows from this, together with the fact that $(uv)\theta = 1$ and the induction hypothesis, that $w\varphi \subseteq T$, and the claim is proved.

Now $1\tau = T$ so that τ^{-1} is T-pure, and we have a T-unitary cover for S over G. □

We note that the T-cover we have constructed is infinite, even if S is finite, because the group involved is infinite.

3 E-dense semigroups

The concept of an E-dense (or E-inversive) semigroup was introduced in [42] and developed by several authors including Mitsch [31]. The latter provides several examples of E-dense semigroups and notes, in particular, that regular, eventually regular and periodic semigroups, in particular, finite semigroups are all E-dense.

For the definition of E-dense semigroup, we introduce the notion of a weak inverse. For an element a of a semigroup S, put

$$W(a) = \{x \in S : xax = x\}.$$

An element $x \in W(a)$ is a *weak inverse* of a. We say that S is E-dense (or E-inversive) if $W(a) \neq \emptyset$ for all $a \in S$. For a subset A of S, we put

$$W(A) = \{x \in S : x \in W(a) \text{ for some } a \in A\}.$$

An E-dense semigroup in which the idempotents form a commutative subsemigroup is said to be *E-commutative dense*.

In the next result we list some conditions equivalent to being E-dense.

Proposition 3.1. *For a semigroup S with $E = E(S)$, the following conditions are equivalent:*

(1) *S is E-dense,*
(2) *for every $a \in S$, there are elements b, c of S such that $ba \in E$ and $ac \in E$,*
(3) *for every $a \in S$, there is an element b of S such that $ab \in E$ and $ba \in E$,*
(4) *for every $a \in S$, there is an element c of S such that $ac \in E$,*
(5) *for every $a \in S$, there is an element d of S such that $da \in E$.*

Proof. The equivalence of (2) to (5) can be found in [2] or [31] and the equivalence of (1) with the rest is in [6], but for completeness we give a short proof.

If a' is a weak inverse of a, then aa' and $a'a$ are idempotent and so (1) implies (3). Clearly, (3) implies (2) and (2) implies (4) and (5). By symmetry, it is enough to show that (4) implies (1). Let $a \in S$ and let $c \in S$ be such that $ac \in E$. Then clearly, cac is a weak inverse of a, proving (1). □

Let S be an E-dense semigroup and U a subset of S; U is *weakly self conjugate* (or *closed under weak conjugation*) if, for each $a \in S$ and $a' \in W(a)$,

$$aUa' \cup a'Ua \subseteq U.$$

The set U is *full* if $E(S) \subseteq U$. Of particular interest to us is the least full weakly self conjugate subsemigroup of S, that is, the intersection of all such subsemigroups. This is called the *weakly self conjugate core* of S, and is denoted by $D(S)$. We can give a 'constructive' definition of $D(S)$ as follows. Put $D_0(S) = \langle E(S) \rangle$ (the subsemigroup generated by $E(S)$), and

$$D_{i+1}(S) = \langle axb, bxa : x \in D_i(S), a \in S^1, b \in W(a) \rangle.$$

Then it is straightforward to show that $\bigcup_{i \geq 0} D_i(S)$ is a weakly self conjugate subsemigroup, and hence that $D(S) = \bigcup_{i \geq 0} D_i(S)$.

Our next objective is to show that if S is an E-dense semigroup, then $D(S)$ is E-dense. We require some preliminary results on weak inverses valid in any semigroup. The first two are in the dissertation of Weipoltshammer [45]; they also appear in [5], the first is also given in [11] and a variant occurs in [1].

Proposition 3.2. *If $a_1, \ldots, a_n$ are elements of a semigroup S, then*

$$W(a_1 \ldots a_n) \subseteq W(a_n) \ldots W(a_1).$$

Proof. If $a' \in W(a_1 \ldots a_n)$, then, for each $i = 1, \ldots n$, define
$$a'_i = a_{i+1} \ldots a_n a' a_1 \ldots a_{i-1}.$$
It is readily verified that $a'_i \in W(a_i)$ for each i, and that $a' = a'_n \ldots a'_1$. □

Proposition 3.3. *If a is an element of a semigroup S, then*
$$W(W(a)) \subseteq E(S)aE(S).$$

Proof. If $b \in W(a)$ and $c \in W(b)$, then $c = (cb)a(bc) \in E(S)aE(S)$. □

Proposition 3.4. *Let S be a semigroup. Then $W(\langle E(S) \rangle) \subseteq \langle E(S) \rangle$. In particular, if S is an E-dense semigroup, then $\langle E(S) \rangle$ is E-dense.*

Proof. If $b \in W(e)$ where $e \in E(S)$, then $b = beb = (be)(eb) \in \langle E(S) \rangle$. Let $e_1, \ldots, e_n$ be idempotents and assume inductively that
$$W(e_1 \ldots e_{n-1}) \subseteq \langle E(S) \rangle.$$
Let $x \in W(e_1 \ldots e_n)$. Then $e_n x \in W(e_1 \ldots e_{n-1})$ and $xe_1 \ldots e_n \in E(S)$ so that, by the induction hypothesis,
$$x = (xe_1 \ldots e_n)(e_n x) \in \langle E(S) \rangle.$$
The result follows. □

We can now prove the following stronger result.

Theorem 3.5. *Let S be a semigroup. Then $W(D(S)) \subseteq D(S)$. In particular, if S is an E-dense semigroup, then $D(S)$ is E-dense.*

Proof. In view of Proposition 3.4 and the recursive definition of $D(S)$, it is enough to prove that for all non-negative integers i, if $W(D_i(S)) \subseteq D_i(S)$, then $W(D_{i+1}(S)) \subseteq D_{i+1}(S)$. Let $c \in S$ and $q \in cD_i(S)W(c)$. Then, using Propositions 3.2 and 3.3, we have
$$\begin{aligned} W(q) &\subseteq W(W(c))W(D_i(S))W(c) \\ &\subseteq E(S)cE(S)D_i(S)W(c) \\ &\subseteq E(S)(cD_i(S)W(c)) \\ &\subseteq E(S)D_{i+1}(S) \\ &\subseteq D_{i+1}(S). \end{aligned}$$
Similarly, if $q \in W(c)D_i(S)c$, then $W(q) \subseteq D_{i+1}(S)$.

Now, if $a \in D_{i+1}(S)$, then $a = q_1 \ldots q_n$ for some q_j where each q_j is in $c_j D_i(S) W(c_j)$ or $W(c_j) D_i(S) c_j$ for some $c_j \in S$. Hence, by Proposition 3.2, $W(a) \subseteq D_{i+1}(S)$, and the proof is complete. □

If T is a full weakly self conjugate subsemigroup of an E-dense semigroup S, then T is strongly dense in S because $\emptyset \neq W(a) \subseteq W_T(a)$ for all $a \in S$. In particular, $D(S)$ is strongly dense in S. It follows from Proposition 2.3 that S has a T-cover.

Now, for an E-dense semigroup S, we choose T to be $D(S)$, and we want to find an E-dense $D(S)$-cover $\widehat{S}$ with $\widehat{D(S)} = D(\widehat{S})$. To do this we use the construction of Section 2 with $\gamma_T(a) = W(a)$. We continue to use the notation introduced in the proof of Proposition 2.3 so that G is the free group on S, and X^* is the free monoid on $S \cup \overline{S}$. We constructed a relational morphism $\tau : G \dashrightarrow S$ by using the natural homomorphism $\theta : X^* \to G$, a relational morphism $\varphi : X^* \dashrightarrow S$, and putting $\tau = \theta^{-1} \varphi$. The cover $\widehat{S}$ of S over G is $\mathrm{gr}(\tau)$, that is, $\{(a, g) \in S \times G : a \in g\tau\}$. We have

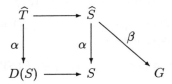

and $\widehat{T} = 1\beta^{-1} = \{(a, 1) : a \in 1\tau\} = \{(a, 1) : a \in D(S)\}$.

Lemma 3.6. *If $(d, g) \in \widehat{S}$ and $c \in W(d)$, then $(c, g^{-1}) \in \widehat{S}$.*

Proof. If $g = 1$, then $d \in 1\tau$, that is, $d \in D(S)$, and so, by Theorem 3.5, $c \in D(S)$. Hence $c \in 1\tau$ and $(c, 1) \in \widehat{S}$.

Now suppose that $g \neq 1$. Since

$$g\tau = \bigcup \{w\varphi : w\theta = g\},$$

we have $d \in w\varphi$ for some $w \in X^*$ with $w\theta = g$. Let $w = x_1 \ldots x_k$ where $x_i \in X$. Then $w\varphi = (x_1\varphi) \ldots (x_k\varphi)$ and so, by Proposition 3.2, $c \in W(x_k\varphi) \ldots W(x_1\varphi)$.

If $x_j = a$ where $a \in S$, then, by definition, $x_j\varphi = D(S)^1 a D(S)^1$, and so, by Proposition 3.2 and Theorem 3.5, we have

$$W(x_j\varphi) \subseteq D(S)^1 W(a) D(S)^1 = \overline{x}_j\varphi.$$

On the other hand, if $x_j = \overline{a}$ where $a \in S$, then $x_j\varphi = D(S)^1 W(a) D(S)^1$ and

170

so, by Propositions 3.2 and 3.3, and Theorem 3.5, we have
$$W(x_j\varphi) \subseteq D(S)^1 W(W(a)) D(S)^1 \subseteq D(S)^1 E(S) a E(S) D(S)^1$$
$$\subseteq D(S)^1 a D(S)^1 = \overline{x}_j\varphi.$$
Thus $c \in (\overline{x}_k\varphi)\ldots(\overline{x}_1\varphi) = \overline{w}\varphi$. Now $\overline{w}\theta = (w\theta)^{-1} = g^{-1}$ so that $c \in g^{-1}\tau$ as required. □

As an immediate consequence we have the following corollary.

Corollary 3.7. *The semigroup $\widehat{S}$ is E-dense.*

Lemma 3.8. $\widehat{T} = D(\widehat{S})$.

Proof. Let $(a,g) \in \widehat{S}$. Then any weak inverse of (a,g) has the form (b,g^{-1}) where $b \in W(a)$. Since $D(S)$ is weakly self conjugate in S, it follows that $\widehat{T} = \{(a,1) : a \in D(S)\}$ is weakly self conjugate in $\widehat{S}$. As $\widehat{T}$ is clearly full, we have $D(\widehat{S}) \subseteq \widehat{T}$.

Now let $(a,1) \in \widehat{T}$. Then $a \in D(S)$ so that $a \in D_i(S)$ for some i. Hence it is enough to show that, for all non-negative integers i,
$$a \in D_i(S) \text{ implies } (a,1) \in D(\widehat{S}). \tag{†}$$

If $i = 0$, then $a = e_1\ldots e_k$ for some idempotents $e_1,\ldots,e_k$. Certainly, $(e_j,1) \in D(\widehat{S})$ for each j since $D(\widehat{S})$ is full, and so $(a,1) \in D(\widehat{S})$.

Assume inductively that (†) holds for i, and let $a \in D_{i+1}(S)$. Then $a = a_1\ldots a_m$ for some $a_j = c_j b_j d_j$ where $b_j \in D_i(S)$, $c_j, d_j \in S$ and one of $c_j \in W(d_j)$ or $d_j \in W(c_j)$ holds. In either case, by Lemma 3.6, there is an element g of G such that $(c_j,g),(d_j,g^{-1}) \in \widehat{S}$. By the induction assumption, we have $(b_j,1) \in D(\widehat{S})$ and so $(a_j,1) = (c_j,g)(b_j,1)(d_j,g^{-1}) \in D(\widehat{S})$. Hence $(a,1) \in D(\widehat{S})$.

It follows that $\widehat{T} = D(\widehat{S})$, as required. □

An E-dense semigroup S is said to be *D-unitary* if $D(S)$ is a unitary subset of S. Putting together the preceding results and Lemma 2.1, we have now proved the following theorem.

Theorem 3.9. *Every E-dense semigroup has a D-unitary E-dense cover over a group.*

The theorem gives an alternative way of looking at $D(S)$ in an E-dense semigroup S. Let $\tau : S \multimap G$ be a relational morphism into a group G. If $e \in E(S)$, then $e\tau$ is a subsemigroup of G, but, in general, it need not be a submonoid. We say that τ is *full* if $E(S) \subseteq 1\tau^{-1}$, so that $e\tau$ is a submonoid for every idempotent e of S. Note that, since subsemigroups of finite groups are subgroups, every relational morphism into a finite group is full. Let $K(S)$ be the intersection of all subsemigroups K of S such that $K = 1\tau^{-1}$ for some full relational morphism τ into a group. It is easy to see that $K(S)$ is a full weakly self conjugate subsemigroup of S, so that $D(S) \subseteq K(S)$. On the other hand, by the theorem, $D(S) = 1\tau^{-1}$ for the relational morphism used in the proof of the theorem, so that $K(S) \subseteq D(S)$. Hence we have the following.

Corollary 3.10. *If S is an E-dense semigroup, then $D(S) = K(S)$.*

On any E-dense semigroup there is a minimum group congruence σ. The existence of σ was noted by Hall and Munn [18], and an explicit description was given by Mitsch [31]. We simply require that σ exists, and that $D(S)$ is contained in a σ-class. This follows from Corollary 3.10 since the natural homomorphism associated with σ is a full relational morphism. We now use these properties to give alternative criteria for an E-dense semigroup to be D-unitary.

Lemma 3.11. *For an E-dense semigroup S, the following conditions are equivalent:*

(1) S is D-unitary,

(2) there is a surjective homomorphism $\beta : S \to G$ onto a group G with $D(S) = 1\beta^{-1}$,

(3) $D(S)$ is a σ-class.

Proof. Suppose that $D(S)$ is unitary in S. As S is E-dense, $D(S)$ is certainly dense in S. Suppose that $x, y \in S$ with $xy \in D(S)$. Since S is E-dense, $W(x) \neq \emptyset$. Now, $D(S)$ is weakly self conjugate, and so if $b \in W(x)$, then $bxyx \in D(S)$. Since $bx \in E(S) \subseteq D(S)$ and $D(S)$ is unitary in S, we have $yx \in D(S)$ so that $D(S)$ is reflexive. Condition (2) now follows from Theorem 2.2.

If condition (2) holds, then $\beta \circ \beta^{-1}$ is a group congruence on S, and so $\sigma \subseteq \beta \circ \beta^{-1}$. Hence, if $a \in b\sigma$ where $b \in D(S)$, then $a\beta = 1$, and so $a \in D(S)$. Since $D(S)$ is contained in a σ-class, condition (3) follows.

If condition (3) holds, then clearly, $D(S)$ is the identity of S/σ. Let $a \in S, b \in D(S)$ with $ab \in D(S)$. Then

$$a\sigma = (a\sigma)(b\sigma) = (ab)\sigma = 1,$$

and hence $a \in D(S)$. Thus S is D-unitary. □

It is worth noting the next result which is an immediate consequence of the lemma.

Corollary 3.12. *Let P, S be E-dense semigroups, and $\alpha : P \to S$ be a surjective homomorphism such that $\alpha|_{D(P)}$ maps $D(P)$ isomorphically onto $D(S)$. If P is D-unitary, then P is a D-unitary cover of S over some group.*

We comment briefly on two special cases: regular semigroups and semigroups in which the idempotents form a subsemigroup. First, let S be a regular semigroup. Then every element a of S has an inverse, that is, an element a' such that $aa'a = a$ and $a'aa' = a'$. In particular, a' is a weak inverse of a. As usual, we denote the set of all inverses of a in S by $V(a)$.

A subsemigroup T of a regular semigroup S is said to be *self conjugate* if $aTa' \subseteq T$ for all $a \in S$ and $a' \in V(a)$. Obviously, a weakly self conjugate subsemigroup of S is self conjugate. The converse is also true if T is full, as we show in the next lemma.

Lemma 3.13. *Let T be a full subsemigroup of a regular semigroup S. Then T is self conjugate if and only if it is weakly self conjugate.*

Proof. Suppose that T is self conjugate, and let $t \in T$, $a \in S$, $b \in W(a)$ and $a' \in V(a)$. Then

$$atb = atbab = atbaa'ab = a(tba)a'(ab).$$

Now $E(S) \subseteq T$, so $ab \in T$ and $tba \in T$. But T is self conjugate, and so it follows that $atb \in T$. Similarly, $bta \in T$, and so T is weakly self conjugate. □

Thus $D(S)$ is the least self conjugate full subsemigroup of a regular semigroup S. We also note that it is immediate from Theorem 3.5 that $D(S)$ is regular.

Given a regular semigroup S, we can use the construction above to obtain a D-unitary E-dense cover $\widehat{S}$ of S over a group. It follows from Lemma 3.6 that the regularity of S implies that $\widehat{S}$ is regular. Thus we have the following result which was first made explicit by Trotter in [43]. Of course, our proof gives an infinite cover, but one of the two proofs in [43] (based on results in [29]) gives a finite regular cover for a finite regular semigroup S.

Corollary 3.14. *A regular semigroup has a D-unitary regular cover over a group.*

It is perhaps worth mentioning that, in the regular case, our construction of a cover can be modified by using the set $V(a)$ of inverses of a rather than the set $W(a)$ of weak inverses of a.

We now consider the case where the set of idempotents is a subsemigroup of S.

Proposition 3.15. *Let S be an E-dense semigroup. If $E(S)$ is a subsemigroup of S, then $E(S) = D(S)$.*

Proof. Certainly, $E(S) \subseteq D(S)$ so that it is enough to prove that $E(S)$ is closed under weak conjugation. Let $a \in S$, $b \in W(a)$ and $e \in E$. Then

$$(aeb)^2 = aebaeb = aebae(bab) = a(eba)(eba)b = aebab = aeb,$$

and, similarly, $bea \in E(S)$. □

We say that an E-dense semigroup S is *E-unitary* if $E(S)$ is a unitary subset of S.

Lemma 3.16. *If S is an E-unitary E-dense semigroup, then $E(S)$ is a subsemigroup of S.*

Proof. If $e, f \in E(S)$, then $efb \in E(S)$ for some $b \in S$ since S is E-dense. But S is E-unitary so that $fb \in E(S)$, and consequently, $b \in E(S)$. From $efb \in E(S)$ and $b \in E(S)$ we get $ef \in E(S)$. □

It follows from the lemma and Proposition 3.15 that if S is an E-dense semigroup, then it is E-unitary if and only if it is D-unitary and $E(S)$ is a subsemigroup. In view of this and Proposition 3.15, we have the following results as corollaries of Theorem 3.9 and Corollary 3.14 respectively.

Corollary 3.17. *Let S be an E-dense semigroup in which $E(S)$ is a subsemigroup. Then S has an E-unitary E-dense cover $\widehat{S}$ over a group. Moreover, if S is E-commutative dense, then so is $\widehat{S}$.*

Recall that an *orthodox* semigroup is a regular semigroup in which the idempotents form a subsemigroup.

Corollary 3.18. *An orthodox semigroup S has an E-unitary orthodox cover $\widehat{S}$ over a group. Moreover, if S is inverse, then so is $\widehat{S}$.*

The general result in the first of these corollaries was proved independently by Almeida, Pin and Weil in [2] and Zhonghao in [46]. The E-commutative dense result is due to the author [10].

The result for orthodox semigroups was proved independently by three authors: McAlister, Szendrei and Takizawa in [28], [40] and [41] respectively.

Their proofs give finite covers for finite orthodox semigroups whereas we have already pointed out that our construction always yields an infinite cover. As mentioned in the introductory remarks, the inverse result was the first of the covering theorems and goes back to unpublished work of Tilson in the finite case and the seminal papers [26], [27] of McAlister in the general case. We give alternative proofs of the orthodox and inverse cases in the following sections because of the desirability of having finite covers of finite semigroups.

4 Inverse semigroups

We use factorisable inverse monoids, to give an alternative proof that every inverse semigroup S has an E-unitary inverse cover which is finite if S is finite, and describe the cover constructed as P-semigroup. This is followed by an account of results of McAlister and Reilly [30] in which E-unitary covers are characterised using idempotent pure prehomomorphisms, and certain dual prehomomorphisms. We conclude the section by illustrating the use of the covering theorem and the P-theorem to obtain Reilly's structure theorem for bisimple inverse ω-semigroups.

4.1 Factorisable monoids

An element a of a monoid M is *unit regular* if $a = aua$ for some unit u of M.

Lemma 4.1. *Let a be an element of a monoid M. Then the following are equivalent:*

(1) a is unit regular,

(2) $a = eg$ for some idempotent e and unit g,

(3) $a = hf$ for some idempotent f and unit h.

Proof. If (1) holds, let u be a unit such that $aua = a$. Then au and ua are idempotents, u^{-1} is a unit, and $a = auu^{-1} = u^{-1}ua$.

If (2) holds, then $a = ea = ag^{-1}a$ so that a is unit regular. Similarly, (3) implies (1). □

A monoid M is *factorisable* (or *unit regular*) if $M = GE$ where G is a subgroup of M and $E = E(M)$ is the set of idempotents of M.

Lemma 4.2. *A monoid M is factorisable if and only if every element of M is unit regular.*

Proof. If $M = GE$ for some subgroup G, then $1 = ge$ for some $g \in G$ and idempotent e, so $e = (ge)e = ge = 1$. Hence $1 = g \in G$. Thus 1 is the identity of G and so G consists of units. (In fact, G is the group of units.) If $a \in M$, then $a = ue$ for some unit u and idempotent e, so that a is unit regular by Lemma 4.1.

The converse is immediate by Lemma 4.1. □

Examples of factorisable monoids include the following. For a finite set X, the symmetric inverse monoid $I(X)$ on X; the full transformation monoid $T(X)$ on X; the monoid $PT(X)$ of all partial transformations on X; and the multiplicative monoid $M_n(F)$ of all $n \times n$ matrices over a field F.

The significance of factorisable monoids in the inverse case arises from the following two results.

Proposition 4.3. *If S is an inverse semigroup, then S can be embedded in a factorisable inverse monoid.*

Proof. By the Vagner-Preston theorem (see [21], [22] or [32]), S can be embedded in the symmetric inverse monoid $I(S)$ on S. If S is finite, $I(S)$ is factorisable and we have the desired embedding.

If S is infinite, let S' be a set disjoint from S and having the same cardinality as S. Clearly, $I(S)$ (and hence S) can be embedded in $I(S \cup S')$. Moreover, if $\alpha \in I(S)$, then $\alpha = \theta|_{\text{dom } \alpha}$ for some θ in the group of units of $I(S \cup S')$. Hence $\alpha \in F$ where

$$F = \{\gamma \in I(S \cup S') : \gamma \leqslant \theta \text{ for some unit } \theta\}.$$

Thus we see that S is embedded in F and that F is a factorisable inverse submonoid of $I(S \cup S')$. □

Proposition 4.4. *Let F be a factorisable inverse monoid with set of idempotents E and group of units G. Then G acts (by automorphisms) on E by conjugation, and $E \rtimes G$ is an E-unitary inverse cover of F.*

Proof. The claim about the action is clear, so that we can form the semidirect product $E \rtimes G$ with multiplication $(e,g)(f,h) = (e(gfg^{-1}), gh)$. It is straightforward to verify that $E \rtimes G$ is E-unitary inverse, and that $E(E \rtimes G) = E \times \{1\}$. Now $\alpha : E \rtimes G \to F$ defined by $(e,g)\alpha = eg$ is a surjective homomorphism which restricts to an isomorphism from $E(E \rtimes G)$ to $E(F)$. □

The finite versions of these two results tell us about the complexity of finite inverse semigroups. A *pseudovariety* of semigroups is a class of semigroups closed under finite direct product, homomorphic images and subsemigroups. Let **A** denote the pseudovariety of all finite *aperiodic* semigroups, that is, finite semigroups on which Green's relation $\mathcal{H}$ is trivial, and let **G** denote the pseudovariety of all finite groups. Let $\mathbf{V}_0 = \mathbf{A}$ and, for $n \geq 0$, let $\mathbf{V}_{n+1} = \mathbf{A} * \mathbf{G} * \mathbf{V}_n$. Here $\mathbf{V} * \mathbf{W}$ is the least pseudovariety which contains all semidirect products $S \rtimes T$ with $S \in \mathbf{V}$ and $T \in \mathbf{W}$. It follows from the Krohn-Rhodes Decomposition Theorem that every semigroup S is in $\mathbf{V}_n$ for some n. The least such n is called the *(group) complexity* of S. If S is an inverse semigroup, then it follows from Propositions 4.3 and 4.4 that $S \in \mathbf{V}_1$ (more precisely, it is in $\mathbf{A} * \mathbf{G}$), and we have the following result of Tilson.

Proposition 4.5. *An inverse semigroup has complexity at most 1.*

Several results on the complexity of various classes of regular semigroups were given by Trotter [43] using the finite version of Corollary 3.14.

4.2 E-unitary covers and the P-theorem

We use Propositions 4.3 and 4.4 to give another proof of the covering theorem, and then show that the cover is a *P*-semigroup. First, we note some elementary variations on the condition that the covering map restricts to an isomorphism between the semilattices of idempotents. In proving the equivalence of the various conditions, we make use of Lallement's lemma. As we will use it several times, we record it here.

Lemma 4.6 (Lallement's Lemma). *Let S, T be regular semigroups and let $\theta : S \to T$ be a surjective homomorphism. If $f \in E(T)$, then there is an idempotent e in S such that $e\theta = f$.*

Equivalently, if ρ is a congruence on a regular semigroup S, and if $a\rho$ is an idempotent in S/ρ, then there is an idempotent e in S such that $a\rho = e\rho$.

A proof of the lemma can be found, for example, in [21, Chapter 2].

Recall that a homomorphism $\alpha : S \to T$ of semigroups is *idempotent separating* if its restriction to $E(S)$ is one-one.

Lemma 4.7. *Let $\alpha : P \to S$ be a surjective homomorphism of inverse semigroups, and $a, b \in P$. Then the following are equivalent:*

(1) $\alpha|_{E(P)}$ is an isomorphism from $E(P)$ onto $E(S)$,

(2) α is idempotent separating,

(3) $a\alpha \mathcal{L} b\alpha$ implies $a \mathcal{L} b$,

(4) $a\alpha \mathcal{R} b\alpha$ implies $a \mathcal{R} b$.

Proof. If (1) holds, then *a fortiori*, (2) holds. If (2) holds and $a\alpha \mathcal{L} b\alpha$, then $(a^{-1}a)\alpha = (a\alpha)^{-1}(a\alpha) = (b\alpha)^{-1}(b\alpha) = (b^{-1}b)\alpha$. Hence $a^{-1}a = b^{-1}b$ since α is idempotent separating, and so $a \mathcal{L} b$. Thus (3) holds. Similarly, (2) implies (4).

If (3) holds, and $e, f \in E(P)$ with $e\alpha = f\alpha$, then, by assumption, $e \mathcal{L} f$ so that $e = f$ and $\alpha|_{E(P)}$ is one-one. Moreover, by Lallement's lemma (Lemma 4.6), we have that $\alpha|_{E(P)}$ maps $E(P)$ onto $E(S)$. Similarly, (4) implies (1). □

We now give a second proof of the existence of E-unitary inverse covers for an inverse semigroup.

Proposition 4.8. *An inverse semigroup S has an E-unitary inverse cover. Moreover, if S is finite, then the cover can be chosen to be finite.*

Proof. By Lemmas 3.12 and 4.7, it is enough to find an E-unitary inverse semigroup P and an idempotent separating homomorphism from P onto S.

By Proposition 4.3, we can embed S in a factorisable inverse monoid F with group of units G and $E = E(F)$. It follows from Proposition 4.4 that the semidirect product $E \rtimes G$ is an E-unitary inverse cover for F over G. Let $\alpha : E \rtimes G \to F$ be the covering homomorphism $((e, g)\alpha = eg)$.

Then $S\alpha^{-1}$ is an inverse subsemigroup of $E \rtimes G$, and hence it is E-unitary. Also, α restricted to $S\alpha^{-1}$ must be idempotent separating, and so $S\alpha^{-1}$ is the desired cover. □

As before, we let β be the projection of $E \rtimes G$ onto G. Then the group over which $S\alpha^{-1}$ is a cover is the image H of β restricted to $S\alpha^{-1}$. Thus

$$H = \{h \in G : eh \in S \text{ for some } e \in E(F)\}.$$

We also note that if $e \in E(F)$ and $eh \in S$ for some $h \in G$, then $e \in S$. For, $(e, h) \in S\alpha^{-1}$, and $S\alpha^{-1}$ is an inverse subsemigroup of $E \rtimes G$, so that $(e, 1) = (e, h)(h^{-1}eh, h^{-1}) = (e, h)(e, h)^{-1} \in S\alpha^{-1}$ and hence $e = (e, 1)\alpha \in S$.

The cover $S\alpha^{-1}$ can be described as another type of "semidirect product" known as a *P-semigroup*, a concept we now define.

Let G be a group, X a poset such that G acts on X by order automorphisms, and Y be a subset of X. Suppose that

(1) Y is an order ideal of X and a meet semilattice under the induced ordering,

(2) $G \cdot Y = X$,

(3) $g \cdot Y \cap Y \neq \emptyset$ for all $g \in G$.

Then we put

$$P = P(G, X, Y) = \{(y, g) \in Y \times G : g^{-1}y \in Y\}$$

with multiplication

$$(y, g)(y', g') = (y \wedge g \cdot y', gg').$$

The proof of the following is straightforward.

Proposition 4.9. *With the above notation, P is an E-unitary inverse semigroup with $E(P) = \{(y, 1) : y \in Y\} \cong Y$, and maximum group homomorphic image G. Moreover, for $(y, g), (z, h) \in P$,*

(1) $(y, g) \mathcal{R} (z, h)$ if and only if $y = z$,

(2) $(y, g) \mathcal{L} (z, h)$ if and only if $g^{-1} \cdot y = h^{-1} \cdot z$.

The importance of P-semigroups arises from the following theorem of McAlister [27] which gives the converse of Proposition 4.9.

Theorem 4.10. *If S is an E-unitary inverse semigroup, then S is isomorphic to a P-semigroup.*

It follows that if S is an inverse semigroup, then the E-unitary cover $S\alpha^{-1}$ in the proof of Proposition 4.8 is a P-semigroup. We could realise it as such by using the subdirect product description or the 'semigroupoid semidirect product' description, but it is easy to do it directly.

Let $Y = E(S)$ and H be the group $\{h \in G : eh \in S \text{ for some } e \in E(F)\}$ considered above. Recall that G (and hence H) acts on $E(F)$ by conjugation. Using this action, put $X = \{h \cdot e : h \in H, e \in E(S)\}$ so that X is a subset of $E(F)$ (and hence inherits the partial order of $E(F)$). It is easy to verify that H, X, Y provide the data for a P-semigroup, and that this semigroup actually is the cover $S\alpha^{-1}$.

The E-unitary cover $S\alpha^{-1}$ was constructed from an embedding of S in a factorisable inverse monoid. McAlister and Reilly proved [30] proved that every E-unitary cover arises from a certain kind of embedding. An embedding $\iota : S \to F$ of an inverse semigroup S into a factorisable inverse monoid F is said to be *strict* if, for each unit g of F, there is an element s of S such that $\iota(s) \leqslant g$.

Theorem 4.11. *Every E-unitary inverse cover of an inverse semigroup S over a group G is isomorphic to one constructed (as in Proposition 4.8) from a strict embedding of S into a factorisable inverse monoid with group of units G.*

A discussion of this theorem in terms of enlargements of groupoids is given in Chapter 8 of [22].

4.3 Prehomomorphisms and dual prehomomorphisms

In [30] a number of different ways in which E-unitary covers of inverse semigroups can arise are discussed. We explain two of these approaches using prehomomorphisms and dual prehomomorphisms.

Let S, T be inverse semigroups and $\theta : S \to T$ be a function. We say that

(1) θ is a *prehomomorphism* if $(ab)\theta \leqslant (a\theta)(b\theta)$ for all $a, b \in S$;

(2) θ is a *dual prehomomorphism* if $(ab)\theta \geqslant (a\theta)(b\theta)$ for all $a, b \in S$ and $(a\theta)^{-1} = a^{-1}\theta$ for all $a \in S$.

Here we follow the terminology of [22]; we note that in [32] what we have called a dual prehomomorphism is known as a prehomomorphism.

Let S be an inverse semigroup and G be a group. If S has an E-unitary inverse cover over G, then by Propositions 1.3 and 1.2, there is an $E(S)$-pure, that is, *idempotent pure* surjective relational morphism $\tau : S \relbar\joinrel\twoheadrightarrow G$ such that the cover is isomorphic to $\mathrm{gr}(\tau)$. Since $\mathrm{gr}(\tau)$ is inverse, so is τ. Let $s \in S$. If $g \in s\tau$, then $s\tau = gH$ where $H = g^{-1}(s\tau)$ is a subgroup of G.

Now the set $K(G)$ of all cosets is an inverse monoid under the operation $\cdot$ where $aH \cdot bK$ is the smallest coset in G which contains the set product $(aH)(bK)$. The set of idempotents of $K(G)$ is the set of subgroups of G, and the identity is the trivial subgroup. The natural partial order on $K(G)$ is given by $aH \leqslant bK$ if and only if $aH \supseteq bK$.

Thus τ is an idempotent pure prehomomorphism from S to $K(G)$ such that, for all $g \in G$, there is an element $s \in S$ with $g \in s\tau$. Conversely, such a prehomomorphism is clearly an idempotent pure surjective relational morphism from S to G.

If we now consider the inverse relational morphism $\tau^{-1} : G \relbar\joinrel\twoheadrightarrow S$, then it is not difficult to verify that for each element g of G, we have:

(1) $g\tau^{-1}$ is an order ideal of S (in the natural partial order),

(2) if $a, b \in g\tau^{-1}$, then $a^{-1}b, ab^{-1} \in E(S)$.

180

A subset of an inverse semigroup S satisfying the second condition is said to be *compatible*, and a compatible order ideal (in the natural partial order) is said to be *permissible*. The set of all permissible subsets is denoted by $C(S)$. Under multiplication of subsets, $C(S)$ is an inverse monoid which was studied by Schein [37] (see also [22]). We mention that if H is a permissible subset of S, then its inverse as an element of $C(S)$ is just $H^{-1} = \{h^{-1} : h \in H\}$, and H is idempotent if and only if $H \subseteq E(S)$. We also note that the natural partial order is given by $H \leqslant K$ if and only if $H \subseteq K$.

Thus τ^{-1} maps G into $C(S)$. Now $(g\tau^{-1})^{-1} = g^{-1}\tau^{-1}$, and $1\tau^{-1} = E(S)$. Hence it is clear from the definition of surjective relational morphism that τ^{-1} is a dual prehomomorphism from G to $C(S)$ satisfying $\bigcup g\tau^{-1} = S$. Conversely, any such dual prehomomorphism gives an idempotent pure surjective relational morphism $G \multimap S$, and thus we have the following result.

Proposition 4.12. *Let S be an inverse semigroup, and G be a group. Then the following are equivalent:*

(1) there is an E-unitary inverse cover of S over G,

(2) there is an idempotent pure prehomomorphism $\theta : S \to K(G)$ such that, for each $g \in G$, there is an element s of S with $g \in s\theta$.

(3) there is a dual prehomomorphism from G to $C(S)$ such that $\bigcup g\tau^{-1} = S$.

4.4 Bisimple inverse ω-semigroups

We illustrate the use of covers and the P-theorem by giving a proof of a structure theorem due to Reilly [35]. Recall that a *bisimple ω-inverse semigroup* S is an inverse semigroup on which Green's relation $\mathcal{D}$ is the universal relation, and in which the idempotents form an ω-chain, that is, $E(S)$ is order isomorphic to the chain of negative integers. We remark that such an inverse semigroup is actually a monoid. Let K be a group, and α be an endomorphism of K. On the set $\mathbb{N} \times K \times \mathbb{N}$ we define a binary operation by the rule:

$$(m, a, n)(p, b, q) = (m - n + t, (a\alpha^{n-t})(b\alpha^{p-t}), q - p + t),$$

where $t = \max\{n, p\}$ and α^0 is the identity map on K. The set $\mathbb{N} \times K \times \mathbb{N}$ together with this operation is a monoid denoted by $B(K, \alpha)$ and called a *Reilly monoid*. It is straightforward to show that $B(K, \alpha)$ is an inverse monoid with identity $(0, 1, 0)$, and semilattice of idempotents $\{(m, 1, m) : m \in \mathbb{N}\}$ with $(m, 1, m) \leqslant (n, 1, n)$ if and only if $m \geqslant n$. Further, the elements (m, a, n) and (p, b, q) are $\mathcal{R}$-related if and only if $m = p$, and $\mathcal{L}$-related if and only if $n = q$ so that $B(K, \alpha)$ is clearly bisimple. Our aim is to prove the following theorem.

Theorem 4.13. *Every bisimple inverse ω-semigroup is isomorphic to a Reilly monoid.*

We start with the following simple lemma.

Lemma 4.14. *Let $\theta : P \to S$ be an idempotent separating surjective homomorphism of inverse semigroups. If S is bisimple, then P is bisimple.*

Proof. If $a, b \in P$, then $a\theta, b\theta$ are $\mathcal{D}$-related in S, and, since θ is onto, there is an element c in P such that $a\theta \, \mathcal{L} \, c\theta \, \mathcal{R} \, b\theta$. By Lemma 4.7, $a \, \mathcal{L} \, c \, \mathcal{R} \, b$ so that all elements of P are $\mathcal{D}$-related, and P is bisimple. $\square$

Lemma 4.15. *If $\theta : B(K, \alpha) \to S$ is an idempotent separating homomorphism from a Reilly monoid onto an inverse monoid S, then S is isomorphic to a Reilly monoid $B(H, \beta)$ where H is the group of units of S.*

Proof. The group of units of $B(K, \alpha)$ is $U = \{(0, k, 0) : k \in K\}$. Since θ is surjective and idempotent separating, $H = U\theta$. Define $\varphi : K \to H$ by $k\varphi = (0, k, 0)\theta$. Then φ is a surjective homomorphism; moreover, if $k \in \ker \varphi$, then $(0, k, 0)\theta = 1$ and hence

$$(0, k\alpha, 0)\theta = ((0, 1, 1)(0, k, 0)(1, 1, 0))\theta = (0, 1, 1)\theta(1, 1, 0)\theta$$
$$= ((0, 1, 1)(1, 1, 0))\theta = 1,$$

so $k\alpha \in \ker \varphi$. Thus the function $\beta : H \to H$, defined by $h\beta = k\alpha\varphi$ where $k\varphi = h$ is well defined. Also, β is an endomorphism so that we have a Reilly monoid $B(H, \beta)$.

If we now define $\psi : B(K, \alpha) \to B(H, \beta)$ by $(m, k, n)\psi = (m, k\varphi, n)$, then ψ is a surjective homomorphism, and $\theta \circ \theta^{-1} = \psi \circ \psi^{-1}$. Hence

$$B(H, \beta) \cong B(K, \alpha)/\psi \circ \psi^{-1} = B(K, \alpha)/\theta \circ \theta^{-1} \cong S.$$

$\square$

It follows from Lemmas 4.14 and 4.15 that to prove Theorem 4.13, it is enough to prove it for E-unitary bisimple inverse ω-semigroups. Let Q be such a semigroup. By Theorem 4.10, Q is isomorphic to a P-semigroup, say $Q \cong P = P(G, X, Y)$. Thus

$$P = \{(e, g) \in Y \times G : g^{-1}e \in Y\},$$

and $E(P) \cong Y$, so that Y is an ω-chain, say $Y = \{e_0, e_1, e_2, \ldots\}$ with the order given by $e_0 > e_1 > e_2 > \ldots$. Put

$$K = \text{stab}_G(e_0) = \{g \in G : g^{-1} \cdot e_0 = e_0\}.$$

Then the group of units of P is $\{(e_0, g) \in P : g \in K\}$ which is isomorphic to K.

Next, we claim that if $k \in K$, then $k \cdot e_i = e_i$ for all i. Certainly, $k \cdot e_0 = e_0$, and, if $k \cdot e_j = e_j$ for some j, then since $e_{j+1} < e_j$ and G acts by order automorphisms, $k \cdot e_{j+1} < e_j$ and $k^{-1} \cdot e_{j+1} < e_j$. As Y is an order ideal, $k \cdot e_{j+1}, k^{-1} \cdot e_{j+1} \in Y$. Hence $k \cdot e_{j+1} \leqslant e_{j+1}$ and $k^{-1} \cdot e_{j+1} \leqslant e_{j+1}$ so that

$$k \cdot e_{j+1} \leqslant e_{j+1} = 1 \cdot e_{j+1} = (kk^{-1}) \cdot e_{j+1} = k \cdot (k^{-1} \cdot e_{j+1}) \leqslant k \cdot e_{j+1}.$$

Thus $k \cdot e_{j+1} = e_{j+1}$, and, by induction, the claim is true.

For any $i, j \in \mathbb{N}$ the pairs $(e_i, 1), (e_j, 1)$ are elements of P, and so they must be $\mathcal{D}$-related. Hence there is an element (e, g) of P with $(e_i, 1) \mathcal{R} (e, g) \mathcal{L} (e_j, 1)$. By Proposition 4.9, $g^{-1} \cdot e = e_j$ and $e = e_i$, and hence $g \cdot e_j = e_i$. In particular, we can choose an element $h \in G$ such that $h \cdot e_1 = e_0$. An induction argument similar to the above shows that $h^{-i} \cdot e_0 = e_i$ for all $i \in \mathbb{N}$.

For $k \in K$, we have $hkh^{-1} \cdot e_0 = hk \cdot e_1 = h \cdot e_1 = e_0$, so we can define an endomorphism $\alpha : K \to K$ by $k\alpha = hkh^{-1}$, and form the Reilly monoid $B(K, \alpha)$.

Finally, we define a mapping $\theta : P \to B(K, \alpha)$ by

$$(e_i, g)\theta = (i, h^i g h^{-j}, j)$$

where $g^{-1} \cdot e_i = e_j$.

We claim that θ is an isomorphism. First, note that $h^i g h^{-j} \cdot e_0 = h^i g \cdot e_j = h^i \cdot e_i = e_0$ so that $(i, h^i g h^{-j}, j)$ is an element of $B(K, \alpha)$.

Clearly, θ is one-one. If $(i, k, j) \in B(K, \alpha)$, put $g = h^{-i} k h^j$. Then $(e_i, g) \in P$ and $(e_i, g)\theta = (i, k, j)$ so that θ is onto.

Let $(e_i, g), (e_m, b) \in P$ with $g^{-1} \cdot e_i = e_j$. Straightforward calculations, considering the two cases $m \leqslant j$ and $m > j$ show that $((e_i, g)(e_m, b))\theta = (e_i, g)\theta(e_m, b)\theta$ so that θ is a homomorphism. Hence $P \cong B(K, \alpha)$ and the proof of Theorem 4.13 is complete.

Other inverse semigroup structure theorems can be proved using the same approach, for example, in unpublished notes, Victoria Gould has used the method to obtain new proofs of the Munn-Kochin result describing the structure of simple inverse ω-semigroups.

5 Orthodox semigroups

This section is based on McAlister's paper [28]. We show that an orthodox semigroup S has an E-unitary orthodox cover which is finite if S is finite. This result is used to obtain a characterisation of those orthodox semigroups in the pseudovariety $\mathbf{A} \vee \mathbf{G}$. Recall that $\mathbf{A}$ is the pseudovariety of all finite aperiodic semigroups, and $\mathbf{G}$ is the pseudovariety of all finite groups. In [28], McAlister proves the following result.

Theorem 5.1. *Let S be a finite orthodox semigroup. Then $S \in \mathbf{A} \vee \mathbf{G}$ if and only if $\mathcal{H}$ is a congruence on S.*

The proof of the "only if" part does not involve covers, and we do not present this. Our aim is to show how the covering theorem is used to prove the "if" part of the theorem.

For basic results on orthodox semigroups, see [21, Chapter 6]. We will need the following easy lemma.

Lemma 5.2. *If e is an idempotent in an orthodox semigroup S, then*
$$V(e) \subseteq E(S).$$

Proof. Let $x \in V(e)$. Then $xex = x$ so that $xe, ex \in E(S)$, and hence the product $(xe)(ex) \in E(S)$. But $x = xex = (xe)(ex)$. □

We also use three special congruences on an orthodox semigroup: the maximum idempotent separating congruence μ, the minimum group congruence σ and the minimum inverse semigroup congruence $\mathcal{Y}$. We have already introduced σ in the more general context of E-dense semigroups. There is an explicit description of μ (see, for example, [20, Theorem VI.1.17]), but we simply need its existence and the fact that it is the largest congruence contained in $\mathcal{H}$ (see, for example, [21, Proposition 2.4.5] where these facts are proved for regular semigroups). However, we do want the following explicit description of $\mathcal{Y}$ due to Hall [16] (see also [21, Theorem 6.2.5]).

Proposition 5.3. *Let S be an orthodox semigroup, and let $\mathcal{Y}$ be the relation defined by*
$$a \, \mathcal{Y} \, b \text{ if and only if } V(a) = V(b).$$
Then $\mathcal{Y}$ is the minimum inverse semigroup congruence on S.

We also use a corollary of the following general result from [17], the proof of which we leave as an exercise.

Lemma 5.4. *Let ρ be a congruence on a semigroup S with $\rho \subseteq \mathcal{L}$. Then $(a, b) \in \mathcal{L}$ in S if and only if $(a\rho, b\rho) \in \mathcal{L}$ in S/ρ.*

Corollary 5.5. *If S is a regular semigroup, then the maximum idempotent separating congruence $\mu_{S/\mu}$ on S/μ is trivial.*

Proof. For some congruence ρ on S, we have $\mu_{S/\mu} = \rho/\mu$. If $a, b \in S$ and $a\rho b$, then $(a\mu, b\mu) \in \rho/\mu$. Hence $(a\mu, b\mu) \in \mathcal{H}$, and so by the lemma and its right-left dual, $a\ \mathcal{H}\ b$ in S. Thus $\rho \subseteq \mathcal{H}$ and so $\rho \subseteq \mu$. Hence $\mu_{S/\mu}$ is trivial. □

As a further corollary, we have the following.

Corollary 5.6. *Let S, T be regular semigroups with maximum idempotent separating congruence μ_S, μ_T respectively, and let $\theta : S \to T$ be a surjective idempotent separating homomorphism. Then $S/\mu_S \cong T/\mu_T$.*

Proof. Let $\varphi : T \to T/\mu_T$ be the natural homomorphism, and let $\rho = \theta\varphi(\theta\varphi)^{-1}$. Then ρ is idempotent separating, so $\rho \subseteq \mu_S$ and μ_S/ρ is idempotent separating on S/ρ. But $S/\rho \cong T/\mu_T$, so that by Corollary 5.5, μ_S/ρ is trivial. Hence $\rho = \mu$. □

Lemma 5.7. *If S is an orthodox semigroup, then $\mathcal{H} \cap \mathcal{Y} = \iota$.*

Proof. Suppose that $a, b \in S$ with $a\ \mathcal{H}\ b$ and $a\ \mathcal{Y}\ b$ and let $a' \in V(a)$. From the proof of Proposition 2.4.1 of [21], $a'a = b'b$ for some $b' \in V(b)$. But $V(a) = V(b)$, so that $b' \in V(a)$, and, by the same argument, there is an inverse b'' of b such that $ab' = bb''$. Now we have

$$a = aa'a = ab'b = bb''b = b.$$

□

If S is E-unitary, we have a stronger result [28].

Lemma 5.8. *If S is an E-unitary orthodox semigroup, then $\mathcal{H} \cap \sigma = \iota$ where σ is the minimum group congruence on S.*

Proof. Suppose that $a, b \in S$ with $a\ \mathcal{H}\ b$ and $a\sigma b$. From $a\ \mathcal{H}\ b$, it follows by Proposition 2.4.1 of [21], that there are inverses a', b' of a and b respectively such that $aa' = bb'$ and $a'a = b'b$. Hence $a'\ \mathcal{H}\ b'$, and since $L_a \cap R_{a'}$ contains an idempotent, we have $ab'\ \mathcal{H}\ aa'$.

From $a\sigma b$, we get $ab'\sigma bb'$. By Lemma 3.11, $E(S)$ is a σ-class, and so ab' is idempotent. Hence $ab' = aa'$. Similarly, $b'a = b'b$. Thus

$$a = aa'a = ab'a = ab'b = aa'b = bb'b = b.$$

□

Next, we note that if $\theta : T \to S$ is an idempotent separating surjective homomorphism of orthodox semigroups, then by Lallement's lemma, $\theta|_{E(T)}$ is an isomorphism from $E(T)$ onto S. Thus to prove that an E-unitary orthodox semigroup T is an E-unitary cover of an orthodox semigroup S, it is enough to show that there is an idempotent separating homomorphism from T onto S.

Theorem 5.9. *If S is a (finite) orthodox semigroup, then there is a (finite) E-unitary orthodox cover $\widehat{S}$ of S.*

Proof. Let $I = S/\mathcal{Y}$ and let $\gamma : S \to I$ be the natural homomorphism. By Proposition 4.8, I has an E-unitary inverse cover P with covering homomorphism $\alpha : P \to I$; moreover, if S is finite, then I is finite, and P can be chosen to be finite. Let

$$\widehat{S} = \{(s\mu, p) \in S/\mu \times P : s \in S \text{ and } p\alpha = s\gamma\}.$$

Clearly, if S is finite, then so is $\widehat{S}$. It is straightforward to verify that $\widehat{S}$ is a regular subsemigroup of the direct product $S/\mu \times P$.

We claim that if $(s\mu, p) \in \widehat{S}$ and $p \in E(P)$, then $s \in E(S)$.

To prove the claim, note that under the hypotheses, $s\gamma = p\alpha \in E(I)$. By Lallement's lemma, $s\gamma = e\gamma$ for some idempotent e of S. By the definition of γ, this gives $V(s) = V(e)$, and hence by Lemma 5.2, $V(s) \subseteq E(S)$. Hence s is an inverse of an idempotent, so that, again by Lemma 5.2, $s \in E(S)$.

It follows that

$$E(\widehat{S}) = \{(e\mu, f) : e \in E(S), f \in E(P) \text{ and } e\gamma = f\alpha\},$$

and hence that $\widehat{S}$ is orthodox.

Next we show that $\widehat{S}$ is E-unitary. Since P is E-unitary, there is, by Lemma 3.11, a homomorphism $\beta : P \to G$ onto a group G with $E(P) = 1\beta^{-1}$. Define $\psi : \widehat{S} \to G$ by the rule:

$$(s\mu, p)\psi = p\beta.$$

If $(s\mu, p)\psi = 1$, then $p \in 1\beta^{-1} = E(P)$, and so, by the claim, s is idempotent. Hence $E(\widehat{S}) = 1\psi^{-1}$, and so, by Lemma 3.11, $\widehat{S}$ is E-unitary.

To see that $\widehat{S}$ is a cover, define $\theta : \widehat{S} \to S$ by $(s\mu, p)\theta = s$. If $s\mu t$ and $(t\mu, p) \in \widehat{S}$, then $s\gamma = p\alpha = t\gamma$ so that $(s, t) \in \mu \cap \gamma\gamma^{-1} = \mu \cap \mathcal{Y}$. Hence, by Lemma 5.7, $s = t$, and θ is well defined. Clearly, it is a surjective homomorphism. Suppose that $(e\mu, p), (f\mu, q)$ are idempotents of $\widehat{S}$ with $(e\mu, p)\theta = (f\mu, q)\theta$. Then $e = f$, that is, $p\alpha = q\alpha$. But $p, q \in E(P)$ and α is idempotent separating, so $p = q$. Thus θ is idempotent separating, and $\widehat{S}$ is an E-unitary cover of S. $\square$

We now prove the "if" part of Theorem 5.1.

Proposition 5.10. *Let S be a finite orthodox semigroup. If $\mathcal{H}$ is a congruence on S, then $S \in \mathbf{A} \vee \mathbf{G}$.*

Proof. By Theorem 5.9, S has a finite E-unitary orthodox cover $\widehat{S}$. Let σ be the minimimum group congruence on $\widehat{S}$, and let $G = \widehat{S}/\sigma$. By Lemma 5.8, $\mathcal{H} \cap \sigma = \iota$ so that $\mu \cap \sigma = \iota$, and hence, by Proposition 1.1, $\widehat{S}$ can be embedded (as a subdirect product) in $\widehat{S}/\mu \times G$.

Now S is an idempotent separating homomorphic image of $\widehat{S}$, and so, by Corollary 5.6, $\widehat{S}/\mu \cong S/\mu$. By assumption, $\mu = \mathcal{H}$ on S, so $\widehat{S}/\mu \cong S/\mathcal{H}$, and thus $\widehat{S}$ can be embedded in $S/\mathcal{H} \times G$. Since $\mathcal{H}$ is a congruence on S, it follows from Lemma 5.4 that $\mathcal{H}$ is trivial on $S/\mathcal{H}$, so $S/\mathcal{H} \in \mathbf{A}$. Hence $S/\mathcal{H} \times G \in \mathbf{A} \vee \mathbf{G}$, and as pseudovarieties are closed under taking subsemigroups and homomorphic images, we have $S \in \mathbf{A} \vee \mathbf{G}$ as required. □

In [29] McAlister generalised Theorem 5.1 to get the following result for regular semigroups.

Theorem 5.11. *Let S be a finite regular semigroup. Then $S \in \mathbf{A} \vee \mathbf{G}$ if and only if $D(S) \in \mathbf{A}$ and $\mathcal{H}$ is a congruence on S.*

6 Generalisations and Related Topics

We conclude the paper by giving a brief account of some other work related to covers. We do not give proofs but point the reader to some of the relevant literature. We start by describing some aspects of covers over monoids other than groups. We then mention the work of Auinger and Trotter on covers over groups belonging to a given variety of groups. Finally, we discuss finite covers of finite semigroups.

6.1 Left ample and weakly left ample semigroups

For a set X, we define the operation $^+$ on the monoid $PT(X)$ of all partial transformations on X by taking α^+ to be the identity mapping on the domain of α. Let S be a semigroup with a unary operation $^+$ such that $e = e^+$ for every idempotent e of S. Then S is said to be *weakly left ample* if there is a $(2, 1)$-algebra embedding of S into $PT(X)$ for some set X.

Note that $I(X)$, the symmetric inverse monoid on X, is a $(2, 1)$-subalgebra of $PT(X)$ and that $\alpha^+ = \alpha\alpha^{-1}$ for all $\alpha \in I(X)$. If S is a semigroup with a unary operation $^+$ and there is a $(2, 1)$-algebra embedding of S into $I(X)$, we say that S is *left ample*. It is easy to see that left ample semigroups are weakly left ample. On an inverse semigroup S, we can define a unary operation $^+$ by $a^+ = aa^{-1}$ for $a \in S$, and then the Vagner-Preston representation shows that S is left ample.

In these definitions we assume that the partial transformations are written on the right of their arguments. By using the dual monoids with the partial transformations written on the left of their arguments, we get definitions of weakly right ample and right ample semigroups. In this case, the unary operation is written as *.

It is immediate from the definition that, in a weakly left ample semigroup, the idempotents commute with each other, and so $E(S)$ is a subsemilattice of S.

On a weakly left ample semigroup S, there is a least congruence σ such that S/σ is a unipotent monoid, that is, the identity is the only idempotent of S/σ. If S is actually left ample, then S/σ is a right cancellative monoid. In both cases (weakly left ample and left ample), S is said to be *proper* if for all elements a and b of S such that $a^+ = b^+$ and $a\sigma b$, we have $a = b$.

It is well known that an inverse semigroup is proper if and only if it is E-unitary (see, for example, [21, Proposition 5.9.1]). Example 3 of [9] shows that the corresponding statement does not hold for left ample semigroups.

Let S be a left ample semigroup and T be a right cancellative monoid. We say that a left ample semigroup P is a *proper cover* of S (*over* T) if P is proper and there is a surjective $(2,1)$-algebra homomorphism from P onto S which maps $E(P)$ isomorphically onto $E(S)$ (and is such that $P/\sigma \cong T$). A *proper cover* of a weakly left ample semigroup over a unipotent monoid is defined similarly.

The existence of a proper cover for a right ample monoid S was established in [9]; the result is easily extended to the case where S is a semigroup, and, of course, the left ample results are simply the duals. Moreover, the cover is finite if S is finite. Thus we have the following analogue for left ample semigroups of Proposition 4.8.

Proposition 6.1. *A left ample semigroup S has a proper left ample cover. Moreover, if S is finite, then the cover can be chosen to be finite.*

In the weakly left ample case, the existence of proper weakly left ample covers was proved in [14]. The fact that a finite weakly left ample semigroup has a finite cover was first shown in [15] using elementary methods, and subsequently obtained, as a consequence of a more general result, in [4] using a sophisticated result of Ash. Moreover, it was shown in [15], that a finite proper weakly left ample semigroup is actually left ample. Combining these results we have the following result.

Proposition 6.2. *A weakly left ample semigroup S has a proper weakly left ample cover. Moreover, if S is finite, then the cover can be chosen to be finite, and is left ample.*

6.2 Covers over group varieties

In Section 3 we showed that an E-dense semigroup S has a D-unitary E-dense cover $\widehat{S}$ over a group G; in fact, G is the maximum group homomorphic image of $\widehat{S}$ and, if $\beta : \widehat{S} \to G$ is the natural homomorphism, $D(S) \cong D(\widehat{S}) = 1\beta^{-1}$. If **H** is a variety of groups, and we want our cover to be over a group in **H**, what subsemigroup do we use instead of $D(S)$, and what relational morphism do we use? These questions have recently been answered by Auinger and Trotter in [5]. We restrict ourselves to describing the subsemigroup and the relational morphism, and stating some of the main theorems. We refer the reader to [5] for proofs.

Let **H** be a variety of groups, S be an E-dense semigroup and X be a countably infinite set. Let $X^{-1} = \{x^{-1} : x \in X\}$ be a set disjoint from X and such that $x \mapsto x^{-1}$ is a bijection. Let $\widetilde{X} = X \cup X^{-1}$ and let $\widetilde{X}^+$ be the free semigroup on $\widetilde{X}$.

For a function $\varphi : X \to S$ and element x of X, put
$$x\overline{\varphi} = \{x\varphi\} \text{ and } x^{-1}\overline{\varphi} = W(x\varphi),$$
and, for $x_1, \ldots, x_n \in \widetilde{X}$, put
$$(x_1 \ldots x_n)\overline{\varphi} = x_1\overline{\varphi} \ldots x_n\overline{\varphi}.$$
For $w \in \widetilde{X}^+$, say that $w\overline{\varphi}$ is the *set of values of w under the substitution φ*. If $w \simeq 1$ is a law in **H**, write $\mathbf{H} \models w \simeq 1$. Now put
$$C_{\mathbf{H}}(S) = \bigcup \{w\overline{\varphi} : w \in \widetilde{X}^+, \mathbf{H} \models w \simeq 1 \text{ and } \varphi : X \to S \text{ is a substitution}\},$$
that is, $C_{\mathbf{H}}(S)$ is the set of all values in S of all words w for which $w \simeq 1$ is a law in **H**. It turns out that $C_{\mathbf{H}}(S)$ is a full weakly self conjugate subsemigroup of S. Moreover, the analogue of Theorem 3.5 holds, that is, $C_{\mathbf{H}}(S)$ contains every weak inverse of each of its elements, so that, in particular, $C_{\mathbf{H}}(S)$ is E-dense, and it is regular if S is regular.

We now describe the relational morphism used to get the covering result. Given an E-dense semigroup S, let $X_S = \{x_s : s \in S\}$ be a copy of S disjoint from S, and $X_S^{-1} = \{x^{-1} : x \in X_S\}$ be a copy of X_S disjoint from $X_S \cup S$.

Let $\widetilde{X_S} = X_S \cup X_S^{-1}$, and $\rho_{\mathbf{H}}$ be the congruence on $\widetilde{X_S}^+$ such that $\widetilde{X_S}^+/\rho_{\mathbf{H}}$ is the relatively free group $F_{\mathbf{H}}(X_S)$ in $\mathbf{H}$ on X_S. Finally, let $\varphi_S : X_S \to S$ be the substitution given by $x_s\varphi_S = s$, and for each $g \in F_{\mathbf{H}}(X_S)$, define

$$g\tau_{\mathbf{H}} = \bigcup\{w\overline{\varphi}_S : w \in \widetilde{X_S}^+ \text{ and } w\rho_{\mathbf{H}} = g\}.$$

Then $\tau_{\mathbf{H}} : F_{\mathbf{H}}(X_S) \multimap S$ is a surjective relational morphism such that $1\tau_{\mathbf{H}} = C_{\mathbf{H}}(S)$, and so $\tau_{\mathbf{H}}^{-1} : S \multimap F_{\mathbf{H}}(X_S)$ is a $C_{\mathbf{H}}(S)$-pure surjective relational morphism, and this is the relational morphism which is used to prove the following result.

Theorem 6.3. *An E-dense semigroup S, has an E-dense $C_{\mathbf{H}}(S)$-unitary cover $\widehat{S}$ over the relatively free group $F_{\mathbf{H}}(X_S)$ with $\widehat{C_{\mathbf{H}}(S)} = C_{\mathbf{H}}(\widehat{S})$. Moreover, if S is regular, then so is $\widehat{S}$.*

When we consider the variety $\mathbf{G}$ of all groups, we have $C_{\mathbf{G}}(S) = D(S)$, and so Theorem 3.9 is a special case of Theorem 6.3.

6.3 Finite semigroups

As we have repeatedly stressed, our general construction of covers always gives an infinite cover. We have, however, also shown that finite inverse and orthodox semigroups have finite E-unitary covers, and mentioned that finite regular semigroups have finite regular D-unitary covers. The existence of a finite D-unitary cover for an arbitrary finite semigroup is a much deeper result as we now explain.

First, for a *finite* semigroup S, we redefine $K(S)$ to be the intersection of all subsemigroups K of S such that $K = 1\tau^{-1}$ for some relational morphism τ into a *finite* group. It is shown in [33] that there is such a τ with $K(S) = 1\tau^{-1}$, and as $E(S) \subseteq 1\tau^{-1}$ for any such τ, we see that $K(S)$ is a full weakly self conjugate subsemigroup of S so that $D(S) \subseteq K(S)$. We now give the straightforward connection with covers, quoting from [44] and [12].

Proposition 6.4. *For a finite semigroup S, the following conditions are equivalent:*

(1) S has a finite D-unitary cover;

(2) $D(S) = K(S)$.

Proof. If (1) holds, then the cover must be over a finite group G, and by Proposition 1.2, $D(S) = 1\tau^{-1}$ for some relational morphism τ into G. Hence $K(S) \subseteq D(S)$ and (2) holds.

If (2) holds, then $D(S) = K(S) = 1\tau^{-1}$ for some relational morphism τ into a finite group, and by Proposition 1.2, $\mathrm{gr}(\tau)$ is a D-unitary cover for S, so that (1) holds. □

The conjectured truth of (2) was one of the major open problems during the 1970s and 1980s, known as the *type II conjecture*. It was finally proved by Ash [3], as a consequence of a much more general result. Shortly afterwards Ribes and Zalesskiĭ [36] proved that finite products of finitely generated subgroups of a free group are closed in the profinite topology of the group. That this result implied the type II conjecture had already been shown by Pin and Reutenauer [34]. More recently, Herwig and Lascar [19] obtained a result on the extendability of partial automorphisms of a relational structure to automorphisms of a containing structure, a consequence of which is the Ribes-Zalesskiĭ theorem. Thus this gives a third proof of the type II conjecture. The methods used in the three proofs have been significantly developed, and the articles by Almeida, Ribes and Coulbois in this volume provide excellent introductions to these topics.

Acknowledgements

I would like to thank the organisers (Gracinda Gomes, Jean-Éric Pin and Pedro Silva) of the Thematic Term on Semigroups, Algorithms, Automata and Languages for inviting me to speak at one of the summer schools at the Centro Internacional de Mátematica in Coimbra.

References

[1] J. Almeida and A. Escada, On the equation $\mathbf{V} * \mathbf{G} = \mathcal{E}\mathbf{V}$, *J. Pure Appl. Alg.* **166** (2002), 1–28.
[2] J. Almeida, J.-É. Pin, and P. Weil, Semigroups whose idempotents form a subsemigroup, *Math. Proc. Camb. Phil. Soc.* **111** (1992), 241–253.
[3] C. J. Ash, Inevitable graphs: A proof of the type II conjecture and some related decision procedures, *Internat. J. Algebra Comput.* **1** (1991), 127–146.
[4] K. Auinger, G. M. S. Gomes, V. Gould and B. Steinberg, An application of a theorem of Ash to finite covers, preprint, http://www-users.york.ac.uk/~varg1/
[5] K. Auinger and P. G. Trotter, A syntactic approach to covers for E-dense semigroups over group varieties, 2000, preprint.

[6] F. Catino and M. M. Miccoli, On semidirect product of semigroup, *Note Mat.* **9** (1989), 189–194.

[7] P. Dubreil, Contribution à la théorie des demigroupes, *Mém. Acad. Sci. Inst. France* **63** (1941), no. 3, 52pp.

[8] S. Eilenberg, *Automata, Languages and Machines* Vol B, Academic Press, 1976.

[9] J. Fountain, A class of right PP monoids, *Quart. J. Math. Oxford* **28** (1977), 285–300.

[10] J. Fountain, E-unitary dense covers of E-dense monoids, *Bull. London Math. Soc.* **22** (1990), 353–358.

[11] J. Fountain and A. Hayes, E^*-dense semigroups whose idempotents form a subsemigroup, preprint 2002.

[12] J. Fountain, J.-É. Pin and P Weil, Covers for monoids, *J. Algebra*, to appear.

[13] G. M. S. Gomes, A characterization of the group congruences on a semigroup, *Semigroup Forum* **46** (1993), 48–53.

[14] G. M. S. Gomes and V. Gould, Proper weakly left ample semigroups, *Internat. J. Algebra Comput.* **9** (1999), 721–739.

[15] G. M. S. Gomes and V. Gould, Finite proper covers in a class of finite semigroups with commuting idempotents, *Semigroup Forum*, to appear.

[16] T. E. Hall, On regular semigroups whose idempotents form a subsemigroup, *Bull. Austral. Math. Soc.* **1** (1969), 195–208.

[17] T. E. Hall, Congruences and Green's relations on regular semigroups, *Glasgow Math. J.* **20** (1972), 167–175.

[18] T. E. Hall and W. D. Munn, The hypercore of a semigroup, *Proc. Edinburgh Math. Soc.* **28** (1985), 107–112.

[19] B. Herwig and D. Lascar, Extending partial automorphisms and the profinite topology on free groups, *Trans. Amer. Math. Soc.* **352** (2000), 1985–2021.

[20] J. M. Howie, *An introduction to semigroup theory*, Academic Press, 1976.

[21] J. M. Howie, *Fundamentals of semigroup theory*, Oxford University Press, 1995.

[22] M. V. Lawson, *Inverse semigroups*, World Scientific, 1998.

[23] F. W. Levi, On semigroups, *Bull. Calcutta Math. Soc.* **36** (1944), 141–146.

[24] F. W. Levi, On semigroups II, *Bull. Calcutta Math. Soc.* **38** (1946), 123–124.

[25] S. W. Margolis and J.-É. Pin, Inverse semigroups and extensions of groups by semilattices, *J. Algebra* **110** (1987), 277–297.

[26] D. B. McAlister, Groups, semilattices and inverse semigroups, *Trans. Amer. Math. Soc.* **192** (1974), 227–244.
[27] D. B. McAlister, Groups, semilattices and inverse semigroups, II, *Trans. Amer. Math. Soc.* **196** (1974), 351–370.
[28] D. B. McAlister, On a problem of M.P.Schützenberger, *Proc. Edinb. Math. Soc.* **23** (1980), 243–247.
[29] D. B. McAlister, Regular semigroups, fundamental semigroups and groups, *J. Austral. Math. Soc. Ser. A* **29** (1980), 475–503.
[30] D. B. McAlister and N. R. Reilly, E-unitary covers for inverse semigroups, *Pacific. J. Math.* **68** (1977), 161–174.
[31] H. Mitsch, Subdirect products of E-inversive semigroups, *J. Austral. Math. Soc.* **48** (1990), 66–78.
[32] M. Petrich, *Inverse semigroups*, Wiley, New York, 1984.
[33] J.-É. Pin, On a conjecture of Rhodes, *Semigroup Forum*, **39** (1989), 1–15.
[34] J.-É. Pin and Ch. Reutenauer, A conjecture on the Hall topology for the free group, *Bull. London Math. Soc.* **23** (1991), 356–362.
[35] N. R. Reilly, Bisimple ω-semigroups, *Proc. Glasgow Math. Assoc.* **7** (1965), 183–187.
[36] L. Ribes and Zalesskiĭ, On the profinite topology on the free group, *Bull. London Math. Soc.* **25** (1993), 37–43.
[37] B. M. Schein, Completions, translational hulls, and ideal extensions of inverse semigroups, *Czech. J. Math.* **23** (1973), 575–610.
[38] B. Steinberg, Semidirect products of categories and applications, *J. Pure Appl. Alg.* **142** (1999), 153–182.
[39] B. Steinberg and B. Tilson, Categories as algebra II, preprint, http://www.fc.up.pt/cmup/cmup-eng.html.
[40] M. B. Szendrei, On a pull-back diagram for orthodox semigroups, *Semigroup Forum* **20** (1980), 1–10, corr. **25** (1982), 311–324.
[41] K. Takizawa, Orthodox semigroups and E-unitary regular semigroups, *Bull. Tokyo Gakugei Univ., Series IV* **31** (1979), 41–43.
[42] G. Thierrin, Demigroupes inversés et rectangulaires, *Bull. Cl. Sci. Acad. Roy. Belgique* **41** (1955), 83–92.
[43] P. G. Trotter, Covers for regular semigroups and an application to complexity, *J. Pure Appl. Alg.* **105** (1995), 319–328.
[44] P. G. Trotter and Zhonghao Jiang, Covers for regular semigroups, *Southeast Asian Bull. Math.* **18** (1995), 319–328.
[45] B. Weipoltshammer, *Kongruenzen auf E-inversiven E-Halbgruppen*, Dissertation, Universität Wien, 2000.
[46] Zhonghao Jiang, E-unitary inversive covers for E-inversive semigroups

whose idempotents form a subsemigroup, *Southeast Asian Bull. Math.*
18 (1994), 59–64.

E^*-UNITARY INVERSE SEMIGROUPS

MARK V. LAWSON*

School of Informatics, University of Wales, Bangor, Dean Street, Bangor,
Gwynedd, United Kingdom
E-mail: m.v.lawson@bangor.ac.uk

With each inverse monoid with zero S we associate a category $C(S)$ which is a slight modification of the category Leech associated with an inverse monoid and a special case of the author's 'category action approach' to inverse semigroups. Using this category, we obtain characterisations of E^*-unitary, strongly E^*-unitary and E-unitary inverse monoids. Specifically, S is E^*-unitary if and and only if $C(S)$ is cancellative; S is strongly E^*-unitary if and only if $C(S)$ can be embedded in a groupoid; and, finally, S is E-unitary (with a zero adjoined) if and only if $C(S)$ has a groupoid of fractions. We also introduce the 'universal group of an inverse semigroup' and show how this group can be used to characterise strongly E^*-unitary semigroups.

1 Introduction

We shall assume that the reader is familiar with the basic theory of inverse semigroups. We use one piece of idiosyncratic notation: if s is an element of an inverse semigroup then $\mathbf{d}(s)$ denotes $s^{-1}s$ and $\mathbf{r}(s)$ denotes ss^{-1}. For undefined terms and standard theorems see [17].

Let S and T be semigroups with zero. A function $\theta\colon S \to T$ is said to be *0-restricted* if $\theta^{-1}(0) = \{0\}$. It is said to be a *0-morphism* if it is 0-restricted and $\theta(ab) = \theta(a)\theta(b)$ when $ab \neq 0$. It is said to be a *prehomomorphism* if $\theta(ab) \leq \theta(a)\theta(b)$. The function θ is *idempotent pure* if $\theta(s)$ an idempotent implies that s is an idempotent. A *0-group* is simply a group with a zero adjoined; if G is a group then G^0 is the group with a zero adjoined.

An inverse semigroup with zero S is said to be E^*-*unitary* if $0 \neq e \leq s$, where e is an idempotent, implies that s is an idempotent. An inverse semigroup S is said to be *strongly* E^*-*unitary* if there is a 0-restricted idempotent pure prehomomorphism from S to a 0-group. It is easy to check that strongly

---

*THE AUTHOR WOULD LIKE TO ACKNOWLEDGE THE FINANCIAL SUPPORT OF FUNDAÇÃO CALOUSTE GULBENKIAN (FCG), FUNDAÇÃO PARA A CIÊNCIA E A TECNOLOGIA (FCT), FACULDADE DE CIÊNCIAS DA UNIVERSIDADE DE LISBOA (FCUL) AND REITORIA DA UNIVERSIDADE DO PORTO."

E^*-unitary semigroups are E^*-unitary. An inverse semigroup S is said to be *E-unitary* if $e \leq s$ where e is an idempotent implies s is an idempotent. Every E-unitary inverse semigroup is strongly E^*-unitary. An inverse semigroup is F^*-*inverse* if every non-zero element is beneath a unique maximal element. It is easy to check that every F^*-inverse semigroup is E^*-unitary. An F^*-inverse semigroup which is also strongly E^*-unitary will be called *strongly F^*-inverse*. An F^*-inverse semigroup generated by its maximal elements is said to be *max-generated*. Mária Szendrei [34] introduced E^*-unitary semigroups, and Meakin and Sapir pointed out that the polycyclic monoids were E^*-unitary in [26]. Strongly categorical E^*-unitary inverse semigroups were described in [5], but the real breakthrough came with the introduction of strongly E^*-unitary inverse semigroups independently by Bulman-Fleming, Fountain and Gould [2] and by the author [19]. The class of F^*-inverse semigroups was introduced by Nica [28].

Examples 1.1.

(1) **Toeplitz inverse semigroups**
This construction is due to Nica [28]. Let G be a non-trivial group, and let P be an arbitrary submonoid of G. For each $g \in G$, we may define a bijection $\lambda_g \colon g^{-1}P \cap P \to gP \cap P$ by $\lambda_g(x) = gx$. The *Toeplitz inverse semigroup* $\mathcal{T}_{(G,P)}$ is the inverse submonoid of the symmetric inverse monoid $I(P)$ generated by the λ_g. Nica proves that these semigroups are F^*-inverse (Lemma 3.2 [28]), and it is easy to verify that they are in fact max-generated and strongly F^*-inverse. Stuart Margolis and the author also showed that they are 0-simple.

(2) **Graph inverse semigroups**
In [18], the author showed how to construct an inverse semigroup from a directed graph, called a 'graph inverse semigroup'. Subsequently, we learnt that this construction had first been introduced by Ash and Hall [1] in the case of directed graphs in which there was at most one arrow from one vertex to another. More recently, Lenz [22] showed how work on a class of groupoid C^*-algebras by Kumjian et al [16] could be interpreted in terms of graph inverse semigroups. They are defined as follows. Let G be a directed graph and let G^* be the free category generated by G. This is a cancellative category, and for each pair $a, b \in G^*$, it is easy to check that

$$G^*a \cap G^*b \neq \emptyset \Rightarrow G^*a \subseteq G^*b \text{ or } G^*b \subseteq G^*a.$$

By the construction of [18], which is a slight extension of the construction we describe in Section 3, we may construct the inverse semigroup with zero $\mathcal{S}(G) = C(G^*)$. These semigroups are max-generated and strongly F^*-inverse. In addition, they are combinatorial[a] and unambiguous except at zero. By generalising a theorem of Ash and Hall [1], it is possible to show that a graph inverse semigroup $\mathcal{S}(G)$ is congruence-free if and only if G is strongly connected and no vertex of G has out-degree equal to 1. This generalises the classical result that whereas the bicyclic monoid has congruences the polycyclic monoids are congruence-free [29]; in terms of our result, we can see that this is because the bicyclic monoid is the graph inverse semigroup of the graph with one vertex and one loop, whereas the polycyclic monoids are the graph inverse semigroups of the graphs with one vertex and more than one loop.

(3) **Tiling semigroups**
Kellendonk [12] showed how to associate an inverse semigroup, called the tiling semigroup, with every tiling in $\mathbb{R}^n$. He did this by making precise the naive idea of constructing a semigroup from a tiling in which the semigroup elements would be the patterns and where multiplication would be 'joining patterns together'. Kellendonk was motivated by some specific questions concerned with the physics of quasi-crystals. Tiling semigroups are always strongly E^*-unitary, and the tiling semigroups of 1-dimensional tilings are strongly F^*-inverse. The structure of 1-dimensional tiling semigroups can be analysed in terms of an old paper of Don McAlister's [24]. Specifically, each 1-dimensional tiling semigroup S is a semigroup of strong quotients of a subsemigroup C, which is 0-cancellative and unambiguous except at zero. The embedding of C in S is such that S is separated over C by the inclusion, and this inclusion is the initial object in the category whose objects consist of all homomorphisms $\theta: C \to T$ to inverse semigroups T such that T is separated over C by θ.

Thus in the 1-dimensional case, tiling semigroups can be described in terms of much simpler kinds of semigroups. However, this description does not appear to generalise to higher dimensions.

The examples above demonstrate the importance of strongly E^*-unitary inverse semigroups. However, arbitrary E^*-unitary inverse semigroups are also interesting. Kellendonk [13] showed how to associate a topological groupoid with each inverse semigroup. As a topological space it is always T_1 but in the case of E^*-unitary inverse semigroups it is hausdorff.

The remainder of this paper deals with the following topics.

---

[a] Aperiodic

In Section 2, we shall show how to construct the universal group of an inverse semigroup, and investigate some of the implications of this result for studying E^*-unitary semigroups.

In Section 3, we shall use some simple category theory to gain some insight into the relationships between E^*-unitary, strongly E^*-unitary and E-unitary monoids.

Finally, in Section 4, we survey some of the structural results on strongly E^*-unitary semigroups.

2 The universal group

The congruence σ is defined on the inverse semigroup S by
$$a\,\sigma\,b \Leftrightarrow \exists c \leq a, b.$$
It is well-known that σ is the minimum group congruence on an inverse semigroup. Amongst its applications, it is useful in studying E-unitary inverse semigroups because of the following well-known result.

Proposition 2.1. *An inverse semigroup semigroup is E-unitary if and only if its minimum group congruence is idempotent pure.*

Observe that if an inverse semigroup S has a zero then the group S/σ is trivial. This means that we cannot hope to find an analogue of this result for strongly E^*-unitary semigroups using congruences and therefore homomorphisms. The following result tells us that to study strongly E^*-unitary semigroups we need to study (0-restricted) prehomomorphisms; these are known to be a reasonable generalisation of homomorphisms [17], and we shall use them to generalise Proposition 2.1.

Lemma 2.2. *Let $\theta\colon S \to G^0$ be a 0-restricted function between an inverse semigroup S and a group with zero G^0. Then θ is a prehomomorphism if and only if $\theta(ab) = \theta(a)\theta(b)$ whenever $ab \neq 0$.*

Proof. Let θ be a prehomomorphism, and let $a, b \in S$ such that $ab \neq 0$. By definition, $\theta(ab) \leq \theta(a)\theta(b)$ and $\theta(ab) \neq 0$. Thus $\theta(ab), \theta(a)\theta(b) \in G$. Hence $\theta(ab) = \theta(a)\theta(b)$, and so θ is a 0-morphism. Conversely, suppose $\theta(ab) = \theta(a)\theta(b)$ whenever $ab \neq 0$. Let $ab = 0$. Then $\theta(ab) = 0$. If the product $\theta(a)\theta(b) = 0$ then $\theta(ab) = \theta(a)\theta(b)$; if the product $\theta(a)\theta(b) \neq 0$ then $\theta(ab) \leq \theta(a)\theta(b)$. It follows that θ is a prehomomorphism. □

Before continuing, we need to deal with the inelegant fashion we are currently handling the zero. This can be done at the cost of working with partial semigroups, in the following way, originally recommended to the author by Johannes Kellendonk.

Let S be a semigroup with zero. If $A \subseteq S$, then define $A^* = A \setminus \{0\}$. The set S^* has a partial multiplication induced by the semigroup operation; we call this structure a *presemigroup*. An inverse presemigroup is what Kellendonk [12] referred to as 'almost groupoid'. An inverse presemigroup S becomes an inverse semigroup with zero S^0 when a zero is adjoined and all undefined products in the presemigroup are defined to be zero. Observe that if S is an inverse semigroup then $S^{*0} = S$. The set of idempotents $E(S^*)$ is not of course a semilattice; however, it does have the property that any two elements with a lower bound have a greatest lower bound. We shall say that any poset with this property is a *presemilattice*. A *morphism* θ from an inverse presemigroup S to an inverse presemigroup T is a function $\theta\colon S \to T$ such that if $a, b \in S$ and ab is defined then $\theta(a)\theta(b)$ is defined and $\theta(ab) = \theta(a)\theta(b)$. It is obvious that there is a bijection between the set of morphisms from S to T and the set of 0-morphisms from S^0 to T^0. From now on we shall work almost exclusively with inverse presemigroups. Definitions made for 0-morphisms can easily be rephrased as definitions for morphisms. Observe that a morphism between groups is just a homomorphism. A *grading* on an inverse semigroup S is a morphism from S^* to a group G. Finally, we say that two inverse presemigroups S and T are *isomorphic* if their corresponding inverse semigroups S^0 and T^0 are isomorphic. This is equivalent to saying that there is a bijective morphism $\theta\colon S \to T$ such that ab is defined in S if and only if $\theta(a)\theta(b)$ is defined in T. In terms of this terminology, an inverse semigroup S is strongly E^*-unitary iff S^* has an idempotent pure grading.

We can now develop an analogue of the minumum group congruence for groups with zero.

Let S be an inverse semigroup with zero. Let $FG(S^*)$ be the free group on the set S^*; we denote the product in $FG(S^*)$ by $\cdot$. Let $\iota\colon S \to FG(S^*)$ be the function which maps $a \in S^*$ to the reduced string (a). Let $\sim$ be the congruence on $FG(S^*)$ generated by the pairs $((a)\cdot(b), (ab))$ where ab is defined in S^*. We enclose $\sim$-equivalence classes in brackets [,]. Let $FG(S^*)/\sim\, = G(S)$ and let ν be the natural map from $FG(S^*)$ to $G(S)$. Put $\tau = \nu\iota$.

Theorem 2.3. *With the definitions above τ is a morphism. If $\alpha\colon S^* \to H$ is any morphism to a group H then there is a unique homomorphism $\beta\colon G(S) \to H$ such that $\beta\tau = \alpha$.*

Proof. We show first that τ is a morphism. Let $a, b \in S^*$ such that ab is defined. Then by the definition of $\sim$ we have that $(a) \cdot (b) \sim (ab)$, and so $[(a) \cdot (b)] = [(ab)]$ in $G(S)$. However $\tau(c) = [(c)]$ for any $c \in S^*$, and $[(a)][(b)] = [(a) \cdot (b)]$ since $\sim$ is a congruence on $FG(S^*)$. Thus

$$\tau(a)\tau(b) = [(a)][(b)] = [(a) \cdot (b)] = [(ab)] = \tau(ab).$$

Let $\alpha\colon S^* \to H$ be any morphism to a group H. Then by the definition of the free group there is a unique homomorphism $\gamma\colon FG(S^*) \to H$ such that $\gamma\iota = \alpha$. Let $a, b \in S^*$ such that ab is defined. Then $\alpha(ab) = \alpha(a)\alpha(b)$, since α is a morphism. Thus $\gamma(\iota(ab)) = \gamma(\iota(a))\gamma(\iota(b)) = \gamma(\iota(a) \cdot \iota(b))$. It follows that $\sim$ is contained in $\ker\gamma$. There is therefore a unique homomorphism $\delta\colon G(S) \to H$ such that $\gamma = \delta\nu$. It follows that $\delta\tau = \alpha$. In addition, δ is the unique morphism with this property because $G(S)$ is generated by $\tau(S^*)$. □

We call $G(S)$ the *universal group* of the inverse semigroup S, and $\tau\colon S^* \to G(S)$ the corresponding *universal grading*.

The next result tells us that in the case of inverse semigroups without zero the above construction coincides with the minimum group congruence.

Proposition 2.4. *If S is an inverse semigroup without zero then $G(S) = S/\sigma$.*

Proof. Since S does not have a zero, a morphism $\alpha\colon S^* \to H$ is just a homomorphism $\alpha\colon S \to H$. It follows that $\ker\alpha$ is a group congruence on S and so $\sigma \subseteq \ker\alpha$. Thus there is a unique homomorphism β from S/σ to H such that $\beta\sigma^\natural = \alpha$. By standard category theory arguments S/σ is isomorphic to $G(S)$. □

We now give some examples of computing the universal group of some well-known inverse semigroups with zero.

Examples 2.1.

(1) Let B_2 be the Brandt semigroup with two non-zero idempotents e and f and two non-idempotent morphisms u and u^{-1} where $u^{-1}u = e$ and $uu^{-1} = f$. Define $\tau\colon B_2^* \to \mathbb{Z}$ by $\tau(e) = \tau(f) = 0$, and $\tau(u) = 1$ and $\tau(u^{-1}) = -1$. It is easy to check that τ determines a morphism from B_2^* to the group $\mathbb{Z}$. Suppose now that β is any morphism from B_2^* to a group H. Then β is determined by its value on u. Let γ be the function which maps 1 in $\mathbb{Z}$ to $\beta(u)$ in H. Since $\mathbb{Z}$ is the free group on 1, γ extends to a homomorphism $\gamma\colon \mathbb{Z} \to H$. It is evident that $\gamma\tau = \beta$ and that γ is unique satisfying this condition. It follows that $G(B_2) = \mathbb{Z}$.

(2) We can generalise Example 1. Let G be an arbitrary groupoid. Then G^0 is a primitive inverse semigroup, and every primitive inverse semigroup is isomorphic to one constructed in this way (see for example [17] page 95). The inverse presemigroup of G^0 is the groupoid G, and morphisms from G to groups are precisely functors from G to groups. It follows that the universal group of G^0 is the same as the universal group of the groupoid G.

(3) Consider I_2, the symmetric monoid on 2 letters. It is easy to check that the function from I_2^* to $\mathbb{Z}_2$ which takes each non-idempotent element to 1, and each idempotent to 0, determines a morphism. To see that $G(I_2) = \mathbb{Z}_2$, observe that I_2 is factorisable; that is, each element lies beneath an element in the group of units. We now make two observations. Let S be factorisable, and let $\tau\colon S^* \to G(S)$ be the universal group. Then first, τ restricted to the group of units of S is a homomorphism to $G(S)$; the proof of this is immediate. Second, $G(S)$ is a homomorphic image of $U(S)$, the group of units of S; to prove this, note that every non-zero element of S is beneath an element of $U(S)$. Thus the image of τ is determined by the images of the elements of $U(S)$. Returning to our example, there are exactly two bijections in $U(I_2)$ and so $G(I_2)$ will have at most two elements. Thus $G(I_2) = \mathbb{Z}_2$.

(4) Consider now I_n, the symmetric inverse monoid on n letters, for $n \geq 3$. In this case, the universal group is trivial. To see why this is so, we begin with an observation. Let $f \in I_n$ be any non-zero element. Then $f \leq g$ for some bijection g. We consider the element g: either $g(i) = i$ for some i in the set $\{1,\ldots,n\}$ or g fixes no element of $\{1,\ldots n\}$. In the first case, $e \leq g$ where e is the idempotent on the subset $\{i\}$. Therefore we have the following diagram where the dotted line represents the natural partial order:

In the second case, $g(1) = j$ where $j \neq 1$. Let h be the partial bijection $1 \to j$. Then $h \leq g$. Now the partial bijection $h \leq (1j)$, where $(1j)$ is the bijection which swaps 1 and j and fixes all other elements of $\{1,\ldots,n\}$. Because $n \geq 3$, there is a letter $k \neq 1, j$ which is fixed by $(1j)$. Let e be the idempotent on the subset $\{k\}$. Thus in the second case, we have the following

Let $\alpha\colon I_n^* \to G$ be any morphism to a group. By Lemma 2.2, we have that α is a prehomomorphism and so preserves the natural partial order. Let $f \in I_n^*$. Then in each of the two cases above, $\alpha(f) = \alpha(e)$. Thus every non-zero element of I_n is mapped to the identity of G. It follows that $G(I_n)$ is the trivial group for $n \geq 3$, as claimed.

(5) Let S be an inverse semigroup with central idempotents. Then S^* is isomorphic to the inverse presemigroup constructed from a presheaf of groups on the poset $E(S^*)$; that is, a functor F from $E(S^*)$ to the category of groups. We prove below that $G(S)$ is the colimit of F. Without loss of generality, $S^* = \bigcup_{e \in E} F(e)$, a disjoint union; there are group homomorphisms $\phi_{i,j}\colon F(i) \to F(j)$ when $i \geq j$; and the product in S^* is given by

$$st = \phi_{e,ef}(s)\phi_{f,ef}(t)$$

when ef exists and is otherwise undefined. Let $\alpha\colon S^* \to G$ be a morphism. Define $\alpha_e = (\alpha \,|\, F(e))$. This is clearly a homomorphism from $F(e)$ to G. Suppose now that $f \leq e$ so that $\phi_{e,f}\colon F(e) \to F(f)$. Let $x \in F(e)$. Then $\phi_{e,f}(x) \leq x$. Thus $\alpha(\phi_{e,f}(x)) = \alpha(x)$. It follows that $\alpha_f(\phi_{e,f}(x)) = \alpha_e(x)$. Hence, α induces a cone from the base F to the vertex G. Conversely, suppose that we have a cone from the base F to the vertex G. Thus we have a family of group homomorphisms $\tau_e\colon F(e) \to G$ satisfying the following condition: if $f \leq e$ then $\tau_f \phi_{e,f} = \tau_e$. Define $\tau\colon S^* \to G$ by $\tau(x) = \tau_e(x)$ if $x \in F(e)$. We show that τ is a morphism. Let $s, t \in S$ where $s \in F(e)$ and $t \in F(f)$ such that $st \neq 0$. Then $ef \neq 0$ and by definition $st = \phi_{e,ef}(s)\phi_{f,ef}(t)$. Now

$$\tau(st) = \tau_{ef}(st) = \tau_{ef}(\phi_{e,ef}(s))\tau_{ef}(\phi_{f,ef}(t)) = \tau_e(s)\tau_f(t) = \tau(s)\tau(t),$$

using the fact that the τ_i form a cone. Thus τ is a morphism.

We have established a bijection between morphisms from S^* to G, and cones from F to G. The proof of the result is now straightforward.

The examples above suggest that our definition of the universal group of an inverse semigroup with zero is a natural one. The following is now the analogue of Proposition 2.1.

Proposition 2.5. *An inverse semigroup S is strongly E^*-unitary if and only if the universal grading $\tau \colon S^* \to G(S)$ is idempotent pure.*

Proof. Suppose that S is strongly E^*-unitary. Then there exists an idempotent pure morphism $\alpha \colon S^* \to H$ to a group. From the universal property of τ there is a morphism $\beta \colon G(S) \to H$ such that $\beta\tau = \alpha$. Suppose that $\tau(s)$ is an idempotent. Then $\beta(\tau(s)) = \alpha(s)$ is an idempotent. Hence s is an idempotent in S^*. It follows that τ is idempotent pure. The converse is immediate. □

How can we decide if an E^*-unitary semigroup is strongly E^*-unitary? This question is undecidable [33] but that does not stop us looking for sufficient conditions. We shall now give some examples of this based on the ideas of this section which arose from conversations with Stuart Margolis.

Let S be a set equipped with a partial binary operation ∘. Following Jekel [9], we say that $(S, \circ)$ is *group-like* if the following three axioms hold:

(GL1) There is an element $1 \in S$ such that $\exists 1 \circ s$ and $\exists s \circ 1$ for all $s \in S$ and $1 \circ s = s = s \circ 1$.

(GL2) For each $s \in S$ there exists an element $s^{-1} \in S$ such that $\exists s^{-1} \circ s$ and $\exists s \circ s^{-1}$ and both products are equal to 1.

(GL3) If $\exists s \circ t = u$ then $\exists t^{-1} \circ s^{-1}$ and equals u^{-1}.

The *universal group* of the group-like set $(S, \circ)$ is the free group on the set S factored out by the congruence generated by the pairs $(st, s \circ t)$ when $s \circ t$ is defined.

Let S be an F^*-inverse monoid, and let $M(S)$ be the set of maximal elements of S. Define ∘ on $M(S)$ as follows: if $s, t \in M(S)$ and $st \neq 0$ then $s \circ t$ is defined to be the unique maximal element above st. It is easy to check that $(M(S), \circ)$ is a group-like set. The following result is straightforward to prove.

Proposition 2.6. *Let S be an F^*-inverse monoid with $(M, \circ)$ its group-like set of maximal elements. If M can be embedded in its universal group then S is strongly E^*-unitary.*

This result is useful because we need only look at the maximal elements of the inverse semigroup rather than all (non-zero) elements.

A special class of group-like sets are the pregroups of Stallings [31]. Each pregroup can be emebedded in its universal group. It is therefore an immediate consequence of Proposition 2.7, that every F^*-inverse monoid, whose maximal elements form a pregroup, is strongly F^*-inverse.

It is not unnatural to ask if the group-like set of maximal elements of an F^*-inverse does not always form a pregroup. The answer is no, and the counter-example comes from an unexpected quarter. Consider the set S of all real analytic, orientation-preserving homeomorphisms between the open intervals of $\mathbb{R}$. These functions form an inverse presemigroup. Analyticity implies that the semigroup is E^*-unitary. Furthermore, each element of S lies beneath a perforce unique maximal element and so S is F^*-inverse. In [9], Jekel shows that the maximal elements of S do not form a pregroup. However, in [8], Jekel proves that the maximal elements of S satisfy a weaker condition, 'the regularity condition', which implies that the group-like structure of maximal elements of S is embedded in its universal group. Thus S is strongly E^*-unitary. We believe that the ideas contained in Jekel's work [8],[9],[10] may be of great interest in developing further the theory of F^*-inverse semigroups.

3 A category approach

In this section, we shall obtain characterisations of E^*-unitary, strongly E^*-unitary and E-unitary inverse *monoids* in terms of a category which may be associated with any inverse semigroup. This category is a special case of [18] and a slight generalisation of [20]. We first discussed it in [19].

Let S be an inverse semigroup with zero. Put

$$C(S) = \{(e, s) \in E(S)^* \times S^* : ss^{-1} \leq e\}.$$

Define

$$\mathbf{d}(e, s) = (s^{-1}s, s^{-1}s) \text{ and } \mathbf{r}(e, s) = (e, e)$$

[b] and define a partial product on $C(S)$ by

$$(e, s)(f, t) = (e, st) \text{ iff } \mathbf{d}(e, s) = \mathbf{r}(f, t).$$

Theorem 3.1. *Let S be an inverse semigroup with zero. Then $C(S)$ is a left cancellative category. Any two elements of $C(S)$ with a common range which can be completed to a commutative square have a pullback. If S has an identity then $C(S)$ has a weak terminal identity.*

Proof. We show first that $C(S)$ is closed under the partial product. Let

$$(e, s), (f, t) \in C(S)$$

such that $(e, s)(f, t)$ is defined. By definition, $s^{-1}s = f$. We have to show that $(e, st) \in C(S)$. By assumption, $e \neq 0$. Suppose $st = 0$. Then $0 =$

---
[b]This is another meaning for the symbols **d** and **r** and should not cause any confusion.

$s^{-1}st = ft = f(tt^{-1})t = t$, since $tt^{-1} \leq f$. But we are given $t \neq 0$. Thus $st \neq 0$. Furthermore, $st(st)^{-1} = stt^{-1}s^{-1} \leq ss^{-1} \leq e$. Hence $(e, st) \in C(S)$, as required.

It is immediate that **d** and **r** are well-defined maps. The verification that $C(S)$ is a category is now straightforward.

To show that $C(S)$ is left cancellative, let $(e, s)(f, t) = (e, s)(i, u)$. Then $f = s^{-1}s = i$, and so $f = i$, and $st = su$. Thus $s^{-1}st = s^{-1}su$ which gives $ft = iu$ and so $t = u$. Hence $(f, t) = (i, u)$, as required.

Let (e, s) and (e, t) be two elements with a common range. Then $s, t \in eS$. They can be completed to a commutative square iff we can find (i, x) and (j, y) such that
$$(e, s)(i, x) = (e, t)(j, y).$$
It follows that $sx = ty$. That is $sS \cap tS \neq \{0\}$. Conversely, suppose that $sS \cap tS \neq \{0\}$. Then we can find $sx = ty \neq 0$. It is clear that
$$(s^{-1}s, s^{-1}sx), (t^{-1}t, t^{-1}ty) \in C(S)$$
and that
$$(e, s)(s^{-1}s, s^{-1}sx) = (e, t)(t^{-1}t, t^{-1}ty).$$
Thus (e, s) and (e, t) can be completed to a commutative square if and only if $sS \cap tS \neq \{0\}$.

Now $sS \cap tS = ss^{-1}S \cap tt^{-1}S = ss^{-1}tt^{-1}S$. Consider the pair of elements of $C(S)$: $(s^{-1}s, s^{-1}tt^{-1})$ and $(t^{-1}t, t^{-1}ss^{-1})$. Then
$$(e, s)(s^{-1}s, s^{-1}tt^{-1}) = (e, t)(t^{-1}t, t^{-1}ss^{-1}).$$
Let (i, x) and (j, y) be as above. Then $sx = ty \in ss^{-1}tt^{-1}S$. Thus $sx = ty = ss^{-1}tt^{-1}a$ for some a. It follows that $x = s^{-1}tt^{-1}a$ and $y = t^{-1}ss^{-1}a$. Put $e = ss^{-1}tt^{-1}$. Then
$$(i, x) = (s^{-1}s, s^{-1}tt^{-1})(e, ea) \text{ and } (j, y) = (t^{-1}t, t^{-1}ss^{-1})(e, ea).$$
It is easy to check that (e, ea) is unique with these properties. Thus $(s^{-1}s, s^{-1}tt^{-1}), (t^{-1}t, t^{-1}ss^{-1})$ is the pullback of (e, s) and (e, t).

Finally, if S has an identity 1, then for every identity (e, e) in $C(S)$ we have that $(1, e): (e, e) \to (1, 1)$. Thus $(1, 1)$ is a weak terminal identity. □

For convenience we say that a category is a *Clifford category* if it is a left cancellative category in which any two elements with a common range which can be completed to a commutative square have a pullback. It is easy to check that the map C is a functor from the category of inverse presemigroups and morphisms to the category of Clifford categories.

In Theorem 3.1, we constructed a Clifford category with a weak terminal identity from an inverse monoid with zero. We now show how to go the other way. Let C be a Clifford category with weak terminal identity 1. Define $S^*(C)$ as follows. Let
$$U = \{(x,y) \in C \times C \colon \mathbf{r}(x) = 1 = \mathbf{r}(y) \text{ and } \mathbf{d}(x) = \mathbf{d}(y)\}.$$
Define a relation $\sim$ on U by $(x,y) \sim (x',y')$ if and only if there is an isomorphism $u \in C$ such that $x = x'u$ and $y = y'u$ (we write this as $(x,y) = (x',y')u$). It is easy to check that $\sim$ is an equivalence relation on U. We denote the $\sim$-equivalence class containing (x,y) by $[x,y]$. Put $S^*(C) = U/\sim$. Define a partial binary operation on $S^*(C)$ as follows. Let $[x,y], [w,z] \in S^*(C)$. Their product is defined iff y and w are part of a commutative square, in which case, since C is a Clifford category, y and w have a pullback a and b, say. Define $[x,y][w,z] = [xa, zb]$. The following picture shows what is going on:

That this product is well-defined, and the proof of the following can either be generalised from Leech [20], or obtained as a special case of the theory developed in [18].

Theorem 3.2. *Let C be a Clifford category with a weak terminal identity. Then $S(C)$ is an inverse monoid with zero.*

Let S be an inverse monoid with zero. Then it can be shown that the function $\iota \colon S \to S(C(S))$ defined by $\iota(s) = [(1,s), (1, s^{-1}s)]$ is a well-defined isomorphism of inverse monoids. The proof of this can either be generalised from [20] or obtained as a special case of the main result in [18].

With this preparation behind us, we can now prove some results about classes of E^*-unitary inverse semigroups.

Recall that a semigroup S is said to have a property *locally* if each local submonoid eSe ($e^2 = e$) has that property. It is immediate that a semigroup which is E^*-unitary is also locally E^*-unitary, although the converse is not true.

Proposition 3.3. *Let S be an inverse semigroup with zero. Then $C(S)$ is cancellative if and only if S is locally E^*-unitary.*

Proof. Suppose first that $C(S)$ is cancellative; in particular, it is right cancellative. Let $0 \neq f \leq a \in eSe$. We prove that a is an idempotent which will prove that eSe is E^*-unitary. Since $a \in eSe$ we have that $ea = a$ and so $\mathbf{r}(a) \leq e$. Thus $(e, a) \in C(S)$. Also $f \leq a$ implies $f \leq \mathbf{d}(a)$. Hence $(\mathbf{d}(a), f) \in C(S)$. It follows that the product $(e, a)(\mathbf{d}(a), f)$ is defined in $C(S)$ and is equal to (e, af). Next observe that $a \in eSe$ implies $ae = a$ and so $\mathbf{d}(a) \leq e$. Thus $(e, \mathbf{d}(a)) \in C(S)$. The product $(e, \mathbf{d}(a))(\mathbf{d}(a), f)$ is defined in $C(S)$ and is equal to $(e, \mathbf{d}(a)f)$. But $f \leq a$ and so $f = af$. Hence

$$(e, a)(\mathbf{d}(a), f) = (e, \mathbf{d}(a))(\mathbf{d}(a), f).$$

Thus by right cancellation, we have that $(e, a) = (e, \mathbf{d}(a))$, and so $a = \mathbf{d}(a)$, an idempotent as required.

Conversely, suppose that S is locally E^*-unitary. We prove that $C(S)$ is right cancellative. Let $(e, a)(f, b) = (e, c)(f, b)$. Then $a, c \in L_f$, $a, c \in eS$, $\mathbf{r}(b) \leq f$ and $ab = cb$. Put $w = abb^{-1} = cbb^{-1}$. Then $w \leq a, c$. Hence $ww^{-1} \leq ac^{-1} \in eSe$. Observe that if $ww^{-1} = 0$ then $w = 0$ and so $abb^{-1} = 0$ giving $ab = 0$. But $\mathbf{d}(ab) = \mathbf{d}(b) \neq 0$. It follows that $0 \neq ww^{-1} \leq ac^{-1} \in eSe$. By assumption, eSe is E^*-unitary. Thus ac^{-1} is an idempotent. Now $a \mathcal{L} c$ implies $ac^{-1} \mathcal{L} cc^{-1}$. Thus $ac^{-1} = cc^{-1}$. It follows that $ac^{-1}c = c$. But $c^{-1}c = f$ and so $af = c$. Thus $a = c$, since $a \mathcal{L} f$. We have therefore proved that $(e, a) = (e, c)$, as required. $\square$

Our first theorem is an immediate corollary to the above proposition.

Theorem 3.4. *Let S be an inverse monoid. Then $C(S)$ is cancellative if and only if S is E^*-unitary.*

We now want to relate idempotent pure gradings on S^* to the properties of the category $C(S)$.

Lemma 3.5.

(i) *Let S be an inverse semigroup with zero. Let $\theta \colon S^* \to G$ be a grading to a group G. Then $\alpha \colon C(S) \to G$ defined by $\alpha(e, a) = \theta(a)$ is a functor. If θ is idempotent pure then α is faithful.*

(ii) *Let C be a Clifford category with a weak terminal identity. Let $\alpha \colon C \to G$ be a functor to a group. Then $\theta \colon S(C)^* \to G$ defined by $\theta([x, y]) = \alpha(x)\alpha(y)^{-1}$ is a well-defined morphism. In addition, if α is faithful then θ is idempotent pure.*

Proof. (i) It is straightforward to check that α is a functor. Suppose that θ is idempotent pure. Let $(e, s), (e, t) \colon (f, f) \to (e, e)$. Suppose that $\alpha(e, s) =$

$\alpha(e, t)$. Then $\theta(s) = \theta(t)$, and $s^{-1}s = t^{-1}t = f$. But θ is idempotent pure and so $s = t$. Hence $(e, s) = (e, t)$, and we have shown that α is faithful.

(ii) We first show that θ is well-defined. Suppose that $[x, y] = [x', y']$. Then by definition there is an isomorphism $u \in C$ such that $(x, y)u = (x', y')$. By definition
$$\theta([x', y']) = \alpha(x')\alpha(y')^{-1} = \alpha(xu)\alpha(yu)^{-1} = \alpha(x)\alpha(y)^{-1} = \theta([x, y]).$$
Thus $\theta([x, y])$ is independent of the choice of representative of $[x, y]$. Now we can prove that θ is a morphism. Let $[x, y], [w, z] \in S(C)^*$ such that $[x, y][w, z] \neq 0$. Let a and b be a pullback of y and w. Then by definition $[x, y][w, z] = [xa, zb]$. We now calculate
$$\alpha([x, y][w, z]) = \alpha([xa, zb]) = \alpha(xa)\alpha(zb)^{-1} = \alpha(x)\alpha(a)\alpha(b)^{-1}\alpha(z)^{-1}.$$
But $ya = wb$, and
$$\alpha(x)\alpha(a)\alpha(b)^{-1}\alpha(z)^{-1} = \alpha(x)\alpha(y)^{-1}\alpha(y)\alpha(a)\alpha(b)^{-1}\alpha(z)^{-1}$$
and this is equal to
$$\alpha(x)\alpha(y)^{-1}\alpha(w)\alpha(z)^{-1} = \alpha([x, y])\alpha([w, z]).$$
Suppose now that α is faithful. We prove that θ is idempotent pure. Suppose that $\theta([x, y]) = 1$. Then $\alpha(x) = \alpha(y)$. But $\mathbf{d}(x) = \mathbf{d}(y)$ and $\mathbf{r}(x) = \mathbf{r}(y)$. Thus $x = y$, since α is faithful. Hence $[x, y]$ is an idempotent. $\square$

Proposition 3.6. *Let S be an inverse monoid with zero and let G be a group. Then S^* has an idempotent pure grading to G if and only if $C(S)$ has a faithful functor to G. In particular, S is strongly E^*-unitary if and only if $C(S)$ admits a faithful functor to a group.*

Proof. Let $\theta: S^* \to G$ be an idempotent pure grading. Define $\alpha: C(S) \to G$ by $\alpha(e, s) = \theta(s)$. Then α is a faithful functor by Lemma 3.5.

Conversely, let $\alpha: C(S) \to G$ be a faithful functor. Then $\theta: S(C(S))^* \to G$ defined by $\theta([\mathbf{x}, \mathbf{y}]) = \alpha(\mathbf{x})\alpha(\mathbf{y})^{-1}$ is a is an idempotent pure morphism by Lemma 3.5. Define $\theta': S^* \to G$ by $\theta'(s) = \theta(\iota(s)) = \alpha(1, s)\alpha(1, s^{-1}s)^{-1}$ where $\iota(s) = [(1, s), (1, s^{-1}s)]$ is the isomorphism defined following Theorem 3.2. Then θ' is an idempotent pure morphism. The proof of the last assertion is immediate. $\square$

A groupoid $\mathcal{G}$ is said to be *connected* if for each pair of identities e and f there is an arrow $g \in \mathcal{G}$ such that $e \xrightarrow{g} f$. A *local group (based at e)* in a groupoid $\mathcal{G}$ is the group of all arrows which begin and end at e. It is well-known that in a connected groupoid any two local groups are isomorphic. The proofs of the following can all be found in [6].

Proposition 3.7. *Let C be a category.*

(i) *If there is a faithful functor from C to the group G, then C can be embedded in a connected groupoid $\mathcal{G}$ with local group G.*

(ii) *Let $\mathcal{G}$ be a connected groupoid, e any identity in $\mathcal{G}$, and G the local group based at e. Then there is a faithful functor from $\mathcal{G}$ to G.*

(iii) *Every groupoid can be embedded in a connected groupoid.*

(iv) *The category C can be embedded in a groupoid if and only if it admits a faithful functor to a group.*

We now have the following characterisation of strongly E^*-unitary inverse monoids.

Theorem 3.8. *Let S be an inverse monoid with zero. Then S^* admits an idempotent pure grading to the group G if and only if $C(S)$ can be embedded in a connected groupoid with local group G. In particular, S is strongly E^*-unitary if and only if the category $C(S)$ can be embedded in a groupoid.*

Proof. Let $\theta\colon S^* \to G$ be a grading to a group G. By Proposition 3.6, there is a faithful functor from $C(S)$ to G. By Proposition 3.7(i), the category $C(S)$ can be embedded in a connected groupoid with local group G.

Conversely, suppose that $C(S)$ can be embedded in a connected groupoid with local group G. Then by Proposition 3.7 (ii), there is a faithful functor from $C(S)$ to G. Hence by Proposition 3.6, there is a grading from S^* to G.

The last assertion follows from Propositions 3.6 and 3.7(iv). □

There is one final case to be considered. Let S be an inverse semigroup without zero. Then from the proof of Theorem 3.1, we have that $C(S)$ is a Clifford category with a stronger property: any two elements with a common range have a pullback.[c] A cancellative category C is said to satisfy the *Ore condition* if any two morphisms with a common range can be completed to a commutative square. Cancellative categories satisfying the Ore condition have a *groupoid of fractions*. Thus if S is an E-unitary inverse semigroup, then $C(S)$ has a groupoid of fractions. Conversely, suppose that S is an E^*-unitary inverse monoid and $C(S)$ satisfies the Ore condition. Then each pair of morphisms in $C(S)$ has a pullback. It follows that the zero in S is removable and we are therefore essentially dealing with an E-unitary semigroup. These results are proved in more detail in [7].

Theorem 3.9. *Let S be an inverse semigroup with zero. Then S is an E-*

---
[c]This is the case considered by [20].

unitary inverse semigroup with a zero adjoined if and only if $C(S)$ satisfies the Ore condition, and therefore has a groupoid of fractions.

In [7], we showed that McAlister's original proof of the P-theorem [25] can be deduced from the above result by the simple expedient of choosing a coordinate system for the inverse semigroup.

The results of this section suggest that the question of whether an inverse monoid is strongly E^*-unitary is analogous to the question of whether a monoid can be embedded in a group. Perhaps techniques which have been developed for investigating this latter question could be generalised from monoids to categories and so to E^*-unitary semigroups. For example, the methods of Cho and Pride [3]. Here is another example. Von Karger [11] proved that a cancellative category C which satisfies the condition

$$Ca \cap Cb \neq \emptyset \Rightarrow Ca \subseteq Cb \text{ or } Cb \subseteq Ca$$

can be embedded in a groupoid. In [19], we proved that von Karger's result implies that every E^*-unitary inverse semigroup which is unambiguous except at zero is strongly E^*-unitary.

4 Structural results

We conclude this paper with some observations on the structure of strongly E^*-unitary semigroups.

The grading associated with a strongly E^*-unitary inverse semigroup is idempotent pure. This means that the general theory of idempotent pure prehomomorphisms described in [17] can easily be used to describe such semigroups in terms which generalise the classical the classical P-theorem in terms of McAlister triples. This result, proved in a different way, together with other characterisations of strongly E^*-unitary inverse semigroups may be found in [2]. In [15], we showed that Munn's proof of the P-theorem [27] could be proved in terms of globalisations of partial group actions on semilattices; this work is closely related to [30]. This approach can easily be extended to yield another description of strongly E^*-unitary semigroups based on partial group actions.

The final characterisation came about through Kellendonk's work on tiling semigroups. The first step was a formalisation of Kellendonk's construction in terms of groups acting partially and freely on categories [14]. Even a superficial glance at tiling semigroups reveals a similarity with Munn's approach to free inverse semigroups in terms of Munn trees, and the more general results of Margolis and Meakin [23] on constructing E-unitary inverse semigroups from group presentations. These connections were raised by the author at the 'Workshop on Tiling Semigroups' held at the University of Essex during the summer of 1999. Subsequently, a common approach to both tilings and free inverse semigroups was found by Ben Steinberg [32] (together with an important contribution by John Fountain). We shall now describe a reformulation of Ben's result in terms which generalise [14].

A *directed graph* consists of a set of *arrows* C and a set of *vertices* C_o together with two functions $\partial_0, \partial_1 \colon C \to C_o$ where $\partial_0(s)$ is the *base* of the arrow and $\partial_1(s)$ is the *head* of the arrow. Elements with the same base and the same head are called *parallel*. An element whose head and base are the same is called a *loop*.

Suppose that C is both a directed graph and a presemigroup and is such that $\exists ab$ implies $\partial_0(a) = \partial_1(b)$, and $\partial_0(ab) = \partial_0(b)$, and $\partial_1(ab) = \partial_1(a)$. Then we call C a *multiplicative graph*. Observe that the set of loops at a given vertex forms a presemigroup, called a *local presemigroup*. A multiplicative graph is *locally idempotent (resp. locally commutative)* if all the local presemigroups are idempotent (resp. commutative). A multiplicative graph whose presemigroup is inverse is called *inverse*.

Let G be a group and X a set. We say that G *acts partially on* X if there is a partial function $G \times X \to X$ denoted by $(a, x) \mapsto a \cdot x$ satisfying the following three axioms:

(PA1) $\exists 1 \cdot x$ for each $x \in X$ and $1 \cdot x = x$.

(PA2) $\exists g \cdot (h \cdot x)$ implies $\exists (gh) \cdot x$ and $g \cdot (h \cdot x) = (gh) \cdot x$.

(PA3) $\exists g \cdot x$ implies that $\exists g^{-1} \cdot (g \cdot x)$, and $g^{-1} \cdot (g \cdot x) = x$.

We say that the action is *free* if $\exists g \cdot x = x$ implies that $g = 1$.

Let S be a multiplicative graph whose partial product we denote by concatenation and let G be a group. We say that G *acts freely and partially on* S if G acts partially on both S_o and S, and freely on S_o; in addition, the following four axioms are satisfied:

(A1) If $\exists g \cdot x$ then $\exists g \cdot \partial_0(x)$ and $\exists g \cdot \partial_1(x)$ and $\partial_0(g \cdot x) = g \cdot \partial_0(x)$ and $\partial_1(g \cdot x) = g \cdot \partial_1(x)$.

(A2) If $\exists xy$ and $\exists g \cdot x$ and $\exists g \cdot y$ and $\exists (g \cdot x)(g \cdot y)$ then $\exists g \cdot (xy)$ and $g \cdot (xy) = (g \cdot x)(g \cdot y)$.

(A3) If $\exists g \cdot (xy)$ then $\exists g \cdot x$ and $\exists g \cdot y$.

(A4) If $\exists g \cdot (xy)$ then $\exists (g \cdot x)(g \cdot y)$ and $g \cdot (xy) = (g \cdot x)(g \cdot y)$.

Proposition 4.1. *Let S be a multiplicative graph equipped with a free partial action by a group G. Define $\sim$ on S by $x \sim y$ iff there exists $g \in G$ such that $g \cdot x = y$. Then $\sim$ is an equivalence relation. Let $S/G = \{[x]: x \in S\}$ be the set of equivalence classes. Define the following operation on S/G:*

$$[x][y] = [(g \cdot x)(h \cdot y)]$$

if there exist $g, h \in G$ such that $(g \cdot x)(h \cdot y)$ is defined in S. Then with respect to this operation, S/G is a presemigroup.

The map from S to S/G defined by $x \mapsto [x]$ is a surjective idempotent pure morphism.

If the multiplicative graph is inverse then so too is S/G.

Steinberg's main result [32] combined with a result of Fountain (private communication) can be expressed in the following way.

Theorem 4.2. *Let G be a group which acts partially and freely on the multiplicative inverse graph S, where S has the additional property that the local presemigroups of S are idempotent. Then S/G is a strongly E^*-unitary inverse presemigroup and every strongly E^*-unitary inverse presemigroup is isomorphic to one constructed in this way.*

It is quite likely that there is no 'definitive' way of describing strongly E^*-unitary inverse semigroups. Instead, different applications will probably require different ways of describing them.

Acknowledgements

I would like to thank Gracinda Gomes, Jean-Éric Pin and Pedro Silva for all their hard work in organising the thematic term on 'Semigroups, automata algorithms and languages' which provided an ideal environment for putting together the ideas in this paper.

This paper has benefited from conversations with John Fountain, Gracinda Gomes, Johannes Kellendonk, Daniel Lenz, Stuart Margolis, and Ben Steinberg.

The author is also grateful to his graduate student Joe Matthews who worked out Examples 2.5 (1),(2),(3),(4).

References

[1] C. J. Ash, T. E. Hall, Inverse semigroups on graphs, *Semigroup Forum* **11** (1975), 140–145.

[2] S. Bulman-Fleming, J. Fountain, V. Gould, Inverse semigroups with zero: covers and their structure, *J. Austral. Math. Soc.* (Series A) **67** (1999), 15–30.

[3] J. R. Cho, S. J. Pride, Embedding semigroups into groups, and the asphericity of semigroups, *Int. J. Algebra and Computation* **3** (1993), 1–13.

[4] P. M. Cohn, *Universal algebra*, D. Reidel Publishing Company, 1981.

[5] G. M. S. Gomes, J. M. Howie, A P-theorem for inverse semigroups with zero, *Portugal. Math.* **53** (1996), 257–278.

[6] M. Hasse, L. Michler, Über die Einbettbarkeit von Kategorien in Gruppoide, *Math. Nachr.* **25** (1963), 169–177.

[7] H. James, M. V. Lawson, An application of groupoids of fractions to inverse semigroups, *Periodica Math. Hung.* **38** (1999), 43–54.

[8] S. M. Jekel, On two theorems of A. Haefliger concerning foliations, *Topology* **15** (1976), 267–271.

[9] S. M. Jekel, Simplicial $K(G,1)$'s, *Manuscripta Math.* **21** (1977), 189–203.

[10] S. M. Jekel, Pseudogroups and homology, *Lecture Notes, Northeastern University Topology Seminar*, 1985.

[11] B. v. Karger, *Temporal algebra*, Habilitationsschrift, Christian-Albrechts Universität Kiel , 1997.

[12] J. Kellendonk, The local structure of tilings and their integer group of invariants, *Comm. Math. Phys.* **187** (1997), 115–157.

[13] J. Kellendonk, Topological equivalence of tilings, *J. Math. Phys.* **38** (1997), 1823–1842.

[14] J. Kellendonk, M. V. Lawson, Tiling semigroups, *J. Algebra* **224** (2000), 140–150.

[15] J. Kellendonk, M. V. Lawson, Partial actions of groups, Preprint.

[16] A. Kumjian, D. Pask, I. Raeburn, J. Renault, Graphs, groupoids and Cuntz-Krieger algebras, *J. Funct. Anal.* **44** (1997), 505–541.

[17] M. V. Lawson, *Inverse semigroups: the theory of partial symmetries*, World Scientific, Singapore, 1998.

[18] M. V. Lawson, Constructing inverse semigroups from category actions, *J. Pure and Applied Algebra* **137** (1999), 57–101.

[19] M. V. Lawson, The structure of 0-E-unitary semigroups I: the monoid case, *Proc. Edinburgh Math. Soc.* **42** (1999), 497–520.

[20] J. Leech, Constructing inverse monoids from small categories, *Semigroup Forum* **36** (1987), 89–116.

[21] J. Leech, On the foundations of inverse monoids and inverse algebras, *Proc. Edinburgh Math. Soc.* **41** (1998), 1–21.

[22] D. H. Lenz, On an order based construction of a groupoid from an inverse semigroup, Preprint.

[23] S. W. Margolis, J. C. Meakin, E-unitary inverse monoids and the Cayley graph of a group presentation, *J. Pure and Applied Algebra* **58** (1989), 45–76.

[24] D. B. McAlister, Inverse semigroups which are separated over a subsemigroup, *Trans. Amer. Math. Soc.* **182** (1973), 85–117.

[25] D. B. McAlister, Groups, semilattices and inverse semigroups II, *Trans. Amer. Math. Soc.* **196** (1974), 351–370.

[26] J. Meakin, M. Sapir, Congruences on free monoids and submonoids of polycyclic monoids, *J. Austral. Math. Soc. (Ser. A)* **54** (1993), 236–253.

[27] D. Munn, A note on E-unitary inverse semigroups, *Bull. London Math. Soc.* **8** (1976), 71–76.

[28] A. Nica, On a groupoid construction for actions of certain inverse semigroups, *Int. J. Mathematics* **5** (1994), 349–372.

[29] M. Nivat, J.-F. Perrot, Une généralisation du monoïde bicyclique, *C. R. Acad. Sci. Paris* **271** (1970), 824–827.

[30] M. Petrich, N. R. Reilly, A representation of E-unitary inverse semigroups, *Quart. J. Math. (2)* **30** (1979), 339–350.

[31] J. R. Stallings, Group theory and three-dimensional manifolds, Yale Monographs **4**, (1971).

[32] B. Steinberg, Building inverse semigroups from group actions, *Preprint*.

[33] B. Steinberg, The uniform word problem for groups and finite Rees quotients of E-unitary inverse semigroups, *Preprint*.

[34] M. B. Szendrei, E-unitary regular semigroups, *Proc. Roy. Soc. Edinburgh* **106A** (1987), 89–102.

SOME RESULTS ON SEMIGROUP-GRADED RINGS

W. D. MUNN*
Department of Mathematics, University of Glasgow, Glasgow G12 8QW, Scotland, UK
E-mail:wdm@maths.gla.ac.uk

A survey is given of some results on semigroup-graded rings concerning the properties of semiprimitivity (Jacobson semisimplicity) and primitivity. Emphasis is placed on rings graded by inverse semigroups and new material in this area is incorporated.

1 Introduction

All rings discussed below are associative, but the existence of unity elements is not assumed.

Let R be a ring and S a semigroup. Then R is said to be S-*graded* (equivalently, *graded by* S) if and only if the following two conditions hold:

(i) $(R, +)$ is the direct sum of a family R_x ($x \in S$) of (abelian) subgroups;
(ii) for all $x, y \in S$, $R_x R_y \subseteq R_{xy}$.

We call the subgroups R_x the *homogeneous components* of R. Various familiar mathematical structures may be viewed as special cases of semigroup-graded rings: some of these appear in Examples 2.1 to 2.5 below.

Suppose that R is such a ring. For a nonempty subset T of S we write $R_T := \oplus_{x \in T} R_x$ and observe that if T is a subsemigroup of S then R_T is a subring of R. In particular, for an idempotent $e \in S$, we have that R_e and R_{H_e} are subrings of R, where H_e denotes the maximal subgroup of S with identity e.

Following Cohen and Montgomery [2] and Kelarev [5], we say that R is *faithful* (equivalently, *faithfully graded by* S) if and only if, for all $x, y \in S$,

$$a \in R_x \backslash 0 \quad \Rightarrow \quad aR_y \neq 0 \text{ and } R_y a \neq 0.$$

This condition ensures that if $R \neq 0$ then each $R_x \neq 0$ and there is a nontrivial linkage between any two homogeneous components. A weaker condition, depending on Green's equivalence $\mathcal{D}$ on S, was introduced by the author in

---

*I WISH TO ACKNOWLEDGE THE FINANCIAL SUPPORT OF FUNDAÇÃO CALOUSTE GULBENKIAN (FCG), FUNDAÇÃO PARA A CIÊNCIA E A TECNOLOGIA (FCT), FACULDADE DE CIÊNCIAS DA UNIVERSIDADE DE LISBOA (FCUL) AND REITORIA DA UNIVERSIDADE DO PORTO.

[11]. We say that R is $\mathcal{D}$-*faithful* (equivalently, $\mathcal{D}$-*faithfully graded by S*) if and only if, for all $x, y \in S$ such that x, y and xy lie in the same $\mathcal{D}$-class,

$$a \in R_x \backslash 0 \Rightarrow aR_y \neq 0 \quad \text{and} \quad b \in R_y \backslash 0 \Rightarrow R_x b \neq 0.$$

Many types of semigroup-graded rings, including semigroup rings and contracted semigroup rings over integral domains, turn out to be $\mathcal{D}$-faithful.

Attention will be confined to the ring-theoretic properties of semiprimitivity and primitivity. These are defined in Section 3. In Section 4, we establish some results from [11] for a ring that is $\mathcal{D}$-faithfully graded by a semigroup S containing $\mathcal{D}$-equivalent idempotents e and f. It is shown in Theorem 4.4 that if R_e [resp. R_{H_e}] is semiprimitive then R_f [resp. R_{H_f}] is semiprimitive and that if R_e [resp. R_{H_e}] is primitive then R_f [resp. R_{H_f}] is primitive.

The remainder of the paper is devoted to an account of rings $\mathcal{D}$-faithfully graded by *inverse* semigroups. Section 5 gives a brief historical survey of results under this heading. In Section 6, it is shown that if R is a ring $\mathcal{D}$-faithfully graded by an inverse semigroup S and if, for each $\mathcal{D}$-class D of S, $R_{G(D)}$ is semiprimitive for one maximal subgroup $G(D)$ in D, then R itself is semiprimitive. Moreover, if R is a ring $\mathcal{D}$-faithfully graded by a 0-bisimple inverse semigroup S with zero z, if $R_z = 0$ and if R_G is primitive for some nonzero maximal subgroup G of S then R is primitive. These results comprise Theorem 6.4 and, between them, generalise all the earlier results quoted in Section 5.

The author's strategy in this account has been to deal with semiprimitivity and primitivity in tandem. An alternative (shorter) proof of the semiprimitivity part of Theorem 6.4 may, however, be of some interest and is provided in Section 7.

We adhere to the standard notation for semigroup theory, as presented in [1] and [4], with one exception: the symbol R_x will here always denote a homogeneous component of a semigroup-graded ring and never an $\mathcal{R}$-class of a semigroup.

2 Some examples of semigroup-graded rings

We start by listing some examples that illustrate the concepts introduced above. The term 'homogeneous component' had its origin in the first example.

Example 2.1. Let $R := A[x_1, x_2, \ldots, x_k]$, the ring of polynomials in the commuting indeterminates $x_1, x_2, \ldots, x_k$ over an integral domain A. Let S denote the additive semigroup of nonnegative integers and, for each $n \in S$, let R_n denote the set of all polynomials in R that are homogeneous of degree n. Clearly, R_n is a subgroup of $(R, +)$, $R = \oplus_{n \in S} R_n$ and, for all $m, n \in S$, $R_m R_n \subseteq R_{m+n}$. Thus R is S-graded; further, being an integral domain, R is faithful.

Example 2.2. Let $R := A[S]$, the semigroup ring of a semigroup S over an integral domain A. The elements of R are the formal sums

$$\sum_{x \in S} \alpha_x x \quad (\alpha_x \in A),$$

with at most finitely many nonzero coefficients α_x; further, for all $x \in S$, we identify $1_A x$ with x, where 1_A is the unity of A. Addition and multiplication in R are according to the rules

$$\sum_x \alpha_x x + \sum_x \beta_x x = \sum_x (\alpha_x + \beta_x) x,$$

$$(\sum_x \alpha_x x)(\sum_y \beta_y y) = \sum_{x,y} \alpha_x \beta_y (xy).$$

For each $x \in S$, let R_x denote Ax $(= \{\alpha x : \alpha \in A\})$. Then $R = \oplus_{x \in S} R_x$ and, for all $x, y \in S$, $R_x R_y \subseteq Axy = R_{xy}$. Since A has no nontrivial zero divisors, it is easily seen that, as an S-graded ring, R is faithful.

Notes.
(i) In Example 2.2, if $e = e^2 \in S$ then $R_e \cong A$.
(ii) Examples 2.1 and 2.2 together show that the same ring can be graded by different semigroups; for we can regard R in Example 2.1 as $A[S]$, where S is the free commutative monoid of rank k.

Example 2.3. Let S be a nontrivial semigroup with zero z and let A be an integral domain. The *contracted semigroup ring* $A_0[S]$ of S over A is defined by $A_0[S] := A[S]/Az$, where Az is the ideal $\{\alpha z : \alpha \in A\}$ of $A[S]$. Write $R := A_0[S]$. Then R can be viewed as an S-graded ring with $R_x = Ax$ ($x \in S \backslash z$) and $R_z = 0$. Since there exists $x \in S \backslash z$, we have that $x \in R_x \backslash 0$; but $xR_z = 0$ and so R is *not* faithful. However, R is $\mathcal{D}$-faithful, as we now demonstrate. Let D be any $\mathcal{D}$-class of S other than $\{z\}$ and let $x, y \in D$ be such that $xy \in D$. Let $a \in R_x \backslash 0$. Then $a = \alpha x$ for some $\alpha \in A \backslash 0$. Hence $ay = \alpha(xy) \neq 0$ and so $aR_y \neq 0$. Similarly, if $b \in R_y \backslash 0$ then $R_x b \neq 0$.

The following special case of Example 2.3 is noteworthy. For a given positive integer n, let S_n denote the semigroup of $n \times n$ matrix units: that is, $S_n := \{e_{ij} : 1 \leq i, j \leq n\} \cup \{z\}$, with multiplication such that z is a zero and

$$e_{ij}e_{kl} = \begin{cases} e_{il} & \text{if } j = k, \\ z & \text{otherwise.} \end{cases}$$

Then, for an integral domain A, $A_0[S_n] \cong M_n(A)$, the ring of $n \times n$ matrices over A, with the usual matrix operations. Thus $M_n(A)$ can be viewed as a $\mathcal{D}$-faithful S_n-graded ring.

Example 2.4. Let $R := M_3(A)$, where A is an integral domain, and let each matrix in R be partitioned into four blocks, with leading block of type 1×1. The rule for block multiplication shows that R is an S_2-graded ring, with homogeneous components

$$R_{e_{11}} = A, \quad R_{e_{12}} = A_{12}, \quad R_{e_{21}} = A_{21}, \quad R_{e_{22}} = M_2(A), \quad R_z = 0,$$

where A_{12} and A_{21} denote respectively, the sets of all 1×2 and 2×1 matrices over A. It can be verified that, as an S_2-graded ring, R is again $\mathcal{D}$-faithful. Now e_{11} and e_{22} are $\mathcal{D}$-equivalent idempotents in S_2; but $R_{e_{11}} \not\cong R_{e_{22}}$. Observe also that (i) $H_{e_{ii}} = \{e_{ii}\}$ ($i = 1, 2$) and (ii) if A is a field then both $R_{e_{11}}$ and $R_{e_{22}}$ are simple rings.

Example 2.5. Let S be a Clifford semigroup (a semilattice of groups); that is, there exists a semilattice E and a family G_e ($e \in E$) of pairwise-disjoint subgroups of S such that $S = \cup_{e \in E} G_e$ and, for all $e, f \in E$, $G_e G_f \subseteq G_{ef}$. Let A be an integral domain and let $R := A[S]$. It is frequently helpful to view R as an E-graded ring, with $R_e = A[G_e]$ for all $e \in E$. Now $\mathcal{D}$ is the identity relation on E; also, for all $e \in E$, $a \in R_e \backslash 0$ implies $aR_e \neq 0$ and $R_e a \neq 0$, since the ring $A[G_e]$ has a unity. Hence R is $\mathcal{D}$-faithfully graded by S. However, R is not always faithfully graded by S. To see this, consider the semigroup $S := \{e, x, z\}$, where e and z are the identity and zero elements, respectively, and $x^2 = e$. Thus we may take $E := \{e, z\}$, $G_e := \{e, x\}$ and $G_z = \{z\}$ above. Let $a \in R_e \backslash 0$ be defined by $a := e - x$. Then $aR_z = 0 = R_z a$.

3 Primitivity and semiprimitivity

This section comprises some definitions and brief comments on those ring- and module-theoretic concepts with which we shall be concerned. All modules considered are assumed to be *right* modules.

Let R be a nonzero ring and V an R-module. We say that V is *irreducible* if and only if $V \neq 0$ and has no submodules other than V and 0. It is convenient here to make a related definition: V is *strictly irreducible* if and only if $V \neq 0$ and, for all $x \in V \backslash 0$, $xR = V$. It is easily seen that if V is strictly irreducible then it is irreducible. However, the converse is not true. If V is irreducible then either V is strictly irreducible or else $(V, +)$ is cyclic of prime power order and $VR = 0$.

Next, we say that V is *faithful* if and only if, for all $a \in R \backslash 0$, $Va \neq 0$. From the remarks above, it is clear that if V is faithful and irreducible then it is faithful and strictly irreducible. (No confusion should arise between this use of the adjective 'faithful' and its use in the context of Section 1.)

Definition. A ring R is *primitive* if and only if $R \neq 0$ and there exists a faithful irreducible R-module.

Again let R be a nonzero ring. A family of R-modules V_λ ($\lambda \in \Lambda$) is termed *faithful* if and only if, for all $a \in R \backslash 0$, there exists $\mu \in \Lambda$ such that $V_\mu a \neq 0$. Observe that if V_λ ($\lambda \in \Lambda$) is a faithful family of irreducible R-modules then there exists a nonempty subset Λ' of Λ such that V_λ ($\lambda \in \Lambda'$) is a faithful family of strictly irreducible R-modules.

Definition. A ring R is *semiprimitive* if and only if either $R = 0$ or there exists a faithful family of strictly irreducible R-modules.

Clearly every primitive ring is semiprimitive.

It is often convenient to characterise these properties of a ring R internally. Observe first that if M is a right ideal of R then the additive group R/M is an R-module under the action defined by

$$(a + M)b = ab + M \quad (a, b \in R).$$

The following result is easily established.

Lemma 3.1. *Let R be a nonzero ring.*
  (i) *If R is primitive then there exists a proper right ideal M of R such that R/M is a faithful irreducible R-module.*
  (ii) *If R is semiprimitive then there exists a nonempty family M_λ ($\lambda \in \Lambda$) of proper right ideals of R such that R/M_λ ($\lambda \in \Lambda$) is a faithful family of strictly irreducible R-modules.*

A different approach to semiprimitivity, using the Jacobson radical, is illustrated in Section 7. (See [6].)

4 Certain subrings of a $\mathcal{D}$-faithful semigroup-graded ring

In this section, for a ring R that is $\mathcal{D}$-faithfully graded by a semigroup S, we establish relationships between certain subrings of R that are naturally associated with a regular $\mathcal{D}$-class of S.

For convenience, we begin by recalling a few definitions and restating some basic properties of semigroups: for the details, see [1] or [4].

The set of all idempotents of a semigroup S is denoted by $E(S)$ and, for $x \in S$, the $\mathcal{H}$-class containing x is denoted by H_x. For all $e \in E(S)$, H_e is the maximal subgroup of S with identity e. A $\mathcal{D}$-class D of S is termed *regular* if and only if $D \cap E(S) \neq \emptyset$, this being equivalent to the condition that every element $x \in D$ is regular in the sense that $x \in xSx$. Elements $p, q \in S$ are said to be *mutually inverse* if and only if $pqp = p$ and $qpq = q$. For such a pair of elements, we have that

- $pq, qp \in E(S)$,
- p, q, pq and qp are $\mathcal{D}$-equivalent,
- $pH_{qp} = H_p$ and $H_p q = H_{pq}$.

Moreover, if $e, f \in E(S)$ and $e \mathcal{D} f$ then there exist mutually inverse elements $x, y \in S$ with $xy = e$, $yx = f$.

Now let R be a ring $\mathcal{D}$-faithfully graded by S and let $a \in R$. For $x \in S$, we denote the R_x-component of a by a_x and we define the *support* of a, $\operatorname{supp}(a)$, to be $\{x \in S : a_x \neq 0\}$. Observe that $\operatorname{supp}(a)$ is a finite set and is empty if and only if $a = 0$. For $X \subseteq S$, we write

$$a_X := \begin{cases} \sum_{x \in X} a_x & \text{if } X \neq \emptyset, \\ 0 & \text{otherwise.} \end{cases}$$

The set-difference $\{x \in X : x \notin Y\}$ of two sets X and Y is denoted by $X \backslash Y$.

We proceed via several lemmas.

Lemma 4.1. *Let R be a ring $\mathcal{D}$-faithfully graded by a semigroup S that has a regular $\mathcal{D}$-class D. If there exists $y \in D$ such that $R_y \neq 0$ then, for all $x \in D$, $R_x \neq 0$.*

Proof. Assume that there exists $y \in D$ such that $R_y \neq 0$. Let $x \in D$. Since D is regular, there exist $e, f \in E(S)$ such that $e \mathcal{R} x$ and $f \mathcal{L} y$. Also, since $e \mathcal{D} f$ there exist mutually inverse elements $p, q \in D$ such that $pq = e$ and $qp = f$.

Let $a \in R_y \setminus 0$. We use the $\mathcal{D}$-faithfulness condition four times to prove that $R_x \neq 0$. First, since $yf = y$, $aR_f \neq 0$ and so there exists $b \in R_f \setminus 0$. Then, since $pf = p$, $R_p b \neq 0$ and so there exists $c \in R_p \setminus 0$. Next, since $ep = p$, $R_e c \neq 0$ and so there exists $d \in R_e \setminus 0$. Finally, since $ex = x$, $dR_x \neq 0$ and so $R_x \neq 0$. □

Lemma 4.2. *Let R be a ring $\mathcal{D}$-faithfully graded by a semigroup S that contains mutually inverse elements p and q. Let $e := pq$ and $f := qp$. Then, for all $a \in R_{H_f} \setminus 0$, there exist $b \in R_p$ and $c \in R_q$ such that $bac \in R_{H_e} \setminus 0$.*

Proof. Let $a \in R_{H_f} \setminus 0$ and let $k \in \mathrm{supp}(a)$. Let D denote the $\mathcal{D}$-class of S containing p, q, e and f. Then $pk \in pH_f = H_p \subseteq D$ and $pkq \in H_p q = H_e \subseteq D$. Hence, by the $\mathcal{D}$-faithfulness of R, since $p, k, pk \in D$ there exists $b \in R_p$ such that $ba_k \neq 0$ and so, since $pk, q, pkq \in D$, there exists $c \in R_q$ such that
$$ba_k c \in R_{pkq} \setminus 0. \tag{1}$$
Now
$$bac = ba_k c + \sum_{h \in \mathrm{supp}(a) \setminus k} ba_h c \tag{2}$$
(with the convention that a sum over the empty set is taken to be 0). But, for all $h \in \mathrm{supp}(a)$, $ba_h c \in R_{phq}$; and, if $phq = pkq$ then $h = k$, since $qp = f$ and $h, k \in H_f$. Hence, from (1) and (2), $bac \neq 0$. Further, $bac \in R_{H_e}$, since $pH_f q = H_e$. Thus $bac \in R_{H_e} \setminus 0$. □

Lemma 4.3. *Let R be a ring $\mathcal{D}$-faithfully graded by a semigroup S that contains mutually inverse elements p and q. Let $e := pq$ and $f := qp$, let T and U be submonoids of the groups H_e and H_f, respectively, and let $T = pUq$.*
  (i) *If R_T is semiprimitive then R_U is semiprimitive.*
  (ii) *If R_T is primitive then R_U is primitive.*

Proof. Note first that, since $pq = e$ and $qp = f$,
$$Tp = pU, \quad qT = Uq, \quad qTp = U. \tag{1}$$

(i) Assume that R_T is semiprimitive. If $R_T = 0$ then, by Lemma 4.1, $R_U = 0$ and so R_U is semiprimitive. Hence we assume that $R_T \neq 0$. Thus, by Lemma 3.1(ii), there exists a nonempty family M_λ ($\lambda \in \Lambda$) of proper right ideals of R_T such that R_T/M_λ ($\lambda \in \Lambda$) is a faithful family of strictly irreducible R_T-modules. We construct a corresponding faithful family of strictly irreducible R_U-modules.

Consider the subgroup $R_T R_{Tp}$ of $(R, +)$. Since, by (1), $TpU = T^2p \subseteq Tp$, we have that

$$(R_T R_{Tp}) R_U \subseteq R_T R_{Tp}. \tag{2}$$

Hence $R_T R_{Tp}$ is an R_U-module under the multiplication induced by that in R.

For $\lambda \in \Lambda$, write

$$N_\lambda := \{a \in R_T R_{Tp} : a R_{qT} \subseteq M_\lambda\}.$$

It is easily verified that N_λ is a subgroup of $(R_T R_{Tp}, +)$. Also, from (1), $UqT \subseteq qT^2 \subseteq qT$ and so, for all $a \in N_\lambda$ and all $r \in R_U$,

$$(ar) R_{qT} \subseteq a R_U R_{qT} \subseteq a R_{UqT} \subseteq a R_{qT} \subseteq M_\lambda.$$

Hence $ar \in N_\lambda$. Consequently, N_λ is an R_U-submodule of $R_T R_{Tp}$.

Let $\Lambda^* := \{\lambda \in \Lambda : N_\lambda \neq R_T R_{Tp}\}$. We show that $\Lambda^* \neq \emptyset$ and that $R_T R_{Tp}/N_\lambda$ ($\lambda \in \Lambda^*$) is a faithful family of irreducible R_U-modules.

By Lemma 4.1, $R_U \neq 0$, since $R_T \neq 0$. Let $a \in R_U \backslash 0$. Then, by Lemma 4.2, there exist $b \in R_p$ and $c \in R_q$ such that $bac \neq 0$. Further, $bac \in R_{pUq} = R_T$; thus $bac \in R_T \backslash 0$. Since the family R_T/M_λ ($\lambda \in \Lambda$) is faithful, there exist $\mu \in \Lambda$ and $r \in R_T$ such that

$$rbac \notin M_\mu. \tag{3}$$

But $R_p \subseteq R_{Tp}$, since $p = ep$ and $e \in T$. Thus $rb \in R_T R_{Tp}$ and so, by (2), $rba \in R_T R_{Tp}$. Also, $R_q \subseteq R_{qT}$, since $q = qe$ and $e \in T$. Hence $c \in R_{qT}$. It follows from (3) that $rba \notin N_\mu$ and so $(rb + N_\mu)a \neq N_\mu$. This shows that $\mu \in \Lambda^*$ and that the family $R_T R_{Tp}/N_\lambda$ ($\lambda \in \Lambda^*$) is faithful.

We now prove that each module in this family is irreducible.

Let $\lambda \in \Lambda^*$ and let $u \in R_T R_{Tp} \backslash N_\lambda$. Then there exists $a \in R_{qT}$ such that

$$ua \notin M_\lambda. \tag{4}$$

For an arbitrary element $v \in R_T R_{Tp}$ we have that $v = \sum_{i=1}^n r^{(i)} d^{(i)}$, for some positive integer n, some $r^{(i)} \in R_T$ and some $d^{(i)} \in R_{Tp}$ ($i = 1, 2, \ldots, n$). Now $ua \in R_T R_{Tp} R_{qT} \subseteq R_T$, since $T^2 pqT \subseteq T$. Hence from (4), since R_T/M_λ is strictly irreducible, there exist $c^{(i)} \in R_T$ such that $uac^{(i)} - r^{(i)} \in M_\lambda$ ($i = 1, 2, \ldots, n$). Thus $uac^{(i)} d^{(i)} - r^{(i)} d^{(i)} \in M_\lambda R_{Tp}$ ($i = 1, 2, \ldots, n$) and so

$$uab - v \in M_\lambda R_{Tp},$$

where $b := \sum_{i=1}^{n} c^{(i)} d^{(i)}$ ($\in R_T R_{Tp}$). Also, since $TpqT \subseteq T$,
$$(uab - v) R_{qT} \subseteq M_\lambda R_{Tp} R_{qT} \subseteq M_\lambda R_T \subseteq M_\lambda$$
and so $uab - v \in N_\lambda$. Further, $ab \in R_{qT} R_T R_{Tp} \subseteq R_U$, since $qT^3 p \subseteq U$, by (1). Hence $(u + N_\lambda) ab = v + N_\lambda$. This shows that $R_T R_{Tp}/N_\lambda$ is strictly irreducible. Thus the family $R_T R_{Tp}/N_\lambda$ ($\lambda \in \Lambda^*$) is faithful and irreducible and so R_U is semiprimitive.

(ii) Suppose that R_T is primitive. This corresponds to the case where $|\Lambda| = 1$ above (see Lemma 3.1(i)). Hence $\Lambda^* = \{\mu\}$, say, and $R_T R_{Tp}/N_\mu$ is a faithful irreducible R_U-module. Thus R_U is primitive. □

The first main theorem now follows easily from Lemma 4.3. It should be noted that, in [11], semiprimitivity is discussed from the standpoint of the Jacobson radical and primitivity is handled independently.

Theorem 4.4 [11]. *Let R be a ring $\mathcal{D}$-faithfully graded by a semigroup S and let e and f be $\mathcal{D}$-equivalent idempotents of S.*
  (i) *If R_e [resp. R_{H_e}] is semiprimitive then R_f [resp. R_{H_f}] is semiprimitive.*
  (ii) *If R_e [resp. R_{H_e}] is primitive then R_f [resp. R_{h_f}] is primitive.*

Proof. First observe that there exist mutually inverse elements p and q in S, with $pq = e$ and $qp = f$. We may therefore apply Lemma 4.3 with suitable choices for T and U. On the one hand, if we take $T = \{e\}$ and $U = \{f\}$ we get the results connecting R_e and R_f; on the other, if we take $T = H_e$ and $U = H_f$ then, noting that $pH_f q = H_e$, we get the results connecting R_{H_e} and R_{H_f}. □

In [11], analogous results are also established for semiprimeness and primeness.

5 Rings graded by inverse semigroups: historical notes

For the remainder of the paper, we consider rings graded by inverse semigroups. This subject has a long history, going back almost fifty years, and the purpose of the present section is to outline some key stages in its development.

Recall that a semigroup S is *inverse* if and only if, for all $x \in S$, there exists a unique $x^{-1} \in S$ (called the 'inverse' of x) such that $xx^{-1}x = x$ and $x^{-1}xx^{-1} = x^{-1}$. Examples include groups, semilattices and semigroups of

$n \times n$ matrix units. For a detailed account of the elementary properties of such semigroups, see [1] or [4].

The earliest result ([12],[13],[7]) concerns the semigroup ring of a *finite* inverse semigroup over a field and dates from the 1950s.

Result 5.1 (Ponizovskiĭ, Oganesyan, Munn). *Let F be a field and S a finite inverse semigroup. If, for all $e \in E(S)$, $F[H_e]$ is semiprimitive then $F[S]$ is semiprimitive.*

In fact, 5.1 was not originally stated in this form. Here semiprimitivity is equivalent to the classical notion of semisimplicity; and Maschke's theorem shows that, for all $e \in E(S)$, $F[H_e]$ is semisimple if F has characteristic 0 or a prime not dividing the order of H_e. The combined result leads to a representation theory of finite inverse semigroups, extending the classical theory of representations of finite groups. The problem is to construct the representations of S from those of the groups H_e.

The notion of a semilattice-graded ring makes its first appearance in a paper by Weissglass [14] in 1973, where such a system is described as a 'supplementary semilattice sum of rings'. The result below is established under a somewhat weaker hypothesis than the existence of a unity in each R_e; and properties other than semiprimitivity are also considered.

Result 5.2 (Weissglass). *Let R be a ring graded by a semilattice E. If, for all $e \in E$, R_e is semiprimitive and has a unity then R is semiprimitive.*

Using this, Weissglass deduces a result analogous to 5.1 for the case where S is an arbitrary Clifford semigroup. (See Example 2.5.)

A significant advance in the study of inverse semigroup rings was made by Domanov [3] in 1976. The omission of the finiteness assumption on the semigroup calls for a more sophisticated argument than that used to prove 5.1. Again, the statement below is paraphrased from the original.

Result 5.3 (Domanov). *Let F be a field and S an inverse semigroup. If, for all $e \in E(S)$, $F[H_e]$ is semiprimitive then $F[S]$ is semiprimitive.*

A semigroup is said to be *bisimple* if and only if it consists of a single $\mathcal{D}$-class. As was observed by Ponizovskiĭ, inspection of Domanov's proof of 5.3 yields the following sufficient condition for primitivity.

Result 5.4. *Let F be a field and S a bisimple inverse semigroup. If there exists $e \in E(S)$ such that $F[H_e]$ is primitive then $F[S]$ is primitive.*

The first general study of rings graded by inverse semigroups was made by Kelarev [5] in 1998. In particular, he obtained the result below.

Result 5.5 (Kelarev). *Let R be a ring faithfully graded by an inverse semigroup S. If, for all $e \in E(S)$, R_{H_e} is semiprimitive then R is semiprimitive.*

Finally, we mention another result on primitivity. This was obtained by the author [10] in 2000.

Result 5.6 (Munn). *Let R be a ring faithfully graded by a bisimple inverse semigroup S. If there exists $e \in E(S)$ such that R_{H_e} is primitive and if, for all $a \in R_{H_e}$, $a \in aR_{H_e}$ then R is primitive.*

It is clear that 5.5 generalises 5.3 (which, in turn, generalises 5.1) and also that 5.6 generalises 5.4. However, it is not true that 5.5 generalises 5.2; for, as was seen in Example 2.5, a ring graded by a semilattice need not be faithful.

6 Rings graded by inverse semigroups: further results

This section is devoted to establishing a two-part theorem which contains, as special cases, all the results mentioned in Section 5.

First, we have another lemma concerning $\mathcal{D}$-faithfulness.

Lemma 6.1. *Let R be a ring $\mathcal{D}$-faithfully graded by an inverse semigroup S and let A be a nonzero ideal of R. Then there exist $f \in E(S)$ and $a \in A\backslash 0$ such that $f \in \mathrm{supp}(a) \subseteq fSf$.*

Proof. Let $b \in A\backslash 0$. Choose f maximal in $\{xx^{-1} : x \in \mathrm{supp}(b)\}$, under the natural partial ordering of $E(S)$, and choose $y \in \mathrm{supp}(b)$ such that $yy^{-1} = f$.

Since $b_y \neq 0$ and $f \mathcal{D} y = fy$, there exists $c \in R_f$ such that $cb_y \neq 0$. Further, since $cb_y \neq 0$ and $y \mathcal{D} y^{-1} \mathcal{D} f$, there exists $d \in R_{y^{-1}}$ such that $cb_y d \neq 0$. Thus

$$cb_y d \in R_f \backslash 0. \tag{1}$$

Write $a := cbd$. Then $a \in A$ and $\mathrm{supp}(a) \subseteq \{fxy^{-1} : x \in \mathrm{supp}(b)\} \subseteq fSf$. By (1), to show that $f \in \mathrm{supp}(a)$ it suffices to prove that if $fxy^{-1} = f$ for $x \in \mathrm{supp}(b)$ then $x = y$. Let $x \in \mathrm{supp}(b)$ be such that $fxy^{-1} = f$.

Then $y^{-1}(fx)y^{-1} = y^{-1}f = y^{-1}$; also, $(fx)y^{-1}(fx) = f^2x = fx$. Hence $fx = (y^{-1})^{-1} = y$. Consequently, $fxx^{-1} = fx(fx)^{-1} = yy^{-1} = f$; that is, $f \le xx^{-1}$. Thus $f = xx^{-1}$, by the choice of f, and so $fx = x$. Since $fx = y$, we have that $x = y$. □

A stronger form of this lemma is stated in Section 7.

Next, we collect together several elementary observations on inverse semigroups. These will be used in the ensuing lemma. For a subset X of an inverse semigroup S we write $X^{-1} := \{x^{-1} : x \in X\}$. Note that if X is an $\mathcal{R}$-class of S then X^{-1} is an $\mathcal{L}$-class in the same $\mathcal{D}$-class and $X \cap X^{-1}$ is a maximal subgroup of S.

Lemma 6.2. *Let S be an inverse semigroup, let G be a maximal subgroup of S and let T denote the $\mathcal{R}$-class of S containing G. The following statements hold for all $x, y, z \in S$.*

(1) *If $x, xyz \in T$ then $xy \in T$.*
(2) *If $x \in G$, $y \in T$ and $xyz \in T$ then $yz \in T$.*
(3) *If $xyz^{-1} \in G$ and $z \in T$ then $yz^{-1} \in T^{-1}$.*
(4) *If $x, xy^{-1} \in T$ then $x^{-1}x \le y^{-1}y$.*
(5) *If $x, y \in T$ then $xy^{-1} \in G$ if and only if $x^{-1}x = y^{-1}y$.*
(6) *If $x \in G$ and $y, z \in T$ then $xyz^{-1} \in G$ if and only if $yz^{-1} \in G$.*

Proof. Denote the identity of G by e.

(i) Let $x, xyz \in T$. Then $xS = xyzS \subseteq xyS \subseteq xS$ and so $xS = xyS$. Thus $xy \in T$.

(ii) Let $x \in G$, $y \in T$ and $xyz \in T$. Then $yz = eyz = x^{-1}(xyz) \in GT = T$.

(iii) Let $xyz^{-1} \in G$ and $z \in T$. Then $zy^{-1}x^{-1} = (xyz^{-1})^{-1} \in G \subseteq T$ and so, by (i), $zy^{-1} \in T$, which gives $yz^{-1} \in T^{-1}$.

(iv) Let $x, xy^{-1} \in T$. Then $xy^{-1}(xy^{-1})^{-1} = e = xx^{-1}$ and so $x^{-1}(xy^{-1}yx^{-1})x = x^{-1}(xx^{-1})x$; that is, $(x^{-1}x)(y^{-1}y) = x^{-1}x$. Thus $x^{-1}x \le y^{-1}y$.

(v) Let $x, y \in T$. Assume that $xy^{-1} \in G$. By (iv), $x^{-1}x \le y^{-1}y$. Also, $yx^{-1} = (xy^{-1})^{-1} \in G$ and so, by (iv), $y^{-1}y \le x^{-1}x$. Hence $x^{-1}x = y^{-1}y$. Conversely, assume that $x^{-1}x = y^{-1}y$. Then $(xy^{-1})(xy^{-1})^{-1} = xy^{-1}yx^{-1} = x(x^{-1}x)x^{-1} = xx^{-1} = e$ and, similarly, $(xy^{-1})^{-1}xy^{-1} = e$. Hence $xy^{-1} \in G$.

(vi) Let $x \in G$ and $y, z \in T$. If $xyz^{-1} \in G$ then, by (ii) and (iii), $yz^{-1} \in T \cap T^{-1} = G$. Conversely, if $yz^{-1} \in G$ then clearly $xyz^{-1} \in G$. □

Lemma 6.3. *Let R be a ring graded by an inverse semigroup S, let D be a $\mathcal{D}$-class of S, let G be a maximal subgroup of S contained in D and let T be the $\mathcal{R}$-class of S containing G. Further, let M be a right ideal of R_G and let*

$$N := \{u \in R_G R_T : (uR_{T^{-1}})_G \subseteq M\}.$$

Then

(1) *the operation $\circ : R_T \times R \to R_T$ defined by $a \circ b = (ab)_T$ ($a \in R_T$, $b \in R$) is an action of R on R_T;*
(2) *$R_G R_T$ is a submodule of R_T;*
(3) *N is a submodule of $R_G R_T$ and so $R_G R_T / N$ is an R-module, with action $\star$ given by $(u + N) \star a = (u \circ a) + N$ ($u \in R_G R_T$, $a \in R$);*
(4) *if $N \neq R_G R_T$ and the R_G-module R_G/M is strictly irreducible then the R-module $R_G R_T/N$ is strictly irreducible.*

Proof. (i) Note first that if $x \in T$ and $y \in S$ then, for $a \in R_x$ and $b \in R_y$,

$$a \circ b = \begin{cases} ab & \text{if } xy \in T, \\ 0 & \text{otherwise.} \end{cases}$$

Since, for all $r, s \in R$, $(r + s)_T = r_T + s_T$, we see that

$$(\forall a, b \in R_T)(\forall c \in R) \quad (a + b) \circ c = a \circ c + b \circ c \tag{1}$$

and

$$(\forall a \in R_T)(\forall c, d \in R) \quad a \circ (c + d) = a \circ c + a \circ d, \tag{2}$$

it remains to show that

$$(\forall a \in R_T)(\forall c, d \in R) \quad (a \circ c) \circ d = a \circ (cd). \tag{3}$$

Let $x \in T$ and $y, z \in S$. In view of (1) and (2), it is sufficient to show that (3) holds for $a \in R_x$, $c \in R_y$ and $d \in R_z$. Let a, c, d be such elements. Then

$$(a \circ c) \circ d = \begin{cases} (ac)d & \text{if } xy \in T \text{ and } xyz \in T, \\ 0 & \text{otherwise,} \end{cases}$$

while

$$a \circ (cd) = \begin{cases} a(cd) & \text{if } xyz \in T, \\ 0 & \text{otherwise.} \end{cases}$$

But, by Lemma 6.2(i), if $xyz \in T$ then $xy \in T$. Hence $(a \circ c) \circ d = a \circ (cd)$. This establishes (3).

(ii) Note that $R_G R_T \subseteq R_T$ and that it is a subgroup of $(R_T, +)$. By (1), to prove that $R_G R_T$ is a submodule of R_T it is sufficient to show that, for all $a \in R_G$, $b \in R_T$ and $c \in R$, $(ab) \circ c \in R_G R_T$. Let $x \in G$, $y \in T$, $z \in S$ and let $a \in R_x$, $b \in R_y$, $c \in R_z$. By (1) and (2), it is enough to show that, for these elements, $(ab) \circ c \in R_G R_T$. If $xyz \in T$ then, by Lemma 6.2(ii), $yz \in T$ and so

$$(abc)_T = abc = a(bc)_T \in R_G R_T,$$

while if $xyz \notin T$ then $(abc)_T = 0$. The result follows.

(iii) It is readily verified that N is a subgroup of $(R_G R_T, +)$. Let $u \in N$ and $a \in R$. We have to show that $u \circ a \in N$.

Let $y \in S$, $z \in T$ and let $b := a_y$, $c \in R_{z^{-1}}$. We prove first that

$$((u \circ b)c)_G \in M. \tag{4}$$

Since this holds if $((u \circ b)c)_G = 0$, we assume that $((u \circ b)c)_G \neq 0$. Now observe that, for all $x \in T$, $((u_x b)_T c)_G = (u_x bc)_G$; for if $xyz^{-1} \in G$ then $xy \in T$, by Lemma 6.2(i), while if $xyz^{-1} \notin G$ then each side is 0. Thus $0 \neq ((u \circ b)c)_G = \sum_{x \in T}((u_x \circ b)c)_G = \sum_{x \in T}(u_x bc)_G$ and so there exists $x \in T$ such that $(u_x bc)_G \neq 0$. This implies that $xyz^{-1} \in G$ and hence that $yz^{-1} \in T^{-1}$, by Lemma 6.2(iii). Thus $bc \in R_{T^{-1}}$. Consequently,

$$((u \circ b)c)_G = \sum_{x \in T}(u_x bc)_G = (ubc)_G \in (uR_{T^{-1}})_G \subseteq M.$$

Hence (4) holds. It follows that $((u \circ a)c)_G \subseteq M$ and so

$$((u \circ a)R_{T^{-1}})_G \subseteq M.$$

Thus $u \circ a \in N$.

(iv) It may happen that $N = R_G R_T$ (as, for example, when $R_G = 0$). Assume that $N \neq R_G R_T$ and that the R_G-module R_G/M is strictly irreducible.

Let $u \in R_G R_T \setminus N$. Denote the elements of $\{x^{-1}x : x \in \operatorname{supp}(u)\}$ by $f_1, f_2, \ldots, f_n$ and, for all $i \in \{1, 2, \ldots, n\}$, let $u^{(i)}$ denote the sum of all those u_x for which $x \in \operatorname{supp}(u)$ and $x^{-1}x = f_i$. Thus

$$u = u^{(1)} + u^{(2)} + \cdots + u^{(n)}.$$

Note also that each $u^{(i)} \in R_G R_T$, since each $u_x \in R_G R_T$. Since $u \notin N$, $(uR_{T^{-1}})_G \not\subseteq M$ and so $\sum_{i=1}^n (u^{(i)} R_{T^{-1}})_G \not\subseteq M$. Assume that the f_i are numbered so that, for some $r \in \{1, 2, \ldots, n\}$,

$$(u^{(i)} R_{T^{-1}})_G \not\subseteq M \quad \Leftrightarrow \quad 1 \leq i \leq r.$$

Write $u' := u^{(1)} + u^{(2)} + \cdots + u^{(r)}$ and $u'' := u - u'$. Then, since $(u''R_{T^{-1}})_G \subseteq M$, we have that

$$u'' \in N. \tag{5}$$

Now assume also that f_1 is minimal in $\{f_1, f_2, \ldots, f_r\}$ and observe that, by Lemma 6.2(v),

$$(\forall x, y \in T) \quad xy^{-1} \in G \Leftrightarrow x^{-1}x = y^{-1}y. \tag{6}$$

Since $(u^{(1)}R_{T^{-1}})_G \not\subseteq M$ and $x^{-1}x = f_1$ for all $x \in \mathrm{supp}(u^{(1)})$, we see from (6) that there exist $y \in T$ and $a \in R_{y^{-1}}$ such that $y^{-1}y = f_1$ and $(u^{(1)}a)_G = u^{(1)}a \in R_G \backslash M$. Thus $u^{(1)} \circ a \in R_G \backslash M$. Further, if $2 \leq i \leq r$ then, for all $x \in \mathrm{supp}(u^{(i)}) \ (\subseteq T)$, we have that $x^{-1}x = f_i \not\leq f_1 = y^{-1}y$ and so, by Lemma 6.2(iv), $xy^{-1} \notin T$, which gives $u^{(i)} \circ a = 0$. Hence

$$u' \circ a \in R_G \backslash M. \tag{7}$$

Let $v \in R_G R_T$. Then $v = \sum_{j=1}^{k} r^{(j)} d^{(j)}$ for some positive integer k, some $r^{(j)} \in R_G$ and some $d^{(j)} \in R_T$ ($j = 1, 2, \ldots, k$). Since R_G/M is strictly irreducible, it follows from (7) that there exists $c^{(j)} \in R_G$ such that

$$(u' \circ a)c^{(j)} - r^{(j)} \in M \quad (j = 1, 2, \ldots, k).$$

Thus $(u' \circ a)c^{(j)}d^{(j)} - r^{(j)}d^{(j)} \in MR_T$ ($j = 1, 2, \ldots, k$) and so

$$(u' \circ a)b - v \in MR_T,$$

where $b := \sum_{j=1}^{k} c^{(j)} d^{(j)}$. Now, by Lemma 6.2(vi), for all $m \in M$, $p \in R_T$ and $q \in R_{T^{-1}}$ we have that $(mpq)_G = m(pq)_G$. Hence

$$(((u' \circ a)b - v)R_{T^{-1}})_G \subseteq (MR_T R_{T^{-1}})_G \subseteq MR_G \subseteq M$$

and so $(u' \circ a)b - v \in N$. Further,

$$(u' \circ a)b = ((u' \circ a)b)_T = (u' \circ a) \circ b = u' \circ (ab).$$

Consequently,

$$u' \circ (ab) - v \in N. \tag{8}$$

But, from (5), $u'' \circ (ab) \in N$. Hence, from (8),

$$u \circ (ab) - v = (u' \circ (ab) - v) + u'' \circ (ab) \in N$$

and so $(u + N) \star (ab) = v + N$. Thus $R_G R_T / N$ is strictly irreducible. □

Before stating the second main theorem we make a further definition. Let R be a ring graded by a nontrivial semigroup S with zero z. We say that R is *0-restricted* if and only if $R_z = 0$. The case where S is a 0-bisimple inverse semigroup (that is, an inverse semigroup with exactly two $\mathcal{D}$-classes, namely $S\backslash z$ and $\{z\}$) features in part (ii) below.

Theorem 6.4. *Let R be a ring $\mathcal{D}$-faithfully graded by an inverse semigroup S.*
 (1) *If, for each $\mathcal{D}$-class D of S, $R_{G(D)}$ is semiprimitive for some maximal subgroup $G(D)$ of S contained in D then R is semiprimitive.*
 (2) *If S is 0-bisimple, if R is 0-restricted and if R_G is primitive for some nonzero maximal subgroup G of S then R is primitive.*

Proof. (i) Assume that, for each $\mathcal{D}$-class D, there exists a maximal subgroup $G(D)$ in D such that $R_{G(D)}$ is semiprimitive. If $R_{G(D)} = 0$ for all choices of D then, by Lemma 4.1, $R = 0$ and so R is semiprimitive. Hence we assume that $R_{G(D)} \neq 0$ for at least one $\mathcal{D}$-class D. By Lemma 3.1(ii), for any D with $R_{G(D)} \neq 0$ there exists a nonempty family $\mathcal{F}(D)$ of proper right ideals of $R_{G(D)}$ such that $R_{G(D)}/M$ ($M \in \mathcal{F}(D)$) is a faithful family of strictly irreducible $R_{G(D)}$-modules.

For a $\mathcal{D}$-class D of S and a proper right ideal M of R_G, where $G := G(D)$, we denote the R-module $R_G R_T/N$, constructed as in Lemma 6.3, by $V(D, M)$. Let $\mathcal{V}$ denote the family of all modules $V(D, M)$, where D is a $\mathcal{D}$-class of S with $R_{G(D)} \neq 0$ and $M \in \mathcal{F}(D)$. We show that $\mathcal{V}$ is faithful.

Let $A := \{a \in R : \text{for all } V \in \mathcal{V},\ V \star a = 0\}$. Clearly, A is an ideal of R. Suppose that $A \neq 0$. By Lemma 6.1, there exist $f \in E(S)$ and $a \in A\backslash 0$ such that
$$f \in \operatorname{supp}(a) \subseteq H \cup K,$$
where $H := H_f$ and $K := fSf\backslash H$. Then
$$a = a_H + a_K, \quad a_H \neq 0. \tag{1}$$
Let D be the $\mathcal{D}$-class containing f, let $G := G(D)$ and let T be the $\mathcal{R}$-class of S containing G. By Lemma 4.2, there exist $x \in T$, with $x^{-1}x = f$, and $b \in R_x$, $c \in R_{x^{-1}}$ such that
$$ba_H c \in R_G\backslash 0.$$
Hence, since R_G/M ($M \in \mathcal{F}(D)$) is a faithful family, there exist $M \in \mathcal{F}(D)$ and $r \in R_G$ such that $rba_H c \in R_G\backslash M$. But $rba_H \in R_G R_x R_H \subseteq R_T$, since

$xH = H_x = Gx \subseteq GT = T$. Hence $rba_H c = (rba_H)_T c = (rb \circ a_H)c$ and so

$$(rb \circ a_H)c \in R_G \setminus M. \tag{2}$$

Let $k \in K$. Then $xkx^{-1} \notin G$; for otherwise, $k = fkf = x^{-1}(xkx^{-1})x \in x^{-1}Gx = H$, which is false. But, for all $g \in G$, $g^{-1}g = xx^{-1}$ and so $gxkx^{-1} \notin G$. Hence $((rb \circ a_k)c)_G = 0$. This shows that

$$((rb \circ a_K)c)_G = 0. \tag{3}$$

Now, from (1),(2) and (3), we have that $((rb \circ a)c)_G \notin M$ and so, since $c \in R_{x^{-1}} \subseteq R_{T^{-1}}$,

$$((rb \circ a)R_{T^{-1}})_G \not\subseteq M.$$

Thus $rb \circ a \notin N$; that is, $(rb + N) \star a \neq N$. This shows that $V(D, M) \star a \neq 0$ and contradicts the assumption that $a \in A$. Consequently, $A = 0$ and so $\mathcal{V}$ is faithful. Since, by Lemma 6.3(iv), each nonzero V in $\mathcal{V}$ is strictly irreducible, it follows that there exists a faithful family of strictly irreducible R-modules. Hence R is semiprimitive.

(ii) Assume that S is 0-bisimple, that R is 0-restricted and that R_G is primitive for some nonzero maximal subgroup of S. Let D be the nonzero $\mathcal{D}$-class of S. In this case we may take $\mathcal{F}(D) = \{M\}$, where M is a proper right ideal of R_G such that R_G/M is a faithful strictly irreducible R_G-module. The proof of (i) then shows that $V(D, M)$ is a faithful irreducible R-module. Thus R is primitive. □

It is clear that Theorem 6.4(i) generalises Result 5.5: first, $\mathcal{D}$-faithfulness replaces faithfulness and, second, we only assume that R_G is semiprimitive for *one* maximal subgroup G in each $\mathcal{D}$-class of S. In addition, it generalises Result 5.2, as we now demonstrate. Let R be a ring graded by a semilattice E and suppose that, for all $e \in E$, R_e is semiprimitive. Since $\mathcal{D}$ is the trivial relation on E, R is $\mathcal{D}$-faithful if and only if, for all $e \in E$ and all $a \in R_e \setminus 0$, $aR_e \neq 0$ and $R_e a \neq 0$. But it is readily verified that this condition is satisfied since each R_e is semiprimitive. Thus R is $\mathcal{D}$-faithful and Theorem 6.4(i) applies.

Also, Theorem 6.4(ii) generalises Result 5.6. To see this, consider a ring R faithfully graded by a bisimple inverse semigroup S. Then R may be regarded as a 0-restricted ring $\mathcal{D}$-faithfully graded by the 0-bisimple semigroup S^0 obtained by adjoining a zero to S. Hence, if R_G is primitive for some maximal subgroup of S then, by 6.4(ii), R itself is primitive.

Thus each of the Results 5.1 to 5.6 is a consequence of Theorem 6.4.

7 An alternative proof of Theorem 6.4(i)

Recall that an element a of a ring R is *right quasiregular* if and only if there exists $b \in R$ such that $a + b = ab$. The *Jacobson radical* $J(R)$ of R can then be defined by

$$J(R) := \{a \in R : \text{for all } x \in R, \ ax \text{ is right quasiregular}\}.$$

It is a standard result that $J(R)$ is an ideal of R and that R is semiprimitive if and only if $J(R) = 0$.

Just as a short proof of Theorem 4.4(i) can be obtained using the Jacobson radical (see [11, Theorem 3.3(i)]), so also we can give a short proof of Theorem 6.4(i). This is provided below. It should be pointed out, however, that in each case the second half of the theorem has then to be established by a separate argument. Moreover, the alternative proof of 6.4(i) also uses 4.4(i).

The following lemma is required. It had its origin in a result by the author on inverse semigroup rings ([8], [9]) which was subsequently adapted by Kelarev [5] to the case of rings faithfully graded by inverse semigroups.

Lemma 7.1. *Let R be a ring $\mathcal{D}$-faithfully graded by an inverse semigroup S and let A be a nonzero ideal of R. Then there exist $f \in E(S)$ and $a \in A \backslash 0$ such that*

$$f \in \operatorname{supp}(a) \subseteq H_f \cup (fSf \backslash P_f),$$

where P_f denotes $\{x \in fSf : x \mathcal{R} f\}$.

The proof is similar to that of Lemma 6.1, although more elaborate. Let $b \in A \backslash 0$, let f be maximal in $\{xx^{-1} : x \in \operatorname{supp}(b)\}$, let g be minimal in $\{x^{-1}x : x \in \operatorname{supp}(b) \text{ and } xx^{-1} = f\}$ and let $y \in \operatorname{supp}(b)$ be such that $yy^{-1} = f$, $y^{-1}y = g$. Using $\mathcal{D}$-faithfulness, we see that there exist $c \in R_f$ and $d \in R_{y^{-1}}$ such that $cb_y d \in R_f \backslash 0$. Then the element $a := cbd$ can be shown to have the required properties.

Proof of Theorem 6.4(i). Let R be a ring $\mathcal{D}$-faithfully graded by an inverse semigroup S and assume that, for each $\mathcal{D}$-class D of S, there exists a maximal subgroup $G(D)$ in D such that $R_{G(D)}$ is semiprimitive. Suppose that $J(R) \neq 0$. We show that this leads to a contradiction.

By Lemma 7.1, there exist $f \in E(S)$ and $a \in J(R) \backslash 0$ such that

$$f \in \operatorname{supp}(a) \subseteq H \cup K,$$

where $H := H_f$ and $K := fSf \backslash P_f$. Note that $H \cap K = \emptyset$, since $H_f \subseteq P_f$. Then

$$a = a_H + a_K, \quad a_H \neq 0. \tag{1}$$

Let $b \in R_H$. Then $ab \in J(R)$ and so there exists $r \in R$ such that

$$ab + r - abr = 0. \tag{2}$$

Now $a \in R_{fS}$, since $H \subseteq fS$ and $K \subseteq fS$. Hence $ab, abr \in R_{fS}$ and so, from (2), $r \in R_{fS}$. Thus

$$r = r_H + r_L, \tag{3}$$

where $L := fS \backslash H$. We now show that

$$HL \subseteq L \tag{4}$$

and

$$KS \subseteq L \tag{5}$$

(with the appropriate interpretation if $K = \emptyset$ and/or $L = \emptyset$).

Let $x \in H$ and $y \in L$. Then $xy \in fS$. But if $xy \in H$ then $y = fy = (x^{-1}x)y = x^{-1}(xy) \in H$, contrary to the definition of L. Hence $xy \in fS \backslash H = L$, which proves (4).

Next, let $x \in K$ and $y \in S$. Again, $xy \in fS$. Suppose that $xy \in H$. Then $xy(xy)^{-1} = f$. But $fx = x$. Hence $x \mathcal{R} f$ and so $x \in P_f$, contrary to the definition of K. Hence $xy \in fS \backslash H = L$, which proves (5).

From (1), (2) and (3),

$$(a_H + a_K)b + (r_H + r_L) - (a_H + a_K)b(r_H + r_L) = 0$$

and so

$$(a_H b + r_H - a_H br_H) + (r_L - a_H br_L + a_K(b - br)) = 0. \tag{6}$$

But, from (4) and (5),

$$r_L - (a_H b)r_L + a_K(b - br) \in R_L. \tag{7}$$

Also, $H \cap L = \emptyset$ and so $R_H \cap R_L = 0$. Hence, from (6) and (7),

$$a_H b + r_H - (a_H b)r_H = 0.$$

Since b was chosen arbitrarily in R_H, this shows that $a_H \in J(R_H)$. Hence, since $a_H \neq 0$, $J(R_H) \neq 0$. But, by hypothesis, $J(R_{H_e}) = 0$ for some $e \in E(S)$ with $e \mathcal{D} f$ and so, by Theorem 4.4(i), $J(R_H) = 0$, contrary to our previous conclusion. Hence $J(R) = 0$. □

Acknowledgments

I am grateful to my friend and colleague, Dr Michael J. Crabb, for his helpful advice on the preparation of the manuscript. In particular, I am much indebted to him for suggesting the formulation of the module $V(D, M)$ in the proof of Theorem 6.4. This is simpler and more elegant than my original version.

References

[1] A. H. Clifford and G. B. Preston, *The algebraic theory of semigroups*, Math. Surveys 7, vol.1 (Amer. Math. Soc., Providence RI, 1961).

[2] M. Cohen and M. S. Montgomery, Group-graded rings, smash products and group actions, *Trans. Amer. Math. Soc.* **282** (1984), 237–258.

[3] O. I. Domanov, On semisimplicity and identities of inverse semigroup algebras, *Rings and Modules, Mat. Issled. Vyp.* **38** (1976), 123–137.

[4] J. M. Howie, *An introduction to semigroup theory* (Acad. Press, London, 1976).

[5] A. V. Kelarev, Semisimple rings graded by inverse semigroups, *J. Algebra* **205** (1998), 451–459.

[6] N. H. McCoy, *The theory of rings* (Macmillan, New York, 1964).

[7] W. D. Munn, Matrix representations of semigroups, *Proc. Cambridge Philos. Soc.* **53** (1957), 5–12.

[8] W. D. Munn, Nil ideals in inverse semigroup algebras, *J. London Math. Soc.* **35** (1987), 433–438.

[9] W. D. Munn, On contracted semigroup rings, *Proc. Roy. Soc. Edinburgh* **115A** (1990), 109–117.

[10] W. D. Munn, Rings graded by bisimple inverse semigroups, *Proc. Roy. Soc. Edinburgh* **130A** (2000), 603–609.

[11] W. D. Munn, $\mathcal{D}$-faithful semigroup-graded rings, *Proc. Edinburgh Math. Soc.* (to appear).

[12] V. A. Oganesyan, On the semisimplicity of a system algebra, *Akad. Nauk Armyan. SSR Dokl.* **21** (1955), 145–147.

[13] J. S. Ponizovskiĭ, On matrix representations of associative systems, *Mat. Sbornik* **38** (1956), 241–260.

[14] J. Weissglass, Semigroup rings and semilattice sums of rings, *Proc. Amer. Math. Soc.* **39** (1973), 471–478.

PROFINITE GROUPS AND APPLICATIONS TO FINITE SEMIGROUPS

LUIS RIBES[*]

*School of Mathematics and Statistics, Carleton University, Ottawa, ON, Canada
K1S 5B6
E-mail: lribes@math.carleton.ca*

This article is based on lectures given at the Thematic Term 2001 on Algorithmic Aspects of the Theory of Semigroups and Automata in Coimbra in May of 2001, and at the Università di L'Aquila in June of 2001. It is a self contained introduction to the theory of profinite groups and profinite graphs with emphasis on algorithms on subgroups of abstract free groups.

1 Introduction

This article is based on lectures given at the Thematic Term 2001 on Algorithmic Aspects of the Theory of Semigroups and Automata in Coimbra in May 2001, and at the Università di L'Aquila in June 2001.

The purpose of these lectures has been to present sufficient background on the theory of profinite groups and profinite graphs so as to allow a good understanding of the proofs of the results on algorithms on subgroups of abstract free groups that constitute the center of the lectures. The necessary background is presented in Sections 2-5, and the main results in Sections 6 and 7. In Section 8 we give a brief indication of the motivations that originally came from several conjectures on algorithms in finite semigroups.

2 Profinite Groups

Profinite groups are Galois groups of Galois extensions (finite or infinite) K/F of fields. For example,

(i) Every finite group is a profinite group.

---

[*]I WOULD LIKE TO ACKNOWLEDGE THE FINANCIAL SUPPORT OF FUNDAÇÃO CALOUSTE GULBENKIAN (FCG), FUNDAÇÃO PARA A CIÊNCIA E A TECNOLOGIA (FCT), FACULDADE DE CIÊNCIAS DA UNIVERSIDADE DE LISBOA (FCUL) AND REITORIA DA UNIVERSIDADE DO PORTO.

(ii) Let $\mathbb{F}_q$ be a field with q elements, where q is a prime number. Let $\bar{\mathbb{F}}_q$ be the algebraic closure of $\mathbb{F}_q$. Define $\widehat{\mathbb{Z}}$ to be the Galois group $\mathrm{Gal}(\bar{\mathbb{F}}_q/\mathbb{F}_q)$ of the extension $\bar{\mathbb{F}}_q/\mathbb{F}_q$. Then $\widehat{\mathbb{Z}}$ is an infinite profinite group (as we shall see later it is the 'free profinite group of rank one').

(iii) Let p and q be prime numbers and let K be the field consisting of the roots in $\bar{\mathbb{F}}_q$ of all polynomials in $\mathbb{F}_q[X]$ of the form $X^{q^{p^t}} - X$ ($t = 0, 1, 2, \ldots$). Then $K/\mathbb{F}_q$ is an infinite Galois extension whose Galois group $\mathrm{Gal}(K/\mathbb{F}_q)$ is $\mathbb{Z}_p$, the additive group of p-adic integers. It is not hard to see that

$$\widehat{\mathbb{Z}} = \prod \mathbb{Z}_p,$$

where p ranges over all primes p.

We give below an equivalent definition of profinite groups. In fact we wish to be more discriminating, and we shall define the more general concept of pro-$\mathcal{C}$ group.

From now on $\mathcal{C}$ denotes a class of finite groups closed under taking subgroups, quotients and finite direct products. We refer to such a class as a variety of finite groups.

Definition 2.1. A *pro-$\mathcal{C}$ group* G is an inverse limit

$$G = \varprojlim_{i \in I} G_i$$

of groups G_i in $\mathcal{C}$.

Remark 2.2. Recall that in the definition above it is understood that I is a poset (partially ordered set) with a partial ordering that usually is denoted by $\preceq$. The groups G_i form an inverse system $\{G_i, \varphi_{ij}, I\}$, i.e., whenever $i, j \in I$ and $i \succeq j$, there is a homomorphism $\varphi_{ij} : G_i \longrightarrow G_j$ such that $\varphi_{ii} : G_i \longrightarrow G_i$ is the identity homomorphism and $\varphi_{kj}\varphi_{ik} = \varphi_{ij}$ if $i \succeq k \succeq j$.

Then $\varprojlim_{i \in I} G_i$ is the following suset of the direct product $\prod_{i \in I} G_i$:

$$\varprojlim_{i \in I} G_i = \{(x_i) \in \prod_{i \in I} G_i \mid x_i \in G_i, \quad \varphi_{ij}(x_i) = x_j \quad \text{whenever} \quad i \succeq j\}.$$

We think of the finite groups G_i as topological groups with the discrete topology; so that the pro-$\mathcal{C}$ group $G = \varprojlim_{i \in I} G_i$ is a (closed) subspace of the topological space $\prod_{i \in I} G_i$. Using these facts one readily shows the following theorem.

Theorem 2.1. *The group G is pro-$\mathcal{C}$ if and only if G is a compact Hausdorff totally-disconnected topological group such that $G/U \in \mathcal{C}$ whenever U is an open normal subgroup of G.*

Terminology 2.3.
 (a) Any pro-$\mathcal{C}$ group is a *profinite group*.
 (b) If $\mathcal{C}$ is a class of finite solvable groups, then a pro-$\mathcal{C}$ group is called *prosolvable*.
 (c) If $\mathcal{C}$ is a class of finite nilpotent groups, then a pro-$\mathcal{C}$ group is called *pronilpotent*.
 (d) If $\mathcal{C}$ is a class of finite p-groups, then a pro-$\mathcal{C}$ group is called *pro-p*.

Example 2.4.
 (a) It is easy to see that for the group $\widehat{\mathbb{Z}}$ mentioned above, we have

$$\widehat{\mathbb{Z}} = \varprojlim_{i \in I} \mathbb{Z}/n\mathbb{Z}.$$

(Observe that in this case, the poset 'I' is the set of natural numbers $\mathbb{N}$, and $n \succeq m$ means $m \mid n$.)
 (b) Similarly, the additive group of p-adic integers $\mathbb{Z}_p$ is a pro-p group that can be expressed as

$$\mathbb{Z}_p = \varprojlim_{n \in I} \mathbb{Z}/p^n\mathbb{Z}.$$

(Here the poset 'I' is the set of natural numbers $\mathbb{N}$, and $n \succeq m$ means $n \geq m$.)

Completions 2.5. Let G be an abstract group. Denote the set of all normal subgroups of G of finite index whose corresponding quotient groups are in $\mathcal{C}$ by $\mathcal{N}_{\mathcal{C}}$; i.e.,

$$\mathcal{N}_{\mathcal{C}} = \{N \mid N \triangleleft G,\ G/N \in \mathcal{C}\}.$$

Then there is a unique topology on G making it into a topological group where the set $\mathcal{N}_{\mathcal{C}}$ is a fundamental system of neighborhoods of the identity element 1 of G. This topology is called the *pro-$\mathcal{C}$ topology* of G. If $\mathcal{C}$ is the class of all finite groups, all finite p-groups, etc., then correspondingly, the topology of G is called the *profinite topology*, the *pro-p topology*, etc., of G.

Then $\{G/N \mid N \in \mathcal{N}_\mathcal{C}\}$ is naturally an inverse system of finite groups in $\mathcal{C}$. The *pro-$\mathcal{C}$ completion* $G_{\hat{\mathcal{C}}}$ of G is the pro-$\mathcal{C}$ group defined by

$$G_{\hat{\mathcal{C}}} = \varprojlim_{N \in \mathcal{N}_\mathcal{C}} G/N.$$

If $\mathcal{C}$ is the class of all finite groups, we usually denote $G_{\hat{\mathcal{C}}}$ by $\widehat{G}$ and we call it the profinite completion; and if $\mathcal{C}$ is the class of all finite p-groups, we usually denote $G_{\hat{\mathcal{C}}}$ by $G_{\hat{p}}$ and we call it the pro-p completion.

For example $\widehat{\mathbb{Z}}$ is the profinite completion of $\mathbb{Z}$ and $\mathbb{Z}_p$ is the pro-p completion of $\mathbb{Z}$.

Lemma 2.2. *Let G and $\mathcal{C}$ be as above. Then*
(a) *there is a natural homomorphism $\varphi : G \longrightarrow G_{\hat{\mathcal{C}}}$ given by $\varphi(g) = (gN \mid N \in \mathcal{N}_\mathcal{C})$ $(g \in G)$.*
(b) $\operatorname{Ker}(\varphi) = \bigcap_{N \in \mathcal{N}_\mathcal{C}} N.$
(c) *φ is an injection if and only if G is residually $\mathcal{C}$ (recall that G is residually $\mathcal{C}$ means that whenever $1 \neq x \in G$, there exists some $N \triangleleft G$ with $G/N \in \mathcal{C}$ and $x \notin N$).*

Next we recall several results without proofs. The reader may consult the indicated references for details.

Proposition 2.3. [3, Proposition 2.2.4] $\varprojlim$ *is an exact functor.*

Explicitly this means the following:
Let $(I, \preceq)$ be a fixed poset and let

$$\{K_i, \varphi_{ij}, I\}, \quad \{G_i, \psi_{ij}, I\} \quad \text{and} \quad \{H_i, \rho_{ij}, I\}$$

be inverse systems of profinite groups over the poset I. Let

$$1 \longrightarrow \{K_i, \varphi_{ij}, I\} \xrightarrow{\{f_i\}} \{G_i, \psi_{ij}, I\} \xrightarrow{\{g_i\}} \{H_i, \rho_{ij} I\} \longrightarrow 1$$

be an exact sequence of inverse systems. Then

$$1 \longrightarrow \varprojlim_{i \in I} K_i \xrightarrow{f} \varprojlim_{i \in I} G_i \xrightarrow{g} \varprojlim_{i \in I} H_i \longrightarrow 1$$

is an exact sequence of profinite groups, where f and g are the continuous homomorphisms $f = \varprojlim f_i$ and $g = \varprojlim g_i$.

Proposition 2.4. [3, Lemma 3.1.2] *Let G be a group endowed with its pro-$\mathcal{C}$ topology.*

(a) Let U be a subgroup of G of finite index. Then U is open in G if an only if
$$G/U_G \in \mathcal{C},$$
where U_G denotes the core of U in G, i.e., the intersection of all the conjugates of U by elements of G.
(b) A subgroup H of G is closed in G if and only if $H = \bigcap U$, where U ranges over the collection of all open subgroups of G containing H.

In some of the most important applications in this paper we shall require that the class $\mathcal{C}$ be in addition *extension closed*, i.e., that whenever K and H are finite groups in $\mathcal{C}$ and
$$1 \longrightarrow K \longrightarrow G \longrightarrow H \longrightarrow 1$$
is an exact sequence of finite groups, then $G \in \mathcal{C}$. For example the class of all finite groups, or of all finite solvable groups, or of all finite p-groups are extension closed, while the class of all finite nilpotent groups is not extension closed.

Proposition 2.5. [3, Lemma 3.1.4] *Assume that the class $\mathcal{C}$ is extension closed. Let U be an open subgroup of an abstract group G endowed with the pro-$\mathcal{C}$ topology. Then the pro-$\mathcal{C}$ topology of U coincides with the topology induced on U from the pro-$\mathcal{C}$ topology of G.*

Proposition 2.6. [3, Lemma 3.1.6] *Let the group G be the free product of its subgroups H and L:*
$$G = H * L.$$
(a) *The pro-$\mathcal{C}$ topology of H coincides with the topology induced from the pro-$\mathcal{C}$ topology of G.*
(b) *Assume that G is residually finite. Then H is closed in the pro-$\mathcal{C}$ topology of G.*

Proposition 2.7. [3, Lemma 3.2.6] *Let H be a subgroup of a group G. Then the topological closure $\bar{H}$ of H in $G_{\hat{\mathcal{C}}}$ is naturally isomorphic with $H_{\hat{\mathcal{C}}}$ if and only if the pro-$\mathcal{C}$ topology of G induces on H its pro-$\mathcal{C}$ topology.*

3 Free Pro-$\mathcal{C}$ Groups

Definition 3.1.

(a) A topological space X is called a *Boolean space* or a *profinite space* if X is compact Hausdorff and totally disconnected. In this article, we prefer to use the term 'profinite'. The justification for this terminology is that a profinite space can be written as an inverse limit, also called a 'projective limit', of finite discrete spaces.

(b) A *pointed* profinite space $(X, *)$ is simply a profinite space X with a distinguished point $*$.

(c) A *map of pointed spaces* $\varphi : (X, *) \longrightarrow (X', *')$ is a continuous map from X to Y such that $\varphi(*) = *'$.

Definition 3.2. Let X be a profinite space. We say that a pro-$\mathcal{C}$ group $F = F(X)$ is a *free pro-$\mathcal{C}$ group* on X if there is a continuous map $\iota : X \longrightarrow F$ and the following universal property is satisfied:

whenever $\varphi : X \longrightarrow G$ is a continuous map from X into a pro-$\mathcal{C}$ group G, then there exists a unique continuous homomorphism

$$\bar{\varphi} : F \longrightarrow G$$

such that $\bar{\varphi}\iota = \varphi$.

There is an analogous definition of a free pro-$\mathcal{C}$ group on a pointed profinite space: simply replace $(X, *)$ for X and continuous map of pointed spaces for continuous map in the definition above (a group is naturally a pointed space if we take its identity element 1 as the distinguished point).

Remark 3.3. It is not hard to check that

(a) The map $\iota : X \longrightarrow F$ is an injection (hence, we often assume that X is a subset of F).

(b) It suffices to check the universal property in Definition 3.2 for finite groups in $\mathcal{C}$ only.

(c) If a free pro-$\mathcal{C}$ group on X exists, it is unique up to isomorphism (this follows from the universal property).

Theorem 3.1. (Existence of free pro-$\mathcal{C}$ groups) *Given a profinite space X (respectively, a pointed profinite space $(X, *)$), there exists a free pro-$\mathcal{C}$ group $F = F(X)$ (respectively, a free pro-$\mathcal{C}$ group $F = F(X, *)$).*

Proof. Let D be the abstract free group on X (considered as a set). Let
$$\mathcal{N} = \{N \mid N \triangleleft D,\ D/N \in \mathcal{C},\ X \cap dN \text{ is open in } X,\ \forall d \in D\}.$$
Then D becomes a topological group for which $\mathcal{N}$ is a fundamental system of neighborhoods of 1. Define
$$F = F(X) = \varprojlim_{N \in \mathcal{N}} D/N.$$
Then the map
$$\iota: X \hookrightarrow D \to \varprojlim_{N \in \mathcal{N}} D/N = F(X)$$
is continuous, and one checks that $F(X)$ and ι satisfy the required universal property. □

Example 3.4.
(1) $\widehat{\mathbb{Z}}$ is the free profinite group on the space $\{1\}$.
(2) $\widehat{\mathbb{Z}}_p$ is the free pro-p group on the space $\{1\}$.
(3) (Douady [22]) Let $X = \mathbf{C} \cup \{*\}$ be the one point compactification of the space of all complex numbers with the discrete topology. Let $G = \text{Gal}(\overline{\mathbf{C}(t)}/\mathbf{C}(t))$ be the absolute Galois group of the field $\mathbf{C}(t)$ of rational functions over the complex numbers. Then G is the free profinite group on the pointed space $(X, *)$.

The following proposition gives an alternative way of characterizing free pro-$\mathcal{C}$ groups on spaces which are one-point compatifications of discrete spaces. The proof is easy and we leave it to the reader.

Proposition 3.2. *Let Y be a discrete topological space and let $X = Y \cup \{*\}$ be its one-point compactification. Then*
(1) *X is a profinite space.*
(2) *The free pro-$\mathcal{C}$ group $F = F(X, *)$ on the pointed space $(X, *)$ can be characterize by the following universal property:*

(i) *$\iota : Y \longrightarrow F$ is a map converging to 1, i.e., every neighborhood U of 1 in F contains almost all the points of $\iota(Y)$: in other words, $\iota(Y) - U$ is finite;*

(ii) $\iota(Y)$ generates F as a topological group;

(iii) whenever G is a pro-$\mathcal{C}$ group and $\varphi : Y \longrightarrow G$ is a map converging to 1, then there exists a unique continuous homomorphism $\bar{\varphi} : F \longrightarrow G$ such that $\bar{\varphi}\iota = \varphi$.

Terminology: When F is as in the proposition above, one says that F is a *free pro-$\mathcal{C}$ group on the set Y converging to 1*; sometimes this group is denoted by $F(Y)$, although this might lead to confusion unless one qualifies this notation.

The following result reduces the study of free pro-$\mathcal{C}$ groups over a general profinite space to that of free pro-$\mathcal{C}$ groups over finite spaces.

Proposition 3.3. *Let $\{X_i, \varphi_{ij}, I\}$ be an inverse system of finite discrete spaces X_i. Let*
$$X = \varprojlim_{i \in I} X_i.$$
Then
$$F(X) = \varprojlim_{i \in I} F(X_i).$$

[A similar result holds for free pro-$\mathcal{C}$ groups over pointed spaces.]

Proof. (Sketch)

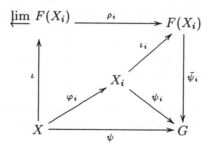

We have to show that $\varprojlim_{i \in I} F(X_i)$ satisfies the universal property of a free pro-$\mathcal{C}$ group. Let $\psi : X \longrightarrow G$ be a continuous map into a group G in $\mathcal{C}$. Since G is finite, there exists some $i \in I$ such that ψ factors through X_i, i.e., there is a continuous map $\psi_i : X_i \longrightarrow G$ with $\psi = \psi_i \varphi_i$, where $\varphi_i : X \longrightarrow X_i$ is the canonical projection. Then ψ_i induces a continuous homomorphism
$$\bar{\psi}_i : F(X_i) \longrightarrow G.$$

Let $\rho_i : \varprojlim_{i \in I} F(X_i) \longrightarrow F(X_i)$ denote the projection, and set $\bar{\psi} = \bar{\psi}_i \rho_i$. Define
$$\iota : X \longrightarrow \varprojlim_{i \in I} F(X_i)$$
to be the map induced by the $\iota_i \varphi_i$. Then clearly $\bar{\psi}\iota = \psi$. $\square$

4 Profinite Graphs

Definition 4.1. A *profinite graph* Γ consists of a profinite space Γ with a distinguished closed subspace $V = V(\Gamma)$ (the space of *vertices*) and two continuous maps
$$d_0, d_1 : \Gamma \longrightarrow V$$
such that $d_{0|V}$ and $d_{1|V}$ are the identity maps on V.

Terminology 4.2.
(a) The space $E = E(\Gamma) = \Gamma - V(\Gamma)$ is called the space of *edges* of Γ.
(b) If $e \in E$, then $d_0(e)$ and $d_1(e)$ are the *initial* and *terminal* vertices of e, respectively.

Definition 4.3. Let Γ and Γ' be profinite graphs. A *morphism*
$$\alpha : \Gamma \longrightarrow \Gamma'$$
is a continuous map such that
$$\alpha d_j = d_j \alpha, \quad (j = 0, 1).$$

Example 4.4.
(a) Every finite graph is a profinite graph.
(b) Let $\Gamma = \mathbb{N} \cup \{*\}$ be the one-point compactification of the space of natural numbers $\mathbb{N} = \{0, 1, 2, \ldots\}$ with the discrete topology. Put $V(\mathbb{N}) = \{*\}$ and define
$$d_0, d_1 : \Gamma \longrightarrow V(\Gamma)$$
to be the constant maps that send every element of Γ to $*$. Then Γ is a profinite graph (a *bouquet*). Observe that in this case, $E(\Gamma)$ is not a profinite space.
(c) Let G be a profinite group and let X be a closed subset of G. Define the *Cayley graph* $\Gamma = \Gamma(G, X)$, associated with G and X, as follows:
$$V(\Gamma) = G; \quad E(\Gamma) = G \times X \quad \text{and} \quad \Gamma = V(\Gamma) \cup E(\Gamma);$$

$$d_0(g,x) = g, \quad d_1(g,x) = gx, \quad d_0(g) = d_1(g) = g, \quad \forall g \in G, \forall x \in X.$$

Definition 4.5. Let Δ be a closed subgraph of a profinite graph Γ. Define the *quotient graph* $\tilde{\Gamma} = \Gamma/\Delta$ as follows: as a space, $\tilde{\Gamma}$ is the quotient space Γ/Δ; let $q : \Gamma \longrightarrow \tilde{\Gamma} = \Gamma/\Delta$ be the quotient map, and define $V(\tilde{\Gamma}) = q(V(\Gamma))$ and $d_j(q(x)) = q(d_j(x))$, $\forall x \in \Gamma$. We say that $\tilde{\Gamma}$ is *the result of collapsing Δ to a vertex*.

Note that in the notation above, q becomes a morphism of profinite graphs. In this case, if Δ has edges and $e \in E(\Delta)$, then $q(e)$ is a vertex. In contrast, a general morphism of graphs sends vertices to vertices but, as the above case shows, edges can be sent to vertices.

Theorem 4.1. *Let Γ be a profinite graph. Then Γ can be written as an inverse limit of finite graphs.*

Proof. (Sketch) Since Γ is a profinite space, it can be written as

$$\varprojlim_R \Gamma/R$$

where R ranges over the set $\mathcal{R}$ of relations on Γ whose equivalence classes xR are clopen. For each $R \in \mathcal{R}$, choose $R' \in \mathcal{R}$ with $R' \succeq R$ (i.e., each equivalence class of R' is contained in some equivalence class of R) such that for each $x \in \Gamma$, there are some $y, z \in \Gamma$ with $d_0(xR') \subseteq yR'$ and $d_1(xR') \subseteq zR'$. Then Γ/R' is a finite quotient graph of Γ and

$$\Gamma = \varprojlim_{R'} \Gamma/R'$$

(since the R' form a cofinal subset of $\mathcal{R}$). $\square$

Definition 4.6. We say that a profinite graph is *connected* if all its finite quotient graphs are connected (as abstract graphs).

Theorem 4.2. *Let Γ be a profinite graph.*
  (a) *If D is an abstract subgraph of Γ, then $\bar{D}$ (the topological closure of D) is a profinite graph.*
  (b) *If $\{\Delta_i \mid i \in I\}$ is a collection of closed connected subgraphs of Γ and $\bigcap \Delta_i \neq \emptyset$, then*

$$\overline{\bigcup_{i \in I} \Delta_i}$$

is connected.

(c) *Let*
$$\Delta_1, \ldots, \Delta_t$$
be a finite collection of closed connected subgraphs of Γ *such that* $\Delta_i \cap \Delta_{i+1} \neq \emptyset$ $(i = 1, \ldots, t-1)$. *Then*
$$\Delta_1 \cup \cdots \cup \Delta_t$$
is a connected profinite graph.

Proof. These assertions follow immediately from the definitions. □

Remark that a connected profinite graph need not be connected when viewed as an abstract graph. For example, define a graph Γ as the one-point compactification $\mathbb{N} \cup \mathbb{N} \cup \{v\}$ of $\mathbb{N} \cup \mathbb{N}$ (here, the first copy of $\mathbb{N}$, together with v is the set of vertices of the graph):

$$v_0 \longrightarrow v_1 \longrightarrow v_2 \longrightarrow \cdots \qquad v$$

This graph can be expressed as an inverse limit of finite connected graphs as follows. Let Γ_n be the finite graph

$$v_0 \longrightarrow v_1 \longrightarrow v_2 \longrightarrow \cdots \longrightarrow v_n \ .$$

Observe that we may think of Γ_n as the quotient graph of Γ_{n+1} obtained by collapsing its subgraph $v_n \longrightarrow v_{n+1}$ to the vertex v_n of Γ_n. Define $\Gamma_{n+1} \longrightarrow \Gamma_n$ to be the canonical quotient morphism. Then

$$\Gamma_1 \longleftarrow \Gamma_2 \longleftarrow \cdots$$

is an inverse system of finite connected graphs and $\Gamma = \varprojlim \Gamma_n$. Therefore Γ is connected as a profinite graph; but v is not a vertex for any edge of Γ, and so Γ is not connected as an abstract graph.

5 Pro-$\mathcal{C}$ Trees

Remark that $\widehat{\mathbb{Z}}$, $\mathbb{Z}_p$ and $\mathbb{Z}_{\hat{\mathcal{C}}}$ (the pro-$\mathcal{C}$ completion of $\mathbb{Z}$) are rings. Note that

$$\mathbb{Z}_{\hat{\mathcal{C}}} = \prod_{C_p \in \mathcal{C}} \mathbb{Z}_p.$$

If X is a profinite space, denote by $[\![\mathbb{Z}_{\hat{\mathcal{C}}} X]\!]$ the free abelian pro-$\mathcal{C}$ group on X, or, to be consistent with the terminology introduced above, the free pro-$\mathcal{C}'$ group on X, where $\mathcal{C}'$ is the variety of those finite groups in $\mathcal{C}$ which are abelian. The following result gives an explicit description of free abelian pro-$\mathcal{C}$ groups.

Lemma 5.1.

(1) *If X is finite,*
$$[\![\mathbb{Z}_{\hat{c}} X]\!] = \bigoplus_{|X|} \mathbb{Z}_{\hat{c}}$$

(the direct sum of $|X|$ copies of $\mathbb{Z}_{\hat{c}}$).

(2) *In general, if*
$$X = \varprojlim_{i \in I} X_i,$$

then
$$[\![\mathbb{Z}_{\hat{c}} X]\!] = \varprojlim_{i \in I} [\![\mathbb{Z}_{\hat{c}} X_i]\!].$$

Let Γ be a profinite graph such that $V(\Gamma)$ is clopen (or equivalently, $E(\Gamma)$ is clopen). Consider the short sequence of profinite abelian groups and continuous homomorphisms

$$0 \longrightarrow [\![\mathbb{Z}_{\hat{c}}(E(\Gamma))]\!] \xrightarrow{d} [\![\mathbb{Z}_{\hat{c}}(V(\Gamma))]\!] \xrightarrow{\epsilon} \mathbb{Z}_{\hat{c}} \longrightarrow 0, \tag{1}$$

where d and ϵ are the continuous homomorphisms determined by

$$d(e) = d_1(e) - d_0(e), \quad \forall e \in E(\Gamma)$$

and

$$\epsilon(v) = 1 \quad \forall v \in V(\Gamma).$$

The homomorphism ϵ is called the *augmentation*; observe that ϵ is always a surjection.

Theorem 5.2. *Let Γ be a profinite graph. Then Γ is connected if and only if the sequence (1) is exact at $[\![\mathbb{Z}_{\hat{c}}(V(\Gamma))]\!]$.*

Proof. Write Γ as an inverse limit
$$\Gamma = \varprojlim_{i \in I} \Gamma_i$$

of finite graphs Γ_i. For each $i \in I$ we have a sequence

$$0 \longrightarrow [\![\mathbb{Z}_{\hat{c}}(E(\Gamma_i))]\!] \xrightarrow{d} [\![\mathbb{Z}_{\hat{c}}(V(\Gamma_i))]\!] \xrightarrow{\epsilon} \mathbb{Z}_{\hat{c}} \longrightarrow 0, \tag{i}$$

Note that the sequence (1) is the inverse limit of the the sequences (i). Hence, by Proposition 2.3, it suffices to prove the theorem for finite Γ.

To do this, assume first that Γ is connected. Observe that in any case $\epsilon d = 0$, so that $\text{Im}(d) \leq \text{Ker}(\epsilon)$. Now, let $\epsilon(\sum_{i=1}^t n_i v_i) = \sum_{i=1}^t n_i = 0$, $(v_1, \ldots, v_t \in V(\Gamma); n_1, \ldots, n_t \in \mathbb{Z}_{\hat{\mathcal{C}}})$. Fix $v_0 \in V(\Gamma)$. Then $\sum_{i=1}^t n_i v_i = \sum_{i=1}^t n_i(v_i - v_0)$; hence it suffices to check that for every pair of distinct vertices v, w of Γ, there exists some $c \in [\![\mathbb{Z}_{\hat{\mathcal{C}}}(V(\Gamma))]\!]$ with $d(c) = v - w$. One can verify this as follows. Let

$$w = w_0, \ldots, w_s = v$$

be distinct vertices of Γ such that for each $i = 0, \ldots, s-1$, there exists an edge e_i whose vertices are w_i and w_{i+1}. Then define

$$c = \sum_{i=1}^s (-1)^{\alpha_i} e_i,$$

where $\alpha_i = \pm 1$ are chosen conveniently (i.e., $+1$ if $d_1(e_i) = w_{i+1}$ and -1 if $d_1(e_i) = w_i$). Now

$$d(c) = w_s - w_0 = v - w.$$

Hence the sequence is exact at $[\![\mathbb{Z}_{\hat{\mathcal{C}}}(V(\Gamma))]\!]$.

Assume next that the sequence is exact at $[\![\mathbb{Z}_{\hat{\mathcal{C}}}(V(\Gamma))]\!]$. Let $v' \in V(\Gamma)$ and let Γ' be the connected component of v' in Γ. Suppose that $\Gamma' \neq \Gamma$ and let Γ'' be the complement of Γ' in Γ; then Γ'' is a subgraph of Γ. Choose $v'' \in V(\Gamma'')$. Clearly $v' - v'' \in \text{Ker}(\epsilon)$. Then there exists $\sum_{i=1}^s n_i e_i \in [\![\mathbb{Z}_{\hat{\mathcal{C}}}(E(\Gamma))]\!]$ ($e_i \in E(\Gamma), n_i \in \mathbb{Z}_{\hat{\mathcal{C}}}, i = 1, \ldots, s$) such that $d(\sum_{i=1}^s n_i e_i) = v' - v''$. We may assume that v' is a vertex of e_1 and $e_1, \ldots, e_t \in \Gamma'$, while $e_{t+1}, \ldots, e_s \in \Gamma''$ and v'' is a vertex of e_s. Clearly $d([\![\mathbb{Z}_{\hat{\mathcal{C}}}(E(\Gamma'))]\!]) \leq [\![\mathbb{Z}_{\hat{\mathcal{C}}}(V(\Gamma'))]\!]$, $d([\![\mathbb{Z}_{\hat{\mathcal{C}}}(E(\Gamma''))]\!]) \leq [\![\mathbb{Z}_{\hat{\mathcal{C}}}(V(\Gamma''))]\!]$ and $[\![\mathbb{Z}_{\hat{\mathcal{C}}}(V(\Gamma))]\!] = [\![\mathbb{Z}_{\hat{\mathcal{C}}}(V(\Gamma'))]\!] \oplus [\![\mathbb{Z}_{\hat{\mathcal{C}}}(V(\Gamma''))]\!]$. Therefore $d(\sum_{i=1}^t n_i e_i) = v'$. However, $v' \notin \text{Ker}(\epsilon)$, a contradiction. Thus $\Gamma = \Gamma'$, i.e., Γ is connected. □

Remark that in the theorem above the use of $\mathbb{Z}_{\hat{\mathcal{C}}}$ may be misleading because the statement seems to depend on the variety $\mathcal{C}$, when in fact the definition of connectedness given in 5.7 is independent of $\mathcal{C}$. As one can easily check, the proof of this theorem is valid if one replaces any given profinite ring R for $\mathbb{Z}_{\hat{\mathcal{C}}}$ and one thinks in terms of free profinite R-modules rather than free abelian pro-$\mathcal{C}$ groups. For details see [6, Proposition 2.2.4].

Definition 5.1. We say that a profinite graph is a *pro-$\mathcal{C}$ tree* if the sequence (1) is exact.

Lemma 5.3. *A finite graph Γ is a pro-$\mathcal{C}$ tree if and only if it is a tree as an abstract graph. (So, in this case, being a tree is independent of the class $\mathcal{C}$.)*

Proof. If $\mathcal{C}$ has a circuit c then $d(c) = 0$. Conversely, if Γ is a tree, define a homomorphism

$$\sigma : [\![\mathbb{Z}_{\hat{\mathcal{C}}}(V(\Gamma))]\!] \longrightarrow [\![\mathbb{Z}_{\hat{\mathcal{C}}}(E(\Gamma))]\!]$$

as follows: fix $v_0 \in V(\Gamma)$; then set

$$\sigma(v) = \begin{cases} 0, & \text{if } v = v_0; \\ \sum_{i=1}^{t} \epsilon_i e_i, & \text{if } v \neq v_0, \end{cases} \text{ where } e_1^{\epsilon_1}, \ldots, e_t^{\epsilon_t} \text{ is the geodesic from } v_0 \text{ to } v.$$

Then $\sigma d = id_{|\mathbb{Z}_{\hat{\mathcal{C}}}(E(\Gamma))}$ since for $e \in V(\Gamma)$,

$$\sigma d(e) = \sigma(d_1(e) - d_0(e)) = \sum_{i=1}^{t} \epsilon_i e_i - \sum_{i=1}^{s} \delta_i f_i = e,$$

where $e_1^{\epsilon_1}, \ldots, e_t^{\epsilon_t}$ and $f_1^{\delta_1}, \ldots, f_s^{\delta_s}$ are the geodesics from v_0 to $d_1(e)$ and from v_0 to $d_0(e)$, respectively. From this we deduce that d is an injection, and hence that the sequence (1) is exact. $\square$

Definition 5.2. Let G be a profinite group and let R be a commutative profinite ring (e.g., $R = \mathbb{Z}_{\hat{\mathcal{C}}}$).
(a) Define the *complete group ring* $[\![RG]\!]$ by

$$[\![RG]\!] = \varprojlim_{U} [\![R(G/U)]\!],$$

where U ranges over the open normal subgroups of G, and $[\![R(G/U)]\!]$ denotes the usual group ring.
(b) The *augmentation ideal* $((IG))$ of $[\![RG]\!]$ is the kernel of the ring homomorphism

$$[\![RG]\!] \longrightarrow R$$

that sends every element $g \in G$ to the identity element 1 of the ring R.

Theorem 5.4. *Assume that $\mathcal{C}$ is an extension closed variety of finite groups. Let $F = F(X)$ be the free pro-$\mathcal{C}$ group on a profinite space X.*
(a) *The Cayley graph $\Gamma(F, X)$ is a pro-$\mathcal{C}$ tree.*
(b) *(Partial converse:) Assume that the groups in $\mathcal{C}$ are solvable. Let G be a pro-$\mathcal{C}$ group and let X be a compact subset of G such that $\Gamma(G, X)$ is a pro-$\mathcal{C}$ tree. Then G is the free pro-$\mathcal{C}$ group on X.*

Proof. Here we give a proof of part (a); for a proof of part (b) see [5, Theorem 1.7]. We have to show that

$$0 \longrightarrow [\![\mathbb{Z}_{\hat{\mathcal{C}}}(F \times X)]\!] \xrightarrow{d} [\![\mathbb{Z}_{\hat{\mathcal{C}}}F]\!] \xrightarrow{\epsilon} \mathbb{Z}_{\hat{\mathcal{C}}} \longrightarrow 0 \qquad (2)$$

is an exact sequence. First observe that if G is a profinite group and X is a compact subset of G, then the Cayley graph is connected if and only if X generates G as a topological group: this is obvious if G is finite, and the general case follows by an inverse limit argument; indeed, the sequence

$$[\![\mathbb{Z}_{\hat{\mathcal{C}}}(G \times X)]\!] \xrightarrow{d} [\![\mathbb{Z}_{\hat{\mathcal{C}}}G]\!] \xrightarrow{\epsilon} \mathbb{Z}_{\hat{\mathcal{C}}} \longrightarrow 0$$

is the inverse limit of the sequences

$$[\![\mathbb{Z}_{\hat{\mathcal{C}}}(G/U \times X_U)]\!] \xrightarrow{d} [\![\mathbb{Z}_{\hat{\mathcal{C}}}(G/U)]\!] \xrightarrow{\epsilon} \mathbb{Z}_{\hat{\mathcal{C}}} \longrightarrow 0,$$

where U ranges over the open normal subgroups of G, and where X_U is the image in G/U of X; therefore the assertion follows from Proposition 2.3. This proves that the sequence (2) is exact at $[\![\mathbb{Z}_{\hat{\mathcal{C}}}F]\!]$.

Hence it only remains to prove that the map d is an injection. Observe that (2) is not only a sequence of abelian pro-$\mathcal{C}$ groups but also of $[\![\mathbb{Z}_{\hat{\mathcal{C}}}F]\!]$-modules ($F$ acts on $[\![\mathbb{Z}_{\hat{\mathcal{C}}}(F \times X)]\!]$ by multiplication on the first component of $F \times X$; see [3, Section 5.7]. According to [3, Proposition 5.7.1], $[\![\mathbb{Z}_{\hat{\mathcal{C}}}(F \times X)]\!]$ is a free $[\![\mathbb{Z}_{\hat{\mathcal{C}}}F]\!]$-module on $1 \times X$. On the other hand, $d(1,x) = x-1$ ($x \in X$); hence, $\text{Im}(d) = ((IF))$. Thus, showing that d is an injection is equivalent to showing that $((IF))$ is a free $[\![\mathbb{Z}_{\hat{\mathcal{C}}}F]\!]$-module on the subspace

$$X - 1 = \{x - 1 \mid x \in X\}$$

(observe that the space $X - 1$ is homeomorphic to X). To see this, we show that $((IF))$ satisfies the universal property of a free module:

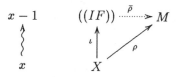

whenever M is an $[\![\mathbb{Z}_{\hat{\mathcal{C}}}F]\!]$-module and $\rho : X \longrightarrow M$ is a continuous map, then there exists a unique $[\![\mathbb{Z}_{\hat{\mathcal{C}}}F]\!]$-module homomorphism $\bar{\rho} : ((IF)) \longrightarrow M$ such that $\bar{\rho}\iota = \rho$, where $\iota : X \longrightarrow ((IF))$ is the map $\iota(x) = x - 1$. Since a module can be expressed as an inverse limit of finite modules [3, Theorem 5.1.1], we may assume that M is a finite $[\![\mathbb{Z}_{\hat{\mathcal{C}}}F]\!]$-module.

Consider the the semidirect product $M \rtimes F$ using the action of F on M. Then $M \rtimes F$ is also a pro-$\mathcal{C}$ group since the class $\mathcal{C}$ is extension closed. Now,

249

since $F = F(X)$ is a free pro-$\mathcal{C}$ group on X, the map ρ induces a continuous homomorphism of pro-$\mathcal{C}$ groups

$$\rho_1 : F \longrightarrow M \rtimes F$$

determined by $\rho_1(x) = (\rho(x), x)$. Define $\delta : F \longrightarrow M$ to be the composite map

$$F \xrightarrow{\rho_1} M \rtimes F \xrightarrow{\pi} M.$$

Then one checks that δ is a derivation, i.e., if $f, f' \in F$, one has

$$\delta(ff') = \delta(f) + f\delta(f')$$

(see [3, Lemma 9.3.6]).

Now, (see [3, Lemma 6.8.3])

$$\mathrm{Der}(F, M) \cong \mathrm{Hom}_{[\![\mathbb{Z}_{\hat{\mathcal{C}}}F]\!]}(((IF)), M).$$

Under this isomorphism δ corresponds to a $[\![\mathbb{Z}_{\hat{\mathcal{C}}}F]\!]$-homomorphism $\bar{\delta} : ((IF)) \longrightarrow M$ such that

$$\bar{\delta}(f - 1) = \delta(f).$$

Then

$$\bar{\delta}\iota(x) = \bar{\delta}(x - 1) = \delta(x) = \rho(x),$$

i.e., $\bar{\delta}\iota = \rho$. □

Remark 5.3.

(a) Jorge Almeida and Pascal Weil [18] pointed out that the proof of the result above does not use the full assumption that the class $\mathcal{C}$ is extension closed; it is only required that $\mathcal{C}$ be closed under extensions with abelian kernel, i.e., that whenever

$$1 \longrightarrow K \longrightarrow G \longrightarrow H \longrightarrow 1$$

is an exact sequence of finite groups such that $K, H \in \mathcal{C}$ and K is abelian, then $G \in \mathcal{C}$. They also proved that this condition is necessary for $\Gamma(F, X)$ to be a pro-$\mathcal{C}$ tree.

(b) The second part of the theorem provides examples to show that the concept of pro-$\mathcal{C}$ tree really depends on the class $\mathcal{C}$. Indeed, if p is a prime number and F is a free pro-p group on a space X, then according to part (a) of this theorem, $\Gamma(F, X)$ is a pro-p tree but, according to part (b), it is not a prosolvable tree.

Lemma 5.5. *Let Γ be a pro-$\mathcal{C}$ tree.*

(a) *A connected subgraph of Γ is a pro-$\mathcal{C}$ tree.*

(b) *If $\{\Delta_i \mid i \in I\}$ is a collection of pro-$\mathcal{C}$ subtrees of Γ, then $\bigcap \Delta_i$ is either empty or a pro-$\mathcal{C}$ tree.*

(c) *If Δ_1 and Δ_2 are pro-$\mathcal{C}$ subtrees of Γ and $\Delta_1 \cap \Delta_2 \neq \emptyset$, then $\Delta_1 \cup \Delta_2$ is a pro-$\mathcal{C}$ subtree.*

Proof. Parts (b) and (c) follow from (a). For part (a) the only thing required to prove is that d in the corresponding sequence (1) is an injection (see Theorem 5.2). But this is obviously the case for any subgraph of Γ since Γ is a pro-$\mathcal{C}$ tree. □

Definition 5.4. Let Γ be a pro-$\mathcal{C}$ tree and let $v, w \in V(\Gamma)$. Define the *chain* $[v, w]$ to be the intersection of all pro-$\mathcal{C}$ subtrees of Γ containing v and w.

By Lemma 5.5, $[v, w]$ is a pro-$\mathcal{C}$ tree. Observe that if $[v, w]$ is finite, then one can identify it with the geometric path joining v and w.

6 The Fundamental Group of a Graph

Definition 6.1. Let Γ be an abstract oriented graph.

(a) If $e \in E(\Gamma)$, we define $e^1 = e$ and we denote by e^{-1} a 'formal' edge ($\neq e$) associated with e (the 'opposite' of e) with $d_0(e^{-1}) = d_1(e)$ and $d_1(e^{-1}) = d_0(e)$. We refer to e^{-1} also as an edge of Γ.

(b) If $v, w \in V(\Gamma)$, a *path* $p_{v,w}$ of length t from v to w is a sequence

$$e_1, \ldots, e_t$$

of edges of Γ such that $d_0(e_1) = v$, $d_0(e_{i+1}) = d_1(e_i)$ ($i = 1, \ldots, t-1$) and $d_1(e_t) = w$. We say that the path $p_{v,w}$ is *reduced* if $e_i \neq e_{i+1}^{-1}$ for all i.

(c) The pairs $\{e, e^{-1}\}$ are called the *geometric edges* of Γ.

(d) A reduced path of the form $p_{v,v}$ is called a *cycle*.

Assume that Γ is a connected abstract graph. Choose a fixed vertex v_0 (the *base* vertex). Let $P(\Gamma, v_0)$ be the set of all cycles p_{v_0, v_0} in Γ. Define a multiplication on $P(\Gamma, v_0)$ by concatenation. We say that two paths in $P(\Gamma, v_0)$ are *equivalent* if one can obtain one of them from the other one by a finite number of additions or deletions of subsequences of the form $e^{\epsilon} e^{-\epsilon}$ where $\epsilon = \pm 1$; this is an equivalence relation on $P(\Gamma, v_0)$ compatible with the multiplication defined above. Let $\pi_1(\Gamma, v_0)$ be the set of all equivalence classes of $P(\Gamma, v_0)$. Then $\pi_1(\Gamma, v_0)$ is a group. This group is independent of the chosen vertex v_0 in the sense that if we choose a different base vertex w_0, then there is an isomorphism $\pi_1(\Gamma, v_0) \cong \pi_1(\Gamma, w_0)$; so, one usually denotes by $\pi_1(\Gamma)$ the group $\pi_1(\Gamma, v_0)$ where v_0 is any (but fixed) vertex of Γ.

Definition 6.2. Let Γ be a connected abstract graph. The group $\pi_1(\Gamma)$ is called the *fundamental group* of Γ.

Theorem 6.1. *Let Γ be a connected abstract graph.*

(a) *$\pi_1(\Gamma)$ is a free group.*

(b) *If Δ is a connected subgraph of Γ, then $\pi_1(\Delta)$ is a free factor of $\pi_1(\Gamma)$.*

Proof. (Sketch) (a) Let T be a maximal tree of Γ. Then $\pi_1(\Gamma)$ is the free group on the set

$$\{c(e) \mid e \in E(\Gamma) - E(T)\},$$

where $c(e)$ is the unique cycle formed as follows: take the unique reduced path from v_0 to $d_0(e)$ in T followed by e followed by the reduced path from $d_1(e)$ to v_0 in T.

(b) Choose a maximal tree T' of Δ. Extend T' to a maximal tree T of Γ. Then if we proceed as in the proof of part (a), we clearly have that the basis obtain for $\pi_1(\Delta)$ is a subset of the basis obtained for $\pi_1(\Gamma)$; therefore $\pi_1(\Delta)$ is a free factor of $\pi_1(\Gamma)$. □

Theorem 6.2. *Let $F = F(X)$ be a free abstract group and let $\Gamma = \Gamma(F, X)$ be the corresponding Cayley graph. If G is a subgroup of F, there is a natural isomorphism*

$$\pi_G : G \longrightarrow \pi_1(G\backslash\Gamma),$$

where $G\backslash\Gamma$ is the quotient graph of Γ under the action of G.

Proof. Denote by

$$\sigma : \Gamma \longrightarrow G\backslash\Gamma$$

the natural epimorphism of graphs. Given $g \in G$, define

$$\pi_G(g) = \sigma(p_{1,g}),$$

where 1 is the vertex of Γ corresponding to the identity element of F and where $p_{1,g}$ is the unique reduced path from 1 to g in Γ (a tree!). We shall show that π_G is an isomorphism of groups.

If $v \in V(\Gamma)$, denote by $Star_\Gamma(v)$ the set of geometric edges of Γ for which v is a vertex. Then σ restricted to $Star_\Gamma(v)$ is a bijection, for each v. From this one sees that every path in $G\backslash\Gamma$ starting at 1 can be lifted to a unique path starting at 1 of Γ. So π_G is onto. It is easy to see that it is also an injection. □

It is worth noticing that the Nielsen-Schreier subgroup theorem, that asserts that a subgroup of an abstract free group is also free, is an immediate consequence of Theorems 6.3 and 6.4.

Next we prove a generalization of a theorem of M. Hall [23].

Theorem 6.3. *Let F be an abstract free group endowed with the pro-$\mathcal{C}$ topology. Let K be a closed subgroup of F and assume that H is a finitely generated subgroup of F which is a free factor of K. Then there exists an open subgroup U of F containing K such that H is a free factor of U.*

Proof. Let Δ be a finite subgraph of the quotient graph $K\backslash\Gamma$. We claim that there exists an open subgroup U of F containing K such that the natural projection of graphs

$$\varphi : K\backslash\Gamma \longrightarrow U\backslash\Gamma$$

restricted to Δ is an injection. To prove this it suffices to show that there is some open subgroup U containing K such that φ is injective on the set of vertices of Δ (since the edges are of the form $(f, x) : f \longrightarrow fx$). Now, assume that $Kf \neq Kf'$ ($f, f' \in F$), i.e., $f' \notin Kf$. Then, since K is closed, there exists some open subgroup V of F containing K such that $f' \notin Vf$. Taking the intersection of the subgroups V corresponding to all pairs of vertices of Δ, we get the desired U. This proves the claim.

We know that $H = \pi_1(H\backslash\Gamma)$ (see Theorem 6.2). Since H is finitely generated, there exist a finite number of cycles in $H\backslash\Gamma$ of the form $p_{1,1}$ representing the generators of H. Denote by Δ the union of those cycles. Then Δ is a finite subgraph of $H\backslash\Gamma$ and

$$\pi_1(\Delta) = \pi_1(H\backslash\Gamma) = H.$$

Let Δ_1 be the image of Δ in $K\backslash\Gamma$ under the mapping $H\backslash\Gamma \longrightarrow K\backslash\Gamma$. Since H is naturally a subgroup of $\pi_1(\Delta_1)$ and since by assumption H is a free factor of K, it follows from the Kurosh subgroup theorem that H is a free factor of $\pi_1(\Delta_1)$. By the claim above, there exists an open subgroup of F containing K such that Δ_1 maps injectively into $U\backslash\Gamma$. So we may consider Δ_1 as a subgraph of $U\backslash\Gamma$. By Theorem 6.1(b), $\pi_1(\Delta_1)$ is a free factor of $\pi_1(U\backslash\Gamma) = U$. Thus H is a free factor of U. □

Corollary 6.4. *Let $\mathcal{C}$ be an extension closed variety of finite groups. Let F be an abstract free group and H a finitely generated closed subgroup of F. Then the topology induced on H by the pro-$\mathcal{C}$ topology of F is the pro-$\mathcal{C}$ topology of H.*

Proof. By the theorem above, H is a free factor of an open subgroup U of F; say $U = H * L$. It is known that the topology induced on U by the pro-$\mathcal{C}$ topology of F is the pro-$\mathcal{C}$ topology of U (see Proposition 2.5). The result follows now from Proposition 2.6. □

Definition 6.3. Let G_1 and G_2 be pro-$\mathcal{C}$ groups. A pro-$\mathcal{C}$ group G together with homomorphisms $\iota_i : G_i \longrightarrow G$ ($i = 1, 2$) is called a *free pro-$\mathcal{C}$ product* of G_1 and G_2 (we write $G = G_1 \amalg G_2$) if it satisfies the following universal property: whenever $\varphi_i : G_i \longrightarrow H$ ($i = 1, 2$) are continuous homomorphisms into a pro-$\mathcal{C}$ group H, then there exists a unique continuous homomorphism $\varphi : G \longrightarrow H$ such that $\varphi \iota_i = \varphi_i$ ($i = 1, 2$).

The free pro-$\mathcal{C}$ product of pro-$\mathcal{C}$ groups G_1 and G_2 always exists and the groups G_1 and G_2 can be canonically embedded in it (cf. [3, Proposition 9.1.2 and Corollary 9.1.4]). From Proposition 2.6(a) we deduce the following

Proposition 6.5. *Let $G = H * L$ be the (abstract) free product of abstract groups H and L. Then*

$$G_{\hat{\mathcal{C}}} = H_{\hat{\mathcal{C}}} \amalg L_{\hat{\mathcal{C}}},$$

where the symbol $\amalg$ denotes 'free pro-$\mathcal{C}$ product'.

Let F be an abstract free group endowed with its pro-$\mathcal{C}$ topology, and let H be a subgroup of F. Denote by $Cl(H)$ the topological closure of H in F. Recall that $\bar{H}$ denotes the topological closure of H in $F_{\hat{\mathcal{C}}}$. Clearly one has that

$$H \leq Cl(H) \leq \bar{H} = \overline{Cl(H)}.$$

Proposition 6.6. *Let $\mathcal{C}$ be an extension closed variety of finite groups. Let F be an abstract free group and H a finitely generated closed subgroup of F. Then (note that $Cl(H)$ is an abstract free group by the Nielsen-Schreier theorem)*

$$\mathrm{rank}(Cl(H)) \leq \mathrm{rank}(H).$$

Proof. Assume to the contrary that $\mathrm{rank}(Cl(H)) > \mathrm{rank}(H)$. Let $x_1, \ldots, x_n$ be a finite subset of a basis of the abstract free group $Cl(H)$ with $n > \mathrm{rank}(H)$.

254

Put $K = \langle x_1, \ldots, x_n \rangle$. By Theorem 6.3 there exists an open subgroup U of F with $U \geq Cl(H)$ such that $U = K * R$ where R is some subgroup of U. Then

$$K_{\hat{\mathcal{C}}} = \bar{K} \leq \bar{H} = \overline{Cl(H)} \leq \bar{U} = U_{\hat{\mathcal{C}}} = K_{\hat{\mathcal{C}}} \amalg R_{\hat{\mathcal{C}}}.$$

Hence the composite map

$$\bar{H} \hookrightarrow \bar{U} \longrightarrow K_{\hat{\mathcal{C}}}$$

is an epimorphism. Therefore, $\text{rank}(H) \geq n$. But this contradicts the assumption $n > \text{rank}(H)$. Thus $\text{rank}(Cl(H)) \leq \text{rank}(H)$. □

7 Algorithmic Problems in Groups

In this section we shall describe some algorithms related to subgroups of abstract free groups.

Definition 7.1. Let n be a natural number. An abstract group G (endowed with the pro-$\mathcal{C}$ topology) is called *n-product subgroup separable* if whenever $H_1, \ldots, H_n$ are finitely generated closed subgroups of G, then their product

$$H_1 \cdots H_n$$

is a closed subset of G.

Remark 7.2. Note that '1-product subgroup separable' is an empty statement. It should not be confused with the concept of G being *subgroup separable* (or LERF), which means that any finitely generated subgroup of G is closed in the profinite topology of G.

Lemma 7.1. *Let $\mathcal{C}$ be an extension closed variety of finite groups. A group G is n-product subgroup separable if and only if it contains an open subgroup which is n-product subgroup separable.*

Proof. Let U be an open subgroup of G. Since $\mathcal{C}$ is extension closed, the pro-$\mathcal{C}$ topology of U is precisely the topology induced by the pro-$\mathcal{C}$ topology of G (see Proposition 2.10). If G is n-product subgroup separable, then plainly so is U. Conversely assume that U is n-product subgroup separable. By the above, the core U_G of U in G is n-product subgroup separable as well. Hence, replacing U by U_G, if necessary, we may assume that U is open and normal in G. Let $H_1, \ldots, H_n$ be finitely generated closed subgroups of G. We prove by induction on the number of H_i which are not contained in U that $H_1 \cdots H_n$ is closed in the pro-$\mathcal{C}$ topology of G. If $H_i \leq U$ for all $i = 1, \ldots, n$, the result is clear. Since each H_i is finitely generated and $U \cap H_i$ has finite index in

H_i, we have that $U \cap H_i$ is also finitely generated. Pick $H_t \not\leq U$. Write $H_t = \bigcup_j h_j(U \cap H_t)$, $(h_j \in H_t)$. Therefore

$$H_1 \cdots H_n = \bigcup_j h_j H_1^{h_j} \cdots H_{t-1}^{h_j}(U \cap H_t)H_{t+1} \cdots H_n \quad \text{(a finite union)}.$$

By the induction hypothesis, $H_1^{h_j} \cdots H_{i-1}^{h_j}(U \cap H_i)H_{i+1} \cdots H_n$ is closed in G. Thus $H_1 \cdots H_n$ is closed in G. □

Theorem 7.2. (L. Ribes and P. Zalesskii [27, 28]) *Let $\mathcal{C}$ be an extension closed variety of finite groups. Assume that G is a group that contains an abstract free group F which is open in the pro-$\mathcal{C}$ topology of G. Then G is $(n+1)$-product subgroup separable, for any $n = 1, 2, \ldots$, i.e.,*

$$H_1 \cdots H_n K$$

is a closed subset of G whenever $H_1, \ldots, H_n$ and K are finitely generated closed subgroups of G.

Proof. By Lemma 7.1 we may assume that $G = F$ is an abstract free group. Note that since the groups $H_1, \ldots, H_n$ and K are finitely generated, they are contained in a finitely generated free factor of F. Hence from now on we will assume that F has finite rank. By Theorem 6.3 there exists an open subgroup U of F such that $U = K * L$. So, by Lemma 7.1 again, we may assume that K is a free factor of F. Hence, by Proposition 6.5,

$$F_{\hat{\mathcal{C}}} = \bar{K} \amalg \bar{L},$$

since $\bar{K} = K_{\hat{\mathcal{C}}}$ and $\bar{L} = L_{\hat{\mathcal{C}}}$.

Let Y be a basis of K as an abstract free group and let X be a basis of F containing Y. Let $\Gamma(F) = \Gamma(F, X)$ and $\Gamma(K) = \Gamma(K, Y)$ denote the abstract Cayley graphs corresponding to (F, X) and (K, Y), respectively; and let $\Gamma(F_{\hat{\mathcal{C}}}) = \Gamma(F, X)$ and $\Gamma(\bar{K}) = \Gamma(\bar{K}, Y)$ denote the profinite Cayley graphs corresponding to $(F_{\hat{\mathcal{C}}}, X)$ and $(\bar{K}, Y)$, respectively. Then we have a commutative diagram of graphs

$$\begin{array}{ccc} \Gamma(F) & \hookrightarrow & \Gamma(F_{\hat{\mathcal{C}}}) \\ \uparrow & & \uparrow \\ \Gamma(K) & \hookrightarrow & \Gamma(\bar{K}) \end{array}$$

Now, proving that $H_1 \cdots H_n K$ is closed in the pro-$\mathcal{C}$ topology of F is equivalent to proving that

$$\bar{H}_1 \cdots \bar{H}_n \bar{K} \cap F = H_1 \cdots H_n K.$$

Hence, if $h_i \in \bar{H}_i$, $k \in \bar{K}$ and
$$h_1 \cdots h_n k \in F,$$
we need to prove that $h_1 \cdots h_n k \in H_1 \cdots H_n K$.

We do this by induction on n. If $n = 0$, this means that $\bar{K} \cap F = K$, which is just the assumption that K is closed in the pro-$\mathcal{C}$ topology of F.

Assume that $n \geq 1$ and that the result holds when the number of factors of the form h_i is less than n. Since $h_1 \cdots h_n k \in F$, the chain $[1, h_1 \cdots h_n k]$ in $\Gamma(F_{\hat{\mathcal{C}}})$ is finite, and hence so is $[h_n^{-1} \cdots h_1^{-1}, k]$. Let v be the first vertex of the chain $[h_n^{-1} \cdots h_1^{-1}, k]$ which is in $\Gamma(\bar{K})$. If $v = h_n^{-1} \cdots h_1^{-1}$, then $h_n^{-1} \cdots h_1^{-1} \in \bar{K}$. So $h_1 \cdots h_n k \in \bar{K} \cap F = K$, and we are done.

So, assume that $[h_n^{-1} \cdots h_1^{-1}, v]$ has length at least 1. Observe that $v = kr$ for some $r \in K$ (since $[v, k]$ is finite). Hence $h_1 \cdots h_n v \in H_1 \cdots H_n K$ if and only if $h_1 \cdots h_n k \in H_1 \cdots H_n K$. Therefore, from now on we may assume that $k = v$.

Next note that $v \in [h_n^{-1} \cdots h_1^{-1}, 1]$. Indeed, let e be the edge of $[h_n^{-1} \cdots h_1^{-1}, v]$ such that $e \notin \Gamma(\bar{K})$ but v is a vertex of e. Then $e \in [h_n^{-1} \cdots h_1^{-1}, 1]$, for otherwise collapsing the subtree
$$[h_n^{-1} \cdots h_1^{-1}, 1] \cup [1, v]$$
of the tree
$$[h_n^{-1} \cdots h_1^{-1}, 1] \cup [1, v] \cup [h_n^{-1} \cdots h_1^{-1}, v]$$
to a point, we get a loop corresponding to $[h_n^{-1} \cdots h_1^{-1}, v]$ since e is not in the subtree. But this contradicts the fact that collapsing a subtree of a tree produces a tree.

Hence from now on we shall assume that
$$v = k \in [h_n^{-1} \cdots h_1^{-1}, 1].$$
For each $i = 1, \ldots n$ define a pro-$\mathcal{C}$ subtree
$$D_i = \bigcup_j \bar{H}_i[1, r_j]$$
of $\Gamma(F_{\hat{\mathcal{C}}})$, where the collection of the r_j is a finite set of generators of H_i. This is a tree since it is a connected subgraph of $\Gamma(F_{\hat{\mathcal{C}}})$; indeed, observe that
$$\bar{H}_i[1, r_j] = \varprojlim_U (\bar{H}_i/\bar{H}_i \cap U)[1U, r_j U],$$
where U ranges over the open subgroups of $F_{\hat{\mathcal{C}}}$, and $(\bar{H}_i/\bar{H}_i \cap U)[1U, r_j U]$ is a finite connected subgraph of $\Gamma(F_{\hat{\mathcal{C}}}/U, X_U)$. Note that D_i is the smallest pro-$\mathcal{C}$

$\bar{H}_i$-invariant subtree of $\Gamma(F_{\hat{C}})$ containing 1; moreover, $\bar{H}_i\backslash D_i$ is a finite graph since it is a quotient of the finite graph $\bigcup_j [1, r_j]$. We deduce from Lemma 5.5 that

$$D = h_n^{-1}\cdots h_2^{-1}D_1 \cup h_n^{-1}\cdots h_3^{-1}D_2 \cup \cdots \cup D_n$$

is a pro-$\mathcal{C}$ subtree of $\Gamma(F_{\hat{C}})$. Moreover,

$$k \in [h_n^{-1}\cdots h_1^{-1}, 1] \subseteq D.$$

Case 1. $k \in h_n^{-1}\cdots h_{i+1}^{-1}D_i$, $i > 1$ (convention: if $i = n$, $h_n^{-1}\cdots h_{i+1}^{-1} = 1$). Then the result follows by induction. Indeed, we assumed that the result is true if the total number of factors is less than $n + 1$. Now, we have

$$h_{i+1}\cdots h_n k = h'_i f \quad (\text{some } h'_i \in \bar{H}_i, f \in F).$$

So

$${h'_i}^{-1} h_{i+1}\cdots h_n k \in F \cap \bar{H}_i \cdots \bar{H}_n \bar{K}.$$

Since $i > 1$, the number of factors is less that $n + 1$, and so by induction

$${h'_i}^{-1} h_{i+1}\cdots h_n k \in H_i H_{i+1} \cdots H_n K.$$

Next, since

$$h_1 \cdots h_n k = h_1 \cdots h_i h'_i {h'_i}^{-1} h_{i+1}\cdots h_n k \in F,$$

we have that

$$h_1 \cdots h_i h'_i \in F.$$

And again by the induction hypothesis, we have $h_1 \cdots h_i h'_i \in H_1 \cdots H_i$. Thus,

$$h_1 \cdots h_n k \in H_1 \cdots H_n K.$$

Case 2. $k \in h_n^{-1}\cdots h_2^{-1}D_1$. We claim that $[k, 1] \cap h_n^{-1}\cdots h_2^{-1}D_1 \cap (h_n^{-1}\cdots h_3^{-1}D_2 \cup \cdots \cup D_n) \neq \emptyset$. To see this first observe that by the definition of the D_i, one has that $h_n^{-1}\cdots h_2^{-1}D_1 \cap (h_n^{-1}\cdots h_3^{-1}D_2 \cup \cdots \cup D_n) \neq \emptyset$, and therefore $h_n^{-1}\cdots h_2^{-1}D_1 \cup (h_n^{-1}\cdots h_3^{-1}D_2 \cup \cdots \cup D_n)$ is a pro-$\mathcal{C}$ tree (see Lemma 5.8); hence, since $k \in h_n^{-1}\cdots h_2^{-1}D_1$ and $1 \in D_n$, one has that $[k, 1] \subseteq h_n^{-1}\cdots h_2^{-1}D_1 \cup (h_n^{-1}\cdots h_3^{-1}D_2 \cup \cdots \cup D_n)$. Now,

$$[k, 1] \cap h_n^{-1}\cdots h_2^{-1}D_1 \cap (h_n^{-1}\cdots h_3^{-1}D_2 \cup \cdots \cup D_n) =$$

$$([k, 1] \cap h_n^{-1}\cdots h_2^{-1}D_1) \cap ([k, 1] \cap (h_n^{-1}\cdots h_3^{-1}D_2 \cup \cdots \cup D_n)).$$

Finally, observe that this intersection is nonempty, because otherwise there would exist an epimorphism of $[k, 1]$ to a disconnected graph consisting of two different vertices obtained by collapsing $[k,1] \cap h_n^{-1} \cdots h_2^{-1} D_1$ and $[k, 1] \cap (h_n^{-1} \cdots h_3^{-1} D_2 \cup \cdots \cup D_n)$, and this would contradict the connectivity of $[k, 1]$. This proves the claim. Say

$$k' \in [k,1] \cap h_n^{-1} \cdots h_2^{-1} D_1 \cap (h_n^{-1} \cdots h_3^{-1} D_2 \cup \cdots \cup D_n).$$

Then $k' \in \bar{K}$, since $[k,1] \subseteq \Gamma(\bar{K})$. Then for some $i \geq 2$, we have

$$k' \in h_n^{-1} \cdots h_{i+1}^{-1} D_i.$$

Put

$$\bar{H} = h_n^{-1} \cdots h_2^{-1} \bar{H}_1 h_2 \cdots h_n.$$

Then $\bar{H}$ acts on $\tilde{D}_1 = h_n^{-1} \cdots h_2^{-1} D_1$. So $\bar{H} \cap \bar{K}$ acts on the pro-$\mathcal{C}$ tree $T = \tilde{D}_1 \cap \Gamma(\bar{K})$.

Note that $\bar{H} \backslash \tilde{D}_1$ is finite. Consider the map of graphs

$$\pi : \tilde{D}_1 \longrightarrow \bar{H} \backslash \tilde{D}_1.$$

It induces an epimorphism of graphs

$$\tilde{\pi} : \bar{H} \cap \bar{K} \backslash T \longrightarrow \pi(T).$$

We claim that $\tilde{\pi}$ is an isomorphism. Indeed, if $x \in \bar{H}$ and t and t' are either both in $V(T)$ or both in $E(T)$ with $xt = t'$, then $x \in \bar{K}$. Thus $\bar{H} \cap \bar{K} \backslash T$ is finite. Hence the morphism of graphs

$$T \longrightarrow \bar{H} \cap \bar{K} \backslash T$$

has a transversal Σ with k' as a vertex. Since $k \in T$, there exists $g \in \bar{H} \cap \bar{K}$ such that

$$gk \in \Sigma.$$

Since Σ is finite and $T \subseteq \Gamma(\bar{K})$, there exists $r \in K$ with $k'r = gk$. Then

$$h_1 \cdots h_n k = h_1 \cdots h_n g^{-1} gk = h_1 \cdots h_n g^{-1} k' r.$$

Therefore, $h_1 \cdots h_n g^{-1} k' \in F$; and since $g \in \bar{H}$, say

$$g = h_n^{-1} \cdots h_2^{-1} h_1' h_2 \cdots h_n,$$

for some $h_1' \in \bar{H}_1$, we have

$$h_1 \cdots h_n g^{-1} k' = h_1 \cdots h_n (h_n^{-1} \cdots h_2^{-1} h_1' h_2 \cdots h_n)^{-1} k' = h_1'' h_2 \cdots h_n k' \in F,$$

where $h_1'' \in \bar{H}_1$. So by Case 1, $h_1 \cdots h_n g^{-1} k' \in H_1 \cdots H_n K$, since $k' \in h_n^{-1} \cdots h_{i+1}^{-1} D_i$ where $i > 1$. Thus

$$h_1 \cdots h_n k = h_1 \cdots h_n g^{-1} k' r \in H_1 \cdots H_n K.$$

□

The following partial generalizations of this result have been obtained by Shihong You and by Thierry Coulbois.

Theorem 7.3. (S. You [32])

(a) *Let F be an abstract free group. Then the group*

$$F \times \mathbb{Z}$$

is subgroup separable and n-product subgroup separable (for any fixed natural number n) with respect to the profinite topology.

(b) *Any amalgamated free product of two cyclic groups over a common subgroup*

$$\mathbb{Z} *_H \mathbb{Z}$$

is subgroup separable and n-product subgroup separable (for any fixed natural number n) with respect to the profinite topology.

Theorem 7.4. (T. Coulbois [20], [21])

(a) *n-product subgroup separability (with respect to the profinite topology) is a property which is closed under taking finite free products of groups that are subgroup separable: if the abstract groups G_1 and G_2 are subgroup separable and n-product subgroup separable with respect to the profinite topology, then so is*

$$G = G_1 * G_2.$$

(b) *Let F_1 and F_2 be abstract free groups with a common cyclic subgroup H. Then their amalgamated free product*

$$F_1 *_H F_2$$

is subgroup separable and n-product subgroup separable (for any fixed natural number n) with respect to the profinite topology.

We turn now to some algorithmic questions related to the pro-p topology of an abstract free group. Recall (see Section 6) that if F is an abstract free group and X is a subset of F, then $Cl(X)$ denotes the topological closure of X in F endowed with the pro-p topology; while $\bar{X}$ denotes the topological closure of X in $F_{\hat{p}}$.

Theorem 7.5. (L. Ribes and P. Zalesskii [28]) *Let p be a fixed prime number. Let H be a finitely generated subgroup of a free group F. Then $Cl(H)$ is also a finitely generated subgroup of F and there exists an algorithm to compute $Cl(H)$.*

Proof. The fact that $Cl(H)$ is finitely generated was shown in Proposition 6.6. Since H is finitely generated, it is contained in a finitely generated free factor of F. Thus, using Proposition 2.6 and Corollary 6.4, we may assume that F has finite rank. Fix a basis X of F. Since F is finitely generated, it has only finitely many open subgroups of a given index (cf. [3, Proposition 2.5.1]). For a natural number m, let $U(m)$ denote the intersection of all open subgroups of F of index at most m containing H; then $U(m)$ is open. Assume that H is generated by $h_1, \ldots, h_r$. Denote by

$$p_{1,h_i}$$

the reduced path in $\Gamma = \Gamma(F, X)$ from 1 to h_i. Consider the subgraph

$$T = \bigcup_{i=1}^{r} p_{1,h_i}$$

of Γ. Let Γ_1, Γ_2 and T_m denote the images of T in the quotient graphs $H\backslash\Gamma$, $Cl(H)\backslash\Gamma$ and $U(m)\backslash\Gamma$, respectively:

$$\begin{array}{ccccccc}
\Gamma & \longrightarrow & H\backslash\Gamma & \stackrel{\tau}{\longrightarrow} & Cl(H)\backslash\Gamma & \stackrel{\tau_m}{\longrightarrow} & U(m)\backslash\Gamma \\
\downarrow & & \downarrow & & \downarrow & & \downarrow \\
T & \longrightarrow & \Gamma_1 & \longrightarrow & \Gamma_2 & \longrightarrow & T_m
\end{array}$$

The images $p_{1,h_i}(m)$ of p_{1,h_i} in $U(m)\backslash\Gamma$ are cycles, and so they represent elements of $\pi_1(T_m)$. Observe that $\pi_1(T_m)$ is a free factor of $\pi_1(U(m)\backslash\Gamma) = U(m)$ (see Theorem 6.1). Clearly (see Proposition 2.4)

$$Cl(H) = \bigcap_m U(m).$$

Note that the graph Γ_1 is finite and $\pi_1(\Gamma_1) = \pi_1(H\backslash\Gamma) = H$.

$$\begin{array}{ccccc}
\pi_1(\Gamma_1) = \pi_1(H\backslash\Gamma) & \stackrel{\bar\tau}{\longrightarrow} & \pi_1(Cl(H)\backslash\Gamma) & \stackrel{\tau_m}{\longrightarrow} & \pi_1(U(m)\backslash T) \\
\uparrow \cong & & \uparrow \cong & & \uparrow \cong \\
H & \hookrightarrow & Cl(H) & \hookrightarrow & U(m)
\end{array}$$

Since Γ_2 is finite, there exists some natural number m such that τ_m is injective on Γ_2 (see the proof of Theorem 6.3). Therefore,
$$\pi_1(\Gamma_2) = \pi_1(T_m).$$
Since
$$H \leq \pi_1(T_m) = \pi_1(\Gamma_2) \leq \pi_1(\Gamma_2) \leq Cl(H)$$
and since $\pi_1(T_m)$ is closed in U, and so in F, it follows that
$$\pi_1(T_m) = Cl(H).$$
Observe that in the considerations above the prime p does not intervene in an essential manner: they are valid (after some obvious changes) for the pro-$\mathcal{C}$ topology as long as the class of finite groups $\mathcal{C}$ is extension closed. However to finish the proof of the theorem we have to show that the natural number m can be found algorithmically.

To find m we have to check that the elements of $\pi_1(T_m)$ represented by the $p_{1,h_i}(m)$ are not contained in an open subgroup of $\pi_1(T_m)$. It is here where we use the fact that we are dealing with the pro-p topology of F. The maximal open subgroups of $\pi_1(T_m)$ (in the pro-p topology) have index p (since the maximal subgroups of a finite p-group have index p, cf. [23]). Hence to find the correct m, we only have to decide whether the elements of $\pi_1(T_m)$ represented by the $p_{1,h_i}(m)$ are in an open subgroup of $\pi_1(T_m)$ of index p. □

The result above has been proved to be valid when one replaces the pro-p topology by the pronilpotent topology by S. Margolis, M. Sapir and P. Weil:

Theorem 7.6. (S. Margolis, M. Sapir and P. Weil [25]) *Let F be an abstract free group and let H be a subgroup of F of finite rank. Then there exists an algorithm to compute the topological closure of H in the pronilpotent topology of F.*

8 Algorithmic Problems in Semigroups

The results in Section 7 were originally motivated by several inter-related conjectures in the theory of finite semigroups. The connection between those conjectures and questions posed in the language of groups and, more specifically, in the language of profinite topologies on groups appeared in papers of J.-É. Pin and C. Reutenauer (cf. [26], [15]). In turn, the conjectures about semigroups were all related to a problem first posed by J. Rhodes about the computability of the 'kernel of a finite semigroup'. In this article we mention only two of these conjectures; for further details see [14, 15, 26, 17, 11]).

A *relational morphism*

$$\varphi : M \longrightarrow\!\!\!\!\!\!\circ\,\, M'$$

of semigroups M and M' is map $\varphi : M \longrightarrow \mathcal{P}(M')$ from M to the power set of M' such that
  (i) for every $m \in M$, $\varphi(m)$ is a nonempty subset of M';
  (ii) for every $m_1, m_2 \in M$, $\varphi(m_1)\varphi(m_2) \subseteq \varphi(m_1 m_2)$; and
  (iii) if M_1 and M_2 are monoids, $1 \in \varphi(1)$ (1 denotes the identity element of a monoid).

For example, if $\rho : M_2 \longrightarrow M_1$ is an epimorphism of semigroups, then $\rho^{-1} : M_1 \longrightarrow \mathcal{P}(M_2)$ is a relational morphism.

The *kernel* of a finite semigroups M is defined to be the subsemigroup

$$K(M) = \bigcap_\varphi \varphi^{-1}(1),$$

where φ ranges over all the relational morphisms of the form $\varphi : M \longrightarrow\!\!\!\!\!\!\circ\,\, G$ with G a finite group. Define the *p-kernel* $K_p(M)$ similarly using only finite p-groups G.

John Rhodes conjectured that $K(M)$ is computable for every finite semigroup M, i.e., that there is an algorithm to decide which elements of M are in $K(M)$. Let's call this conjecture $C1$. The motivation of this conjecture was an attempt to decide the 'complexity' of finite semigroups. For details see [14].

Next we briefly describe a question in term of Mal'cev products, equivalent to the question posed by Rhodes in the case of monoids. Given a variety of finite monoids $\mathcal{V}$ and a variety of finite groups $\mathcal{G}$, the *Mal'cev product* $\mathcal{V}^{-1}\mathcal{G}$ is defined to be the smallest variety of finite monoids containing the monoids M such that there exists a group $G \in \mathcal{G}$ and a morphism of monoids $\varphi : M \longrightarrow G$ with $\varphi^{-1}(1) \in \mathcal{V}$. One says that a variety of finite monoids $\mathcal{W}$ is *decidable* if there exists an algorithm to decide whether or not a given finite monoid is in the $\mathcal{W}$. Then conjecture $C2$ is the following: if the varieties $\mathcal{V}$ and $\mathcal{G}$ are decidable, then so is their Mal'cev product.

There are several other related problems (for example in terms of automata theory) that imply these conjectures (see [14]). In [26], J.-É. Pin and C. Reutenauer proposed the following conjecture in terms of profinite topologies of groups: If $H_1, \ldots, H_n$ are finitely generated subgroups of an abstract free group F, then the set $H_1 \cdots H_n$ is closed in the profinite topology of F. In the same paper they proved that a positive answer to this conjecture implies a positive answer to the conjecture of Rhodes. The conjecture of Pin and Reutenauer is proven to be correct as a special case of Theorem 7.2. Rhodes' conjecture was independently and using completely different methods proved by C. J. Ash [16]. The corresponding conjecture for the p-kernel of a finite semigroup requires not only Theorem 7.2 but also an application of Theorem 7.5; see [28]. Thus we have

Theorem 8.1. *Let M be a finite semigroup and let p be a prime number.*
(1) *There exists an algorithm to compute the kernel $K(M)$ of M.*
(2) *There exists an algorithm to compute the p-kernel $K_p(M)$ of M.*

There have been recently several papers with reformulations, improvements, applications and new proofs of some of the results in this article. See [19, 18, 20, 29, 30, 24, 31, 32].

Acknowledgments

I am very pleased to thank the organizers of the Thematic Term 2001 on Algorithmic Aspects of the theory of Semigroups and Automata, Gracinda Gomes, Pedro Silva and Jean-Éric Pin, for inviting me to participate to the meeting and for making my stay in Coimbra a very pleasant and scientifically rewarding experience.

It is also a pleasure to thank Carlo M. Scoppola for arranging my visit to L'Aquila and Rome; for being my 'banker' and my guide for a few weeks in Italy; and for his mathematical suggestions.

The referee of this paper has been very helpful and I thank him/her for his/her corrections and suggestions.

References on profinite groups:

[1] J-P. Serre, *Galois Cohomology*, Springer, Berlin - New York, 1996.
[2] J. Dixon, M. du Sautoy, A. Mann and D. Segal, *Analytic pro-p Groups*, Cambridge Univ. Press, Cambridge, 1999.
[3] L. Ribes and P. Zalesskii, *Profinite Groups*, Springer, Berlin - New York, 2000.
[4] J. Wilson, *Profinite Groups*, Clarendon Press, Oxford, 1998.

References on profinite graphs:

[5] D. Gildenhuys and L. Ribes, Profinite groups and Boolean graphs, *J. Pure Appl. Algebra*, **12** (1978) 21-47.
[6] L. Ribes and P. Zalesskii, *Profinite Graphs*, Springer, Berlin - New York, to appear.
[7] P. Zalesskii and O. Melnikov, Subgroups of profinite groups acting on trees, *Math. USSR-Sbornik*, **63** (1989) 405-424.

References on semigroups:

[8] J. Almeida, *Finite Semigroups and Universal Algebra*, World Scientific, Singapore, 1995.
[9] M. Lawson, *Inverse Semigroups*, World Scientific, Singapore, 1998.
[10] M. Howie, *Fundamentals of Semigroup Theory*, Clarendon Press, Oxford, 1995.

References on background:

[11] K. Henckell, S. Margolis, J.-É. Pin and J. Rhodes, Ash type II theorem, profinite groups and Mal'cev products, *Internat. J. Algebra Comput.*, **1** (1991) 411-436.
[12] K. Krohn and J. Rhodes, Algebraic theory of machines, *Trans. Amer. Math. Soc.*, **116** (1965) 450-464.
[13] K. Krohn and J. Rhodes, Complexity of finite semigroups, *Ann. Math* **88** (1968) 128-160.
[14] J.-É. Pin, On a conjecture of Rhodes, *Semigroup Forum*, **39** (1989) 1-15.
[15] J.-É. Pin, A topological approach to a conjecture of Rhodes, *Bull. Austral. Math. Soc.* **38** (1988) 421-431.

References on results and applications:

[16] C. Ash, Inevitable graphs: A proof of the type II conjecture and some related decision procedures, *Internat. J. Algebra Comput.*, **1** (1991) 127-146.
[17] J. Almeida and M. Delgado, Sur certains systèmes d'équations avec contraintes dans un groupe libre, *Portugal. Math.*, **56** (1999) 409-417.
[18] J. Almeida and P. Weil, Reduced factorizations in free profinite groups and join decompositions of pseudovarieties, *Internat. J. Algebra Comput.*, **4** (1994) 375-403.

[19] K. Auinger and B. Steinberg, The geometry of profinite graphs with applications to free groups and finite monoids, preprint.
[20] T. Coulbois, Free products, profinite topology and finitely generated subgroups, *Internat. J. Algebra Comput.*, **11** (2001) 171-184.
[21] T. Coulbois, Propriétés de Ribes-Zalesskiĭ, topologie profinie, produit libre et généralisations, Thèse de Doctorat, Université Paris VII – Denis Diderot, 2000.
[22] A. Douady, Détermination d'un groupe de Galois, *C. R. Acad Sc. Paris*, **258** (1964) 5305-5308.
[23] M. Hall, Coset representations in free groups, *Trans. Amer. Math. Soc.*, **67** (1949) 421-432.
[24] B. Herwig and D. Lascar, Extending partial automorphisms and the profinite topology on free groups, *Trans. Amer. Math. Soc.* **352** (2000) 1985-2021.
[25] S. Margolis, M. Sapir and P. Weil, Closed subgroups in pro-**V** topologies and the extension problem for inverse automata, *Internat. J. Algebra Comput.* **11** (2001) 405-445.
[26] J.-É. Pin and C. Reutenauer, A conjecture on the Hall topology for the free group, *Bull. London Math. Soc.* **23** (1991) 356-362.
[27] L. Ribes and P. Zalesskii, On the profinite topology on a free group, *Bull. London Math. Soc.*, **25** (1993) 37-43.
[28] L. Ribes and P. Zalesskii, The pro-p topology of a free group and algorithmic problems in semigroups, *Internat. J. Algebra Comput.* **4** (1994) 359-374.
[29] B. Steinberg, Inevitable graphs and profinite topologies: Some solutions to algorithmic problems in monoid and automatata theory stemming from group theory, *Internat. J. Algebra Comput.* **11** (2001) 25-71.
[30] B. Steinberg, Finite state automata: a geometric approach, *Trans. Amer. math. Soc.* **353** (2001) 3409-3464.
[31] P. Weil, Computing closures of finitely generated subgroups of the free group, pp 289-307 in: Algorithmic Problems in Groups and Semigroups, J-C. Birget, S. Margolis, J. Meakin and M. Sapir eds., Birkhäuser, Boston, 2000.
[32] S. You, The product separability of the generalized free product of cyclic groups, *J. London Math. Soc.* (2) **56** (1996) 91-103.

Research articles

DYNAMICS OF FINITE SEMIGROUPS

J. ALMEIDA*

*Dep. Matemática Pura, Faculdade de Ciências, Universidade do Porto
R. Campo Alegre, 687, 4169-007 Porto, Portugal
E-mail: jalmeida@fc.up.pt*

Iteration of implicit operators on profinite algebras provides an easy way of producing interesting implicit operations which in turn can be used to construct useful pseudoidentities. This paper reviews this method of producing implicit operations and explores its usage to obtain pseudoidentity characterizations of various pseudovarieties of groups and semigroups. In particular, the Plotkin conjecture for solvable groups and characterizations of finite nilpotent groups are discussed.

1 Introduction

Implicit operations on pseudovarieties of finite algebras (also called profinite words in [4, 5] in the case of semigroups and monoids) have been studied extensively in connection with applications in the theory of finite semigroups [2]. Such operations may be viewed as elements of relatively free profinite algebras but the operation point of view is sometimes very useful. Indeed, we may not only operate with implicit operations using the basic algebraic operations of the algebras in question, but we may also compose them, with proper care since implicit operations depend in general on several variables. In [3], this aspect was explored leading to the notion of implicit operator on a profinite algebra. Such operators turn out to form a profinite monoid on which we may therefore apply monoid implicit operations. In particular, taking the profinite algebra to be an algebra of implicit operations, we may use this idea to exhibit new implicit operations from known ones. An application to the theory

---

*WORK SUPPORTED, IN PART, BY *FUNDAÇÃO PARA A CIÊNCIA E A TECNOLOGIA* (FCT) THROUGH THE *CENTRO DE MATEMÁTICA DA UNIVERSIDADE DO PORTO*, BY THE FCT AND POCTI APPROVED PROJECT POCTI/32817/MAT/2000 WHICH IS COMPARTICIPATED BY THE EUROPEAN COMMUNITY FUND FEDER, AND BY THE INTAS GRANT #99-1224. THIS WORK WAS DONE IN PART WHILE THE AUTHOR WAS VISITING THE *CENTRO INTERNACIONAL DE MATEMÁTICA*, IN COIMBRA, PORTUGAL. FINANCIAL SUPPORT OF *FUNDAÇÃO CALOUSTE GULBENKIAN* (FCG), FCT, *FACULDADE DE CIÊNCIAS DA UNIVERSIDADE DE LISBOA* (FCUL) AND *REITORIA DA UNIVERSIDADE DO PORTO* IS GRATEFULLY ACKNOWLEDGED.

of semigroup pseudovarieties is given in [3].

The aim of this paper is to survey the basic results on implicit operators and to explore new applications. Section 2 presents the basic material on implicit operators. This is developed for general profinite algebras since the theory holds in that context. Section 3 reveals the relevance of the ω-power in terms of dynamical systems. Section 4 introduces some examples in the context of semigroups and monoids.

The remainder of the paper is concerned with pseudoidentities $v_1 = v_2$ built up from equating the two components of the ω-power $(v_1, v_2) = (w_1, w_2)^\omega$ of a binary implicit operator on semigroups. The last two sections contain several new results. Section 5 aims to find criteria for a pseudoidentity $v_1 = v_2$ to hold in pseudovarieties of groups. For this purpose, numerical properties of the frequency matrix of the base operator (w_1, w_2) are considered. This amounts to deducing that certain finite groups satisfy $v_1 = v_2$ from the knowledge that certain finite Abelian groups satisfy $w_1 = w_2$. In Section 6, these results are used as a guide for the choice of interesting specific examples of pseudoidentities. In particular, the Plotkin conjecture on the characterization of finite solvable groups, already introduced in Section 2.3, and characterizations of finite nilpotent groups are discussed.

2 Basics

In this section, wherever the development of the theory is not specific to semigroups, we prefer to consider arbitrary algebras of a finitary type. This means that a set of symbols of operations with corresponding finite arities is fixed. An algebra of this type is a set A endowed with interpretations of the operation symbols f as operations $f_A : A^n \to A$ (often denoted simply f) on A of the arity n associated with the symbol. We normally talk about *the algebra A*, without explicit reference to the interpretation of the operation symbols. If the set A is endowed with a topology with respect to which the operation symbols are interpreted as continuous operations, then we say that the algebra is a *topological algebra*. In case the topology is compact, we talk about a *compact algebra*. Finite algebras are examples of compact algebras under the discrete topology. A topological algebra is said to be *finitely generated* if there is a finite subset generating a dense subalgebra.

2.1 Profinite algebras and implicit operations

By a *profinite algebra* we mean a compact algebra A which is residually finite as a topological algebra in the sense that, given two distinct elements a and b

of A, there is a continuous homomorphism $\varphi : A \to F$ into a finite algebra such that $\varphi(a) \neq \varphi(b)$. Such algebras embed in a direct product of finite algebras and, therefore, they are totally disconnected (the connected components are singletons) or, equivalently, for a compact space, they are zero-dimensional (the clopen subsets form a basis of the topology). For certain types of compact algebras, such as semigroups, monoids, groups, rings, and lattices, residual finiteness is equivalent to zero-dimensionality [1], but this is not true in general as an unpublished example of K. Auinger shows.

Assume that A is a finitely generated profinite algebra. We endow A with a distance function defined, for $a \neq b$, by $d(a,b) = 2^{-r(a,b)}$ where $r(a,b)$ is the cardinality $|F|$ of the smallest finite algebra F for which there exists a continuous homomorphism $\varphi : A \to F$ such that $\varphi(a) \neq \varphi(b)$, and letting $d(a,a) = 0$. This is an ultrametric which induces the topology of A.

A *pseudovariety* is a class of finite algebras which is closed under taking homomorphic images, subalgebras and finite direct products. Let V be a pseudovariety. By a *pro-V algebra* we mean a profinite algebra whose finite continuous homomorphic images belong to V. The *free pro-V algebra on the set X* is the unique (up to continuous isomorphism) pro-V algebra $\overline{\Omega}_X V$ endowed with a mapping $\iota : X \to \overline{\Omega}_X V$ such that, given any mapping $\varphi : X \to A$ into a pro-V algebra A, there is a unique continuous homomorphism $\hat{\varphi} : \overline{\Omega}_X V \to A$ such that $\hat{\varphi} \circ \iota = \varphi$.

The existence of the free pro-V algebra on A may be established by considering the projective limit of all X-generated algebras from V. We denote by $\overline{\Omega}_X$, without mention of a specific pseudovariety, the absolutely free profinite algebra on the set X, that is the free pro-V algebra on X where V is the pseudovariety of all finite algebras of the given type.

The elements of $\overline{\Omega}_X V$ may be naturally interpreted as X-*ary operations* on profinite algebras: for $w \in \overline{\Omega}_X V$ and $\varphi \in A^X$, let $w_A(\varphi) = \hat{\varphi}(w)$. This interpretation defines an *implicit operation* since it commutes with continuous homomorphisms in the sense that the following diagram commutes for every continuous homomorphism $h : A \to B$:

$$\begin{array}{ccc} A^X & \xrightarrow{w_A} & A \\ {\scriptstyle h^X}\downarrow & & \downarrow{\scriptstyle h} \\ B^X & \xrightarrow{w_B} & B \end{array}$$

The restriction of the interpretation to members of V is a bijection from $\overline{\Omega}_X \mathsf{V}$ onto the set of all X-ary implicit operations on V. We view the elements of $\overline{\Omega}_X \mathsf{V}$ as X-ary implicit operations on V.

The elements of X correspond to the component projections. In case $X = \{x_1, \ldots, x_n\}$, the projection x_i is given by

$$(x_i)_A : (a_1, \ldots, a_n) \mapsto a_i.$$

We then identify $\overline{\Omega}_n \mathsf{V}$ with $\overline{\Omega}_X \mathsf{V}$. The subalgebra $\Omega_X \mathsf{V}$ of $\overline{\Omega}_X \mathsf{V}$ generated by the component projections consists of the X-*ary explicit operations on* V. The algebra $\Omega_X \mathsf{V}$ is the V-free algebra on the set X.

Implicit operations may be composed as usual operations to produce again implicit operations, a composition which may be described more shortly as follows. If $v_1, \ldots, v_m$ are n-ary implicit operations on V and w is an m-ary implicit operation on V, then the composite $w(v_1, \ldots, v_m)$ is the n-ary implicit operation on V determined by $w_{\overline{\Omega}_n \mathsf{V}}(v_1, \ldots, v_m)$.

A *pseudoidentity* over a pseudovariety V is a formal equality $u = v$ between implicit operations over V of the same arity. It is said to *hold* in a pro-V algebra A if $u_A = v_A$. Since implicit operations commute with continuous homomorphisms, this property is respected by taking continuous homomorphic images, subalgebras and direct products of finitely many factors. The class $[\![\Sigma]\!]$ of all finite members of V satisfying all pseudoidentities from a set Σ of pseudoidentities over V is therefore a (sub)pseudovariety of V. By a theorem of Reiterman [15], every subpseudovariety of V is of this form.

For more information on profinite algebras and implicit operations, see [2, 6].

2.2 Implicit operators

An *n-ary implicit operator* on a pro-V algebra A is a transformation $T : A^n \to A^n$ of A^n into itself whose components $T_i : A^n \to A$ are determined by (n-ary) implicit operations. The set of all n-ary implicit operators on A is denoted $\mathcal{O}_n(A)$. This is clearly a monoid under composition. The metric on

$\overline{\Omega}_n$ induces a metric on $\mathcal{O}_n(A)$:

$$d(T, U) = \sum_{i=1}^{n} \inf\{d(v, w) : v, w \in \overline{\Omega}_n, v_A = T_i, w_A = U_i\}.$$

Proposition 2.1 ([3]). *The correspondence*

$$\mathcal{O}_n : A \mapsto \mathcal{O}_n(A)$$

defines a functor from the category of profinite algebras with onto continuous homomorphisms as morphisms into the category of profinite monoids.

The proof of Proposition 2.1 involves showing that, if A is the projective limit of a projective family (A_i, φ_{ij}) of onto homomorphisms between finite algebras, then there is an associated projective family $(\mathcal{O}_n(A_i), \mathcal{O}_n(\varphi_{ij}))$ whose projective limit is isomorphic with $\mathcal{O}_n(A)$. This fact immediately yields the following lemma which will be of use in Section 3.

Lemma 2.2. *The evaluation mapping*

$$\mathcal{O}_n(A) \times A^n \to A^n$$
$$(f, P) \mapsto f(P)$$

is continuous.

We say that V *requires bounded memory for the computation of explicit operations* if, for every $n \geq 1$, there exists $k \geq 0$ such that, for all $w \in \Omega_n\mathsf{V}$, w may be computed from the projections $x_1, \ldots, x_n$ using at each step either a projection or one of the basic operations applied to memorized values and storing in the process values in at most k memory cells. Examples of such pseudovarieties are given by semigroups, monoids, groups, and rings [3]. Denote by M the pseudovariety of all finite monoids.

Proposition 2.3 ([3]). *Let* V *be a pseudovariety of finite finitary type which requires bounded memory for the computation of explicit operations. Then every n-ary implicit operation on* V *is a component of an implicit operator of the form*

$$\pi_M(\rho_1, \ldots, \rho_r)$$

where $r \geq 1$, $\rho_i \in \Omega_m\mathsf{V}$, $M = \mathcal{O}_r(\overline{\Omega}_m\mathsf{V})$, $\pi \in \overline{\Omega}_r\mathsf{M}$, *and the integers r and m depend only on n and not on the implicit operation.*

Proposition 2.3 confirms a central role for implicit operations on finite monoids in the theory of finite algebras.

2.3 Examples of implicit operators

Denote by x^ω the unary implicit operation on finite semigroups (and monoids) which associates to each element s of a finite semigroup S the unique idempotent power s^ω of s. Being an implicit operation on finite semigroups, x^ω has a natural interpretation on every profinite semigroup. It is easy to see that
$$x^\omega = \lim_{n \to \infty} x^{n!}.$$
For a unary semigroup implicit operation v, we can consider its ω-power v^ω or the unique component of the ω-power $(v)^\omega$ of the unary implicit operator it defines. Since the notation is ambiguous and the two constructions lead to different results, we will denote the latter by $v^{\circ \omega}$ to emphasize the semigroup operation being considered is composition.

For n-ary implicit operations (on V) $w_1, \ldots, w_n$ and a pro-V algebra A, denote by $(w_1, \ldots, w_n)$ the implicit operator
$$A^n \to A^n$$
$$(a_1, \ldots, a_n) \mapsto ((w_1)_A(a_1, \ldots, a_n), \ldots, (w_n)_A(a_1, \ldots, a_n))$$
Denote composition of operators by concatenation: $(v_1, \ldots, v_n)(w_1, \ldots, w_n)$ has component i determined by the operation $v_i(w_1, \ldots, w_n)$.

Remark. This remark is not necessary for understanding the rest of the paper. We claim that $\mathcal{O}_n(A)$ may be viewed as a monoid of endomorphisms of A^n in a suitable clone category in the sense of [12, Subsection 3.6], a book to which we refer the reader for undefined terms involving clones. To prove the claim, let $\mathbf{C}_\mathsf{V}$ denote the clone of the monogenic relatively free profinite algebra $\overline{\Omega}_1 \mathsf{V}$ in the opposite category of pro-V algebras: $\mathbf{C}_\mathsf{V}$ is the full subcategory whose objects are the free products of n copies of $\overline{\Omega}_1 \mathsf{V}$ in the category of pro-V algebras, that is the finitely generated relatively free pro-V algebras $\overline{\Omega}_n \mathsf{V}$ with $n \geq 1$. As suggested by one of the referees, following [12, Exercise 3.16.23] we call this clone the *clone of* V. On the other hand, for a profinite algebra A, let $\mathbf{C}_A$ denote the clone of A in the category of profinite algebras, that is the full subcategory whose objects are the spaces A^n with $n \geq 1$. In case A is a pro-V algebra, its structure as such induces a clone-map $\mathbf{C}_\mathsf{V} \to \mathbf{C}_A$, namely the functor $F : \mathbf{C}_\mathsf{V} \to \mathbf{C}_A$ which sends $\overline{\Omega}_n \mathsf{V}$ to A^n and each continuous homomorphism $\varphi : \overline{\Omega}_n \mathsf{V} \leftarrow \overline{\Omega}_m \mathsf{V}$ to the continuous homomorphism $F\varphi : A^n \to A^m$ whose components are the operations $(\varphi x_1)_A, \ldots, (\varphi x_m)_A$. The monoid $\mathcal{O}_n(A)$ is then precisely the monoid of endomorphisms of A^n in the image category $F\mathbf{C}_\mathsf{V}$. Moreover, [12, Exercise 3.16.23] may be transposed to the context of profinite algebras and pseudovarieties by stating that a profinite algebra A is pro-V if and only if there exists a clone-map $\mathbf{C}_\mathsf{V} \to \mathbf{C}_A$.

Implicit operators serve as a means of constructing implicit operations. The following are examples of monoid implicit operations constructed as components of implicit operators.

The equality
$$x^\omega = x_1(x_1 x_2, x_2)^\omega (1, x)$$
is a simple illustration of how the ω-power of implicit operators produces implicit operations. The same idea can be used to obtain many other implicit operations. For instance, another rather important unary implicit operation is the $(\omega - 1)$-power which may be defined by
$$x^{\omega-1} = x_1(x_1 x_2, x_3, x_3)^\omega (1, 1, x)$$
which is just a variation of the preceding equality, obtained by adding the gimmick of delaying the start of the action of the first component. The operation $x^{\omega-1}$ provides a *weak inverse* of x in the sense that $x^{\omega-1} = x^{\omega-1} x x^{\omega-1}$. In particular, over profinite groups, $x^{\omega-1}$ is the inverse of x.

Another useful unary implicit operation is defined by
$$x^{n^\omega} = (x^n)^{\circ \omega} = \lim_{k \to \infty} x^{n^{k!}}.$$
Consider the *commutator* $[x, y] = x^{\omega-1} y^{\omega-1} xy$ and let
$$[x, {}_k y] = x_1([x_1, x_2], x_2)^k (x, y).$$
The right hand side makes sense for any unary monoid operation x^k. In particular, taking $k = \omega$, we obtain
$$[x, {}_\omega y] = \lim_{n \to \infty} [x, {}_{n!} y].$$
The identities $[x, {}_n y] = 1$ are known as the *(right) Engel identities*.

Zorn (see [21, 16]) proved that a finite group is nilpotent if and only if it satisfies an Engel identity. This result may be expressed by saying that the pseudovariety G_{nil} of all finite nilpotent groups is defined by the pseudoidentity $[x, {}_\omega y] = 1$.

Consider next the binary operations defined by
$$u_n = x_1([[x_1, x_2], [x_1, x_3]], x_2, x_3)^n ([x, y], x, y).$$
In particular, $u_\omega = \lim_{n \to \infty} u_{n!}$. B. Plotkin (see [10]) has conjectured that a finite group is solvable if and only if it satisfies some identity of the form $u_n = 1$ with $n \geq 1$. In other words, the conjecture is that the pseudovariety G_{sol} of all finite solvable groups is defined by the pseudoidentity $u_\omega = 1$. Interpreting $[\ ,\ ]$ as the Lie bracket in a Lie algebra, we have the following related result which adds some evidence to Plotkin's conjecture.

Theorem 2.4 ([10]). *A finite-dimensional Lie algebra over an infinite field of characteristic $p > 5$ is solvable if and only if it satisfies an identity of the form $u_n = 0$.*

The validity of Plotkin's conjecture would entail immediately a proof of a result of Thompson [19] according to which a finite group is solvable if and only if all its 2-generated subgroups are solvable. The original proof of this result depends on Thompson's complete classification of finite simple groups whose proper subgroups are solvable which extends over more than 400 published pages. A much shorter proof was given by Flavell [8].

To prove the conjecture, since obviously every finite solvable group satisfies $u_\omega = 1$, it suffices to show that every non-solvable group fails $u_\omega = 1$. Since $\mathsf{G}_{\mathsf{sol}}$ is closed under extensions, it suffices to consider finite non-Abelian simple groups. If a simple group has a proper non-solvable subgroup, then it suffices to show that the subgroup fails $u_\omega = 1$. Hence it suffices to consider finite non-Abelian simple groups all of whose proper subgroups are solvable. The list of all such groups is the result of Thompson's classification mentioned above and consists of the following groups:

$$\mathrm{PSL}(2,p),\ \mathrm{PSL}(2,2^q),\ \mathrm{PSL}(2,3^r),\ \mathrm{PSL}(3,3),\ \mathrm{Sz}(2^r) \tag{1}$$

where p, q and r are primes such that p is different from 3 and is either 5 or congruent with $\pm 2 \bmod 5$, and r is odd. Here, $\mathrm{PSL}(n, q^l)$ denotes the projective special linear group in dimension n over the Galois field $\mathrm{GF}(q^l)$ and $\mathrm{Sz}(2^r)$ denotes the Suzuki group over $\mathrm{GF}(2^r)$ (see [9]).

Calculations on any specific such group (with not too large parameters, so as to be feasible) show that it fails $u_\omega = 1$ but no pattern for a good evaluation seems to emerge. Such calculations add further evidence for Plotkin's conjecture but fall short of proving it.

2.4 Invertible implicit operators

We conclude this section with a remark on the invertibility of implicit operators.

In a profinite monoid, note that an element f is invertible if and only if $f^\omega = 1$. In the case of a profinite monoid of implicit operators over a free profinite algebra $\overline{\Omega}_n \mathsf{V}$ over a pseudovariety, since a pseudoidentity on n variables is valid in $\overline{\Omega}_n \mathsf{V}$ if and only if it is valid in every member of V, invertibility of an implicit operator over $\overline{\Omega}_n \mathsf{V}$ is therefore equivalent to its invertibility over every member of V. In this case, we will simply say that the operator is *invertible over* V. The following result is the generalization of an idea which was used in [3].

Lemma 2.5. *An implicit operator $(w_1, \ldots, w_n)$ is invertible over a pseudovariety V if and only if the subalgebra $\langle w_1, \ldots, w_n \rangle$ generated by its components is dense in $\overline{\Omega}_n \mathsf{V}$.*

Proof. An inverse of the operator $(w_1, \ldots, w_n)$ is another operator $(\pi_1, \ldots, \pi_n)$ such that the composite $(\pi_1, \ldots, \pi_n)(w_1, \ldots, w_n)$ is the identity operator $(x_1, \ldots, x_n)$. Hence, the given operator is invertible if and only if the generators x_i of $\overline{\Omega}_n \mathsf{V}$ are of the form $\pi_i(w_1, \ldots, w_n)$ for some implicit operation π_i. Now, the generators have this property if and only if every element of $\overline{\Omega}_n \mathsf{V}$ has it. Since the closure of the subalgebra $\langle w_1, \ldots, w_n \rangle$ consists precisely of the elements of $\overline{\Omega}_n \mathsf{V}$ of the form $\pi(w_1, \ldots, w_n)$, the result follows. □

3 Dynamics of implicit operators

In general, given n-ary implicit operations $w_1, \ldots, w_n$, and a profinite algebra A, we have a dynamical system (A^n, f), where $f = (w_1, \ldots, w_n)$, consisting of a compact space together with a continuous transformation of this space into itself. If A is finite, of course there are some $k \geq 0$ and $\ell > 0$ such that, for any point $P \in A^n$, $f^{k+\ell}(P) = f^k(P)$, that is the dynamics is eventually periodic.

If A is infinite, there are in general no repetitions in each *orbit* $\{f^k(P) : k \geq 0\}$. Denote by $\mathrm{Im}_\infty(f)$ the intersection of all $\mathrm{Im}(f^k)$ ($k \geq 1$). The reader can easily verify the following facts:

- $\mathrm{Im}_\infty(f) = \mathrm{Im}(f^\omega)$.

- f acts as a permutation on $\mathrm{Im}_\infty(f)$ (with inverse $f^{\omega-1}$).

- f^ω is the identity mapping 1_{A^n} on A^n if and only if there are implicit operations $v_1, \ldots, v_n$ such that $g = (v_1, \ldots, v_n)$ satisfies $gf = 1_{A^n}$.

Recall a point $P \in A^n$ of the dynamical system (A^n, f) is said to be

- *recurrent* if, for every neighbourhood U of P and every k, there is $\ell \geq k$ such that $f^\ell(P) \in U$;

- *uniformly recurrent* if there exists m such that, for every neighbourhood U of P and every k, there is $\ell \in \{k+1, \ldots, k+m\}$ such that $f^\ell(P) \in U$.

Proposition 3.1. *Let A be a finitely generated profinite algebra and let f be an n-ary implicit operator on A. Then every recurrent point of A^n under the action of f is uniformly recurrent and the set of all such points is $\mathrm{Im}(f^\omega)$.*

Proof. Suppose first that $P \in \mathrm{Im}(f^\omega)$. We show that P is uniformly recurrent.

Let U be a neighbourhood of P. Since A is zero-dimensional, we may assume that U is of the form $U = U_1 \times \cdots \times U_n$ where the U_i are clopen subsets of A. Since A is residually finite, there exists a continuous homomorphism $\varphi : A \to F$ into a finite algebra whose kernel saturates each U_i.

Consider the following commutative diagram where g is the implicit operator on F defined by $(w_1, \ldots, w_n)$ and $\varphi^{(n)}$ is φ on each component:

$$\begin{array}{ccc} A^n & \xrightarrow{f} & A^n \\ {\scriptstyle \varphi^{(n)}}\downarrow & & \downarrow{\scriptstyle \varphi^{(n)}} \\ F^n & \xrightarrow{g} & F^n \end{array}$$

By construction, we have $(\varphi^{(n)})^{-1}(Q) \subseteq U$ where $Q = \varphi^{(n)}(P)$. Since $\varphi^{(n)}(\mathrm{Im}(f^\omega)) \subseteq \mathrm{Im}(g^\omega)$, Q belongs to $\mathrm{Im}(g^\omega)$.

Let $m = |F|^n$. Every orbit of the action of g on F^n has length at most m. Hence, for every $k \geq 0$ at least one of the points

$$g^{k+1}(Q), g^{k+2}(Q), \ldots, g^{k+m}(Q)$$

must coincide with Q. Hence, for every $k \geq 0$, at least one of the points

$$f^{k+1}(P), f^{k+2}(P), \ldots, f^{k+m}(P)$$

must lie in U. Hence P is a uniformly recurrent point with respect to f.

Conversely, suppose that P is a recurrent point of A^n with respect to the action of f. Given a positive integer r, for every $k \geq 0$ there exists $\ell \geq k$ such that $f^\ell(P)$ lies at a distance at most $1/r$ from P. Taking $k = r!$, we deduce that there exists some point Q_r such that

$$d(f^{r!}(Q_r), P) < \frac{1}{r}$$

Since A^n is compact, there is some convergent subsequence $(Q_{r_m})_m$, say with limit Q. In view of Lemma 2.2, we deduce that

$$P = \lim_{m \to \infty} f^{r_m!}(Q_{r_m}) = f^\omega(Q)$$

which completes the proof. □

4 Examples in the realm of groups and semigroups

In this section, we present some examples of implicit operators on semigroups. In these examples we consider an n-tuple $(w_1, \ldots, w_n)$ of implicit operations on the n-letter alphabet $\{x_1, \ldots, x_n\}$. For simplicity, we may sometimes write $a = x_1$, $b = x_2$, ...

In the case where the w_i are words, such an n-tuple may be represented by a labeled directed graph $\Gamma(w_1, \ldots, w_n)$ whose vertices are the letters and whose labeled edges are those of the form

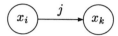

if the letter in position $j + 1$ of the word w_i is x_k. The *label* of a path in $\Gamma(w_1, \ldots, w_n)$ is the concatenation of the labels in its successive edges. In the general case, where the w_i are not necessarily words, we may still construct the underlying directed graph by drawing an edge from x_i to x_k if w_i depends on x_k. But the labeling makes no sense in this case. The set of all x_k such that w depends on x_k is called the *content* of w and is denoted $c(w)$.

Let $\mathsf{K} = [\![x^\omega y = x^\omega]\!]$, the pseudovariety of all finite semigroups whose idempotents are left zeros. The free pro-K semigroup $\overline{\Omega}_X \mathsf{K}$ may be realized as the free semigroup X^+ with the set X^∞ of all right infinite words on X adjoined as left zeros. Multiplication of a finite word by an infinite word is obtained by concatenation. The topology induced on X^+ is discrete while on X^∞ it is the topology of the Cantor set. A metric is roughly but suggestively described by saying that two words (finite or infinite) are *close* if they have a common *long* prefix.

In $\overline{\Omega}_X \mathsf{K}$ every infinite product $w_1 w_2 \cdots$ converges. Thus $\overline{\Omega}_X \mathsf{K}$ has a natural structure as an ω-semigroup (see [13]). It is in fact the free ω-semigroup on the set X.

For words $w_1, \ldots, w_n$ such that w_1 starts with x_1, if

$$(v_1, \ldots, v_n) = (w_1, \ldots, w_n)^\omega$$

then v_1 on $\overline{\Omega}_n \mathsf{K}$ may be identified with an infinite word.

Example 4.1. Let $\phi_n = \phi_n(a, b) = x_1(ab, a)^n$. The so-called *Fibonacci infinite word* is defined as the value of the binary operation ϕ_ω on K. The graph of the operator generating ϕ_ω by iteration is $\Gamma(ab, a)$:

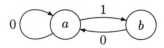

We will call ϕ_ω the *Fibonacci implicit operation*. For finite n, the number of occurrences of the letter b in $\phi_n(a,b)$ is the nth Fibonacci number.

For a word u, denote the reverse word by $\bar{u}$. If $c = |w_i|$ $(i = 1, \ldots, n)$ is constant, then there is a path

with label u if and only if the letter in position $\bar{u}_{(c)}$ (the integer which is represented by $\bar{u}$ in base c) of $x_i(w_1, \ldots, w_n)^{|u|}$ is x_k, where we start counting from the left of a word at position 0.

Example 4.2. The *Prouhet-Thue-Morse infinite word* is defined to be the value of the binary operation $\tau(a,b) = x_1(ab, ba)^\omega$ on K. The graph of the operator generating $\tau(a,b)$ by iteration is $\Gamma(ab, ba)$:

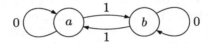

Since both components have the same length, the letters in each position are given by the process described above. For instance, the letter in position $6 = \overline{011}_{(2)}$ of the first component of the iterate $(ab, ba)^3$ is the letter at the end vertex of the path labeled 011 starting at the vertex a, namely an a. We will call $\tau(a,b)$ the *Prouhet-Thue-Morse implicit operation*.

Let $w_1, \ldots, w_n \in \overline{\Omega}_n S$ be implicit operations, where $n > 1$, and let

$$(v_1, \ldots, v_n) = (w_1, \ldots, w_n)^\omega$$

in $\mathcal{O}_n(\overline{\Omega}_n S)$. The following results concerning the relative location of the implicit operations v_i with respect to Green's relations on the free profinite semigroup $\overline{\Omega}_n S$ are taken from [5] and are easily established using standard compactness arguments.

Proposition 4.1. *a) If all the w_i have full content then the v_i are $\mathcal{J}$-equivalent regular members of $\overline{\Omega}_n S$.*

b) If $x_i \in c(w_i)$ for some i then the v_i are $\mathcal{J}$-equivalent regular operations if and only if the graph $\Gamma(w_1, \ldots, w_n)$ is strongly connected.

c) If all the w_i have full content and they all have a common prefix x_j, then the v_i are $\mathcal{R}$-equivalent.

d) If all the w_i have full content and they all have a common nontrivial *prefix and a common nontrivial suffix, then the v_i are $\mathcal{H}$-equivalent group members of* $\overline{\Omega}_n S$.

For instance, consider the implicit operators
$$(\phi_r, \phi_{r-1}) = (ab, a)^r.$$
Since $f^\omega = (f^2)^\omega$, we also have $(\phi_\omega, \phi_{\omega-1}) = (aba, ab)^\omega$. Hence ϕ_ω and $\phi_{\omega-1}$ are $\mathcal{R}$-equivalent regular implicit operations. Moreover, since
$$\phi_{r+2} = \phi_{r+1}\phi_r = \phi_r\phi_{r-1}\phi_r, \qquad (2)$$
$\phi_{r+2} \leq_{\mathcal{L},\mathcal{R}} \phi_r$ so that, in a finite semigroup, the ϕ_{2n} (resp. the ϕ_{2n-1}) are eventually all in the same $\mathcal{H}$-class. Taking $r = n!$ in (2) and letting $n \to \infty$, we obtain the equality $\phi_{\omega+2} = \phi_\omega\phi_{\omega-1}\phi_\omega$ from which, together with the fact that $\phi_{\omega+2} \mathcal{H} \phi_\omega$, it follows that ϕ_ω and $\phi_{\omega-1}$ are $\mathcal{R}$-equivalent group elements. They are not $\mathcal{H}$-equivalent: on right-zero semigroups, $(ab, a) = (b, a)$ so that ϕ_ω and $\phi_{\omega-1}$ end with different letters (respectively a and b). Since the implicit operator (ab, a) is invertible on profinite groups, ϕ_ω and $\phi_{\omega-1}$ are two $\mathcal{R}$-equivalent group members of $\overline{\Omega}_2 S$ which give respectively the projections a and b on profinite groups. In particular, $\mathsf{I} = [\![\phi_\omega = 1]\!]$ and $\mathsf{R} = [\![\phi_\omega = \phi_{\omega-1}]\!]$ are respectively the trivial pseudovariety, consisting of singleton semigroups, and the pseudovariety of all finite $\mathcal{R}$-trivial semigroups.

The components of the operator
$$(u, v) = (a^2 b, ab)^\omega$$
give two $\mathcal{H}$-equivalent group elements of $\overline{\Omega}_2 S$. Since the implicit operator $(ab^{\omega-1}, (ab^{\omega-1})^{\omega-1}b)$ is an inverse of (a^2b, ab) on finite groups, the implicit operator $(a^2b, ab)^\omega$ is the identity on profinite groups. Thus, given two $\mathcal{H}$-equivalent group elements s and t of a profinite semigroup S, denoting by G the common $\mathcal{H}$-class, which is a profinite group, we have $u_S(s, t) = u_G(s, t) = s$ and $v_S(s, t) = v_G(s, t) = t$. In other words, the values of the implicit operations u and v describe two arbitrary $\mathcal{H}$-equivalent group elements of any profinite semigroup. This may be used to give bases of pseudoidentities for pseudovarieties of the form $\overline{\mathsf{H}}$, consisting of all finite semigroups whose subgroups lie in a given pseudovariety of groups H. For example, for the pseudovariety Ab of all finite Abelian groups, using the above implicit operations u and v, we have
$$\overline{\mathsf{Ab}} = [\![uv = vu]\!].$$
More generally, we could take the components of the implicit operator
$$(x_1^2 x_2 \cdots x_n, x_1 x_2^2 x_3 \cdots x_n, \ldots, x_1 \cdots x_{n-2} x_{n-1}^2 x_n, x_1 \cdots x_n)^\omega \qquad (3)$$

to obtain n implicit operations whose values describe n arbitrary $\mathcal{H}$-equivalent group elements of profinite semigroups. Moreover, the above operator is the identity mapping on G^n for every profinite group G. Given a basis of pseudoidentities for a pseudovariety H of groups, the components of such an operator may be used to exhibit a basis of pseudoidentities for $\overline{\mathsf{H}}$ in the way described above for Ab. This was first done by Schützenberger [17] (see also Pin [14, Chapter 3, Section 4.6]) in the language of the "ultimate definition by sequences of equations" by considering, in the language of this paper, the operator

$$(x_1^{\omega+1} x_2^{\omega} \cdots x_n^{\omega}, x_1^{\omega} x_2^{\omega+1} x_3^{\omega} \cdots x_n^{\omega}, \ldots, x_1^{\omega} \cdots x_{n-1}^{\omega} x_n^{\omega+1})^{\omega}. \qquad (4)$$

See [5] for further details and applications.

5 Some remarks on pseudoidentities built from binary implicit operators

Given a finite semigroup S and a regular $\mathcal{D}$-class D, let $\equiv$ denote the smallest equivalence relation on group elements of D which identifies $\mathcal{R}$-equivalent elements and $\mathcal{L}$-equivalent elements. A *block* of D is the Rees quotient of the subsemigroup generated by an $\equiv$-class modulo the ideal (which may be empty) consisting of those elements which do not lie in D. For a pseudovariety V, BV is the class of all finite semigroups whose blocks belong to V. It is easy to show that BV is a pseudovariety.

Proposition 5.1. *Let (w_1, w_2) be a binary implicit operator and let $(v_1, v_2) = (w_1, w_2)^{\omega}$. Suppose $\Gamma(w_1, w_2)$ is strongly connected and the implicit operations w_1 and w_2 start (and, respectively, end) with different letters. Then*

$$\mathsf{B}(\llbracket v_1 = v_2 \rrbracket \cap \mathsf{G}) = \llbracket v_1 = v_2 \rrbracket.$$

Proof. By induction on k, we see that the components of every finite power $(w_1, w_2)^k$ retain the property of starting (respectively ending) with different letters. Hence a finite semigroup S satisfying the pseudoidentity $v_1 = v_2$ can not have any distinct $\mathcal{R}$- or $\mathcal{L}$-equivalent idempotents. Hence $S \in \mathsf{BG}$. Since the subgroups of S also satisfy $v_1 = v_2$, S belongs to the pseudovariety $\mathsf{B}(\llbracket v_1 = v_2 \rrbracket \cap \mathsf{G})$.

Conversely, suppose that $S \in \mathsf{B}(\llbracket v_1 = v_2 \rrbracket \cap \mathsf{G})$ and $a, b \in S$. By Proposition 4.1(a), $v_1(a, b)$ and $v_2(a, b)$ are $\mathcal{J}$-equivalent regular elements. Since (v_1, v_2) is an idempotent operator, we have

$$v_i(a, b) = v_i(v_1(a, b), v_2(a, b)) \quad (i = 1, 2). \qquad (5)$$

Taking into account these equalities, since the blocks of S are groups, we conclude that $v_1(a,b)$ and $v_2(a,b)$ are elements of the same subgroup. Since this subgroup satisfies the pseudoidentity $v_1 = v_2$, using (5) we obtain

$$v_1(a,b) = v_1(v_1(a,b), v_2(a,b)) = v_2(v_1(a,b), v_2(a,b)) = v_2(a,b),$$

which shows that S satisfies $v_1 = v_2$. □

Consider the implicit operator

$$(u,v) = (a^2b, ba)^\omega.$$

On profinite Abelian groups, the operator (a^2b, ba) behaves precisely like the invertible operator (a^2b, ab). Hence (u,v) is the identity operator on profinite Abelian groups. In particular, the only finite cyclic group satisfying the pseudoidentity $u = v$ is the trivial group. Since every nontrivial finite group contains a nontrivial finite cyclic group, it follows that no nontrivial finite group verifies the pseudoidentity $u = v$. By noting that $(a^2b, ba)^2 = (a^2ba^2b^2a, ba^3b)$, we deduce that $u \in a(\overline{\Omega}_2 S)a$ and $v \in b(\overline{\Omega}_2 S)b$, and therefore Proposition 5.1 shows that

$$[\![u = v]\!] = \mathsf{BI} = \mathsf{A} \cap \mathsf{BG}$$

where A denotes the pseudovariety of all finite semigroups with no nontrivial subgroups.

More generally, let $A(w_1, \ldots, w_n)$ denote the $n \times n$ matrix whose (i,j) entry is the *number* $|w_i|_{x_j}$ *of occurrences of the letter* x_j *in* w_i, which we call the *frequency matrix*. For a word w, the number $|w|_{x_j}$ is obtained precisely by counting how many times the letter x_j occurs in w. Hence, for words $w_1, \ldots, w_n$, the frequency matrix is the adjacency matrix of the graph $\Gamma(w_1, \ldots, w_n)$. For an arbitrary n-ary implicit operation w, $|w|_{x_j}$ is constructed by first taking the unary implicit operation on G (on the single variable x_j) obtained by evaluating to 1 all other variables and then identifying all these generators of different free procyclic groups $\overline{\Omega}_{\{x_j\}}$G. Thus $|w|_{x_j}$ is viewed as an element of $\overline{\Omega}_1$G. Note that $\overline{\Omega}_1$G is a profinite ring with multiplication of implicit operations as addition and composition as the ring multiplication. This ring is in fact the completion $\hat{\mathbb{Z}}$ of the ring $\mathbb{Z} \simeq \Omega_1$G of integers with respect to the metric of Subsection 2.1, also known as its *profinite completion*. We observe that the correspondence between integers and implicit operations sends n to $x^{\omega+n}$.

As an example, we compute the frequency matrix

$$A(a^{\omega-1}ba, ab^{\omega-1}) = \begin{pmatrix} 0 & 1 \\ 1 & -1 \end{pmatrix}$$

where the result is justified since a^ω is the neutral element of the group $\overline{\Omega}_{\{a\}}\mathsf{G}$ and $b^{\omega-1}$ is the inverse of b in the group $\overline{\Omega}_{\{b\}}\mathsf{G}$. The determinant and the trace of the frequency matrix will play an important role. In our example the determinant and the trace are both -1.

Let $\mathfrak{A}_m$ denote the pseudovariety of all finite Abelian groups of exponent m. For a prime p, let G_p denote the pseudovariety of all finite p-groups.

Proposition 5.2. *The following conditions are equivalent for an implicit operator $f = (w_1, \ldots, w_n)$ and a prime p:*
(1) f is invertible on G_p;
(2) f is invertible on $\mathfrak{A}_p$;
(3) $\det A(w_1, \ldots, w_n) \not\equiv 0 \pmod{p}$.

Proof. (1)$\Rightarrow$(2) is trivial. Since elements of $\mathfrak{A}_p$ are just finite-dimensional vector spaces over $\mathrm{GF}(p)$, the equivalence between (2) and (3) follows by elementary linear algebra. It remains to show that (2)$\Rightarrow$(1).

Suppose f is not invertible over G_p. Then, by Lemma 2.5, the closed subgroup H of $\overline{\Omega}_n \mathsf{G}_p$ generated by $\{w_1, \ldots, w_n\}$ is a proper subgroup. Since $\overline{\Omega}_n \mathsf{G}_p$ is a profinite group, there exists a continuous homomorphism $\varphi : \overline{\Omega}_n \mathsf{G}_p \to F$ onto a finite p-group such that φH is a proper subgroup of F. Since F is a finite p-group, φH is contained in a subgroup K of F of index p, which is normal in F. Let $\eta : F \to F/K$ be the natural homomorphism. Then $\eta \varphi H$ is trivial, which shows that f is not invertible on $\mathfrak{A}_p$. $\square$

Corollary 5.3. *The following conditions are equivalent for an implicit operator $f = (w_1, \ldots, w_n)$:*
(1) f is invertible on $\mathsf{G}_{\mathsf{nil}}$;
(2) f is invertible on Ab;
(3) $\det A(w_1, \ldots, w_n) = \pm 1$.

Proof. The result follows immediately from Proposition 5.2 since finite nilpotent groups are direct products of subgroups of prime-power exponent. $\square$

In particular, a necessary condition for the explicit operator $(w_1, \ldots, w_n)$ to be invertible on G is that its frequency matrix be invertible over $\mathbb{Z}$. For example, the ω-power of the operator

$$(x_1^2 x_2 \cdots x_n, x_1 x_2^2 x_3 \cdots x_n, \ldots, x_1 \cdots x_{n-2} x_{n-1}^2 x_n, x_1 \cdots x_{n-1} x_n^2)^\omega$$

is not invertible over G for $n \geq 1$ since the determinant of the frequency matrix of the base operator is $n+1$. Compare with the operators (3) and (4) at the end of Section 4.

For a fairly detailed study of implicit operators which are invertible on the pseudovariety of all finite groups, see [5]. For the remainder of this section, we consider binary implicit operators of the form $(v_1, v_2) = (w_1, w_2)^\omega$. Let $d = \det A(w_1, w_2)$, $t = \operatorname{tr} A(w_1, w_2)$, and take π to be the set of all integer primes which divide t (in the profinite ring $\overline{\Omega}_1 \mathsf{G}$). The fact that the operator (v_1, v_2) is idempotent gives us the following formulas which will be crucial in the sequel:

$$v_i(x, y) = v_i(v_1(x, y), v_2(x, y)) \quad (i = 1, 2). \tag{6}$$

We are interested in the question of determining which finite groups verify the pseudoidentity $v_1 = v_2$. If $d = \pm 1$ then no nontrivial finite cyclic group satisfies the pseudoidentity $v_1 = v_2$ by Corollary 5.3 and therefore no nontrivial finite group satisfies it. If $\mathsf{G}_{\mathsf{nil}}$ satisfies the pseudoidentity $v_1 = v_2$, then so does G_p for every prime p so that $d = 0$ by Proposition 5.2.

Lemma 5.4. *Let G be a finite group and H a normal subgroup such that G/H satisfies $v_1 = 1 = v_2$.*
(a) If H satisfies $v_1 = 1 = v_2$ then so does G.
(b) If H satisfies $v_1 = v_2$ then so does G.

Proof. Given $a, b \in G$, we know that $v_1(a, b), v_2(a, b) \in H$ since the quotient group G/H satisfies $v_1 = 1 = v_2$. In view of (6), if H satisfies $v_1 = v_2$, then

$$v_1(a, b) = v_1(v_1(a, b), v_2(a, b)) = v_2(v_1(a, b), v_2(a, b)) = v_2(a, b),$$

which shows that G also satisfies the pseudoidentity $v_1 = v_2$. This proves (b) and (a) is handled similarly. $\square$

Lemma 5.5. *If $d = 0$, then every finite Abelian group satisfies the pseudoidentities $v_i = w_i^{t^{\omega-1}}$ $(i = 1, 2)$.*

Proof. Let $M = A(w_1, w_2)$. By the Cayley-Hamilton formula, we know that $M^2 = tM$. Hence $M^n = t^{n-1}M$ for every $n \geq 1$. Thus, in a finite Abelian group G, $(w_1, w_2)^n = (w_1^{t^{n-1}}, w_2^{t^{n-1}})$. Taking $n = k!$ and letting $k \to \infty$, we conclude that, in G, $v_i = w_i^{t^{\omega-1}}$. $\square$

Let G_π denote the pseudovariety consisting of all finite π-groups. Define also the pseudovarieties

$$\mathsf{G}_{\mathsf{nil},\pi} = \mathsf{G}_{\mathsf{nil}} \cap \mathsf{G}_\pi$$
$$\mathsf{G}_{\mathsf{sol},\pi} = \mathsf{G}_{\mathsf{sol}} \cap \mathsf{G}_\pi$$

For two pseudovarieties V and W, denote by V∗W their *semidirect product* which is generated by all semidirect products $S * T$ with $S \in \mathsf{V}$ and $T \in \mathsf{W}$.

The above simple lemmas allow us to show that many finite groups satisfy pseudoidentities of the form $v_1 = v_2$.

Proposition 5.6. *Suppose $d = 0$. Then the pseudovariety $\mathsf{G}_{\mathsf{sol},\pi}$ satisfies the pseudoidentities $v_1 = 1 = v_2$.*

Proof. By Lemma 5.5, for $p \in \pi$, we have $v_i = 1$ in every $G \in \mathsf{G}_p \cap \mathsf{Ab}$. The result now follows from Lemma 5.4(a) since $\mathsf{G}_{\mathsf{sol},\pi}$ is the closure under semidirect product of the join of the pseudovarieties $\mathfrak{A}_p$ with $p \in \pi$. □

Assuming that certain Abelian groups satisfy $v_1 = v_2$, we may deduce that much more general groups also satisfy this pseudoidentity.

Proposition 5.7. *If $\mathsf{G}_p \cap \mathsf{Ab}$ satisfies the pseudoidentity $v_1 = v_2$, then so does G_p.*

Proof. Suppose that G is a finite p-group which is a minimal counter-example for the pseudoidentity $v_1 = v_2$. Let $a, b \in G$ be such that $v_1(a, b) \neq v_2(a, b)$ and let Z be the center of G. By minimality of G, the pseudoidentity $v_1 = v_2$ must hold in the quotient group G/Z. Hence $v_2(a, b) = v_1(a, b)z$ for some $z \in Z$. Let $g = v_1(a, b)$. Using (6), we deduce that

$$v_i(a, b) = v_i(v_1(a, b), v_2(a, b)) = v_i(g, gz) \quad (i = 1, 2).$$

But, since the subgroup generated by g and z is Abelian, the pseudoidentity $v_1 = v_2$ is valid in this subgroup and, therefore,

$$v_1(a, b) = v_1(g, gz) = v_2(g, gz) = v_2(a, b),$$

in contradiction with the assumption. □

A more satisfying hypothesis would be for Abelian groups to satisfy the pseudoidentity $w_1 = w_2$. The relationship with $v_1 = v_2$ is considered in the next result.

Proposition 5.8. *If the pseudovariety Ab satisfies the pseudoidentity $w_1 = w_2$ then it also satisfies $v_1 = v_2$. Conversely, if Ab satisfies $v_1 = v_2$ and p is a prime not in π, then every finite Abelian p-group satisfies $w_1 = w_2$.*

Proof. From $w_1 = w_2$ we deduce that $w_1(w_1, w_2) = w_2(w_1, w_2)$. Iterating this argument shows that the pseudoidentity $v_1 = v_2$ is a consequence of $w_1 = w_2$.

Suppose for the remainder of the proof that Ab satisfies $v_1 = v_2$. If $d \neq 0$, then there would be some prime q not dividing d. By Proposition 5.2, the implicit operator (v_1, v_2) would be the identity operator (a, b) on $\mathfrak{A}_q$, in contradiction with the hypothesis that $\mathfrak{A}_q$ satisfies $v_1 = v_2$. Hence $d = 0$.

By Lemma 5.5, it follows that Ab satisfies the pseudoidentities $v_i = w_i^{t^{\omega-1}}$. Substituting in $v_1 = v_2$ and raising both sides to the power t, this yields

$$w_1^{t^\omega} = w_2^{t^\omega}. \tag{7}$$

Let G be a finite Abelian p-group, where p is a prime number not in π. Then G satisfies $x^{t^\omega} = x$ and so the validity of the pseudoidentity $w_1 = w_2$ in G is an immediate consequence of (7). □

To show that the hypothesis in Proposition 5.8 that the prime p is not in π can not be removed, consider the following example. Let $w_1 = a^{2^\omega+1}b^{2^\omega+1}$ and $w_2 = a^2 b^2$. Then, in the profinite ring $\overline{\Omega}_1 G$, where each x^ν is identified with ν, $d = 0$ and $t = 2^{\omega+1} + 2 = 2(2^\omega + 1)$ so that the prime 2 is a member of π. We claim that the pseudovariety Ab satisfies $v_1 = v_2$. Indeed, over Ab,

$$v_1 = a^{2^{\omega-1}(2^\omega+1)^{\omega-1}2^{\omega+1}}b^{2^{\omega-1}(2^\omega+1)^{\omega-1}2^{\omega+1}} = a^{2^\omega(1-2^\omega)}b^{2^\omega(1-2^\omega)}$$
$$v_2 = a^{2^{\omega-1}(2^\omega+1)^{\omega-1}2}b^{2^{\omega-1}(2^\omega+1)^{\omega-1}2} = a^{2^\omega(1-2^\omega)}b^{2^\omega(1-2^\omega)}.$$

On the other hand Ab fails $w_1 = w_2$: for instance, the cyclic group of order 4 fails the pseudoidentity $a^{2^{\omega+1}} = a^2$ which is a consequence of $w_1 = w_2$.

Proposition 5.9. *If Ab satisfies the pseudoidentity $w_1 = w_2$, then $\mathsf{G}_{\mathsf{nil}} * \mathsf{G}_{\mathsf{sol},\pi}$ satisfies $v_1 = v_2$.*

Proof. By Proposition 5.8, Ab satisfies the pseudoidentity $v_1 = v_2$. Hence we may apply Proposition 5.7 to conclude that $\mathsf{G}_{\mathsf{nil}}$ satisfies $v_1 = v_2$.

By substituting in turn 1 for each variable, we deduce that $A(w_1, w_2)$ has equal rows and so, in particular, $d = 0$. By Proposition 5.6, it follows that $\mathsf{G}_{\mathsf{sol},\pi}$ satisfies the pseudoidentities $v_1 = 1 = v_2$. By Lemma 5.4(b), we conclude that indeed $\mathsf{G}_{\mathsf{nil}} * \mathsf{G}_{\mathsf{sol},\pi}$ satisfies $v_1 = v_2$. □

For the next three results, we assume that w_1 and w_2 are explicit operators, that is semigroup words or, more generally, monoid words. Then the entries of the matrix $A(w_1, w_2)$ are non-negative integers and an entry is zero if and only if the word corresponding to that row does not depend on the variable corresponding to that column.

Proposition 5.10. *Suppose w_1 and w_2 are words and the matrix $A(w_1, w_2)$ has a zero column. Let δ be the absolute value of the difference of the entries* *in the other column. Then*

$$[\![v_1 = v_2]\!] \cap \mathsf{G}_{\mathsf{nil}} = \mathsf{G}_{\mathsf{nil},\pi} \vee ([\![x^\delta = 1]\!] \cap \mathsf{G}_{\mathsf{nil}}). \tag{8}$$

Proof. Without loss of generality, we may assume $(w_1, w_2) = (a^t, a^k)$ for some k. Then $v_1 = v_2$ is the pseudoidentity $a^{t^\omega} = a^{t^{\omega-1}k}$ which, for groups, is equivalent to $a^{t^{\omega-1}\delta} = 1$ since $\delta = |t-k|$. Raising to the power t, we obtain the equivalent pseudoidentity $a^{t^\omega \delta} = 1$. This pseudoidentity is obviously true in π-groups and also in groups satisfying the pseudoidentity $x^\delta = 1$. Conversely, if a finite group G satisfies the pseudoidentity $x^{t^\omega \delta} = 1$ then, for every $g \in G$, either some $p \in \pi$ divides the order of g or $g^{t^\omega} = g$, so that $g^\delta = 1$. Hence, if G is nilpotent, then G must lie in the join in (8). □

Allowing for monoid implicit operations, so that w_1 and w_2 may be 1, we could also consider the case when $A(w_1, w_2)$ has a zero row. Again, without loss of generality, we may then assume that the second row is zero, that is $w_2 = 1$. Then a simple inductive calculation yields $(w_1, w_2)^k = (w_1^{t^{k-1}}, 1)$. It follows that $v_1 = w_1^{t^\omega - 1}$, with $v_1(a, 1) = a^{t^\omega}$, while $v_2 = 1$. Hence the pseudoidentity $v_1 = v_2$ holds in a finite group if and only if the prime factors of the orders of its elements divide t, that is the group is a π-group. This proves the following result.

Proposition 5.11. *Suppose w_1 and w_2 are monoid words and the matrix $A(w_1, w_2)$ has a zero row. Then*

$$[\![v_1 = v_2]\!] \cap \mathsf{G} = \mathsf{G}_\pi.$$

The cases of word operators (w_1, w_2) which remain to consider with $d = 0$ are those with no zero entries. We further reduce this to the case of equal rows for nilpotent groups.

Proposition 5.12. *Suppose w_1 and w_2 are monoid words and the second row of the matrix $A(w_1, w_2)$ is λ times the first row. Then*

$$[\![v_1 = v_2]\!] \cap \mathsf{G}_{\mathsf{nil}} = \mathsf{G}_{\mathsf{nil},\rho} \vee ([\![x^{\lambda-1} = 1]\!] \cap \mathsf{G}_{\mathsf{nil}}) \tag{9}$$

where ρ is the set of all prime divisors of both elements of the main diagonal of $A(w_1, w_2)$.

Proof. Let $G \in \mathsf{G}_{\mathsf{nil}}$. Assume first that G verifies the pseudoidentity $v_1 = v_2$. Then, substituting 1 for one of the variables we obtain the pseudoidentities

$$x^{\alpha^\omega(\lambda-1)} = 1 \quad \text{and} \quad x^{\lambda^{\omega-1}\beta^\omega(\lambda-1)} = 1, \tag{10}$$

where α and $\lambda\beta$ are the diagonal entries of the matrix $A(w_1, w_2)$. Let H be one of the Sylow subgroups of G and suppose H is a p-group. If $p \notin \rho$, then $x^{\alpha^\omega} = x^{\beta^\omega} = x$ so that each of the pseudoidentities (10) yields that H satisfies the identity $x^{\lambda-1} = 1$. The inclusion from left to right in (9) follows since G is the direct product of its Sylow subgroups.

Conversely, since $\rho \subseteq \pi$, if $G \in \mathsf{G}_{\mathsf{nil},\rho}$, then G satisfies $v_1 = v_2$ by Proposition 5.6. On the other hand, if G satisfies $x^{\lambda-1} = 1$, then, over G, the operator (w_1, w_2) is equivalent to an operator (w_1', w_2') whose frequency matrix has equal rows. By Proposition 5.8, every finite nilpotent group satisfies the corresponding pseudoidentity $v_1' = v_2'$. Hence G satisfies $v_1 = v_2$. □

6 Further examples and applications

The results of Section 5 should be viewed as a guide to restrictions that implicit operators should satisfy in order to obtain interesting pseudoidentities for finite groups. It should however be emphasized that we are basically just inferring consequences of the hypothesis that some Abelian groups satisfy certain pseudoidentities. In the most interesting situations, where all finite Abelian groups satisfy the pseudoidentity $w_1 = w_2$, a little perturbation of the initial operator (w_1, w_2) by mere commutation of factors may transform an interesting pseudoidentity $v_1 = v_2$ into a trivial one as the following example shows. A deeper investigation of these pseudoidentities should be able to distinguish just which perturbations lead to interesting pseudoidentities.

As an example, consider the implicit operator

$$(w_1, w_2) = (ab^{\omega-1}, b^{\omega-1}a). \tag{11}$$

Its frequency matrix is $\begin{pmatrix} 1 & -1 \\ 1 & -1 \end{pmatrix}$. The pseudovariety Ab satisfies the pseudoidentity $w_1 = w_2$. Since $t = 0$, π is the set of all primes and so, by Proposition 5.6, $\mathsf{G}_{\mathsf{sol}} \subseteq [\![v_1 = 1 = v_2]\!] \cap \mathsf{G}$. But it turns out that exhaustive computer calculations using GAP [18] give that the simple group $\mathrm{PSL}(2, 23)$ satisfies the pseudoidentities $v_1 = 1 = v_2$, 23 being the smallest prime greater than 3 for which this property holds. On the other hand, if we take the operator $(ab^{\omega-1}, ab^{\omega-1})$, which coincides with (11) on Ab and, therefore, has the same frequency matrix, then the associated pseudoidentity $v_1 = v_2$ is trivial.

Another interesting example is given by the operator

$$(w_1, w_2) = (a^2 b^{\omega-1}, b^{\omega-1} a^2). \tag{12}$$

Again here Ab satisfies $w_1 = w_2$ but now $t = 1$ and so $\pi = \emptyset$. By Proposition 5.9, $\mathsf{G}_{\mathsf{nil}}$ satisfies $v_1 = v_2$. However, the symmetric group $\mathrm{Sym}(3)$ also satisfies $v_1 = v_2$.

The operator

$$(w_1, w_2) = (bab^{\omega-1}, a) \tag{13}$$

also has Ab satisfying $w_1 = w_2$ and frequency matrix with trace 1. Again by Proposition 5.9, G$_{\text{nil}}$ satisfies $v_1 = v_2$. The operator (13) is particularly interesting since, in all its finite powers, the components may be written, over groups, as products in which the letters a and b with exponent ± 1 alternate. Alternatively, we could consider the operator

$$(dab, a, dcb, c)^\omega \tag{14}$$

and evaluate c to $a^{\omega-1}$ and d to $b^{\omega-1}$ to obtain the values v_1 and v_2, respectively, for the first and second components of (14). One may also consider the related operator $(v_1', v_2') = (b^{\omega-1}ab, a)^\omega$. Since

$$(b^{\omega-1}ab, a) = (a^{\omega-1}, b^{\omega-1})(bab^{\omega-1}, a)(a^{\omega-1}, b^{\omega-1})$$

over finite groups and the operator $(a^{\omega-1}, b^{\omega-1})$ is an involution on finite groups, we conclude that

$$(v_1', v_2') = (a^{\omega-1}, b^{\omega-1})(v_1, v_2)(a^{\omega-1}, b^{\omega-1})$$

on finite groups and so the pseudoidentities $v_1' = v_2'$ and $v_1 = v_2$ are equivalent over G.

The pseudoidentity $v_1 = v_2$ provides yet another definition of nilpotency for finite groups. The proof of the following result will appear elsewhere.

Theorem 6.1. *Let* $(v_1, v_2) = (b^{\omega-1}ab, a)^\omega$. *Then a finite group satisfies the pseudoidentity* $v_1 = v_2$ *if and only if it is nilpotent.*

Taking into account that the components of each finite power of the implicit operator (13) start with different letters and also end with different letters, the following corollary is an immediate application of Proposition 5.1.

Corollary 6.2. *Let* $(v_1, v_2) = (b^{\omega-1}ab, a)^\omega$. *Then* BG$_{\text{nil}} = [\![v_1 = v_2]\!]$.

For the Prouhet-Thue-Morse implicit operation $\tau(a, b) = x_1(ab, ba)^\omega$, we have $(\tau(a, b), \tau(b, a)) = (ab, ba)^\omega$. Since the operator $(ab, ba)^\omega$ is idempotent, the equality

$$\tau(a, b) = \tau(\tau(a, b), \tau(b, a)) \tag{15}$$

holds.

The original proof of the following result is based on a theorem of M. Hall on the existence of p-complements in finite groups [11, Theorem 14.4.7]. A proof using Thompson's list (1) of finite simple groups whose proper subgroups are solvable was given by Boffa and Point [7].

Theorem 6.3 (Širšov [20]). *The pseudovariety* $\mathsf{G}_{\mathsf{nil}} * \mathsf{G}_2$ *consists of all finite groups which satisfy the pseudoidentity* $\tau(a, b) = \tau(b, a)$.

Using Proposition 5.1, we immediately obtain the following corollary.

Corollary 6.4. $\mathsf{B}(\mathsf{G}_{\mathsf{nil}} * \mathsf{G}_2) = [\![\tau(a,b) = \tau(b,a)]\!]$.

The square of the operator (11), coincides on groups with the operator $([a^{\omega-1}, b], [a, b^{\omega-1}])$, whose frequency matrix is zero. Perhaps the most promising operator of this kind is the implicit operator

$$(w_1, w_2) = ([a, b], [a^{\omega-1}, b^{\omega-1}]). \tag{16}$$

By Proposition 5.6, $\mathsf{G}_{\mathsf{sol}}$ satisfies $v_1 = v_2$. Computer calculations with the Thompson groups (1) gave that, within 10 random tries, a counter-example for the pseudoidentity $v_1 = v_2$ was found for the groups $\mathrm{PSL}(2, p)$ with $3 < p < 3000$, $\mathrm{PSL}(2, 2^q)$ with $q < 17$, $\mathrm{PSL}(2, 3^r)$ with $r < 11$, $\mathrm{PSL}(3, 3)$, and $\mathrm{Sz}(2^r)$ with $r < 7$. Carrying the calculations further for some of the parameters seems to be beyond the capacities of GAP or the computer. The pseudoidentity is therefore probably a good alternative to the pseudoidentities resulting from the Plotkin conjecture mentioned in Subsection 2.3 in order to define $\mathsf{G}_{\mathsf{sol}}$.

Acknowledgement

The author wishes to thank Antonio Restivo and Véronique Bruyère for bringing to his attention the work of M. Boffa and F. Point on identities built up from Prouhet-Thue-Morse words, Françoise Point for referring earlier work of A. I. Širšov, and Semyon Yakubovich for translating Širšov's paper. Thanks also to the anonymous referees and Alfredo Costa for their helpful comments on an earlier version of this paper.

References

[1] J. Almeida, *Residually finite congruences and quasi-regular subsets in uniform algebras*, Portugal. Math. **46** (1989) 313–328.

[2] ———, *Finite Semigroups and Universal Algebra*, World Scientific, Singapore, 1995. English translation.

[3] ———, *Dynamics of implicit operations and tameness of pseudovarieties of groups*, Trans. Amer. Math. Soc. **354** (2002) 387–411.

[4] J. Almeida and M. V. Volkov, *Profinite methods in finite semigroup theory*, Tech. Rep. CMUP 2001-02, Univ. Porto, 2001.

[5] ————, *Profinite identities for finite semigroups whose subgroups belong to a given pseudovariety*. In preparation.

[6] J. Almeida and P. Weil, *Relatively free profinite monoids: an introduction and examples*, in Semigroups, Formal Languages and Groups, J. B. Fountain, ed., vol. 466, Dordrecht, 1995, Kluwer Academic Publ., 73–117.

[7] M. Boffa and F. Point, *Identités de Thue-Morse dans les groupes*, C. R. Acad. Sci. Paris Sér. I Math. **312** (1991) 667–670.

[8] P. Flavell, *Finite groups in which every two elements generate a soluble group*, Invent. Math. **121** (1995) 279–285.

[9] D. Gorenstein, *Finite Groups*, Harper & Row, New York, 1968.

[10] F. Grunewald, B. Kuniavskii, D. Nikolova, and E. Plotkin, *Two-variable identities in groups and Lie algebras*, Zapiski Nauch. Seminarov POMI **272** (2000) 161–176. To appear also in J. Math. Sciences.

[11] M. Hall, *Theory of Groups*, Chelsea, New York, 1976.

[12] R. McKenzie, G. McNulty, and W. Taylor, *Algebras, Lattices and Varieties*, vol. I, Wadsworth, Mont erey, CA, 1987.

[13] D. Perrin and J.-É. Pin, *Infinite Words*, Academic Press, to appear. (http://www.liafa.jussieu.fr/~jep/Resumes/InfiniteWords.html).

[14] J.-É. Pin, *Variétés de langages et variétés de semigroupes*, Thèse d'état, Univ. Paris 6, 1981.

[15] J. Reiterman, *The Birkhoff theorem for finite algebras*, Algebra Universalis **14** (1982) 1–10.

[16] D. J. S. Robinson, *A Course in Theory of Groups*, no. 80 in Grad. Texts in Math., Springer-Verlag, New York, 1982.

[17] M. P. Schützenberger, *Sur certaines pseudo-variétés de monoïdes finis*, vol. 3 of Cahiers Math. Montpellier, Montpellier, 1974, 317–327.

[18] The GAP Group, *GAP - Groups, Algorithms, and Programming, Version 4.1*, Aachen, St Andrews, 1999. (http://www-gap.dcs.st-and.ac.uk/~gap).

[19] J. G. Thompson, *Non-solvable groups all of whose local subgroups are solvable*, Bull. Amer. Math. Soc. **74** (1968) 383–437.

[20] A. I. Širšov, *On certain near-Engel groups*, Algebra i Logika **2** (1963) 5–18.

[21] M. Zorn, *Nilpotency of finite groups (abstract)*, Bull. Amer. Math. Soc. **42** (1936) 485–486.

GROUP PRESENTATIONS FOR A CLASS OF RADICAL RINGS OF MATRICES

NOELLE ANTONY,* CLARE COLEMAN AND DAVID EASDOWN
School of Mathematics and Statistics, University of Sydney, NSW 2006, Australia
E-mail: noellea@maths.usyd.edu.au, cec@maths.usyd.edu.au, de@maths.usyd.edu.au

In this paper we assemble some facts about radical rings which arise by sandwich multiplication using a Jacobson radical element. We find a group presentation for the Munn ring consisting of $n \times n$ matrices over $\mathbb{Z}_{p^k}$, where p is prime, which employs a scalar sandwich matrix where the scalar is from $p\mathbb{Z}_{p^k}$, regarded as a group with respect to the circle operation.

1 Introduction and preliminaries

Groups of units of rings with identity are well studied. However many rings arise naturally without an identity. For example, nontrivial rings which coincide with their Jacobson radical never have an identity, and one does not typically expect a semigroup ring or algebra to have an identity. Nevertheless, all rings possess groups of *quasi-units*, that is, elements which are invertible with respect to the circle operation ∘ defined by

$$x \circ y = x + y - xy .$$

Consider a ring S, not necessarily with 1, with multiplication denoted by · or juxtaposition. We refer to $(S, \circ)$ as the *circle monoid* of S. Denote by S^1 the result of adjoining 1 to S, which may be done in such a way that the characteristic n of the ring is preserved, so that the prime subring of S^1 is a copy of $\mathbb{Z}_n$ (where we take $\mathbb{Z}_0$ to be $\mathbb{Z}$) (see, for example, [11, Theorem 2.26] or [1]). Then the mapping

$$\widehat{} : (S, \circ) \to (S^1, \cdot), \quad x \mapsto \widehat{x} = 1 - x \quad (x \in S)$$

is a monoid embedding, which is an isomorphism when $S = S^1$. An element $x \in S$ is called *quasi-invertible* if there is an element y such that

$$x \circ y = y \circ x = 0 ,$$

---

*THE FIRST AUTHOR WOULD LIKE TO ACKNOWLEDGE THE FINANCIAL SUPPORT OF FUNDAÇÃO CALOUSTE GULBENKIAN (FCG), FUNDAÇÃO PARA A CIÊNCIA E A TECNOLOGIA (FCT), FACULDADE DE CIÊNCIAS DA UNIVERSIDADE DE LISBOA (FCUL) AND REITORIA DA UNIVERSIDADE DO PORTO.

in which case we call y the *quasi-inverse* of x and write $x' = y$. Put
$$\mathcal{G}(S) = \{\, x \in S \mid x \text{ is quasi-invertible}\,\},$$
called the *group of quasi-units* or the *circle group* of S. When $S = S^1$, denote by $G(S)$ the group of units of $(S, \cdot)$, in which case $\widehat{} : \mathcal{G}(S) \to G(S)$ is a group isomorphism. It is straightforward to check that if $S \ne S^1$ then $\mathcal{G}(S^1) \cong G(\mathbb{Z}_n) \times \mathcal{G}(S)$, where n is the characteristic of S.

The Jacobson radical of S, denoted by $\mathcal{J}(S)$, may be defined to be the largest ideal of S consisting of quasi-invertible elements. We say that a normal subgroup N of a group G is *complemented* in G if there exists a subgroup H of G such that $G = NH$ and $N \cap H = \{e\}$, where e is the identity of the group. We call H the *complement* of N in G and G the *internal semidirect product* of N by H. It is easy to see that any ideal of S contained in $\mathcal{J}(S)$ forms a normal subgroup of $(\mathcal{G}(S), \circ)$. The existence of complements of $\mathcal{J}(S)$ and the nilradical in $\mathcal{G}(S)$ appears to be a delicate issue and is discussed in [1]. For example if S is a finite ring with identity of prime characteristic, then the Jacobson radical is always complemented [17]. However finite rings exist in non-prime characteristic where complementation fails [8].

Call S *radical* if $S = \mathcal{J}(S)$. The circle group of a radical ring has also been called the *adjoint group* [23]. Chick [3] [4] investigates, also with Gardner [5], interesting examples of commutative radical rings S in which $(S, \circ)$ and $(S, +)$ are isomorphic. The question of when an abstract group arises as the circle group of a ring, and the interplay between finite generation, nilpotency of the ring and nilpotency of its circle group have been investigated by a number of authors including Ault, Watters, Kruse, Tahara, Hosomi and Sandling [2] [23] [13] [14] [22] [21]. Membership criteria for the circle groups of band graded rings have been investigated by Kelarev [12].

It should be remarked that many authors use as circle operation $\circ^+$ defined by $x \circ^+ y = x + y + xy$. (We prefer our definition because the quasi-inverse of a nilpotent element is then the negative sum of powers of the element, rather than an alternating sum.) This does not matter in our context, however, because negation is an isomorphism between the monoids $(S, \circ)$ and $(S, \circ^+)$, an instance of a phenomenon remarked upon below. Both $\circ = \circ^{(1)}$ and $\circ^+ = \circ^{(-1)}$ are special cases of the associative derived operation $\circ^{(k)}$, where k is an integer, defined by
$$x \circ^{(k)} y = x + y - kxy.$$

Derived operations are binary operations formed by combinations of the ring operations as polynomial expressions in two noncommuting variables with integer coefficients, and, by a theorem of McConnell and Stokes [16], are associative if and only if they have one of the following forms:

$$x * y = \begin{cases} x + y - kxy & (1) \\ x + y - kyx & (2) \\ kxy & (3) \\ kyx & (4) \\ x & (5) \\ y & (6) \\ 0 & (7) \end{cases}$$

where k is an integer. Observe that (5), (6) and (7) do not produce interesting semigroups, being left-zero, right-zero and null respectively, whilst (2) is dual to (1) and (4) dual to (3). If (3) is regarded as defining a new ring multiplication then all instances of (1) arise as circle monoids. We put this in a more general setting as follows:

Let $(S, +)$ be an abelian group, T a set containing S, and $z \in T$. Suppose that T has an associative partial binary operation, denoted by juxtaposition, satisfying the following conditions:

(1) $(\forall x, y \in S) \qquad xzy \in S$;

(2) $(\forall u, v, w \in S) \quad (u+v)zw = uzw+vzw \quad$ and $\quad uz(v+w) = uzv+uzw$.

Then it is routine to verify that $(S, +, *^{(z)})$ is a ring where

$$x *^{(z)} y = xzy .$$

We refer to $*^{(z)}$ as *sandwich* multiplication. This construction includes for example the Munn ring

$$(\mathsf{Mat}_{m,n}(R), +, *^{(z)})$$

formed by taking T to be the set of all matrices over R^1 for a given ring R, under the partial binary operation of matrix multiplication, $S = \mathsf{Mat}_{m,n}(R)$ to be the subset of $m \times n$ matrices over R, and z an $n \times m$ sandwich matrix. The terminology *Munn ring* is due to McAlister [15], which in turn derives from the notion of *Munn algebra* (see [18], [6, Section 5.2] and [20, Chapter 5]), though in our definition above we allow an unrestricted sandwich matrix P (see also [19]). More generally, one could take T to be the set of all homomorphisms between abelian groups, under the partial binary operation of composition, S to be the set of all homomorphisms from an abelian group A to an abelian group B, regarded itself as an abelian group under pointwise addition, and z some fixed homomorphism from B to A.

A further class of examples is obtained by taking T to be any ring and S an ideal of T. For each $z \in T$ we obtain the ring $(S, +, *^{(z)})$. Denote the circle operation for this ring by $\circ^{(z)}$, so that

$$x \circ^{(z)} y = x + y - xzy.$$

In the special case that $T = S^1$ then, in the McConnell-Stokes classification, each instance of (3) becomes sandwich multiplication $*^{(k)}$ where k may be taken to lie in the prime subring of T, and then each instance of (1) becomes a circle operation.

One may ask what relationship, if any, exists between the monoids $(S, \circ^{(z)})$ and $(S, \circ^{(w)})$ as z and w vary over T. For example, if $S = T = \mathbb{Z}_4$ then one may easily verify that $(S, \circ^{(1)}) \cong (S, \circ^{(3)}) \cong (S, \cdot)$ is not a group (in fact, it is an ideal extension of a two-element null semigroup by a two element group), $(S, \circ^{(2)})$ is a product of two cyclic groups of order 2, and $(S, \circ^{(0)}) = (S, +)$ is a cyclic group of order 4.

It is easy to check that if S is an ideal of a ring T with identity, $z, w \in T$, and u is a unit of T which commutes with w such that $z = uw$, then

$$x \mapsto u^{-1}x \quad \text{for} \quad x \in S$$

yields a ring isomorphism: $(S, +, *^{(z)}) \to (S, +, *^{(w)})$, which in turn induces a monoid isomorphism: $(S, \circ^{(z)}) \to (S, \circ^{(w)})$. In particular, for any ring S, $(S, \circ) \cong (S, \circ^+) \cong (S, \circ^{(k)})$ for any integer k which is invertible modulo the characteristic.

We establish here some notational conventions used throughout the paper. The kth power of an element x with respect to a binary operation $\odot$ is denoted by $x^{\odot k}$. The cyclic group of order n is denoted by C_n, written multiplicatively. If G is a group and $x, y \in G$ then we write

$$x^y = y^{-1}xy \quad \text{and} \quad [x, y] = x^{-1}y^{-1}xy.$$

If Σ is an alphabet and $\mathcal{R}$ a collection of relations then $\langle \Sigma \mid \mathcal{R} \rangle$ denotes a group presentation.

Let S be a ring. Abbreviate to

$$\mathsf{Mat}_n(S) = \mathsf{Mat}_{n,n}(S, +, \cdot),$$

the usual ring of $n \times n$ matrices with entries from a ring $(S, +, \cdot)$. Note that $\mathcal{J}(\mathsf{Mat}_n(S)) = \mathsf{Mat}_n(\mathcal{J}(S))$. Thus if S is radical then so is $\mathsf{Mat}_n(S)$, whence $\mathsf{Mat}_n(S) = \mathcal{G}(\mathsf{Mat}_n(S))$ is a group under $\circ$. Observe further that for $z \in \mathbb{Z}$,

$$\mathsf{Mat}_n(S, +, *^{(z)}) = (\mathsf{Mat}_n(S), +, *^{(z)}) = (\mathsf{Mat}_n(S), +, *^{(zI)}),$$

where I denotes the $n \times n$ identity matrix, the latter being a Munn ring with respect to the scalar sandwich matrix zI. Membership of the Jacobson radical of Munn rings in general is characterised by Munn:

Theorem 1.1. *[19] If S is any ring then a matrix α lies in the Jacobson radical of the Munn ring $(\mathsf{Mat}_{m,n}(S), +, *^{(P)})$ where P is a sandwich matrix over S^1 if and only if all entries of $P\alpha P$ lie in $\mathcal{J}(S)$.*

Using $n = m = 1$ and noting that $\mathcal{J}(S) = S \cap \mathcal{J}(S^1)$, we immediately deduce the following:

Corollary 1.2. *If S is any ring and $z \in \mathcal{J}(S^1)$ then $(S, +, *^{(z)})$ is radical, so that $(S, \circ^{(z)})$ is a group. In particular, $\mathsf{Mat}_n(\mathbb{Z}_{p^n}, \circ^{(q)})$ is a group for any prime p and $q = p^\ell$ for $\ell \geq 1$.*

The remainder of this paper is devoted to finding a group presentation for $\mathsf{Mat}_n(\mathbb{Z}_{p^n}, \circ^{(q)})$.

2 Circle monoids of integers modulo a power of a prime

This section provides detailed analysis of circle monoids of $\mathbb{Z}_{p^n}$ with respect to sandwich multiplication by powers of a prime p, which is interesting in its own right, but aspects of which are needed later in developing certain group presentations. We begin by noting the following, the proof of which is routine:

Lemma 2.1. *If m, n are positive integers such that m divides n, then the mapping $z \mapsto mz$, for $z \in \mathbb{Z}$, induces a ring isomorphism: $(\mathbb{Z}_{n/m}, +, *^{(m)}) \to (m\mathbb{Z}_n, +, \cdot)$.*

Corollary 2.2. *If m, n are positive integers such that m divides n, then the monoids $(\mathbb{Z}_{n/m}, \circ^{(m)})$ and $(m\mathbb{Z}_n, \circ)$ are isomorphic.*

The following is well-known (see, for example [17, Theorem XVI.9]):

Lemma 2.3. *If p is prime and $n \geq 1$ then*

$$G(\mathbb{Z}_{p^n}) \cong \mathcal{G}(\mathbb{Z}_{p^n}) \cong \begin{cases} C_{p-1} \times C_{p^{n-1}} & \text{if } p \text{ is odd or } p = 2 \text{ and } n \leq 2 \, ; \\ C_2 \times C_{2^{n-2}} & \text{if } p = 2 \text{ and } n > 2 \, . \end{cases}$$

Observe that $\mathcal{J}(\mathbb{Z}_{p^n}) = p\mathbb{Z}_{p^n}$ is the Sylow p-subgroup of $\mathcal{G}(\mathbb{Z}_{p^n})$, yielding the following:

Corollary 2.4. *If p is prime and $n \geq 1$ then*
$$(p\mathbb{Z}_{p^n}, \circ) \cong \begin{cases} C_{p^{n-1}} & \text{if } p \text{ is odd or } p = 2 \text{ and } n \leq 2 \text{;} \\ C_2 \times C_{2^{n-2}} & \text{if } p = 2 \text{ and } n > 2 \text{.} \end{cases}$$

Note that 2 has order 2 in the group $(\mathbb{Z}_{2^n}, \circ)$, yet 2 is not a multiple of 2^ℓ if $\ell > 1$. From this, and the fact that subgroups of cyclic groups are cyclic, Corollaries 2.2 and 2.4 yield the following:

Corollary 2.5. *If p is prime, $1 \leq \ell \leq n$, and $\ell > 1$ if $p = 2$, then*
$$(\mathbb{Z}_{p^n}, \circ^{(p^\ell)}) \cong (p^\ell \mathbb{Z}_{p^{n+\ell}}, \circ) \cong (1 + p^\ell \mathbb{Z}_{p^{n+\ell}}, \cdot) \cong C_{p^n} \text{ .}$$

To get more detailed information concerning generators and orders of elements, we first note the following facts about binomial coefficients (proofs of which the reader can supply, or can be found in [8]):

Lemma 2.6. *If p is prime, $t \geq 1$ and $0 \leq i \leq p^{t-1}$ then*
$$\binom{p^t}{pi} \equiv \binom{p^{t-1}}{i} \mod p^t \text{ .}$$

Lemma 2.7. *If p is prime, $t \geq 1$, $0 \leq i < p^{t-1}$ and $1 \leq k \leq p - 1$ then*
$$\binom{p^t}{pi + k} \equiv 0 \mod p^t \text{ .}$$

Some aspects of the following theorem are remarked upon, without proof, in [10].

Theorem 2.8. *Let p be prime and k, ℓ be positive integers such that $\ell \leq k$. Suppose that $\ell > 1$ if $p = 2$. Put $q = p^\ell$. Then $(\mathbb{Z}_{p^k}, \circ^{(q)})$ is a cyclic group and, for each integer x and $n \geq 0$,*
$$x^{\circ^{(q)}n} \equiv \frac{1 - (1 - xq)^n}{q} \mod p^k \text{ .}$$

If $0 < x < p^k$, then, regarded as an element of the group $(\mathbb{Z}_{p^k}, \circ^{(q)})$, the order of x is p^{k-j} where p^j is the highest power of p dividing x. In particular $(\mathbb{Z}_{p^k}, \circ^{(q)})$ is generated by 1, and for all $n \geq 0$,
$$q1^{\circ^{(q)}n} \equiv 1 - (1 - q)^n \mod p^{k+\ell}$$
and
$$q(1^{\circ^{(q)}n})^\dagger \equiv 1 - (1 - q')^n \mod p^{k+\ell} \text{ ,}$$

where † denotes quasi-inversion in $(\mathbb{Z}_{p^k}, \circ^{(q)})$ and ′ denotes usual quasi-inversion in $\mathbb{Z}_{p^k}$.

Proof. By Corollary 2.5, $(\mathbb{Z}_{p^k}, \circ^{(q)})$ is a cyclic group. If x is any integer and $n \geq 0$ then, by Lemma 2.1,
$$qx^{\circ^{(q)}n} \equiv (qx)^{\circ n} \quad \mod p^k q = p^{k+\ell},$$
so that
$$1 - qx^{\circ^{(q)}n} \equiv 1 - (qx)^{\circ n} = (1-qx)^n \quad \mod p^{k+\ell},$$
yielding
$$x^{\circ^{(q)}n} \equiv \frac{1 - (1-qx)^n}{q} \quad \mod p^k.$$

Now suppose $0 < x < p^k$ and write $x = p^j y$ where y is not divisible by p. Note that $0 \leq j \leq k-1$.

Consider s such that $1 \leq s \leq p^{k-j}$. Suppose first that s is not divisible by p. By Lemma 2.7,
$$\binom{p^{k-j}}{s} \equiv 0 \quad \mod p^{k-j},$$
so that
$$\binom{p^{k-j}}{s} x^s q^{s-1} = \binom{p^{k-j}}{s} y^s p^{js+\ell(s-1)} \equiv 0 \quad \mod p^k$$
since $js + \ell(s-1) \geq j$. If, further, $1 < s \leq p^{k-j-1}$ then, by Lemma 2.7,
$$\binom{p^{k-j-1}}{s} \equiv 0 \quad \mod p^{k-j-1},$$
yielding
$$\binom{p^{k-j-1}}{s} x^s q^{s-1} = \binom{p^{k-j-1}}{s} y^s p^{js+\ell(s-1)} \equiv 0 \quad \mod p^k$$
since now $js + \ell(s-1) \geq j+1$.

Suppose secondly that s is divisible by p, say $s = p^m t$ where t is not divisible by p and $m \geq 1$. Note that
$$js + \ell(s-1) \geq j + \ell(p^m - 1) \geq j + m + 1.$$
By Lemmas 2.7 and 2.6,
$$\binom{p^{k-j}}{s} \equiv \binom{p^{k-j-m}}{t} \equiv 0 \quad \mod p^{k-j-m},$$

so that

$$\binom{p^{k-j}}{s}x^s q^{s-1} \equiv \binom{p^{k-j-m}}{t}y^s p^{js+\ell(s-1)} \equiv 0 \mod p^k,$$

since $js + \ell(s-1) \geq j + m$. In the case that $s \leq p^{k-j-1}$ then, again by Lemmas 2.7 and 2.6,

$$\binom{p^{k-j-1}}{s} \equiv \binom{p^{k-j-m-1}}{t} \equiv 0 \mod p^{k-j-m-1},$$

yielding

$$\binom{p^{k-j-1}}{s}x^s q^{s-1} \equiv \binom{p^{k-j-m-1}}{t}y^s p^{js+\ell(s-1)} \equiv 0 \mod p^k,$$

now since $js + \ell(s-1) \geq j + m + 1$.

These congruences tell us, firstly, that

$$x^{o(q)p^{k-j}} \equiv \frac{1-(1-qx)^{p^{k-j}}}{q} = \sum_{s=1}^{p^{k-j}}(-1)^{s-1}\binom{p^{k-j}}{s}x^s q^{s-1} \equiv 0 \mod p^k,$$

so that the order of x divides p^{k-j}, and, secondly, that

$$x^{o(q)p^{k-j-1}} \equiv \frac{1-(1-qx)^{p^{k-j-1}}}{q} = \sum_{s=1}^{p^{k-j-1}}(-1)^{s-1}\binom{p^{k-j-1}}{s}x^s q^{s-1}$$
$$\equiv p^{k-j-1}x = yp^{k-1} \not\equiv 0 \mod p^k,$$

so that the order of x is precisely p^{k-j}. The very last assertion of the theorem follows by noting that $1^\dagger = -(1 + q + q^2 + \ldots)$, so that $1^\dagger q = q'$. □

3 Some technical lemmas

In this section we collect together some observations of a technical nature which will be useful later in deriving group presentations. The proofs of Lemmas 3.1 and 3.2, also noted in [9], are straightforward inductions and left to the reader.

Lemma 3.1. *If G is a group and $x, y, z \in G$ are such that $[x, y] = z$ and $[x, z] = [y, z] = 1$ then $[x^\lambda, y^\mu] = z^{\lambda\mu}$ for all $\lambda, \mu \in \mathbb{Z}^+$.*

Lemma 3.2. *If G is a group and $x, y \in G$ are such that $[x, y] = y^\alpha$ for some $\alpha \in \mathbb{Z}$ then*

$$[x^\lambda, y^\mu] = y^{\mu(1-(1-\alpha)^\lambda)}$$

300

for all $\lambda, \mu \in \mathbb{Z}^+$.

Lemma 3.3. *Let p be prime and k, ℓ be positive integers such that $\ell \leq k$. If $p = 2$ we assume that $\ell > 1$. Put*

$$q = p^\ell \quad \text{and} \quad L = \begin{cases} 0 & \text{if } k - 2\ell \leq 0 \\ p^{k-2\ell} - 1 & \text{if } k - 2\ell > 0 \end{cases}.$$

Suppose G is a group, $x, y, z, w \in G$ are such that x, y, z, w each have order dividing p^k,

$$x^z = x^{1-q}, \; x^w = x^{1-q'}, \; y^z = y^{1-q'}, \; y^w = y^{1-q}, \; [z, w] = 1$$

(all quasi-inversion taking place in $\mathbb{Z}_{p^k}$), and for each $m = 0, \ldots, L$,

$$x^\alpha y = z^{-\beta} y x^\alpha w^\beta$$

where $\alpha = (1 - q)^m q$ and β is the least nonnegative integer such that

$$1^{o^{(q)}\beta} = -\alpha q$$

in $\mathbb{Z}_{p^k}$ (which exists because $(\mathbb{Z}_{p^k}, o^{(q)})$ is generated by 1). Then, for all $\lambda, \mu \in \mathbb{Z}^+$,

$$x^\lambda y^\mu = z^{-\nu} y^\mu x^\lambda w^\nu$$

where ν is the least nonnegative integer such that

$$1^{o^{(q)}\nu} = -\lambda\mu q$$

in $\mathbb{Z}_{p^k}$.

Proof. The case $\lambda = \mu = 1$ is covered by the hypothesis (when $m = 0$), which starts an induction. In the following, since orders divide p^k, we may interpret exponents as elements of $\mathbb{Z}_{p^k}$. Let $\lambda > 1$. By an inductive hypothesis, choosing β so that $1^{o^{(q)}\beta} = -(\lambda - 1)q$,

$$\begin{aligned} x^\lambda y &= xx^{\lambda-1}y = xz^{-\beta}yx^{\lambda-1}w^\beta = z^{-\beta}x^{z^{-\beta}}yx^{\lambda-1}w^\beta \\ &= z^{-\beta}x^{(1-q)^{-\beta}}yx^{\lambda-1}w^\beta. \end{aligned}$$

But β is a multiple of q, since, by Theorem 2.8, $-(\lambda-1)q$ has order dividing $p^{k-\ell}$ and 1 has order p^k in the group $(\mathbb{Z}_{p^k}, o^{(q)})$. Note also, by Theorem 2.8 and Corollary 2.3, that $1 - q$ has order $p^{k-\ell}$ in $G(\mathbb{Z}_{p^k})$. Write $-\beta = Kq$ for some integer K. If $L = 0$ then

$$(1 - q)^{-\beta} = ((1 - q)^q)^K = 1^K = 1,$$

since $p^{k-\ell}$ divides q, yielding, by the hypothesis when $m = 0$,
$$x^{(1-q)^{-\beta}} y = xy = z^{-\gamma} yxw^\gamma = z^{-\gamma} yx^{(1-q)^{-\beta}} w^\gamma$$
where $1^{o^{(q)}\gamma} = -q = -(1-q)^{-\beta} q$. If $L > 0$ then write
$$K = mp^{k-2\ell} + n \quad \text{where} \quad 0 \le n \le L$$
so that
$$(1-q)^{-\beta} = ((1-q)^{p^{k-\ell}})^{mq}(1-q)^{nq} = (1-q)^{nq},$$
yielding
$$x^{(1-q)^{-\beta}} y = z^{-\gamma} yx^{(1-q)^{-\beta}} w^\gamma$$
by the hypothesis where $1^{o^{(q)}\gamma} = -(1-q)^{-\beta} q$. In both cases,
$$x^\lambda y = z^{-\beta} z^{-\gamma} yx^{(1-q)^{-\beta}} w^\gamma x^{\lambda-1} w^\beta = z^{-(\beta+\gamma)} yx^{(1-q)^{-\beta}} (x^{\lambda-1}) w^{-\gamma} w^\gamma w^\beta$$
$$= z^{-(\beta+\gamma)} yx^{(1-q)^{-\beta}} x^{(1-q')^{-\gamma}(\lambda-1)} w^{\gamma+\beta} = z^{-\delta} yx^\lambda w^\delta$$

where $\delta = \alpha + \beta$, after observing that, in $\mathbb{Z}_{p^k}$, using the formulae of Theorem 2.8,

$$(1-q)^{-\beta} + (1-q')^{-\gamma}(\lambda - 1) = (1-q)^{-\beta} + (1-q)^\gamma(\lambda - 1)$$
$$= (1-q)^{-\beta} + (1 - q1^{o^{(q)}\gamma})(\lambda - 1) = \lambda - 1 + (1-q)^{-\beta}(1 + (\lambda - 1)q^2)$$
$$= \lambda - 1 + (1-q)^{-\beta}(1 - q1^{o^{(q)}\beta}) = \lambda - 1 + (1-q)^{-\beta}(1-q)^\beta = \lambda.$$

Further, by Theorem 2.8,
$$1^{o^{(q)}\delta} = 1^{o^{(q)}\beta} \circ^{(q)} 1^{o^{(q)}\gamma} = -(\lambda - 1)q - (1-q)^{-\beta} q(1 - q1^{o^{(q)}\beta})$$
$$= -\lambda q + q - q(1-q)^{-\beta}(1-q)^\beta = -\lambda q.$$

Now let $\mu > 1, \lambda \ge 1$. By an inductive hypothesis, we have, choosing δ such that $1^{o^{(q)}\delta} = -\lambda(\mu - 1)q$,

$$x^\lambda y^\mu = x^\lambda y^{\mu-1} y = z^{-\delta} y^{\mu-1} x^\lambda w^\delta y = z^{-\delta} y^{\mu-1} w^\delta (x^\lambda)^{w^\delta} y$$
$$= z^{-\delta} y^{\mu-1} w^\delta x^{(1-q')^\delta \lambda} y = z^{-\delta} y^{\mu-1} w^\delta z^{-\epsilon} yx^{(1-q')^\delta \lambda} w^\epsilon,$$

choosing ϵ such that $1^{o^{(q)}\epsilon} = -(1-q')^\delta \lambda q$ by the first half, so that, since $[z,w]=1$,

$$\begin{aligned}
x^\lambda y^\mu &= z^{-\delta} y^{\mu-1} z^{-\epsilon} w^\delta y x^{(1-q')^\delta \lambda} w^\epsilon \\
&= z^{-\delta} z^{-\epsilon} (y^{\mu-1})^{z^{-\epsilon}} y^{w^{-\delta}} w^\delta x^{(1-q')^\delta \lambda} w^\epsilon \\
&= z^{-(\delta+\epsilon)} y^{(1-q')^{-\epsilon}(\mu-1)} y^{(1-q)^{-\delta}} (x^{(1-q')^\delta \lambda})^{w^{-\delta}} w^\delta w^\epsilon \\
&= z^{-\sigma} y^{(1-q')^{-\epsilon}(\mu-1)+(1-q)^{-\delta}} x^{(1-q')^{-\delta}(1-q')^\delta \lambda} w^\sigma \\
&= z^{-\sigma} y^\mu x^\lambda w^\sigma
\end{aligned}$$

where $\sigma = \epsilon + \gamma$, after observing that, again by Theorem 2.8,

$$\begin{aligned}
(1-q')^{-\epsilon}(\mu-1) + (1-q)^{-\delta} &= (1-q)^\epsilon (\mu-1) + (1-q)^{-\delta} \\
&= (1 - q 1^{o^{(q)}\epsilon})(\mu-1) + (1-q)^{-\delta} \\
&= \mu - 1 + q(\mu-1)(1-q')^\delta \lambda q + (1-q)^{-\delta} \\
&= \mu - 1 + (1-q)^{-\delta}(1 - q 1^{o^{(q)}\delta}) \\
&= \mu - 1 + (1-q)^{-\delta}(1-q)^\delta \;=\; \mu.
\end{aligned}$$

Finally we observe that

$$\begin{aligned}
1^{o^{(q)}\sigma} &= 1^{o^{(q)}\delta} \, o^{(q)} \, 1^{o^{(q)}\epsilon} \\
&= -\lambda(\mu-1)q - (1-q')^\delta \lambda q(1 - q 1^{o^{(q)}\delta}) \\
&= -\lambda \mu q + \lambda q(1 - (1-q)^{-\delta}(1-q)^\delta) \;=\; -\lambda \mu q.
\end{aligned}$$

$\square$

4 A group presentation

Throughout p denotes a prime and k a positive integer. Our aim is to exhibit a group presentation for any Munn ring over $\mathbb{Z}_{p^k}$ with respect to a scalar sandwich matrix where the scalar is a multiple of p, regarded as a group with respect to the circle operation. Multiplying through by a unit, as explained in the first section, it suffices to assume the scalar is a power of p. Up to isomorphism then, the problem is to find a presentation for the group

$$(\mathrm{Mat}_n(\mathbb{Z}_{p^k}), o^{(q)})$$

where $q = p^\ell$ for some $\ell \leq k$. It is hoped that this will be a stepping stone towards a complete understanding of the circle monoids of all finite Munn rings over $\mathbb{Z}_{p^k}$, since, in the general case, up to isomorphism, one may take the sandwich matrix to be a direct sum of scalar matrices augmented possibly by extra rows or columns of zeros (see Chapters 3 and 6 of [7]).

We begin by reviewing the general theory of group presentations of radical matrix rings initiated by Coleman and Easdown [9], and apply one of their main results in our context. Let S be any radical ring and suppose

$$(S, +) \cong \langle \Gamma^{(+)} \mid \mathcal{R}^{(+)} \rangle, \ (S, \circ) \cong \langle \Gamma^{(\circ)} \mid \mathcal{R}^{(\circ)} \rangle$$

for some alphabet $\Gamma^{(+)}$, $\Gamma^{(\circ)}$ and collections of relations $\mathcal{R}^{(+)}$, $\mathcal{R}^{(\circ)}$ over $\Gamma^{(+)}$, $\Gamma^{(\circ)}$ respectively. We may suppose no generator is redundant, so that there are collections $W^{(+)}$, $W^{(\circ)}$ of words (normal forms), over $\Gamma^{(+)}$, $\Gamma^{(\circ)}$ respectively such that

$$\Gamma^{(+)} \subseteq W^{(+)}, \ \Gamma^{(\circ)} \subseteq W^{(\circ)},$$

and bijections

$$\varphi : W^{(+)} \longrightarrow S, \ \psi : W^{(\circ)} \longrightarrow S$$

whose inverses induce the above isomorphisms. Let n be a fixed positive integer. We now create a new alphabet

$$\Gamma = \{\, \sigma_{ij} \mid i, j \in \{1, \ldots, n\}, \ \sigma \in \Gamma^{(+)} \text{ if } i \neq j, \text{ and } \sigma \in \Gamma^{(\circ)} \text{ if } i = j \,\}.$$

For any $i, j \in \{1, \ldots, n\}$, put

$$1_{ij} = 1$$

where 1 here denotes the empty word, and if $w = \sigma^{(1)} \ldots \sigma^{(m)}$ is any non-empty word where $\sigma^{(1)}, \ldots, \sigma^{(m)}$ are letters, put

$$w_{ij} = \sigma^{(1)}_{ij} \ldots \sigma^{(m)}_{ij},$$

so that, if w is over $\Gamma^{(+)}$ and $i \neq j$, or over $\Gamma^{(\circ)}$ and $i = j$, then w_{ij} is over Γ. For any $i \neq j$, let $\mathcal{R}^{(+)}_{ij}$ denote the collection of relations of the form

$$v_{ij} = w_{ij}$$

where $v = w$ is a relation of $\mathcal{R}^{(+)}$. For any i, let $\mathcal{R}^{(\circ)}_i$ denote the collection of relations of the form

$$v_{ii} = w_{ii}$$

where $v = w$ is a relation of $\mathcal{R}^{(\circ)}$. Now let $\mathcal{R}$ denote the collection of relations of the following types:

(1) $\bigcup_{i \neq j} \mathcal{R}^{(+)}_{ij} \cup \bigcup_i \mathcal{R}^{(\circ)}_i$.

(2) $(\forall i \neq l, j \neq k) \left(\forall a \in \begin{cases} \Gamma^{(+)} & \text{if } i \neq j \\ \Gamma^{(\circ)} & \text{if } i = j \end{cases} \right) \left(\forall b \in \begin{cases} \Gamma^{(+)} & \text{if } k \neq l \\ \Gamma^{(\circ)} & \text{if } k = l \end{cases} \right)$

$$[a_{ij}, b_{kl}] = 1.$$

(3) $(\forall i \neq j \neq k \neq i)(\forall u, v \in W^{(+)})$

$$[u_{ij}, v_{jk}] = ((-(u\varphi)(v\varphi))\varphi^{-1})_{ik}.$$

(4) $(\forall i \neq j)(\forall u \in W^{(\circ)}, v \in W^{(+)})$

$$[u_{ii}, v_{ij}] = (((u\psi)'(v\varphi))\varphi^{-1})_{ij}.$$

(5) $(\forall i \neq j)(\forall u \in W^{(+)}, v \in W^{(\circ)})$

$$[u_{ij}, v_{jj}] = ((-(u\varphi)(v\psi))\varphi^{-1})_{ij}.$$

(6) $(\forall i > j)(\forall u, v \in W^{(+)})$

$$u_{ij}v_{ji} = ((-(v\varphi)(u\varphi))'\psi^{-1})_{jj} v_{ji} u_{ij} ((-(u\varphi)(v\varphi))\psi^{-1})_{ii}.$$

The following is proved in [9]:

Theorem 4.1. *The groups* $(\text{Mat}_n(S), \circ)$ *and* $\langle \Gamma \mid \mathcal{R} \rangle$ *are isomorphic.*

We now apply this theorem to prove the following:

Theorem 4.2. *Let p be prime and k, ℓ positive integers such that $\ell \leq k$. Suppose $\ell > 1$ if $p = 2$. Put*

$$q = p^\ell \quad \text{and} \quad L = \begin{cases} 0 & \text{if } k - 2\ell \leq 0 \\ p^{k-2\ell} - 1 & \text{if } k - 2\ell > 0. \end{cases}$$

Then the ring

$$\text{Mat}_n(\mathbb{Z}_{p^k}, +, *^{(q)})$$

is a group with respect to the circle operation $\circ^{(q)}$ and has the group presentation $\langle A \mid \mathcal{R} \rangle$ where

$$A = \{ a_{ij} \mid i, j \in \{1, \ldots, n\} \}$$

and $\mathcal{R}$ consists of the following relations:

(1) $(\forall i, j) \quad a_{ij}^{p^k} = 1;$

(2) $(\forall i \neq l, j \neq k)$ $[a_{ij}, a_{kl}] = 1$;

(3) $(\forall i \neq j \neq k \neq i)$ $[a_{ij}, a_{jk}] = a_{ik}^{-q}$;

(4) $(\forall i \neq j)$ $[a_{ii}, a_{ij}] = a_{ij}^{q'}$;

(5) $(\forall i \neq j)$ $[a_{ij}, a_{jj}] = a_{ij}^{-q}$;

(6) $(\forall i > j)(\forall m = 0, \ldots, L)$

$$a_{ij}^\alpha a_{ji} = a_{jj}^{-\beta} a_{ji} a_{ij}^\alpha a_{ii}^\beta$$

where $\alpha = (1 - q)^{mq}$ and β is the least nonnegative integer such that

$$1^{o^{(q)}\beta} = -\alpha q$$

in $\mathbb{Z}_{p^k}$ (which exists because $(p\mathbb{Z}_{p^k}, o^{(q)})$ is cyclic generated by 1).

Note that (3) and (5) could be amalgamated. However in practice it is convenient to keep them separate because when $n = 2$ the collection (3) is empty, and also (5) may easily be reformulated as conjugation relations.

Proof. Put $G = (\mathrm{Mat}_n(\mathbb{Z}_{p^k}), o^{(q)})$. If $x \in \mathbb{Z}_{p^k}$ denote the quasi-inverse with respect to o by $'$ and with respect to $o^{(q)}$ by $\dagger$. By Corollary 2.5,

$$(\mathbb{Z}_{p^k}, +) \cong (\mathbb{Z}_{p^k}, o^{(q)}) \cong C_{p^k} \cong \langle a \mid a^{p^k} = 1 \rangle.$$

In the preceding theory, then, we take $\Gamma^{(+)} = \Gamma^{(o)} = \{a\}$,

$$W^{(+)} = W^{(o)} = \{a^i \mid 0 \leq i \leq p^k - 1\},$$

and $\phi : a^i \mapsto i$, $\psi : a^i \mapsto 1^{o^{(q)}i}$ for each i. By Theorem 4.1, and the formulae in Theorem 2.8, a presentation for G uses generators a_{ij} where $i, j \in \{1, \ldots, n\}$ and relations

(1) $(\forall i, j)$ $a_{ij}^{p^k} = 1$;

(2) $(\forall i \neq l, j \neq k)$ $[a_{ij}, a_{kl}] = 1$;

(3) $(\forall i \neq j \neq k \neq i)(\forall \lambda, \mu)$ $[a_{ij}^\lambda, a_{jk}^\mu] = a_{ik}^{-\lambda \mu q}$;

(4) $(\forall i \neq j)(\forall \lambda, \mu)$ $[a_{ii}^\lambda, a_{ij}^\mu] = a_{ij}^{(1^{o^{(q)}\lambda})\dagger \mu q} = a_{ij}^{\mu(1-(1-q')^\lambda)}$;

(5) $(\forall i \neq j)(\forall \lambda, \mu)$ $[a_{ij}^\lambda, a_{jj}^\mu] = a_{ij}^{-\lambda 1^{o^{(q)}\mu} q} = a_{ij}^{-\lambda(1-(1-q)^\mu)}$;

(6) $(\forall i > j)(\forall \lambda, \mu)$

$$a_{ij}^\lambda a_{ji}^\mu = a_{jj}^{-\alpha} a_{ji}^\mu a_{ij}^\lambda a_{ii}^\alpha$$

where $1^{o^{(q)}}\alpha = -\lambda\mu q$.

Collections (1) and (2) are the same as in the statement of this theorem. We may replace (3) by the relations

$$(\forall i \neq j \neq k \neq i) \quad [a_{ij}, a_{jk}] = a_{ik}^{-q},$$

since the rest are implied by these and Lemma 3.1, in the presence of (2). We may replace (4) by the relations

$$(\forall i \neq j) \quad [a_{ii}, a_{ij}] = a_{ij}^{q'},$$

since the rest are implied by these and Lemma 3.2. We may replace (5) by the relations

$$(\forall i \neq j) \quad [a_{ij}, a_{jj}] = a_{ij}^{-q},$$

equivalent to

$$(\forall i \neq j) \quad [a_{jj}, a_{ij}] = a_{ij}^{q},$$

which, by Lemma 3.2, imply, for each λ, μ,

$$[a_{ij}^\lambda, a_{jj}^\mu] = [a_{jj}^\mu, a_{ij}^\lambda]^{-1} = = a_{ij}^{-\lambda(1-(1-q)^\mu)}.$$

Suppose $i \geq j$. Observe that relations above imply

$$a_{ii}^{p^k} = a_{ij}^{p^k} = a_{ji}^{p^k} = a_{jj}^{p^k} = [a_{jj}, a_{ii}] = 1,$$

$$a_{ij}^{a_{jj}} = a_{ij}[a_{ij}, a_{jj}] = a_{ij}^{1-q}, \qquad a_{ij}^{a_{ii}} = a_{ij}[a_{ii}, a_{ij}]^{-1} = a_{ij}^{1-q'},$$

and similarly,

$$a_{ji}^{a_{jj}} = a_{ji}^{1-q'}, \qquad a_{ji}^{a_{ii}} = a_{ji}^{1-q},$$

so that, taking $x = a_{ij}$, $y = a_{ji}$, $z = a_{jj}$ and $w = a_{ii}$, Lemma 3.3 implies all of the relations of (6) may be replaced by the subcollection, for $m = 0, \ldots, L$,

$$a_{ij}^\alpha a_{ji} = a_{jj}^{-\beta} a_{ji} a_{ij}^\alpha a_{ii}^\beta$$

where $\alpha = (1-q)^{mq}$ and $1^{o^{(q)}}\beta = -\alpha q$. $\square$

The previous proof employs the general machinery of [9]. A direct, though very long, argument can be constructed along the following lines, using some counting to exploit finiteness of the groups. First observe, by calculation, that all of the relations hold if we identify the generator a_{ij} with the $n \times n$ matrix consisting of 1 in the (i,j)th position and 0 elsewhere. For example, for (6), if $\alpha = (1-q)^{mq}$ and $1\circ^{(q)}\beta = -\alpha q$ then $a_{jj}^{-\beta}a_{ji}a_{ij}^{\alpha}a_{ii}^{\beta}$ becomes identified with

$$\begin{matrix} & j & i \\ j \\ i \end{matrix} \begin{pmatrix} (-\alpha q)^{\dagger} & 0 \\ 0 & 0 \end{pmatrix} \circ^{(q)} \begin{pmatrix} 0 & 1 \\ 0 & 0 \end{pmatrix} \circ^{(q)} \begin{pmatrix} 0 & 0 \\ \alpha & 0 \end{pmatrix} \circ^{(q)} \begin{pmatrix} 0 & 0 \\ 0 & -\alpha q \end{pmatrix}$$

$$= \begin{pmatrix} (-\alpha q)^{\dagger} & 1 - q(-\alpha q)^{\dagger} \\ 0 & 0 \end{pmatrix} \circ^{(q)} \begin{pmatrix} 0 & 0 \\ \alpha & -\alpha q \end{pmatrix}$$

$$= \begin{pmatrix} (-\alpha q)^{\dagger} - q(1 - q(-\alpha q)^{\dagger})\alpha & 1 - q(-\alpha q)^{\dagger} - q(1 - q(-\alpha q)^{\dagger})(-\alpha q) \\ \alpha & -\alpha q \end{pmatrix}$$

$$= \begin{pmatrix} (-\alpha q) \circ^{(q)} (-\alpha q)^{\dagger} & 1 - q((-\alpha q) \circ^{(q)} (-\alpha q)^{\dagger}) \\ \alpha & -\alpha q \end{pmatrix}$$

$$= \begin{pmatrix} 0 & 1 \\ \alpha & -\alpha q \end{pmatrix} = \begin{pmatrix} 0 & 0 \\ \alpha & 0 \end{pmatrix} \circ \begin{pmatrix} 0 & 1 \\ 0 & 0 \end{pmatrix},$$

which becomes identified with $a_{ij}^{\alpha}a_{ji}$. The next step is to verify that these matrices generate the entire group. For example, when $n = 2$,

$$\begin{pmatrix} x & y \\ z & w \end{pmatrix} = \begin{pmatrix} x & 0 \\ z & 0 \end{pmatrix} \circ^{(q)} \begin{pmatrix} 0 & y(1-xq)^{-1} \\ 0 & w + zqy(1-xq)^{-1} \end{pmatrix}$$

$$= \begin{pmatrix} x & 0 \\ 0 & 0 \end{pmatrix} \circ^{(q)} \begin{pmatrix} 0 & 0 \\ z & 0 \end{pmatrix} \circ^{(q)} \begin{pmatrix} 0 & 0 \\ 0 & w + zqy(1-xq)^{-1} \end{pmatrix}$$

$$\circ^{(q)} \begin{pmatrix} 0 & y(1-xq)^{-1} \\ 0 & 0 \end{pmatrix},$$

which becomes identified with $a_{11}^{\alpha}a_{21}^{z}a_{22}^{\beta}a_{12}^{y(1-xq)^{-1}}$ where $1\circ^{(q)}\alpha = x$ and $1\circ^{(q)}\beta = w + zqy(1-xq)^{-1}$. The previous calculation can be put into a conceptual framework involving semidirect and general products, leading to the following recursive formula for any radical matrix ring S, which is proved in Section 6 of [9]:

$$(\mathsf{Mat}_n(S), \circ) \cong ((S,+)^{n-1} \rtimes (\mathsf{Mat}_{n-1}(S), \circ)) \circledast ((S,+)^{n-1} \rtimes (S, \circ)),$$

where $\rtimes$ and $\circledast$ denote semidirect and general product respectively. Details and background, particularly for general products, can be found in Sections 1 and 6 of [9].

In our sketch of a direct proof of Theorem 4.2 we have reached the stage where $\text{Mat}_n(\mathbb{Z}_{p^k}, \circ^{(q)})$ is a homomorphic image of the group defined by the presentation. To see that the homomorphism is an isomorphism it is sufficient to prove that the group contains at most p^{kn^2} elements, the size of $\text{Mat}_n(\mathbb{Z}_{p^k})$. One may do this by proving that if $w \in W$ and a_{kl} is any generator, where

$$W = \left\{ \prod_{i=1}^{n}\prod_{j=1}^{n} a_{ij}^{\alpha_{ij}} \;\middle|\; \alpha_{ij} \in \{0, \ldots, p^k - 1\} \right\}$$

then the relations can be used to transform the word wa_{kl} into another word from W, and noticing that the size of W is p^{kn^2}. The first step towards a word argument would be to apply Lemmas 3.1, 3.2 and 3.3 (as in the formal proof of Theorem 4.2) to amplify the supply of relations and then use an inductive argument to manipulate words. A general argument and technique along these lines is provided in Lemma 3.5 of [9], which is stated and proved for monoids.

5 Examples

Example 5.1. The group $(\text{Mat}_2(\mathbb{Z}_9), \circ^{(3)})$ has the following presentation:

$$\langle a,b,c,d \mid a^9 = b^9 = c^9 = d^9 = 1, \; ad = da,$$
$$b^a = b^{-d} = b^4, \; c^d = c^{-a} = c^4, \; bc = cba^{-3}d^3 \rangle$$

This follows from Theorem 4.2, putting

$$a = a_{11}, \; b = a_{12}, \; c = a_{21}, \; d = a_{22},$$

rearranging commutator relations as conjugation relations, noting that $3' = 6$ in $\mathbb{Z}_9$, $L = 0$ (so (6) only provides one relation), and observing that a^3 and d^3 are central. By Corollary 2.5, this is also a presentation for $(3\text{Mat}_2(\mathbb{Z}_{27}), \circ)$, and appears as Example 8.3 in [9] when $p = 3$. This group has a decomposition

$$(C_9 \rtimes C_9) \circledast (C_9 \rtimes C_9)$$

with appropriate semidirect and general product actions.

Example 5.2. Applying Theorem 4.2, as in the previous example, but noting $L = 2$ so (6) provides 3 relations, the group $(\text{Mat}_2(\mathbb{Z}_{27}), \circ^{(3)})$ has the following presentation:

$$\langle a,b,c,d \mid a^{27} = b^{27} = c^{27} = d^{27} = 1, \; ad = da, \; b^a = b^{-d} = b^{-2},$$
$$c^d = c^{-a} = c^{-2}, \; bc = d^3cba^{-3}, \; b^{10}c = d^3cb^{10}a^{-3}, \; b^{19}c = d^3cb^{19}a^{-3} \rangle$$

But the matrix $\begin{pmatrix} 0 & 9 \\ 0 & 0 \end{pmatrix}$ clearly commutes under $\circ^{(3)}$ with all matrices, so that b^9 must be central. In particular we may add the relation $b^9 c = cb^9$ and use it to deduce the last two relations in the above presentation, which may then be deleted. Thus we get the more succinct presentation:

$$\langle a,b,c,d \mid a^{27} = b^{27} = c^{27} = d^{27} = 1,\ ad = da,\ b^a = b^{-d} = b^{-2},$$
$$c^d = c^{-a} = c^{-2},\ b^9 c = cb^9,\ bc = d^3 cba^{-3} \rangle$$

(The relation $b^9 c = cb^9$ also follows, of course, by Lemma 3.3, taking $x = b$, $\lambda = 9$, $y = c$, $\mu = 1$, $z = d$, $w = a$ and $\nu = 0$, the least nonnegative integer such that $1\circ^{(3)}\nu = -27 = 0$ in $\mathbb{Z}_{27}$.)

Example 5.3. In the development of the main results of this paper we have been careful to avoid the case $p = 2$, $\ell = 1$. This case is also accounted for, in fact, by observing that, by Corollary 2.2,

$$(\mathbb{Z}_{2^k}, \circ^{(2)}) \cong (2\mathbb{Z}_{2^{k+1}}, \circ),$$

so that

$$(\text{Mat}_n(\mathbb{Z}_{2^k}), \circ^{(2)}) \cong (\text{Mat}_n(2\mathbb{Z}_{2^{k+1}}), \circ),$$

a presentation for which is furnished by Theorem 8.4 of [9].

References

[1] N. Antony, "Radicals of rings and the circle operation," Honours thesis, University of Sydney, 2000.

[2] J.C. Ault and J.F. Watters, "Circle groups of nilpotent rings," *American Math. Monthly* **80** (1973), 48-52.

[3] H.L. Chick, "The properties of some rings having isomorphic additive and circle composition groups," *Austral. Math. Soc. Gaz.* **23** (1996), 112-117.

[4] H.L. Chick, "Rings with isomorphic additive and circle composition groups," *Rings and Radicals* (Shijiazhuang, 1994), Pitman Res. Notes. Ser., 346, Longman, Harlow (1996), 160-169.

[5] H.L. Chick and B.J. Gardner, "Commutative quasiregular rings with isomorphic additive and circle composition groups. II. Rational algebras," *Comm. Algebra* **26** (1998), 657-670.

[6] A.H. Clifford and G.B. Preston, *The Algebraic Theory of Semigroups*, Vol. 1, Math. Surveys of the American Math. Soc. 7, Providence, RI, 1961.

[7] C. Coleman, "Circle decompositions of rings, in particular, Munn rings", Doctoral thesis, University of Sydney, 1999.
[8] C. Coleman and D. Easdown, "Complementation in the group of units of a ring," *Bull. Austral. Math. Soc.* **62** (2000), 183-192.
[9] C. Coleman and D. Easdown, "Decompositions of rings under the circle operation," *Beiträge zur algebra und geometrie*, **43** (2002) 55-88.
[10] I. Fischer and K.E. Eldridge, "Artinian rings with a cyclic quasi-regular group", *Duke Math. J.* **36** (1969), 43-47.
[11] N. Jacobson, *Basic Algebra I*, Freeman, New York, 1985.
[12] A.V. Kelarev, "The groups of units of a commutative semigroup ring," *J. Algebra* **169** (1994), 902-912.
[13] R.L. Kruse, "A note on the adjoint group of a finitely generated radical ring," *J. London Math. Soc.* **1** (1969), 743-744.
[14] R.L. Kruse, "On the circle group of a nilpotent ring," *American Math. Monthly* **77** (1970), 168-170.
[15] D.B. McAlister, "The category of representations of a completely 0-simple semigroup," *J. Austral. Math. Soc.* **12** (1971), 193-210.
[16] N.R. McConnell and T. Stokes, "Generalising quasiregularity for rings," *Austral. Math. Soc. Gaz.* **25** (1998), 250-252.
[17] B. McDonald, *Finite Rings with Identity*, Marcel Dekker, New York, 1974.
[18] W.D. Munn, "On semigroup algebras," *Proc. Camb. Phil. Soc.* **51** (1955), 1-15.
[19] W.D. Munn, "Semigroup rings of completely regular semigroups," *Lattices, Semigroups and Universal Algebras*, Lisbon (1988), 191-210, Plenum, New York, 1990.
[20] J. Okninski, *Semigroup Algebras*, Marcel Dekker, New York, 1991.
[21] R. Sandling, "Group rings of circle and unit groups," *Math. Z.* **140** (1974), 195-202.
[22] K.I. Tahara and A. Hosomi, "On the circle group of finite nilpotent rings," *Groups - Korea 1983*, Eds. A.C. Kim and B.H. Neumann, Springer Lecture Notes in Mathematics 1098 (1984), 161-179.
[23] J.F. Watters, "On the adjoint group of a radical ring," *J. London Math. Soc.* **43** (1968), 725-729.

FINITE SEMIGROUPS
IMPOSING TRACTABLE CONSTRAINTS

ANDREI BULATOV

Department of Mathematics and Mechanics, Ural State University,
620083 Ekaterinburg, RUSSIA
E-mail: Andrei.Bulatov@usu.ru

PETER JEAVONS

Computing Laboratory, University of Oxford
Oxford OX1 3QD, UNITED KINGDOM
E-mail: Peter.Jeavons@comlab.ox.ac.uk

MIKHAIL VOLKOV[*]

Department of Mathematics and Mechanics, Ural State University,
620083 Ekaterinburg, RUSSIA
E-mail: Mikhail.Volkov@usu.ru

The Constraint Satisfaction Problem provides a framework in which a wide variety of combinatorial problems can be expressed in a natural way. In the present paper, after defining the general Constraint Satisfaction Problem and giving several concrete examples, we briefly survey a recent approach to studying constraint satisfaction problems which is based on characterizing the complexity of problem classes via finite algebras and their pseudovarieties. Then we investigate the case when the finite algebra assigned to a subclass of the constraint satisfaction problem with a finite set of values turns out to be a semigroup. Here we show that the subclass is tractable if and only if the assigned semigroup is a block group; otherwise it always contains an NP-complete problem.

---

[*]THE PARTICIPATION OF THE THIRD NAMED AUTHOR IN THE SCHOOL WAS FINANCIALLY SUPPORTED BY FUNDAÇÃO CALOUSTE GULBENKIAN (FCG), FUNDAÇÃO PARA A CIÊNCIA E A TECNOLOGIA (FCT), FACULDADE DE CIÊNCIAS DA UNIVERSIDADE DE LISBOA (FCUL) AND REITORIA DA UNIVERSIDADE DO PORTO. THE RESEARCH PRESENTED IN THE PAPER WAS ALSO SUPPORTED BY RUSSIAN EDUCATION MINISTRY, GRANT E00-1.0-92, BY THE RUSSIAN FOUNDATION OF BASIC RESEARCH, GRANT 01-01-00258, AND BY THE INTAS THROUGH THE NETWORK PROJECT 99-1224 'COMBINATORIAL AND GEOMETRIC THEORY OF GROUPS AND SEMIGROUPS AND ITS APPLICATIONS TO COMPUTER SCIENCE'.

Introduction

Many combinatorial problems can be naturally expressed in the framework of the Constraint Satisfaction Problem (CSP, for short). The aim in a CSP is to assign values to a given set of variables such that certain constraints are satisfied on the values that can be assigned simultaneously to some subsets of variables [31, 26]. A standard satisfiability problem [11], where the variables must be assigned Boolean values, gives a typical example of such problem.

Throughout the paper we assume that P $\neq$ NP. Recall that problems from P are said to be *tractable*. The general CSP is known to be NP-hard, and hence, intractable [26, 31]. However, certain restrictions on the form of constraints may affect the complexity of the corresponding problem class and give rise to a tractable subclass of the general CSP. There is, therefore, a fundamental research direction aiming to recognize tractable subclasses of the CSP. A progress in this direction may provide efficient algorithms for those applications, which fall in one of the known tractable classes. The direction is also very important from a theoretical perspective, as it helps to clarify the boundary between tractability and intractability in a wide range of combinatorial search problems.

For the important special case of Boolean CSPs, Schaefer [38] managed to find the precise boundary between tractability and intractability. Namely, Schaefer established that for Boolean CSPs (which he called "Generalised Satisfiability Problems") there are exactly six different families of tractable constraints, and any problem involving constraints not contained in one of these six families is NP-complete. This important result is known as Schaefer's Dichotomy Theorem for Boolean relations.

Schaefer [38] raised the question of how this result could be generalised to larger domains (that is, sets with more than two elements). Several approaches tackling this problem have appeared recently. For instance, Feder and Vardi [10] used techniques from logic programming and group theory to identify three broad families of tractable constraints which include all of Schaefer's six classes. In a series of papers the second named author and his coauthors have characterized a number of tractable constraint classes over finite domains using invariance properties of relations [21, 16, 18, 19, 20, 17].

In [20] and then in [4, 5] it has been shown that, for any subclass of CSP over a finite set of values, there is a finite algebra such that the complexity of the subclass depends only on the corresponding algebra. The first aim of the present paper is to give a brief overview of this link between computational complexity theory and universal algebra. Our second aim is to analyze finite semigroups from the complexity-theoretical point of view. It is interesting and somewhat unexpected that these complexity-theoretical considerations lead to a class which is known to play a distinguished role in finite semigroup theory: the class of so-called block-groups [35].

The paper is organized as follows. In Section 1 we introduce the CSP, define some connected notions, and provide a few typical examples. Then in Section 2 we present the algebraic approach to the CSP. Section 3 contains a treatment of the semigroup case. We describe all finite semigroups such that the corresponding subclass of the CSP is tractable and deduce a semigroup version of the Dichotomy Theorem: if a subclass defined by a finite semigroup is not tractable, then it always contains an NP-complete problem. In conclusion we compare the semigroup case with the general situation.

The reader is referred to the books [11, 34] and [6, 2] for the basic notions of respectively complexity theory and universal algebra that we use below.

1 Constraint Satisfaction Problem

1.1 Definition

The 'constraint satisfaction problem' was introduced by Montanari [31] in 1974 and has been widely studied [26, 27, 3, 9, 28, 29, 41, 22, 25, 12, 10, 23].

Definition 1.1. The *constraint satisfaction problem (CSP)* is the combinatorial decision problem with

INSTANCE: a triple $(V, A, \mathcal{C})$ where

- V is a set of *variables*;
- A is a set *[domain]*;
- $\mathcal{C}$ is a set of *constraints*.
- Each constraint $C \in \mathcal{C}$ is a pair $\langle s, \varrho \rangle$, where
  - $s = (v_1, \ldots, v_m)$ is a tuple of variables (whose length depends on C), called the *constraint scope*;
  - $\varrho \subseteq A^m$ is an m-ary relation over A called the *constraint relation*.

QUESTION: does there exist a *solution*, i.e. a function $\varphi : V \to A$, such that, for each constraint $\langle s, \varrho \rangle \in \mathcal{C}$, with $s = (v_1, \ldots, v_m)$, the tuple $(\varphi(v_1), \ldots, \varphi(v_m))$ belongs to ϱ?

The size of a problem instance is the length of the encoding of all tuples in all constraints.

For some applications it is necessary to consider a more general version of the CSP, where each variable $v \in V$ has its own domain A_v and, for each constraint $\langle s, \varrho \rangle \in \mathcal{C}$ with the scope $s = (v_1, \ldots, v_m)$, the constraint relation ϱ is a subset in $A_{v_1} \times \cdots \times A_{v_m}$. This general version is usually referred to as the *multi-sorted* CSP in contrast to the *one-sorted* CSP as defined above. Even though we concentrate on the one-sorted version of the CSP in the present paper, we would like to mention that the general version admits a similar treatment.

1.2 Examples

The CSP provides a framework for a wide range of problems from various areas of computer science such as database theory [42, 24], temporal and spatial reasoning [39], machine vision [31], belief maintenance [8], technical design [33], natural language comprehension [1], programming language analysis [32], etc. Here we restrict ourselves to a few examples showing how certain well known problems from logic, graph theory, linear algebra, planning can be naturally expressed as instances of the CSP.

Example 1.1. An instance of the standard propositional SATISFIABILITY problem [11, 34] is specified by giving a formula in conjunctive normal form, that is, a conjunction of clauses, and asking whether there are truth values for the variables which make the formula true.

Suppose that $\Phi = X_1 \wedge \cdots \wedge X_n$ is such a formula, where the X_i are clauses. The satisfiability question for Φ can be expressed as the instance $(V, \{0, 1\}, \mathcal{C})$ of the CSP, where V is the set of all variables appearing in the clauses X_i, and $\mathcal{C}$ is the set of constraints $\{\langle s_1, \varrho_1 \rangle, \ldots, \langle s_n, \varrho_n \rangle\}$, where each constraint $\langle s_k, \varrho_k \rangle$ is constructed as follows:

- $s_k = (x_1^k, \ldots, x_{m_k}^k)$ where $x_1^k, \ldots, x_{m_k}^k$ are the variables appearing in the clause X_k;

- $\varrho_k = \{0, 1\}^{m_k} \setminus \{(a_1, \ldots, a_{m_k})\}$ where $a_i = 1$ if x_i^k is negated in X_k and $a_i = 0$ otherwise (i.e., ϱ_k consists of all m_k-tuples that make X_k true).

The solutions of this CSP instance are exactly the assignments which make the formula Φ true. Hence, any instance of SATISFIABILITY can be expressed as an instance of the CSP.

We notice that Example 1.1 shows that the CSP in general is NP-hard.

Example 1.2. An instance of GRAPH UNREACHABILITY consists of a graph $G = (V, E)$ and a pair of vertices, $v, w \in V$. The question is whether there is no path in G from v to w. This can expressed as the CSP instance $(V, \{0, 1\}, \mathcal{C})$

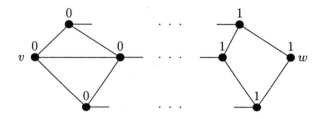

Figure 1. GRAPH UNREACHABILITY as an instance of the CSP

where
$$\mathcal{C} = \{\langle e, \{(0,0), (1,1)\}\rangle \mid e \in E\} \cup \{\langle v, \{0\}\rangle, \langle w, \{1\}\rangle\}.$$
Indeed, this CSP instance has a solution if and only if one can label the vertices of G by 0's and 1's such that v is labelled 0, w is labelled 1 and adjacent vertices are assigned equal labels, see Fig. 1. Clearly, this is possible if and only if there is no path from v to w.

Example 1.3. A system of linear equations over a field F can be expressed as the CSP instance $(V, F, \mathcal{C})$ where V is the set of variables of the system, and each constraint $\langle s_i, \varrho_i \rangle$ from $\mathcal{C}$ corresponds to an equation. Then s_i is the set of variables appearing in the equation, and ϱ_i is the set of solutions of the equation, that is, a hyperplane.

Example 1.4. A graph $G = (V, E)$ is said to be *interval* if its vertices can be represented by intervals of the real line such that two vertices are adjacent if and only if intervals which represent them have a common point, see Fig. 2. This notion plays an important role in planning and temporal reasoning. The question whether a given graph G is interval is expressed by the CSP instance $(V, I(\mathbb{R}), \{\langle e, \varrho\rangle \mid e \in E\} \cup \{\langle (u, v), \tau\rangle \mid u \neq v,\ (u, v) \notin E\})$ where $I(\mathbb{R})$ is the set of real intervals, and ϱ and τ are the binary realtions defined as

$$\varrho = \{((a, b), (c, d)) \mid (a, b) \cap (c, d) \neq \varnothing\},$$
$$\tau = \{((a, b), (c, d)) \mid (a, b) \cap (c, d) = \varnothing\}.$$

All the examples above are essentially one-sorted. Multi-sorted CSPs arise naturally in the study of databases; in fact, even though it uses a rather different vocabulary, the notion of a relational database is easily seen to be equivalent to the notion of an instance of the multi-sorted CSP.

317

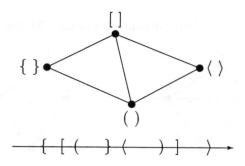

Figure 2. An interval graph

The list of examples can easily be extended. The reader is encouraged to translate his own favorite combinatorial problem to the CSP language, although sometimes this is a nontrivial exercise[a].

1.3 Restricted classes of CSP

As already mentioned, the CSP is NP-hard in general. However, some restrictions on the form of constraints may ensure tractability of a problem class. There are two natural ways to restrict the CSP: first, restrictions on the form of the constraint relations; and second, restrictions on the way in which the constraint scopes may interact. In this paper we adopt the first way; some results and references concerning the second direction can be found in [12].

Definition 1.2. For a given set of relations Γ over a set A, CSP(Γ) is defined to be the decision problem with

INSTANCE: An instance, $P = (V, A, \mathcal{C})$, of the CSP, in which, for each constraint $\langle s, \varrho \rangle$, the relation ϱ is an element of Γ.

QUESTION: Does P have a solution?

The set Γ is said to be *tractable* if, for each finite subset $\Gamma' \subseteq \Gamma$, the class CSP(Γ') is tractable, Γ is said to be *NP-complete* if CSP(Γ') is NP-complete for some finite subset $\Gamma' \subseteq \Gamma$.

Many well-known combinatorial problems can be represented in the form CSP(Γ).

---

[a]Of course, it is known that any problem from NP can be expressed by an instance of SATISFIABILITY, and thus, by an instance of the CSP, see Example 1.1. However, we are seeking for simple and natural reductions only, as in the examples above.

Example 1.5. An instance of GRAPH q-COLORABILITY consists of a graph $G = (V, E)$. The question is whether the vertices of G can be labelled with q colors so that adjacent vertices are assigned different colors, see Fig. 3. This

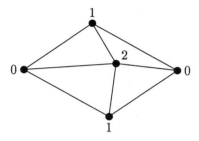

Figure 3. 3-coloring of a graph

can expressed as the CSP instance $(V, A, \{\langle e, \neq_A \rangle \, | \, e \in E\})$ where A is a q-element set (of colors) and $\neq_A = \{(a, b) \in A^2 \, | \, a \neq b\}$ is the disequality relation on A. Observe that this is in fact an instance of CSP($\{\neq_A\}$). Conversely, we can assign a graph G to each instance $P = (V, A, \langle s, \neq_A \rangle)$ of CSP($\{\neq_A\}$) so that P has a solution if and only if G admits a coloring with q colors: let V and the set of all constraint scopes of P be respectively the vertex and the edge sets of G. Thus, we see that GRAPH q-COLORABILITY is precisely CSP($\{\neq_A\}$) with A being a q-element set.

We note that for $q \geq 3$ GRAPH q-COLORABILITY is known to be NP-complete [11, 34].

Example 1.6. The NOT-ALL-EQUAL SATISFIABILITY problem (NAE) [11, 34] is a restricted version of the standard SATISFIABILITY problem which remains NP-complete. An instance of NAE is a conjunction $\Phi = X_1 \wedge \cdots \wedge X_n$ of 3-literal clauses $X_i = z_1^i \vee z_2^i \vee z_3^i$ (recall that a *literal* is either a variable or the negation of a variable). The question is whether there are assignments of truth values for the variables appearing in Φ such that in no clause X_i all three literals become equal in truth value. (Clearly, such assignments make the formula Φ true so that NAE is indeed a version of SATISFIABILITY.) This can be expressed as the instance $(V, \{0,1\}, \mathcal{C})$ of CSP($\{\nu\}$), where ν is the ternary relation $\{(a, b, c) \, | \, \{a, b, c\} = \{0, 1\}\}$ on $\{0, 1\}$, V is the set of all literals appearing in the clauses X_i, and $\mathcal{C}$ consists of the constraints $\langle (z_1^i, z_2^i, z_3^i), \nu \rangle$ for each clause X_i and the constraints $\langle (x, \neg x, \neg x), \nu \rangle$ for each variable x appearing in Φ. Indeed, the latter constraints ensure that each solution of the instance is induced by an assignment of values for the variables while the former constraints guarantee that the values of the literals in X_i are not all equal under the assignment. Conversely, each instance of $\text{CSP}(\{\nu\})$ can be easily converted into an equivalent instance of NAE. Thus, NAE can be identified with $\text{CSP}(\{\nu\})$.

While the examples above give NP-complete classes, the next problem class is tractable in spite of the infinite family of relations involved.

Example 1.7. The family of all linear systems over a field F corresponds to the class $\text{CSP}(H)$ where H is the class of all hyperplanes of finite-dimensional F-vector spaces (compare with Example 1.3).

2 Algebraic structure of CSP

Throughout the rest of the paper we assume the domains to be finite. The relation families, however, may (and often will) be infinite.

2.1 From arbitrary sets of relations to relational clones

The main goal in the study of $\text{CSP}(\Gamma)$ is to find a way to distinguish between tractable and intractable sets of relations. In this connection, it is natural to ask how a tractable family of relations can be extended so that the extended set of relations remains tractable. In other words, which operations on relations do preserve tractability? An important class of such operations has been discovered in [15].

We make use of the standard correspondence between relations and logical predicates: to each m-ary relation $\varrho \subseteq A^m$ we assign the m-ary predicate on A which becomes true precisely on m-tuples from ϱ. Recall that an existential first order formula is said to be *primitive positive* if its quantifier-free part is a conjunction of predicates. These formulas can be used as operations that allow one to derive new relations from given ones.

Example 2.1. The standard product of two binary relations ϱ_1 and ϱ_2 may be defined by the following primitive positive formula:

$$(x, z) \in \varrho_1 \circ \varrho_2 \iff (\exists y)\ (x, y) \in \varrho_1 \wedge (y, z) \in \varrho_2.$$

A set of relations on a set A closed with respect to all operations defined by primitive positive formulas is called a *relational clone*. Given a set Γ of relations, the smallest relational clone containing Γ is said to be *generated by* Γ and denoted by $\langle \Gamma \rangle$.

Theorem 2.1 ([15]). *For any set of relations Γ over a finite set, and any finite set of relations $\Gamma' \subseteq \langle \Gamma \rangle$, there is a polynomial time reduction from $\mathrm{CSP}(\Gamma')$ to $\mathrm{CSP}(\Gamma)$.*

Corollary 2.2. *If a set of relation Γ is tractable, then the relational clone $\langle \Gamma \rangle$ is also tractable.*

In a sense, Corollary 2.2 amounts to saying that, in the study of $\mathrm{CSP}(\Gamma)$, a reasonable strategy is to concentrate on relational clones rather than arbitrary families of relations. In many cases, this strategy has proved to be successful: for example, in [15] it is shown that Schaefer's results [38] on CSPs over a two-element domain can easily be deduced from Theorem 2.1 and the classical description of clones on a two-element set [37].

2.2 From relational clones to finite algebras

An n-ary operation $f : A^n \to A$ on a set A *preserves* an m-ary relation $\varrho \subseteq A^m$ (or ϱ is *invariant* under f) if applying f to the rows of an arbitrary $m \times n$-matrix over A whose columns belong to ϱ, one always gets a column in ϱ:

$$
\begin{array}{cccc}
a_{11} & a_{12} & \cdots & a_{1n} \\
a_{21} & a_{22} & \cdots & a_{2n} \\
\cdots & \cdots & \cdots & \cdots \\
a_{m1} & a_{m2} & \cdots & a_{mn} \\
\mathbin{\rotatebox[origin=c]{90}{$\in$}} & \mathbin{\rotatebox[origin=c]{90}{$\in$}} & & \mathbin{\rotatebox[origin=c]{90}{$\in$}} \\
\varrho & \varrho & & \varrho
\end{array}
\implies
\begin{array}{c}
f(a_{11}, a_{12}, \ldots, a_{1n}) \\
f(a_{21}, a_{22}, \ldots, a_{2n}) \\
\cdots \\
f(a_{m1}, a_{m2}, \ldots, a_{mn}) \\
\mathbin{\rotatebox[origin=c]{90}{$\in$}} \\
\varrho
\end{array}.
$$

This can be also expressed by saying that $(\varrho; f)$ is a subalgebra of the m-th direct power of the algebra $(A; f)$. For a given set of operations, F, the set of all relations invariant under operations from F is denoted by $\mathrm{Inv}\, F$. Conversely, for a set of relations, Δ, the set of all operations preserving relations from Δ is denoted by $\mathrm{Pol}\, \Delta$.

It is well known [36, 40] that, for any finite algebra $\mathbb{A} = (A; F)$, the set $\mathrm{Inv}\, F$, that is, the set of subalgebras of finite direct powers of $\mathbb{A}$, forms a relational clone. Moreover, for any relational clone Δ over a finite set A, there is a finite algebra $\mathbb{A} = (A; F)$ such that $\Delta = \mathrm{Inv}\, F$. In fact, F can be chosen to be $\mathrm{Pol}\, \Delta$.

Thus, every finite algebra $\mathbb{A}$ determines a subclass $\mathrm{CSP}(\mathrm{Inv}\, F)$ of the CSP which will also be denoted by $\mathrm{CSP}(\mathbb{A})$. We say that the algebra $\mathbb{A}$ is *tractable* [*NP-complete*] if the set $\mathrm{Inv}\, F$ is tractable [NP-complete]. Conversely, for any problem class of the form $\mathrm{CSP}(\Gamma)$, where Γ is a family of relations over a finite set A, there is a corresponding finite algebra (for instance, $(A; \text{Pol}\,\Gamma)$), and the tractability [NP-completeness] of the problem class depends only on this algebra.

Example 2.2. Every left zero semigroup $\mathbb{A} = (A; \cdot)$ with $|A| > 1$ is NP-complete.

Recall that the multiplication in left zero semigroups is simply the projection on the left component: $x \cdot y = x$ for all $x, y \in A$. Clearly, this multiplication preserves **every** relation on A. Therefore if $|A| = 2$ the class $\text{CSP}(\mathbb{A})$ contains the NP-complete problem $\text{CSP}(\{\nu\})$, cf. Example 1.6, and for $|A| > 2$ we can refer to Example 1.5 since in this case $\text{CSP}(\mathbb{A})$ contains the NP-complete problem $\text{CSP}(\{\neq_A\})$.

The reader may be surprised by the fact that a fairly obvious algebra such as the two-element left zero semigroup, say, turns out to be intractable in our sense. However, there is no contradiction in that. The correspondence $F \mapsto \text{Inv}\, F$, $\Delta \mapsto \text{Pol}\,\Delta$ between operations and relations is in fact a Galois connection, so 'weak' operations preserve many relations (among which one can easily find NP-complete ones) while 'strong' operations that preserve fewer relations may lead to tractable constraints.

Thus, computational complexity theory suggests a fresh look at finite algebras, and describing algebraic properties which reflect in the tractability or the intractability of the corresponding problem classes has proved to be an interesting task which is highly non-trivial in its general setting. The papers [21, 16, 19, 20, 17, 4, 7, 5] provide a number of partial results going in this direction. In the next section we shall apply some of these results to classify tractable semigroups.

3 Tractable semigroups

Originally the theory and practice of the CSP dealt with sets of relations rather than algebras. It will be fair to say that from this point of view, finite semigroups do not arise naturally: given a set Γ of relations, the requirement that Γ is invariant under a binary associative operation appears to be rather exotic. However, in studying many important problems which can be posed for arbitrary algebras, semigroups constitute a 'golden middle' case that is sufficiently rich and diverse with respect to the problem under consideration and at the same time is not hopelessly difficult as is the general case. Classifying tractable algebras in general and even tractable groupoids seems to be hard enough; on the other hand, Feder and Vardi [10] proved that an arbitrary finite group is tractable. This suggests that it may be reasonable to have a look at the intermediate case of semigroups.

We have seen that non-trivial finite left zero semigroups are NP-complete (Example 2.2). Of course, the same conclusion holds for non-trivial finite right zero semigroups. The following observation from [4] allows one to enlarge the supply of NP-complete semigroups.

Lemma 3.1 ([4], Theorem 3.9). *Let $\mathbb{A}$ be a finite algebra. If $\mathbb{A}$ is tractable then so is every subalgebra of $\mathbb{A}$. If $\mathbb{A}$ has an NP-complete subalgebra then $\mathbb{A}$ is NP-complete itself.*

Corollary 3.2. *If a finite semigroup $\mathbb{S} = (S; \cdot)$ contains two distinct idempotents e and f such that either $ef = e$ and $fe = f$ or $ef = f$ and $fe = e$, then $\mathbb{S}$ is NP-complete.*

Proof. The subsemigroup $(\{e, f\}; \cdot)$ of $\mathbb{S}$ is a non-trivial left or right zero semigroup which is NP-complete. Now Lemma 3.1 applies. □

We conclude that every finite semigroup $\mathbb{S}$ which is not NP-complete must satisfy the following two conditions:

$$ef = e^2 = e \wedge fe = f^2 = f \Rightarrow e = f, \tag{1}$$

$$ef = f^2 = f \wedge fe = e^2 = e \Rightarrow e = f. \tag{2}$$

Finite semigroups satisfying (1) and (2) are called *block groups*. We refer the reader to [35] for a brilliant survey of the theory of block groups which reveals remarkable connections between block groups, progroup topology of the free monoid and recognizable languages. A jewel of the theory is a characterization of block groups in terms of power semigroups of groups due to Henckell and Rhodes [14], see also [13]. As the main result of the present paper shows, from the computational complexity point of view, block groups also play a distinguished role:

Theorem 3.3. *A finite semigroup is tractable if and only if it is a block group.*

Proof. The 'only if' part follows from Corollary 3.2. One of the crucial ingredients for the proof of the 'if' part is a recent result by Dalmau [7] which may be formulated as follows. By a *set function* on a set A we mean an arbitrary mapping from the power set $\mathcal{P}(A)$ to A. A set function $f : \mathcal{P}(A) \to A$ *preserves* an m-ary relation $\varrho \subseteq A^m$ if applying f to the sets of elements in the rows of an arbitrary $m \times n$-matrix over A whose columns belong to ϱ, one always gets a column in ϱ:

$$
\begin{array}{cccc}
a_{11} & a_{12} & \ldots & a_{1n} \\
a_{21} & a_{22} & \ldots & a_{2n} \\
\multicolumn{4}{c}{\ldots\ldots\ldots\ldots\ldots} \\
a_{m1} & a_{m2} & \ldots & a_{mn} \\
\cap & \cap & & \cap \\
\varrho & \varrho & & \varrho
\end{array}
\quad\Longrightarrow\quad
\begin{array}{c}
f(\{a_{11}, a_{12}, \ldots, a_{1n}\}) \\
f(\{a_{21}, a_{22}, \ldots, a_{2n}\}) \\
\ldots \\
f(\{a_{m1}, a_{m2}, \ldots, a_{mn}\}) \\
\cap \\
\varrho
\end{array}
$$

Proposition 3.4 ([7]). *Let Γ be a set of relations on a finite set A. If there is a set function preserving all relations from Γ, then there exists a uniform algorithm that solves arbitrary instance from $\mathrm{CSP}(\Gamma)$ in polynomial time; in particular, Γ is tractable.*

Now let $\mathbb{S} = (S; \cdot)$ be a block group and let $|S| = k$. For arbitrary $s_1, \ldots, s_n \in S$, we consider the element

$$f(s_1, \ldots, s_n) = (s_1^{k!} \cdots s_n^{k!})^{k!}. \tag{3}$$

First we prove that the right-hand side of (3) depends only on the set $\{s_1, \ldots, s_n\}$; in other words, that f is in fact a set function. Let $e_i = s_i^{k!}$; it is known (and easy to check) that each e_i is an idempotent. Then it is easy to realize that in order to show that f is a set function, it suffices to verify that

$$(e_1 \cdots e_n)^{k!} = (e_{1\pi} \cdots e_{n\pi})^{k!} \tag{4}$$

for every permutation π of the set $\{1, \ldots, n\}$. We shall employ the following observation:

Lemma 3.5 ([30], proof of Proposition 2.3). *If $e_1, \ldots, e_n$ are idempotents of a k-element block group. Then for every $i = 1, \ldots, n$*

$$(e_1 \cdots e_n)^{k!} = (e_1 \cdots e_n)^{k!} e_1 \cdots e_i.$$

Now, for an arbitrary permutation π of the set $\{1,\ldots,n\}$, we have

$$\begin{aligned}
(e_1\cdots e_n)^{k!} &= (e_1\cdots e_n)^{k!}e_1\cdots e_{n\pi} && \text{by Lemma 3.5,} \\
&= e_1\cdots e_{n\pi}(e_{n\pi+1}\cdots e_n e_1\cdots e_{n\pi})^{k!} \\
&= (e_{n\pi+1}\cdots e_n e_1\cdots e_{n\pi})^{k!} && \text{by the dual of Lemma 3.5,} \\
&= (e_{n\pi+1}\cdots e_n e_1\cdots e_{n\pi})^{k!}e_{n\pi} && \text{by Lemma 3.5,} \\
&= (e_1\cdots e_n)^{k!}e_{n\pi} \\
&= (e_1\cdots e_n)^{k!}e_{1\pi}\cdots e_{n\pi} && \text{by the same trick applied} \\
& && \text{to } e_{(n-1)\pi},\ldots,e_{1\pi}, \\
&= (e_1\cdots e_n)^{k!}(e_{1\pi}\cdots e_{n\pi})^{k!} \\
&= (e_{1\pi}\cdots e_{n\pi})^{k!} && \text{by an analogous process} \\
& && \text{applied on the left.}
\end{aligned}$$

Thus, the equality (4) is proved, and the formula (3) indeed defines a set function.

Let Γ be the set of all relations on S which are invariant under the multiplication in $\mathbb{S}$; then $\mathrm{CSP}(\Gamma) = \mathrm{CSP}(\mathbb{S})$ and the set function f respects all relations from Γ because by (3) f is expressed in terms of the multiplication. Hence Proposition 3.4 applies. □

Let us list several consequences of Theorem 3.3 and its proof.

A set Γ of relations is said to be *globally tractable* if a uniform algorithm solves arbitrary instances from $\mathrm{CSP}(\Gamma)$ in polynomial time. (This requirement is stronger than the usual tractability which we defined in Subsection 1.3 as the existence, for each finite subset $\Gamma' \subseteq \Gamma$, of a polynomial time algorithm which solves the problem $\mathrm{CSP}(\Gamma')$.) The most well known example of a globally tractable set of relations is provided by linear systems over a field, cf. Example 1.7. We notice that our proof shows that the set of all relations which are invariant under the multiplication in a block group is globally tractable. In particular, constraints imposed by an arbitrary finite group are globally tractable — a conclusion which cannot be extracted from the tractability proof by Feder and Vardi [10].

Combining Theorem 3.3 and Corollary 3.2, we deduce a semigroup analogue of the Dichotomy Theorem: a finite semigroup is either tractable or NP-complete.

Given the multiplication table of a finite semigroup $\mathbb{S} = (S;\cdot)$, one can check if $\mathbb{S}$ satisfies the conditions (1) and (2) in quadratic (of $|S|$) time. Thus, by Theorem 3.3, the problem of recognizing the tractability of a semigroup is tractable.

We conclude with a short comparison between the semigroup case and the general situation. For arbitrary finite algebras, several tempting conjectures have been formulated in [4]. We reproduce here the most important of them.

An algebra $\mathbb{B} = (B; F)$ where F is a collection of permutations of the set B is said to be a *G-set*. A G-set $\mathbb{B}$ is non-trivial if $|B| > 1$. Since permutations preserve the disequality relation $\neq_B$ and also the not-all-equal relation ν (in the case $|B| = 2$), it follows from Examples 1.5 and Example 1.6 that every non-trivial G-set is NP-complete. The main conjecture of [4] basically means that a non-trivial G-set hides behind every NP-complete algebra. Recall that a *factor* (semigroupists would say *divisor*) of an algebra $\mathbb{A}$ is a homomorphic image of a subalgebra of $\mathbb{A}$.

G-set Conjecture. Every NP-complete finite algebra $\mathbb{A}$ has a factor $\mathbb{A}'$ such that $\text{CSP}(\mathbb{A}') = \text{CSP}(\mathbb{B})$ for a non-trivial G-set $\mathbb{B}$.

The next conjecture is inspired by Schaefer's Dichotomy Theorem for Boolean relations (which confirms the conjecture for two-element algebras):

Dichotomy Conjecture. Every finite algebra is either tractable or NP-complete.

It is known that if both the G-set conjecture and the Dichotomy Conjecture hold true, then so does also the following

Algorithmic Conjecture. There exits a algorithm which, given the Cayley tables of a finite algebra $\mathbb{A} = (A; F)$, determines whether $\mathbb{A}$ is tractable in polynomial of $|A|$ time.

Finally, we mention

Local-Global Conjecture. Every tractable finite algebra is globally tractable.

Returning to the semigroup case, we see that Theorem 3.3 and its consequences are consistent with all the above conjectures.

Acknowledgments

This paper was presented in the third named author's lectures at the School on Algorithmic Aspects of the Theory of Semigroups and its Applications held at Centro Internacional de Matemática, Coimbra, Portugal, in May 2001 as a part of the Thematic term on Semigroups, Algorithms, Automata and Languages. He is very much indebted to the organizers of the term: Prof. Gracinda M. S. Gomes, Prof. Jean-Éric Pin and Prof. Pedro V. Silva for the invitation and for great help during his stay in Coimbra.

References

[1] J. F. Allen, Natural Language Understanding, Benjamin Cummings, Menlo Park, 1994.
[2] J. Almeida, Finite Semigroups and Universal Algebra [Ser. in Algebra 3], World Scientific, Singapore–New Jersey–London–Hong Kong, 1995.
[3] W. Bibel, *Constraint satisfaction from a deductive viewpoint*, Artif. Intelligence **35** (1988) 401–413.
[4] A. A. Bulatov, P. G. Jeavons, and A. A. Krokhin, *Constraint satisfaction problems and finite algebras*, in U.Montanari et al. (eds.), Proc. 27th Int. Colloq. on Automata, Languages and Programming — ICALP'00 [Lect. Notes Comp. Sci. **1853**], Springer-Verlag, Berlin–Heidelberg–N.Y., 2000, 272–278.
[5] A. A. Bulatov, P. G. Jeavons, and A. A. Krokhin, *Classifying the complexity of constraints using universal algebras*, SIAM J. Computing (submitted).
[6] S. Burris and H. P. Sankappanavar, A Course in Universal Algebra, Springer-Verlag, Berlin–Heidelberg–N.Y, 1981.
[7] V. Dalmau, Computational Complexity of Problems over Generalised Formulas, PhD thesis, Department LSI, Universitat Politècnica de Catalunya, Barcelona, 2000.
[8] R. Dechter and A. Dechter, *Structure-driven algorithms for truth maintenance*, Artif. Intelligence **82** (1996) 1–20.
[9] R. Dechter and J. Pearl, *Network-based heuristics for constraint satisfaction problems*, Artif. Intelligence **34** (1988) 1–38.
[10] T. Feder and M. Y. Vardi, *The computational structure of monotone monadic SNP and constraint satisfaction: A study through datalog and group theory*, SIAM J. Computing **28** (1998) 57–104.
[11] M. R. Garey and D. S. Johnson, Computers and Intractability: A Guide to the Theory of NP-Completeness, W. H. Freeman and Company, San Francisco, 1979.
[12] M. Gyssens, P. Jeavons, and D. Cohen, *Decomposing constraint satisfaction problems using database techniques* Artif. Intelligence 66 (1994) 57–89.
[13] K. Henckell, S. W. Margolis, J.-É. Pin, and J. Rhodes, *Ash's Type II*

Theorem, profinite topologies and Malcev products, Int. J. Algebra and Computation **1** (1991) 411–436.
[14] K. Henckell and J. Rhodes, *The theorem of Knast, the PG = BG and Type II Conjectures*, in J.Rhodes (ed.), Monoids and Semigroups with Applications, World Scientific, Singapore, 1991, 453–463.
[15] P. G. Jeavons, *On the algebraic structure of combinatorial problems*, Theor. Comp. Sci. **200** (1998) 185–204.
[16] P. G. Jeavons and D. A. Cohen, *An algebraic characterization of tractable constraints*, in Computing and Combinatorics. 1st Int. Conf. — COCOON'95 (Xian, China, August 1995) [Lect. Notes Comput. Sci. **959**], Springer-Verlag, Berlin–Heidelberg–N.Y., 1995, 633–642.
[17] P. G. Jeavons, D. A. Cohen, and M. C. Cooper, *Constraints, consistency and closure*, Artif. Intelligence **101** (1998) 251–265.
[18] P. G. Jeavons, D. A. Cohen, and M. Gyssens, *A unifying framework for tractable constraints*, in Proc. 1st Int. Conf. on Constraint Programming — CP'95 (Cassis, France, September 1995) [Lect. Notes Comput. Sci. **976**], Springer-Verlag, Berlin–Heidelberg–N.Y., 1995, 276–291.
[19] P. G. Jeavons, D. A. Cohen, and M. Gyssens, *Closure properties of constraints*, J. ACM **44** (1997) 527–548.
[20] P. G. Jeavons, D. A. Cohen, and J. K. Pearson, *Constraints and universal algebra*, Ann. Math. Artif. Intelligence **24** (1998) 51–67.
[21] P. G. Jeavons and M. C. Cooper, *Tractable constraints on ordered domains*, Artif. Intelligence **79** (1995) 327–339.
[22] L. Kirousis, *Fast parallel constraint satisfaction*, Artif. Intelligence, **64** (1993) 147–160.
[23] Ph. G. Kolaitis and M. Y. Vardi, *A game-theoretic approach to constraint satisfaction*, in Proc. 17th National Conf. on Artif. Intelligence and 12th Innovative Applications of Artif. Intelligence Conf. — AAAI'00/IAAI'00 (Austin, TX, July 30–August 3, 2000), MIT Press, Cambridge, 2000, 175–181.
[24] Ph. G. Kolaitis and M. Y. Vardi, *Conjunctive-query containment and constraint satisfaction*, J. Comput. Syst. Sci. **61** (2000) 302–332.
[25] P. B. Ladkin and R. D. Maddux, *On binary constraint problems*, J. ACM **41** (1994) 435–469.
[26] A. K. Mackworth, *Consistency in networks of relations*, Artif. Intelligence **8** (1977) 99–118.
[27] A. K. Mackworth, *Constraint satisfaction*, in S.C.Shapiro and D.Eckroth (eds.), Encyclopedia of Artificial Intelligence, Vol.I, John Wiley & Sons, N.Y., 1987, 285–293.
[28] A. K. Mackworth, *The logic of constraint satisfaction*, Artif. Intelligence

58 (1992) 3–20.
[29] A. K. Mackworth and E. C. Freuder, *The complexity of constraint satisfaction revisited*, Artif. Intelligence **59** (1993) 57–62.
[30] S. W. Margolis and J.-É. Pin, *Varieties of finite monoids and topology for the free monoid*, in K.Byleen et al. (eds.), Proc. of the 1984 Marquette Conf. on Semigroups, Marquette Univ., Milwaukee, 1984, 113–130.
[31] U. Montanari, *Networks of constrains: Fundamental properties and applications to picture processing*, Information Sciences **7** (1974) 95–132.
[32] B. A. Nadel, *Constraint satisfaction in Prolog: Complexity and theory-based heuristics*, Information Sciences **83** (1995) 113–131.
[33] B. A. Nadel and J. Lin, *Automobile transmission design as a constraint satisfaction problem: Modelling the kinematics level*, Artif. Intelligence for Engineering Design, Analysis and Manufacturing **5** (1991) 137–171.
[34] C. H. Papadimitriou, Computational complexity, Addison-Wesley Publishing Company, Reading–Menlo Park–N.Y., 1994.
[35] J.-É. Pin, $BG = PG$, *a success story*, in J.Fountain (ed.), Semigroups, Formal Languages and Groups [NATO ASI Ser., Ser. C: Math. Phys. Sci. **466**], Kluwer Academic Publishers, Dordrecht–Boston–London, 1995, 33–47.
[36] R. Pöschel and L.A. Kalužnin, Funktionen- und Relationenalgebren, VEB Deutscher Verlag der Wissenschaften, Berlin 1979; [Math. Reihe **67**] Birkhäuser Verlag, Basel–Stuttgart, 1979.
[37] E. L. Post, The two-valued iterative systems of mathematical logic [Ann. Math. Studies **5**], Princeton Univ. Press, 1941.
[38] T. J. Schaefer, *The complexity of satisfiability problems*, in Proc. 10th Ann. ACM Symp. on Theory of Computing — STOC'78, Association for Computer Machinery, N.Y., 1978, 216–226.
[39] E. Schwalb and L. Vila, *Temporal constraints: a survey*, Constraints **3** (1998) 129–149.
[40] Á. Szendrei, Clones in universal algebra [Séminaire de Mathématiques Supérieurs **99**], Les Presses de l'Université de Montréal, Montréal, 1986.
[41] E. Tsang, Foundations of constraint satisfaction, Academic Press, London, 1993.
[42] M. Y. Vardi, *Constraint satisfaction and database theory: a tutorial*, in Proc. 19th ACM SIGACT-SIGMOD-SIGART Symp. on Principles of Database Systems — PODS'00 (Dallas, TX, May 2000), 2000 (electronic, see http://www.acm.org/sigmod/pods/proc00/).

ON THE EFFICIENCY AND DEFICIENCY OF REES MATRIX SEMIGROUPS

C.M. CAMPBELL, J.D. MITCHELL AND N. RUŠKUC*

Mathematical Institute, North Haugh, St Andrews, Fife KY16 9SS, Scotland
E-mail: cmc@st-and.ac.uk, jamesm@mcs.st-and.ac.uk, nik@mcs.st-and.ac.uk

Let S be a finite simple semigroup represented as a Rees matrix semigroup $\mathcal{M}[G; I, J; P]$ over a group G. We show that if G is efficient (i.e. if it can be defined by a presentation $\langle\, A \mid R \,\rangle$ with $|R| - |A| = \mathrm{rank}(H_2(G))$) then S is also efficient. We also show how to find a minimal presentation for S in this case.

1 Introduction

In this paper we combine results from [1] and [3] in order to determine the relationship between the deficiency of a group G and the deficiency of a Rees matrix semigroup S over G; in particular, we investigate how the efficiency of S depends on the efficiency of G.

We start by defining semigroup and group presentations. Let A be an alphabet. We denote by A^+ the free semigroup on A consisting of all non-empty words over A, and by $F(A)$ the free group of all freely reduced words over $A \cup A^{-1}$ (including the empty word), where A^{-1} is an alphabet whose elements represent the inverses of elements of A. A *semigroup presentation* is an ordered pair $\langle\, A \mid R \,\rangle$, where $R \subseteq A^+ \times A^+$. If both A and R are finite then we have a *finite presentation*. A semigroup S is said to be *defined by the semigroup presentation* $\langle\, A \mid R \,\rangle$ if $S \cong A^+/\rho$, where ρ is the congruence on A^+ generated by R. Replacing A^+ by $F(A)$ in the above definitions yields the notions of a *group presentation* and of a *group defined by a presentation*. For basic facts about semigroup and group presentations see any standard introductory texts on semigroups and groups, such as [6], [8] and [10].

We define the *deficiency* of a finite presentation $\mathcal{P} = \langle A|R\rangle$ to be $|R|-|A|$. The *semigroup deficiency* of a finitely presented semigroup S is the minimum

---

*THE AUTHORS WOULD LIKE TO ACKNOWLEDGE THE FINANCIAL SUPPORT OF FUNDAÇÃO CALOUSTE GULBENKIAN (FCG), FUNDAÇÃO PARA A CIÊNCIA E A TECNOLOGIA (FCT), FACULDADE DE CIÊNCIAS DA UNIVERSIDADE DE LISBOA (FCUL) AND REITORIA DA UNIVERSIDADE DO PORTO. THE THIRD AUTHOR ACKNOWLEDGES THE PARTIAL SUPPORT OF INTAS-99/1224.

deficiency of any semigroup presentation $\mathcal{P}$ defining S:

$\mathrm{def}_S(S) = \min\{\, \mathrm{def}(\mathcal{P}) \mid \mathcal{P}$ is a finite semigroup presentation that defines $S\,\}$.

We define the group deficiency $\mathrm{def}_G(G)$ of a finitely presented group analogously, using finite group presentations:

$\mathrm{def}_G(G) = \min\{\, \mathrm{def}(\mathcal{P}) \mid \mathcal{P}$ is a finite group presentation that defines $G\,\}$.

Hence there are two notions of deficiency for a finitely presented group. We call a presentation $\mathcal{P} = \langle\, A \mid R\,\rangle$, that defines S, *minimal* if $\mathrm{def}(\mathcal{P}) = \mathrm{def}_S(S)$. If S is a finite semigroup and G is a finite group it can be shown that:

$$\mathrm{def}_S(S) \geq 0 \text{ and } \mathrm{def}_G(G) \geq 0,$$

and it can also be shown that $\mathrm{def}_S(G) \geq \mathrm{def}_G(G)$. We call a presentation $\mathcal{P}$ with $\mathrm{def}(\mathcal{P}) = 0$, a *balanced* presentation.

A better bound for the deficiency of a finite semigroup S or a finite group G is given by the rank of the second integral homology of S or G:

$$\mathrm{def}_S(S) \geq \mathrm{rank}(H_2(S)) \text{ and } \mathrm{def}_G(G) \geq \mathrm{rank}(H_2(G)). \qquad (1)$$

The first of these inequalities is due to S.J. Pride (unpublished). The second is a well known result, a proof of which may be found in [9]. A finite semigroup or group is called *efficient* if it attains this lower bound, and is called *inefficient* otherwise. Thus there are two notions of efficiency for finite groups, namely the group and semigroup efficiency. Note that the definition of efficiency is different in the case of infinite semigroups or groups, see Section 4.

It is a well known fact that a completely simple semigroup is a Rees matrix semigroup over a group. Let G be a group, I and J be index sets and P be a $|J| \times |I|$ matrix with entries p_{ji} from G. Then the semigroup of elements from $I \times G \times J$ with the multiplication

$$(i, g, j)(k, h, l) = (i, g p_{jk} h, l)$$

is called a *Rees matrix semigroup*, and is denoted $S = \mathcal{M}[G; I, J; P]$. The matrix P can be chosen to be *normal*, in other words $p_{1i} = p_{j1} = 1_G$, the identity of G, for all $i \in I$ and $j \in J$. For further details see [6].

Example 1.1. The *Fibonacci groups* $F(r, n)$ ($r \geq 2$, $n \geq 1$) are defined by the (balanced) group presentations

$$\langle\, a_1, a_2, \ldots, a_n \mid a_1 a_2 \ldots a_r = a_{r+1},\ a_2 a_3 \ldots a_{r+1} = a_{r+2}, \ldots,$$
$$a_{n-1} a_n a_1 \ldots a_{r-2} = a_{r-1},\ a_n a_1 a_2 \ldots a_{r-1} = a_r\,\rangle.$$

It was shown in [4] that the semigroup $S(r, n)$ defined by this presentation is a disjoint union of $\gcd(r, n)$ copies of $F(r, n)$. In fact, $S(r, n)$ is completely simple, and can be represented as a Rees matrix semigroup $\mathcal{M}[F(r,n); I, J; P]$ where $|J| = 1$ and $|I| = \gcd(r,n)$. Since $S(r,n)$ is defined by a balanced presentation, it is efficient whenever it is finite, which happens, of course, if, and only if, $F(r,n)$ is finite.

Example 1.2. Let S be a finite rectangular band $I \times J$, where $I = \{1,\ldots,m\}$, $J = \{1,\ldots,n\}$ and $(i,j)(k,l) = (i,l)$. Of course, S is a Rees matrix semigroup over the trivial group. In [2] the second homology of S was found to be $H_2(S) = \mathbb{Z}^{(|I|-1)(|J|-1)}$. Also, S has the following presentation:

$\langle\, y_2,\ldots,y_m, z_2,\ldots,z_n \mid z_n y_m y_2 = z_n y_m,\ y_i y_{i+1} = y_i\ (2 \leq i \leq m-1),$
$\qquad\qquad\qquad\qquad z_n y_m z_2 = z_2,\ z_j z_{j+1} = z_{j+1}\ (2 \leq j \leq n-1),$
$\qquad\qquad\qquad\qquad y_m z_n z_n y_m = y_m,$
$\qquad\qquad\qquad\qquad z_j y_i = z_n y_m \qquad\qquad\qquad (2 \leq i \leq m,\ 2 \leq j \leq n)\,\rangle$

in terms of the generators $\{\,(i,1) \mid 2 \leq i \leq m\,\}$ and $\{\,(1,j) \mid 2 \leq j \leq n\,\}$. The deficiency of the above presentation is $(|I|-1)(|J|-1)$, and hence S is efficient.

2 Presentations and homology

We begin by giving a presentation for an arbitrary finite Rees matrix semigroup $S = \mathcal{M}[G; I, J; P]$, where P is normal. Throughout we take $I = \{1, 2, \ldots, m\}$ and $J = \{1, 2, \ldots, n\}$. Let $\langle A \mid R \rangle$ be a minimal semigroup presentation defining G, and let

$$Y = \{\, y_i \mid 2 \leq i \leq m \,\} \qquad\qquad (2)$$
$$Z = \{\, z_j \mid 2 \leq j \leq n \,\}$$

be two new alphabets. It is easy to verify that the set $X = A \cup Y \cup Z$ generates S, where $a \in A$ represents $(1, a, 1) \in S$, $y_i \in Y$ represents $(i, 1_G, 1) \in S$ and $z_j \in Z$ represents $(1, 1_G, j) \in S$. If $e \in A^+$ is any word representing 1_G, then we have the following presentation for S

$$\langle X \mid R, y_i e = y_i, e y_i = e, z_j e = e, e z_j = z_j, z_j y_i = p_{ji}\ (2 \leq i \leq m,\ 2 \leq j \leq n)\,\rangle. \qquad (3)$$

For further details see [7]. Since the presentation for G is minimal, the deficiency of (3) is $\mathrm{def}_S(G) + (|I|-1)(|J|-1) + (|I|-1) + (|J|-1)$. A presentation for S with smaller deficiency was derived from (3) in [1, Proposition 4.5]:

$\langle\, X \mid R,\ e y_2 = e,\ y_i y_{i+1} = y_i\ (2 \leq i \leq m-1),$
$\qquad e z_2 = z_2,\ z_j z_{j+1} = z_{j+1}\ (2 \leq j \leq n-1),$
$\qquad y_m z_n e = y_m,$ $\qquad\qquad\qquad\qquad\qquad\qquad\qquad\qquad\qquad$ (4)
$\qquad z_j y_i = p_{ji} \qquad\qquad (2 \leq i \leq m,\ 2 \leq j \leq n)\,\rangle.$

The deficiency of this presentation is $\mathrm{def}_S(G) + (|I| - 1)(|J| - 1) + 1$. It is important to note that this presentation has only been shown to define S for finite G; see [1] for further details.

We end this section by giving the second integral homology of an arbitrary Rees matrix semigroup $S = \mathcal{M}[G; I, J; P]$. Presentation (3) can be used to obtain a complete rewriting system, which in turn can be used to compute $H_2(S)$; see [12] for further details. This method was employed in the proof of the following proposition.

Proposition 2.1. *The second integral homology of $S = \mathcal{M}[G; I, J; P]$ is given by*

$$H_2(S) = H_2(G) \times \mathbb{Z}^{(|I|-1)(|J|-1)}.$$

The proof of this result may be found in [1].

As a consequence of this result and (1), it follows for a finite semigroup S that:

$$\mathrm{def}_S(S) \geq \mathrm{rank}(H_2(G)) + (|I| - 1)(|J| - 1).$$

3 Deficiency and efficiency

We now consider the following questions: what is the relationship between the group and semigroup deficiency of a finite group G? In other words, under what circumstances does G have a minimal semigroup presentation $\mathcal{P} = \langle A|R \rangle$ where $|R| - |A| = \mathrm{def}_G(G)$? And if G has such a presentation, is it possible to eliminate one relation from presentation (4)? The answer to the first question is given in [3], from which we state some results. We begin by giving a class of semigroup presentations that define groups.

Proposition 3.1. *The semigroup presentation*

$\langle a_1, a_2, \ldots, a_n \mid a_1 = a_2\beta_1 a_2 a_1, a_2 = a_3\beta_2 a_3, a_3 = a_4\beta_3 a_4, \ldots, a_n = a_1\beta_n a_1, R \rangle$

where $n \geq 1$, $\beta_1, \beta_2, \ldots, \beta_n$ are arbitrary words over $\{a_1, a_2, \ldots, a_n\}$ and R is an arbitrary set of relations, defines a group.

The proof of this result may be found in [3].

As a consequence of this result, the semigroup deficiency of a group of non-negative deficiency may be determined.

Proposition 3.2. *Let G be a group of non-negative deficiency. Then there exists a semigroup presentation $\mathcal{P}$ that defines G, such that $\mathrm{def}(\mathcal{P}) = \mathrm{def}_G(G)$. In particular, for finite groups, $\mathrm{def}_G(G) = \mathrm{def}_S(G)$.*

Proof. It was shown in [3] that a minimal group presentation for G could be used to obtain a semigroup presentation of the type given in Proposition 3.1 without increasing the deficiency. □

Corollary 3.3. *Let G be a finite group. There exists a minimal semigroup presentation $\mathcal{P} = \langle\, A \mid R \,\rangle$ that defines G, such that R contains a relation of the form $a = aua$, where $a \in A$ and $u \in A^+$.*

Proof. From the discussion in the proof of Proposition 3.2 there exists a minimal semigroup presentation $\mathcal{Q}$ of the type given in Proposition 3.1. Let $\mathcal{P}$ denote the presentation obtained from $\mathcal{Q}$ by removing the relation $a_1 = a_2\beta_1 a_2 a_1$ and replacing it with the relation

$$a_1 = a_1\beta_n a_1 \beta_{n-1} \ldots \beta_3 a_4 \beta_2 a_3 \beta_1 a_2 a_1.$$

The following sequence of equalities may be obtained as a consequence of the relations in $\mathcal{Q}$, without using the relation $a_1 = a_2\beta_1 a_2 a_1$. Similarly, the sequence can be obtained as a consequence of the relations in $\mathcal{P}$, without using the relation $a_1 = a_1\beta_n a_1 \beta_{n-1} \ldots \beta_3 a_4 \beta_2 a_3 \beta_1 a_2 a_1$:

$$\begin{aligned}(a_1\beta_n a_1)\beta_{n-1} a_n \ldots a_4\beta_3 a_4\beta_2 a_3\beta_1 a_2 a_1 &= (a_n\beta_{n-1} a_n)\ldots a_4\beta_3 a_4\beta_2 a_3\beta_1 a_2 a_1 \\ &= \cdots = (a_4\beta_3 a_4)\beta_2 a_3\beta_1 a_2 a_1 \\ &= (a_3\beta_2 a_3)\beta_1 a_2 a_1 = a_2\beta_1 a_2 a_1.\end{aligned}$$

We see from this that $\mathcal{P}$ and $\mathcal{Q}$ define the same semigroup. We let $a = a_1$ and $u = \beta_n a_1 \beta_{n-1} \ldots \beta_3 a_4 \beta_2 a_3 \beta_1 a_2$. It follows that $\mathcal{P}$ is the desired presentation. □

We now return to a finite Rees matrix semigroup $S = \mathcal{M}[G; I, J; P]$. Modifying the method of [1] we show that it is possible to replace two of the relations in (4) with a single relation, hence reducing the deficiency by one. Let us choose our minimal presentation $\langle\, A \mid R \,\rangle$ in accord with Corollary 3.3. Let X denote the generating set used in (3).

Theorem 3.4. *The presentation*

$$\begin{aligned}\langle\, X \mid &R\backslash\{aua = a\}, \\ &auy_2 z_n a = a,\ y_i y_{i+1} = y_i\ (2 \leq i \leq m-1), \\ &auz_2 = z_2,\ z_j z_{j+1} = z_{j+1}\ (2 \leq j \leq n-1), \\ &y_m z_n au = y_m,\ z_j y_i = p_{ji}\ (2 \leq i \leq m, 2 \leq j \leq n)\,\rangle\end{aligned} \quad (5)$$

defines the finite Rees matrix semigroup $S = \mathcal{M}[G; I, J; P]$ and has deficiency $\mathrm{def}_S(G) + (|I|-1)(|J|-1)$.

Proof. We begin by showing that the relation $auy_2z_na = a$ is a consequence of the relations in (4). First, we let $e = au$ in (4), giving that

$$auy_2 = au, \ auz_2 = z_2, \ y_m z_n au = y_m.$$

Observe that by repeatedly applying the relation $y_iy_{i+1} = y_i$ we obtain

$$y_2 = y_2y_3 = y_2y_3y_4 = \cdots = y_2y_3y_4\cdots y_{m-3}y_{m-2}y_{m-1}y_m$$
$$= y_2y_3y_4\cdots y_{m-3}y_{m-2}y_m = y_2y_3y_4\cdots y_{m-3}y_m = \cdots = y_2y_m.$$

In the same way, by repeatedly applying the relation $z_jz_{j+1} = z_{j+1}$, we obtain the relation $z_n = z_2z_n$. We use these five new relations to show

$$z_na = z_2z_na = auz_2z_na = auz_na = auy_2z_na = auy_2y_mz_na = auy_2y_mz_naua$$
$$= auy_2y_ma = auy_2a = aua = a.$$

It follows that

$$auy_2z_na = auy_2a = aua = a.$$

We have proved that the semigroup defined by (5) is a homomorphic image of S. We now show the converse, in other words that the relations in (4) are consequences of the relations in (5). First we see that

$$auy_2 = auy_2y_m = auy_2y_mz_nau = auy_2z_nau = au.$$

It follows that

$$aua = au(auy_2z_na) = auauz_na = auz_na = auy_2z_na = a.$$

The final part of the result follows immediately from Proposition 3.2. □

As an immediate corollary of the above result we obtain the main result of this paper, which relates the efficiency of G to the efficiency of S.

Theorem 3.5. *Let $S = \mathcal{M}[G;I,J;P]$ be a finite simple semigroup expressed as a Rees matrix semigroup over a group G. If G is efficient then S is efficient.*

4 Open questions and examples

We ask the natural question: does the converse of Theorem 3.5 hold?

Open Question 4.1. *Does there exist an efficient Rees matrix semigroup over an inefficient group?*

Or, does the semigroup deficiency of G provide a better lower bound for the deficiency of a Rees matrix semigroup?

Open Question 4.2. *Does the inequality*
$$\operatorname{def}_S(S) \geq \operatorname{def}_S(G) + (|I| - 1)(|J| - 1)$$
hold for all Rees matrix semigroups?

Note that if the answer to Question 4.2 is yes then the answer to Question 4.1 is no.

In the particular case of $1 \times n$ Rees matrix semigroups we can show that the answer to Question 4.2 is indeed yes.

Proposition 4.3. *Let $S = G \times Z$ be the direct product of a finite group G and a finite right zero semigroup Z. Then $\operatorname{def}_S(S) = \operatorname{def}_G(G)$. In particular, S is efficient if, and only if, G is efficient.*

Proof. Let $\langle\, A \mid R\,\rangle$ be a semigroup presentation for G with $|R| - |A| = \operatorname{def}_S(G) = \operatorname{def}_G(G)$. Assume that $|Z| = n$. Let $X = \{\,x_2, \ldots, x_n\,\}$ be a new alphabet, and let $x_1 \in A^+$ be a word representing 1_G, the identity of G. It is now a routine exercise to show that the presentation:

$$\langle\, A, X \mid R,\ x_1 x_2 = x_2,\ x_2 x_3 = x_3, \ldots,\ x_{n-1} x_n = x_n,\ x_n x_1 = x_1\,\rangle$$

defines S and has deficiency $|R| - |A|$. Thus $\operatorname{def}_S(S) \leq \operatorname{def}_G(G)$. For the reverse inequality it is sufficient to note that any presentation for S, when treated as a group presentation, defines G; this follows from [5]. Finally, by Proposition 2.1, we have $H_2(S) = H_2(G)$, and the proposition follows. □

We conclude the paper by stating some open questions about infinite Rees matrix semigroups. In the case of infinite groups and semigroups the basic bounds for deficiency are:

$$\operatorname{def}_G(G) \geq \operatorname{rank}(H_2(G)) - \operatorname{rank}_{\mathbb{Z}}(H_1(G)) \qquad (6)$$
$$\operatorname{def}_S(S) \geq \operatorname{rank}(H_2(S)) - \operatorname{rank}_{\mathbb{Z}}(H_1(S)). \qquad (7)$$

Here $\operatorname{rank}_{\mathbb{Z}}(A)$ denotes the $\mathbb{Z}$-rank of a finitely generated abelian group A, i.e. the number of infinite cyclic factors in the canonical decomposition of A into the direct sum of cyclic groups. An (infinite) group or semigroup is said to be efficient if the equalities hold in (6) or (7), respectively. Inequality (6) is well known (see [11]); (7) was proved analogously by S.J. Pride (unpublished). Also it is well known that $H_1(G) = G/G'$, the abelianisation of G. If S is defined by a presentation $\langle\, A \mid R\,\rangle$, then $H_1(S) = H_1(K)$ where K is the group defined by $\langle\, A \mid R\,\rangle$; this follows immediately from the resolution used in [12]. Note that K is the largest group homomorphic image of S, if S has one (which, of course, is the case for Rees matrix semigroups).

As observed above, the proof that presentation (4) defines S, relies on the finiteness of G, and this presentation was crucial in the proof of Theorem 3.4. However, neither the presentation (4) nor Theorem 3.4 hold for Rees matrix semigroups over infinite groups. To see this, let G be an infinite group which is efficient both as a group and as a semigroup, and for which $\text{rank}_{\mathbb{Z}}(G/G') > 0$; one such example is the free abelian group of any rank greater than 2. Let P be any normalised $|J| \times |I|$ matrix, the entries of which form a generating set for G, and form the Rees matrix semigroup $S = \mathcal{M}[G; I, J; P]$. Now it is well known (and easy to verify) that the maximal group homomorphic image of S is G/N, where N is the normal subgroup of G generated by the entries of P, and hence it is trivial. Now we have

$$\begin{aligned}\text{def}_S(S) &\geq \text{rank}(H_2(S)) - \text{rank}_{\mathbb{Z}}(H_1(S))) \\ &= \text{rank}(H_2(G)) + (|I|-1)(|J|-1) - 0 \\ &> \text{rank}(H_2(G)) + (|I|-1)(|J|-1) - \text{rank}_{\mathbb{Z}}(G/G') \\ &= \text{def}_S(G) + (|I|-1)(|J|-1),\end{aligned}$$

as required.

This leads us to ask the following question:

Open Question 4.4. *What is the deficiency of a Rees matrix semigroup over an infinite group G?*

A significant first step towards solving this question would be achieved by answering:

Open Question 4.5. *Is it possible to find a presentation $\mathcal{P}$ that defines $S = \mathcal{M}[G; I, J; P]$, for arbitrary G, where $\text{def}(\mathcal{P})$ is smaller than the deficiency of presentation (3)?*

Acknowledgements

The authors wish to thank the two anonymous referees for their helpful suggestions.

References

[1] H. Ayik, C.M. Campbell, J.J. O'Connor and N. Ruškuc, On the efficiency of finite simple semigroups, *Turkish J. Math.* **24**, 129-146 (2000).
[2] H. Ayik, C.M. Campbell, J.J. O'Connor and N. Ruškuc, Minimal presentations and efficiency of semigroups, *Semigroup Forum* **60**, 231-242

(2000).
- [3] C.M. Campbell, J.D. Mitchell and N. Ruškuc, On defining groups efficiently without inverses, *Math. Proc. Cambridge Philos. Soc.*, to appear.
- [4] C.M. Campbell, E.F. Robertson, N. Ruškuc and R.M. Thomas, Fibonacci semigroups, *J. Pure Appl. Algebra* **94**, 49-57 (1994).
- [5] C.M. Campbell, E.F. Robertson, N. Ruškuc and R.M. Thomas, Semigroup and group presentations, *Bull. London Math. Soc.* **27**, 46-50 (1995).
- [6] J.M. Howie, *Fundamentals of Semigroup Theory*, (Clarendon Press, Oxford, 1995).
- [7] J.M. Howie and N. Ruškuc, Constructions and presentations for monoids, *Comm. Algebra* **22**, 6209-6224 (1994).
- [8] G. Lallement, *Semigroups and Combinatorial Applications*, (John Wiley, New York, 1979).
- [9] D.J.S. Robinson, *A Course in the Theory of Groups*, (Springer, New York, 1996).
- [10] J.J. Rotman, *The Theory of Groups: an Introduction*, (Allyn and Bacon, Boston, 1965).
- [11] J.J. Rotman, *An Introduction to Homological Algebra*, (Academic Press, New York, 1979).
- [12] C.C. Squier, Word problems and a homological finiteness condition for monoids, *J. Pure Appl. Algebra* **49**, 201-217 (1987).

SOME PSEUDOVARIETY JOINS INVOLVING GROUPS AND LOCALLY TRIVIAL SEMIGROUPS

JOSÉ CARLOS COSTA [*]

Departamento de Matemática, Universidade do Minho, Campus de Gualtar, 4700-320 Braga, Portugal
E-mail: jcosta@math.uminho.pt

In this paper, we present some computations of joins of the form $\mathcal{L}\mathbf{I} \vee \mathbf{H} \vee \mathbf{V}$ where $\mathcal{L}\mathbf{I}$ is the pseudovariety of locally trivial semigroups, $\mathbf{H}$ is a pseudovariety of groups, and $\mathbf{V}$ is a pseudovariety of semigroups satisfying some pseudoidentities. Similar results are obtained for the pseudovarieties $\mathbf{K}$, of semigroups in which idempotents are left zeros, and its dual $\mathbf{D}$, in place of $\mathcal{L}\mathbf{I}$.

1 Introduction

The problem of calculation of joins of pseudovarieties is in general very difficult and seems to depend greatly on the specific pseudovarieties involved. In fact, Albert, Baldinger and Rhodes [1] gave an example of two decidable pseudovarieties whose join is undecidable. This unexpected result has brought some attention and a new interest to the join operator and revealed the apparent impossibility of finding general results for doing computations. The join operator has received the attention of many authors and several calculations and answers to decision problems are known at the moment, but always involving specific pseudovarieties. In particular, the pseudovariety of groups is one of the most visited, from which the article of Almeida and Weil [6] constitutes a remarkable example. In that paper, some difficult computations involving pseudovarieties of groups are obtained using elaborate techniques based on a study of profinite groups. For other calculations, we must also cite the work of Almeida, Azevedo, Steinberg, Trotter, Volkov, Zeitoun and the

---

[*]THE AUTHOR WOULD LIKE TO ACKNOWLEDGE THE FINANCIAL SUPPORT OF FUNDAÇÃO CALOUSTE GULBENKIAN (FCG), FUNDAÇÃO PARA A CIÊNCIA E A TECNOLOGIA (FCT), FACULDADE DE CIÊNCIAS DA UNIVERSIDADE DE LISBOA (FCUL) AND REITORIA DA UNIVERSIDADE DO PORTO. WORK SUPPORTED, IN PART, BY FCT THROUGH THE *CENTRO DE MATEMÁTICA DA UNIVERSIDADE DO MINHO*, AND BY THE FCT AND POCTI APPROVED PROJECT POCTI/32817/MAT/2000 WHICH IS COMPARTICIPATED BY THE EUROPEAN COMMUNITY FUND FEDER.

author [2, 4, 5, 8, 9, 10, 15, 16, 17, 18]. A survey on the subject is presented in [19].

For a pseudovariety **H** of groups, we denote by $\mathcal{D}\mathbf{ReH}$ the pseudovariety of all finite semigroups S in which each regular $\mathcal{D}$-class is a subsemigroup of S isomorphic to the direct product of a rectangular band by a group of **H**. We notice that a finite semigroup S belongs to $\mathcal{D}\mathbf{ReH}$ if and only if each regular $\mathcal{D}$-class of S is a subsemigroup of S whose idempotents form a subsemigroup and all subgroups of S lie in **H**. We recall that $\mathcal{D}\mathbf{ReG}$ is usually denoted by $\mathcal{D}\mathbf{O}$, where **G** is the pseudovariety of all finite groups.

The purpose of this paper is the computation of some pseudovariety joins involving groups and locally trivial semigroups. To be more precise, if **H** is a pseudovariety of groups and **V** is a subpseudovariety of $(\mathbf{CR}\textcircled{m}\mathbf{N}) \cap \mathcal{D}\mathbf{ReH}$, where $\mathbf{CR}\textcircled{m}\mathbf{N}$ is the Mal'cev product of the pseudovarieties of completely regular semigroups and of nilpotent semigroups, we give a description of the joins of the form $\mathcal{L}\mathbf{I} \vee \mathbf{H} \vee \mathbf{V}$ in terms of a set of pseudoidentities defining **V**. This work constitutes an extension of our paper [11], where we dealt with pseudovariety joins of the form $\mathcal{L}\mathbf{I} \vee \mathbf{V}$. In that paper, we introduced an operator $\mathcal{U}$ defined on the lattice of pseudovarieties as follows: if **V** is a pseudovariety defined by a set Σ of pseudoidentities, then $\mathcal{U}\mathbf{V}$ is the pseudovariety defined by the pseudoidentities of the form $a^\omega bxcd^\omega = a^\omega bycd^\omega$ where $x = y$ is an element of Σ. The pseudovariety $\mathcal{U}\mathbf{V}$ is a natural upper bound of the join $\mathcal{L}\mathbf{I} \vee \mathbf{V}$. In many cases the equality $\mathcal{L}\mathbf{I} \vee \mathbf{V} = \mathcal{U}\mathbf{V}$ holds. This is the case (see [11]) when **V** satisfies a certain cancellation property, which seems to be frequent: it is verified, for instance, by any pseudovariety of groups, by the pseudovariety **Com** of commutative semigroups and by the pseudovariety **J** of $\mathcal{J}$-trivial semigroups. However the equality $\mathcal{U}\mathbf{V} = \mathcal{L}\mathbf{I} \vee \mathbf{V}$ is not valid in general since the equality $\mathcal{L}\mathbf{I} \vee \mathbf{V} = \mathcal{U}\mathbf{V} \cap (\mathbf{CR}\textcircled{m}\mathbf{N})$ is verified by any subpseudovariety **V** of $\mathbf{CR}\textcircled{m}\mathbf{N}$, as shown by the author in [11]. Thus, for example, the pseudovariety $\mathcal{U}\mathbf{Sl}$ constitutes a strict upper bound of $\mathcal{L}\mathbf{I} \vee \mathbf{Sl}$, where **Sl** is the pseudovariety of semilattices.

Here, we introduce a new operator which is related with the joins of the form $\mathcal{L}\mathbf{I} \vee \mathbf{H} \vee \mathbf{V}$. Given a pseudovariety **V** defined by a set Σ of pseudoidentities, we define $\mathcal{U}^\omega \mathbf{V}$ to be the pseudovariety defined by the pseudoidentities $(a^\omega bxcd^\omega)^\omega = (a^\omega bycd^\omega)^\omega$ where $x = y$ is an element of Σ. For each pseudovariety **H** of groups, $\mathcal{U}^\omega \mathbf{V}$ is a natural upper bound of the join $\mathcal{L}\mathbf{I} \vee \mathbf{H} \vee \mathbf{V}$ and it will be used to compute it when **V** is a subpseudovariety of $(\mathbf{CR}\textcircled{m}\mathbf{N}) \cap \mathcal{D}\mathbf{ReH}$.

2 Preliminaries

In this section, we will briefly recall some notions and results concerning the objects that we will be dealing with. We presuppose familiarity with the basic definitions of finite semigroup theory and implicit operations. For a comprehensive treatment of these subjects, the reader is referred to the books of Eilenberg [12], Pin [13] and Almeida [3].

Let $\mathbf{V}$ be a pseudovariety. The semigroup of (n-ary) implicit operations on $\mathbf{V}$ is denoted by $\hat{F}_n(\mathbf{V})$ and the subsemigroup of its explicit elements is denoted by $F_n(\mathbf{V})$. A pseudoidentity is a formal identity $x = y$ of elements of $\hat{F}_n(\mathbf{S})$, where $\mathbf{S}$ denotes the pseudovariety of all finite semigroups. If x and y are explicit operations, $x = y$ is called an identity.

For a pseudovariety $\mathbf{V}$, let $\mathcal{L}\mathbf{V}$ (resp. $\mathcal{L}_l\mathbf{V}$, $\mathcal{L}_r\mathbf{V}$) be the pseudovariety of all finite semigroups S such that eSe (resp. eS, Se) belongs to $\mathbf{V}$ for each idempotent e of S. We notice that $\mathbf{K} = \mathcal{L}_l\mathbf{I}$ and $\mathbf{D} = \mathcal{L}_r\mathbf{I}$, where $\mathbf{I}$ denotes the trivial pseudovariety. According to Pin and Weil [14], $\mathbf{CR}\textcircled{w}\mathbf{N}$ is defined by the pseudoidentity $ab^\omega c = (ab^\omega c)^{\omega+1}$. Thus $\mathbf{CR}\textcircled{w}\mathbf{N} = \mathcal{L}_l\mathbf{CR} \cap \mathcal{L}_r\mathbf{CR}$ since the pseudovarieties $\mathcal{L}_l\mathbf{CR}$ and $\mathcal{L}_r\mathbf{CR}$ are defined, respectively, by the pseudoidentities $a^\omega b = (a^\omega b)^{\omega+1}$ and $ba^\omega = (ba^\omega)^{\omega+1}$.

We recall the following well-known and useful observation about the pseudovarieties $\mathcal{L}\mathbf{I}$, $\mathbf{K}$ and $\mathbf{D}$. See, for instance, [3, pages 88-91] for a proof.

Lemma 2.1. *Let $\mathbf{V}$ be a pseudovariety containing $\mathbf{N}$.*

(1) *The pseudovariety $\mathbf{V}$ does not satisfy any non-trivial identity. Furthermore, if $\mathbf{V}$ satisfies a pseudoidentity $y = u$, with u explicit, then y and u are equal.*

(2) *If $\mathbf{V}$ is the pseudovariety $\mathcal{L}\mathbf{I}$ (resp. $\mathbf{K}$, $\mathbf{D}$) and $\mathbf{V}$ satisfies a pseudoidentity $x = y$, with $x, y \in \hat{F}_n(\mathbf{S})$ not explicit, then there exist $r, s, u, v \in \hat{F}_n(\mathbf{S})$, with r and s not explicit, such that $x = rus$ and $y = rvs$ (resp. $x = ru$ and $y = rv$, $x = us$ and $y = vs$).*

The next result, which is part of Corollary 5.6.2 of [3], is also very useful.

Lemma 2.2. *If $x \in \hat{F}_n(\mathbf{S})$ is a non-explicit operation, then there exist $x_1, x_2, x_3 \in \hat{F}_n(\mathbf{S})$ such that $x = x_1 x_2^\omega x_3$.*

We shall need the following result, proved by Almeida and Azevedo, which can be found in [7].

Proposition 2.3. *Let $\mathbf{V}$ be a subpseudovariety of $\mathcal{D}\mathbf{ReG}$ and let $x, y \in \hat{F}_n(\mathbf{S})$. If x and y are regular when restricted to $\mathbf{V}$, then $\mathbf{V}$ satisfies $x = y$ if* and only if **V** satisfies $x^\omega = y^\omega$ and $\mathbf{V} \cap \mathbf{G}$ satisfies $x = y$.

3 The computations

Let Σ be a set of pseudoidentities. In [11] the author considered the following pseudovarieties

$$\mathbf{U}_\Sigma = [\![a^\omega bxcd^\omega = a^\omega bycd^\omega \mid x = y \in \Sigma]\!]$$
$$\mathbf{U}_{\Sigma,l} = [\![a^\omega bx = a^\omega by \mid x = y \in \Sigma]\!]$$

where, for each $x = y \in \Sigma$ (with $x, y \in \hat{F}_n(\mathbf{S})$ for a certain natural number n), $a, b, c, d \notin \hat{F}_n(\mathbf{S})$. The author showed that, if Σ and Σ' are two sets of pseudoidentities defining the same pseudovariety, then $\mathbf{U}_\Sigma = \mathbf{U}_{\Sigma'}$ and $\mathbf{U}_{\Sigma,l} = \mathbf{U}_{\Sigma',l}$. As a consequence, there were introduced the following operators on the lattice of pseudovarieties: for a pseudovariety $\mathbf{V} = [\![\Sigma]\!]$,

$$\mathcal{U}\mathbf{V} = \mathbf{U}_\Sigma \quad \text{and} \quad \mathcal{U}_l\mathbf{V} = \mathbf{U}_{\Sigma,l}.$$

Let us now define, for a set Σ of pseudoidentities,

$$\mathbf{U}_\Sigma^\omega = [\![(a^\omega bxcd^\omega)^\omega = (a^\omega bycd^\omega)^\omega \mid x = y \in \Sigma]\!]$$
$$\mathbf{U}_{\Sigma,l}^\omega = [\![(a^\omega bx)^\omega = (a^\omega by)^\omega \mid x = y \in \Sigma]\!]$$

where, for each $x = y \in \Sigma$ (with $x, y \in \hat{F}_n(\mathbf{S})$ for a certain natural number n), $a, b, c, d \notin \hat{F}_n(\mathbf{S})$.

Proposition 3.1. *Let Σ and Σ' be two sets of pseudoidentities defining the same pseudovariety, i.e., such that $[\![\Sigma]\!] = [\![\Sigma']\!]$. Then $\mathbf{U}_\Sigma^\omega = \mathbf{U}_{\Sigma'}^\omega$ and $\mathbf{U}_{\Sigma,l}^\omega = \mathbf{U}_{\Sigma',l}^\omega$.*

Proof. Using the fact that the correspondence $x \mapsto x^\omega$, defined on $\hat{F}_n(\mathbf{S})$, is continuous, the proof is a simple adaptation of the proof of Proposition 3.1 in [11]. It suffices to make in that proof the obvious changes, namely the substitution of the implicit operations of the form $a^\omega bzcd^\omega$ by $(a^\omega bzcd^\omega)^\omega$ and the pseudovarieties of the form $\mathbf{U}_\Phi$ by $\mathbf{U}_\Phi^\omega$. □

As a consequence of this result, we may introduce the following two operators on the lattice of pseudovarieties. For a pseudovariety $\mathbf{V} = [\![\Sigma]\!]$, let

$$\mathcal{U}^\omega \mathbf{V} = \mathbf{U}_\Sigma^\omega \quad \text{and} \quad \mathcal{U}_l^\omega \mathbf{V} = \mathbf{U}_{\Sigma,l}^\omega.$$

Notice that, for pseudovarieties $\mathbf{V}$ and $\mathbf{W}$, the following properties hold:

- $\mathcal{L}\mathbf{I} \vee \mathbf{V} \subseteq \mathcal{U}\mathbf{V} \subseteq \mathcal{U}^\omega \mathbf{V}$ and $\mathbf{K} \vee \mathbf{V} \subseteq \mathcal{U}_l \mathbf{V} \subseteq \mathcal{U}_l^\omega \mathbf{V}$;

- $\mathcal{L}\mathbf{I} \vee \mathbf{G} \vee \mathbf{V} \subseteq \mathcal{U}^\omega \mathbf{V}$ and $\mathbf{K} \vee \mathbf{G} \vee \mathbf{V} \subseteq \mathcal{U}_l^\omega \mathbf{V}$;

- if $\mathbf{V} \subseteq \mathbf{W}$, then $\mathcal{U}^\omega \mathbf{V} \subseteq \mathcal{U}^\omega \mathbf{W}$ and $\mathcal{U}_l^\omega \mathbf{V} \subseteq \mathcal{U}_l^\omega \mathbf{W}$.

Remark 3.1. We notice that, if in the definition of the operator $\mathcal{U}^\omega$ one would substitute (only) the ω over parenthesis by an arbitrary positive integer k, one would obtain a new (well-defined) operator. That is, the operators $\mathcal{U}$ and $\mathcal{U}^\omega$ can be obtained as members of a countable family of operators $(\mathcal{U}^k)_{k \in \mathbb{N} \cup \{\omega\}}$, where, for a pseudovariety $\mathbf{V} = [\![\Sigma]\!]$,

$$\mathcal{U}^k \mathbf{V} = [\![(a^\omega bxcd^\omega)^k = (a^\omega bycd^\omega)^k \mid x = y \in \Sigma]\!].$$

We notice also that the operators $\mathcal{U}_l$ and $\mathcal{U}_l^\omega$ admit symmetrical versions $\mathcal{U}_r$ and $\mathcal{U}_r^\omega$: for a pseudovariety $\mathbf{V} = [\![\Sigma]\!]$,

$$\mathcal{U}_r \mathbf{V} = [\![xba^\omega = yba^\omega \mid x = y \in \Sigma]\!]$$

and

$$\mathcal{U}_r^\omega \mathbf{V} = [\![(xba^\omega)^\omega = (yba^\omega)^\omega \mid x = y \in \Sigma]\!].$$

We can now present the announced characterization of the pseudovarieties of the form $\mathcal{L}\mathbf{I} \vee \mathbf{H} \vee \mathbf{V}$, with $\mathbf{V} \subseteq (\mathbf{CR}\text{\textcircled{m}}\mathbf{N}) \cap \mathcal{D}\mathbf{ReH}$. We denote by $\mathbf{K}_1$ (resp. $\mathbf{D}_1$) the pseudovariety of left zero (resp. right zero) semigroups.

Theorem 3.2. *Let* $\mathbf{H}$ *be a pseudovariety of groups and let* $\mathbf{V}$ *be a subpseudovariety of* $\mathcal{L}_l \mathbf{CR} \cap \mathcal{L}_r \mathbf{CR} \cap \mathcal{D}\mathbf{ReH}$. *Then*

$$\mathcal{L}\mathbf{I} \vee \mathbf{H} \vee \mathbf{V} = \mathcal{U}^\omega \mathbf{V} \cap \mathcal{L}_l \mathbf{CR} \cap \mathcal{L}_r \mathbf{CR} \cap \mathcal{D}\mathbf{ReH}$$
$$= \mathcal{U}^\omega \mathbf{V} \cap \mathcal{L}_l(\mathbf{D}_1 \vee \mathbf{H} \vee \mathbf{V}) \cap \mathcal{L}_r(\mathbf{K}_1 \vee \mathbf{H} \vee \mathbf{V}) \cap \mathcal{D}\mathbf{ReH}$$

$$\mathbf{K} \vee \mathbf{H} \vee \mathbf{V} = \mathcal{U}_l^\omega \mathbf{V} \cap \mathcal{L}_l \mathbf{CR} \cap \mathcal{L}_r \mathbf{CR} \cap \mathcal{D}\mathbf{ReH}$$
$$= \mathcal{U}_l^\omega \mathbf{V} \cap \mathcal{L}_l(\mathbf{D}_1 \vee \mathbf{H} \vee \mathbf{V}) \cap \mathcal{L}_r(\mathbf{K}_1 \vee \mathbf{H} \vee \mathbf{V}) \cap \mathcal{D}\mathbf{ReH}$$

$$\mathbf{D} \vee \mathbf{H} \vee \mathbf{V} = \mathcal{U}_r^\omega \mathbf{V} \cap \mathcal{L}_l \mathbf{CR} \cap \mathcal{L}_r \mathbf{CR} \cap \mathcal{D}\mathbf{ReH}$$
$$= \mathcal{U}_r^\omega \mathbf{V} \cap \mathcal{L}_l(\mathbf{D}_1 \vee \mathbf{H} \vee \mathbf{V}) \cap \mathcal{L}_r(\mathbf{K}_1 \vee \mathbf{H} \vee \mathbf{V}) \cap \mathcal{D}\mathbf{ReH}.$$

Proof. The other cases being similar, we show the result only for $\mathcal{L}\mathbf{I}$. Let

$$\mathbf{W} = \mathcal{U}^\omega \mathbf{V} \cap \mathcal{L}_l \mathbf{CR} \cap \mathcal{L}_r \mathbf{CR} \cap \mathcal{D}\mathbf{ReH}.$$

The inclusion $\mathcal{L}\mathbf{I} \vee \mathbf{H} \vee \mathbf{V} \subseteq \mathbf{W}$ is clear. For the proof of the reverse inclusion, consider a pseudoidentity $x = y$, with $x, y \in \hat{F}_n(\mathbf{S})$ for a certain natural number n, and suppose that $\mathcal{L}\mathbf{I} \vee \mathbf{H} \vee \mathbf{V}$ satisfies $x = y$. By Reiterman's Theorem, it suffices to prove that $\mathbf{W}$ satisfies $x = y$.

Since $\mathcal{L}\mathbf{I}$ satisfies $x = y$, by Lemma 2.1 (1) two cases may arise: either x and y are the same word or x and y are both not explicit. In this last case, Lemma 2.1 (2) shows that we can write $x = rus$ and $y = rvs$, for some $r, s, u, v \in \hat{F}_n(\mathbf{S})$ with r and s not explicit. Moreover, from Lemma 2.2, we can write $r = r_1 r_2^\omega r_3$ and $s = s_1 s_2^\omega s_3$, for some $r_1, r_2, r_3, s_1, s_2, s_3 \in \hat{F}_n(\mathbf{S})$. Now, we notice that $\mathcal{L}_l \mathbf{CR} \cap \mathcal{L}_r \mathbf{CR}$ satisfies

$$x = rus = r_1 r_2^\omega r_3 u s_1 s_2^\omega s_3 = (r_1 r_2^\omega)^{\omega+1} r_3 u s_1 (s_2^\omega s_3)^{\omega+1} = (r_1 r_2^\omega)^\omega x (s_2^\omega s_3)^\omega.$$

Analogously, $\mathcal{L}_l \mathbf{CR} \cap \mathcal{L}_r \mathbf{CR}$ satisfies $y = (r_1 r_2^\omega)^\omega y (s_2^\omega s_3)^\omega$. Now since $\mathbf{V}$ satisfies $x = y$ it is clear, from its definition, that $\mathcal{U}^\omega \mathbf{V}$ satisfies $((r_1 r_2^\omega)^\omega x (s_2^\omega s_3)^\omega)^\omega = ((r_1 r_2^\omega)^\omega y (s_2^\omega s_3)^\omega)^\omega$. Therefore, $\mathbf{W}$ satisfies

$$x^\omega = ((r_1 r_2^\omega)^\omega x (s_2^\omega s_3)^\omega)^\omega = ((r_1 r_2^\omega)^\omega y (s_2^\omega s_3)^\omega)^\omega = y^\omega.$$

We notice that, since $\mathbf{W} \subseteq \mathcal{L}_l \mathbf{CR} \cap \mathcal{L}_r \mathbf{CR} = \mathbf{CR}\textcircled{m}\mathbf{N}$, an immediate consequence of Lemma 2.2 is that every non-explicit operation is a regular element when restricted to $\mathbf{W}$. Therefore, since $\mathbf{H} = \mathbf{W} \cap \mathbf{G}$ satisfies $x = y$ and $\mathbf{W} \subseteq \mathcal{D}\mathbf{ReG}$, we deduce from Proposition 2.3 that $\mathbf{W}$ satisfies $x = y$.

To prove the second equality concerning $\mathcal{L}\mathbf{I}$ let us note that the following relations are valid

$$\mathcal{L}\mathbf{I} \vee \mathbf{H} \vee \mathbf{V} \subseteq \mathcal{L}_l(\mathbf{D}_1 \vee \mathbf{H} \vee \mathbf{V}) \subseteq \mathcal{L}_l(\mathcal{L}_l \mathbf{CR}) = \mathcal{L}_l \mathbf{CR}.$$

Analogously, $\mathcal{L}\mathbf{I} \vee \mathbf{H} \vee \mathbf{V} \subseteq \mathcal{L}_r(\mathbf{K}_1 \vee \mathbf{H} \vee \mathbf{V}) \subseteq \mathcal{L}_r \mathbf{CR}$. Now, these inclusions and the first equality concerning $\mathcal{L}\mathbf{I}$ imply the second one. □

4 Conclusion and open questions

Theorem 3.2 presents a description of the pseudovarieties of the form $\mathcal{L}\mathbf{I} \vee \mathbf{H} \vee \mathbf{V}$, with $\mathbf{V} \subseteq (\mathbf{CR}\textcircled{m}\mathbf{N}) \cap \mathcal{D}\mathbf{ReH}$. We must notice that these calculations were obtained in [11] but under a different form. In fact Theorem 4.1 of [11] presents a description of the pseudovarieties $\mathcal{L}\mathbf{I} \vee \mathbf{W}$, with $\mathbf{W} \subseteq \mathbf{CR}\textcircled{m}\mathbf{N}$. Thus, in particular when $\mathbf{W} = \mathbf{H} \vee \mathbf{V}$, Theorem 4.1 of [11] presents a characterization of $\mathcal{L}\mathbf{I} \vee \mathbf{H} \vee \mathbf{V}$, with $\mathbf{V} \subseteq (\mathbf{CR}\textcircled{m}\mathbf{N}) \cap \mathcal{D}\mathbf{ReH}$. However, this characterization is obtained in terms of the pseudovariety $\mathcal{U}\mathbf{W} = \mathcal{U}(\mathbf{H} \vee \mathbf{V})$, and so in terms of a set of pseudoidentities defining $\mathbf{H} \vee \mathbf{V}$. Therefore Theorem 3.2 above gives effectively a new description of the joins $\mathcal{L}\mathbf{I} \vee \mathbf{H} \vee \mathbf{V}$, since it presents them in terms of the pseudovariety $\mathcal{U}^\omega \mathbf{V}$, and so in terms of a set of pseudoidentities defining $\mathbf{V}$. Therefore, the operator $\mathcal{U}^\omega$ seems to be more appropriate to characterize the pseudovarieties $\mathcal{L}\mathbf{I} \vee \mathbf{H} \vee \mathbf{V}$.

As to the pseudovarieties $\mathbf{V}$ not contained in $(\mathbf{CR}\textcircled{m}\mathbf{N}) \cap \mathcal{D}\mathbf{ReH}$, some calculations of the form $\mathcal{L}\mathbf{I} \vee \mathbf{H} \vee \mathbf{V}$ are known from [11], in terms of the operator $\mathcal{U}$. For instance, as a consequence of the considerations of Section 5 in [11], we deduce the equality

$$\mathcal{L}\mathbf{I} \vee \mathbf{G} \vee \mathbf{J} = \mathcal{U}(\mathbf{G} \vee \mathbf{J}).$$

Trotter and Volkov [17] have shown that the pseudovariety $\mathbf{G} \vee \mathbf{J}$ is not finitely based, while Steinberg [16] has proved that $\mathbf{G} \vee \mathbf{J}$ is definable by a recursive set of pseudoidentities. So, $\mathcal{L}\mathbf{I} \vee \mathbf{G} \vee \mathbf{J}$ is also definable by a recursive set of pseudoidentities. We propose the following question, in terms of the operator $\mathcal{U}^\omega$.

Problem 4.1. *Does the equality* $\mathcal{L}\mathbf{I} \vee \mathbf{G} \vee \mathbf{J} = \mathcal{U}^\omega \mathbf{J} \cap \mathcal{D}\mathrm{ReG}$ *hold?*

The positive answer to this question would give a finite basis for $\mathcal{L}\mathbf{I} \vee \mathbf{G} \vee \mathbf{J}$.

References

[1] D. Albert, R. Baldinger and J. Rhodes, *Undecidability of the identity problem for finite semigroups*, J. Symbolic Logic **57** (1992) 179-192.

[2] J. Almeida, *Some pseudovariety joins involving the pseudovariety of finite groups*, Semigroup Forum **37** (1988) 53-57.

[3] J. Almeida, Finite Semigroups and Universal Algebra, World Scientific, Singapore, 1994.

[4] J. Almeida and A. Azevedo, *The join of the pseudovarieties of $\mathcal{R}$-trivial and $\mathcal{L}$-trivial semigroups*, J. Pure and Applied Algebra **60** (1989) 129-137.

[5] J. Almeida, A. Azevedo and M. Zeitoun, *Pseudovariety joins involving $\mathcal{J}$-trivial semigroups and completely regular semigroups*, Int. J. Algebra and Computation **9** (1999) 99-112.

[6] J. Almeida and P. Weil, *Reduced factorizations in free profinite groups and join decompositions of pseudovarieties*, Int. J. Algebra and Computation **4** (1994) 375-403.

[7] A. Azevedo, *Operations preserving homomorphisms on the class of finite semigroups* **DS**, in J. Almeida (Ed.), Actas II Encontro de Algebristas Portugueses, Universidade do Porto, Porto (1987) 33-43.

[8] A. Azevedo, *The join of the pseudovariety* **J** *with permutative pseudovarieties*, in J. Almeida and al (Eds.), Lattices, Semigroups and Universal Algebra, Plenum, New York (1990) 1-11.

[9] A. Azevedo and M. Zeitoun, *Three examples of join computations*, Semigroup Forum **57** (1998) 249-277.

[10] J. Costa, *Free profinite semigroups over some classes of semigroups locally in $\mathcal{D}\mathbf{G}$*, Int. J. Algebra and Computation **10** (2000) 491-537.

[11] J. Costa, *Some pseudovariety joins involving locally trivial semigroups*, Semigroup Forum, to appear.
[12] S. Eilenberg, Automata, Languages and Machines, vol. B, Academic Press, New York, 1976.
[13] J.-É. Pin, Varieties of Formal Languages, Plenum, New York and North Oxford, London, 1986.
[14] J.-É. Pin and P. Weil, *Profinite semigroups, Mal'cev products and identities*, J. Algebra **182** (1996) 604-626.
[15] B. Steinberg, *On pointlike sets and joins of pseudovarieties*, Int. J. Algebra and Computation **8** (1998) 203-231.
[16] B. Steinberg, *On algorithmic problems for joins of pseudovarieties*, Semigroup Forum **62** (2001) 1-40.
[17] P. Trotter and M. Volkov, *The finite basis problem in the pseudovariety joins of aperiodic semigroups with groups*, Semigroup Forum **52** (1996) 83-91.
[18] M. Zeitoun, *On the decidability of the membership problem of the pseudovariety* $\mathbf{J} \vee \mathbf{B}$, Int. J. Algebra and Computation 5 (1995) 47-64.
[19] M. Zeitoun, *On the join of two pseudovarieties*, in J. Almeida and al (Eds.), Semigroups, Automata and Languages, World Scientific, Singapore (1996) 281-288.

PARTIAL ACTION OF GROUPS ON RELATIONAL STRUCTURES: A CONNECTION BETWEEN MODEL THEORY AND PROFINITE TOPOLOGY

THIERRY COULBOIS*

*Équipe de logique, université Paris 7,
2 place Jussieu, 75251, Paris Cédex 05, France
E-mail:* coulbois@logique.jussieu.fr

This note gives an overview of the links between the study of the profinite topology of free groups and the techniques of extension of partial isomorphisms and partial actions of groups. A few open problems are also included.

1 Introduction

This note is the text of a lecture delivered by the author at the workshop "Model Theory, Profinite Topology and Semigroups" held in Coimbra (Portugal) in June 2001. Its aim is to give an overview of the links between the study of the profinite topology of free groups and the techniques of extension of partial isomorphisms and partial actions of groups. There are no proofs in this note since most of the results are published elsewhere as indicated in the text. Moreover, though we give all the definitions used here, it may be useful for the reader to go back to the original articles to get more details and examples.

The second purpose of this note is to explain the perspectives of this area. Therefore we include at the end a few open problems.

The origin of this research comes from the interaction between model theory, profinite topology of groups and formal languages. Indeed it came out from the conjecture of J.-É. Pin and C. Reutenauer [19], [18] on the profinite topology of free groups (which is now known as Ribes-Zalesskiĭ's theorem) that they needed in order to solve the type II conjecture of J. Rhodes. Then B. Herwig and D. Lascar gave another proof of this theorem using some model theory associated with automorphisms of first order structures as it was

---

*THE AUTHOR WOULD LIKE TO ACKNOWLEDGE THE FINANCIAL SUPPORT OF FUNDAÇÃO CALOUSTE GULBENKIAN (FCG), FUNDAÇÃO PARA A CIÊNCIA E A TECNOLOGIA (FCT), FACULDADE DE CIÊNCIAS DA UNIVERSIDADE DE LISBOA (FCUL) AND REITORIA DA UNIVERSIDADE DO PORTO.

done by E. Hrushovski in his proof that the class of graphs has the extension property (see below).

These problems also received attention from people that come from inverse and finite semigroup theory. C. Ash [4] was the first to prove the type II conjecture. For connections with semigroup theory one can refer to articles [8] and [17].

In fact, the question of extending partial isomorphisms of structures and the study of automorphism groups was already popular among model theorists. It is quite natural, while trying to classify first-order structures to look at their automorphism groups. This has been done in various ways. One can refer to the article by D. Lascar [15] to obtain information in this direction and to learn about the small index property.

Section 2 is devoted to defining relational structures and their partial isomorphisms. Then we introduce the extension property as was defined by B. Herwig and D. Lascar [9]. The reader should refer to their work for all that concerns extension property and profinite topology of free groups. In Section 3 we deal with the profinite topology of groups. We describe several properties of free groups and we give examples of groups having the same properties. Partial actions are the core of Section 4. We present there the results that link extension properties and the profinite topology of groups. A precise and detailed presentation of these topics can be found in works by the author [5, 6]. The last two sections present perspectives of this research field. In Section 5, we consider a wide class of extension properties by defining $\mathcal{T}$-free structures. They relate to the profinite topology of groups through left systems. For $\mathcal{T}$-free structures and left systems our main sources are the work of B. Herwig and D. Lascar [9] and that of J. Almeida and M. Delgado [2]. The last section contains some open problems.

2 Relational structures

2.1 Some definitions

A *language* is a (finite) set of symbols together with their arity.

A *relational structure* M is a set endowed with interpretations of relational symbols from a given language. For a given symbol R we denote by R_M or simply R its interpretation in M.

Example 2.1. A *graph* is a relational structure in the language $\mathcal{L} = \{R\}$ where R is a binary relation. With this definition a graph is oriented and it is said to be unoriented if the relation R is symmetric.

A *substructure* A of M is a subset of M where each relation symbol is interpreted as the restriction of its interpretation in M. M is said to be an *extension* of A.

A *partial isomorphism* of a relational structure is an isomorphism between two substructures.

We denote by $\mathrm{PI}(M)$ the inverse monoid of all partial isomorphisms of a relational structure M. This inverse monoid is equipped with the usual partial order, which can be defined alternatively by the inclusion of graphs: we say that q is an *extension* of p if the underlying graph of the function p is a subset of the graph of the function q.

If A is a substructure of a relational structure M and p is in $\mathrm{PI}(M)$ we define the restriction of p to A and we denote it by $p{\restriction}A$ as follows:

$$\forall x, y \in A, \ p{\restriction}A(x) = y \iff p(x) = y.$$

$\mathrm{Aut}(M)$ is the subgroup of $\mathrm{PI}(M)$ of all automorphisms of the relational structure M.

In this note we will use a particular kind of relational structures invented by B. Herwig and D. Lascar: *the n-partitioned structures*. The language for these structures is $\mathcal{L} = \{\longrightarrow, U_1, \ldots, U_n\}$ where $\longrightarrow$ is a binary relation and $U_1, \ldots, U_n$ are unary predicates. An $\mathcal{L}$-structure M is said to be n-partitioned if the predicates U_i define a partition of M and the relation $\longrightarrow$ holds between elements satisfying U_i and elements satisfying U_{i+1} ($i = 1, \ldots, n$, $n+1 = 1$).

A n-partitioned structure is n-*cycle free* if it does not contains n elements $s_1, \ldots, s_n$ such that

$$s_1 \longrightarrow s_2 \longrightarrow \cdots \longrightarrow s_n \longrightarrow s_1.$$

2.2 Extension properties

To link these relational structures and their partial isomorphisms to the profinite topology of free groups, we use the extension property for a class of structures.

A class $\mathcal{C}$ of structures has the *extension property* if given any finite structure A of $\mathcal{C}$ and a finite collection $p_1, \ldots, p_r$ of partial isomorphisms of A, the following properties are equivalent:

(1) there exists an extension M of A (possibly infinite) in $\mathcal{C}$ and automorphisms $\bar{p}_1, \ldots, \bar{p}_r$ of M that extend $p_1, \ldots, p_r$;
(2) there exists a finite extension B of A in $\mathcal{C}$ and automorphisms $\tilde{p}_1, \ldots, \tilde{p}_r$ of B that extend $p_1, \ldots, p_r$.

It is obvious that the class of sets has the extension property. Indeed one can always take B equal to A and extend partial isomorphisms of a finite set to automorphisms.

The second result concerning extension property was proved by E. Hrushovski [10]: the class of graphs has the extension property.

In these first two examples the first condition of the definition is always satisfied, which means that the second is as well.

It is a result of B. Herwig and D. Lascar [9] that the class of n-cycle free n-partitioned structures has the extension property. This is a complicated result which was proved in order to get a new proof of the result of L. Ribes and P. Zalesskiĭ that will be mentioned in the sequel. Unlike what happens in the first two examples, here condition 1 of the definition is not always satisfied.

B. Herwig and D. Lascar [9] obtained some even stronger result for the extension property. They proved that the class of $\mathcal{T}$-free structures (of which the class of n-cycle free n-partitioned structures is a particular case) has the extension property. We will wait until Section 5.2 before giving the definition of $\mathcal{T}$-free structures.

3 Profinite topology

The purpose of this note is to recall the deep link between the profinite topology of groups and extension of partial isomorphisms. We begin with free groups which may be more familiar and easier. Then we will deal with the general case of groups and for that we will introduce the notion of partial action of a group on a structure.

The profinite topology on a group G is the coarsest topology for which any mapping from G into a finite discrete group is continuous. A basis of clopen neighborhoods of the identity element is given by the finite index subgroups.

3.1 Free groups

Many results are known on the profinite topology of free groups. It is known to be Hausdorff. This means that a free group is residually finite, in other words that for any non-trivial element x of a free group F there exists a finite group G and a morphism π from F to G that maps x to a non-trivial element.

It is also known that any finitely generated subgroup of a free group is closed with respect to its profinite topology. This means that a free group is LERF (locally extended residually finite), in other words that for any finitely generated subgroup H of a free group F and any element x of F which is not in H, there exists a finite group G and a morphism π from F to G which separates H and x:

$$\pi(x) \notin \pi(H).$$

The second of these results was obtained in the pioneering article of M. Hall [7] where the profinite topology is introduced.

This second result is tightly linked with the extension property for the class of sets. Indeed let F be a free group on an alphabet Σ. To any subgroup H of F we can associate the set M of left cosets of H and for any letter of the alphabet we define an automorphism of M by left multiplication. If H is finitely generated and x is an element of $F\backslash H$, we can define the finite subset A of M whose elements are the cosets wH where w is a suffix of one of the generators of H or of x. For any letter we can now define a partial isomorphism of A which is the restriction of the automorphism of M.

The extension property for sets gives us a finite set B containing A and for each letter of the alphabet an automorphism of B that extends the corresponding partial isomorphism of A. This is enough to define a morphism π from F into the symmetric group on B which is a finite group. And it is easy to check that

$$\pi(x) \notin \pi(H).$$

Example 3.1. Let $F = <a, b, c>$, $H = <ab, bbb, c^{-1}bc>$ and $x = cba^{-1}$. Figure 1 shows the set A and the partial isomorphisms. It is clear from the figure and from the proof, that we constructed a finite inverse automata. The connection between finitely generated subgroups of free groups and inverse automata or inverse semigroups is well known and detailed in [17].

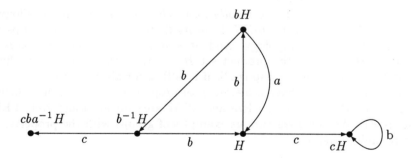

Figure 1. An example of partial isomorphisms associated with a finitely generated subgroup of a free group.

The sketch of the proof that free groups are LERF and Example 3.1 were given here because they illustrate the correspondence between finitely generated subgroups of free groups and partial isomorphisms. They show how extension of partial isomorphisms can be used to prove that some subsets of free groups are closed for the profinite topology.

After these old results on the profinite topology of free groups, R. Gitik and E. Rips proved:

Theorem 1. [12] *Let H and K be finitely generated subgroups of a free group F. The double coset HK is closed for the profinite topology of F.*

We say that a free group is double coset separable or RZ_2.

This result is tightly linked with the extension property for the class of graphs. Indeed it is a consequence of Hrushovski's result.

L. Ribes and P. Zalesskiĭ proved an even stronger result:

Theorem 2. [20] *Let $H_1, \ldots, H_n$ be finitely generated subgroups of a free group F. The product set $H_1 \cdots H_n$ is closed for the profinite topology of F.*

We say that a free group is RZ_n.

In turn this result is tightly linked with the extension property for the class of n-cycle free n-partitioned structures. B. Herwig and D. Lascar used the extension property for n-cycle free n-partitioned structures to give a new proof of Theorem 2.

3.2 Profinite topology of groups

We are here interested in groups whose profinite topology has similar properties to those of free groups.

A group is said to be *residually finite (RF)* if its profinite topology is Hausdorff. It is said to be *LERF* if all its finitely generated subgroups are closed for its profinite topology. And, it is RZ_n if for all finitely generated subgroups $H_1, \ldots, H_n$, the product set $H_1 \cdots H_n$ is closed for the profinite topology. We say that a group is RZ if it is RZ_n for all integer n.

It is clear that we have a hierarchy of properties. Every group which is RZ_{n+1} is also RZ_n. Properties RZ_1 and LERF are equivalent and every LERF group is RF. We give here various examples of groups with the properties of their profinite topology.

Example 3.2.

(1) It is obvious that finite groups and finitely generated abelian groups are RZ.

(2) It is a result of L. Ribes and P. Zalesskiĭ [20] that free groups are RZ.
(3) A consequence (see [5]) of the previous example is that $GL_2(\mathbb{Z})$ and $SL_2(\mathbb{Z})$ are RZ.
(4) As a corollary of results by the author [5] one gets that surface groups are RZ;
(5) J. Lennox and J. Wilson [16] proved that polycyclic-by-finite groups are RZ_2;
(6) The author proved in his thesis [5] that free metabelian groups are LERF;
(7) It is a result of K. Gruenberg [13] that free solvable groups are RF.

4 Partial actions

In the sketch of the proof that free groups are LERF given in Section 3.1, we constructed a morphism from a free group on an alphabet Σ into an automorphism group by giving the image of each of the elements of Σ. In the general case, for a group G with a set of generators S, to define a morphism into an automorphism group it is not enough to give an automorphism for each element of S. It is the purpose of our notion of partial action to make possible similar constructions.

Let $\bar{\varphi}$ be an action of a group G on a relational structure M. Let A be a finite substructure of M and S a finite subset of G. The *partial action* φ of G on A induced by $\bar{\varphi}$ and S is the application φ from G into $PI(A)$ such that for all g in G and for all a, a' in A, we have

$$\varphi(g)(a) = a'$$

if and only if there exists a finite collection $s_1, \ldots, s_r$ in S, $\epsilon_1, \ldots, \epsilon_r$ in $\{\pm 1\}$ and $a_0, \ldots, a_r$ in A such that

$$g = s_1^{\epsilon_1} \cdots s_r^{\epsilon_r}, \ a_0 = a, \ a_r = a' \text{ and, } \bar{\varphi}(s_i^{\epsilon_i})(a_{i-1}) = a_i.$$

Although the definition is technical, it is a rather natural notion as illustrated by the following properties of partial actions.

Property 1. *In the conditions of the previous definition, we have:*
(1) $\forall s \in S, \varphi(s) = \bar{\varphi}(s) \lceil A$;
(2) $\forall g \in G, \varphi(g) \leq \bar{\varphi}(g)$ and $\varphi(g)^{-1} = \varphi(g^{-1})$;
(3) $\forall g, h \in G, \varphi(g) \circ \varphi(h) \leq \varphi(gh)$.

As did J. Kellendonk and M. Lawson [14] this property can be used to define a notion of partial action which is very similar to ours.

Thanks to these three properties it is clear that the *stabilizer* of an element a of A which is defined as

$$\mathrm{Stab}_\varphi(a) = \{g \in G \mid \varphi(g)(a) = a\}$$

is a subgroup of G.

Moreover the requirement that S and A are finite enforces that a stabilizer is a finitely generated subgroup of G.

We can extend the order on the inverse monoid of partial isomorphisms into an order on partial actions. We say that $\tilde\varphi$ is an extension of φ if for all g in G, $\tilde\varphi(g)$ is an extension of $\varphi(g)$ as a partial isomorphism of A.

The partial action of G on A can now also be defined as the smallest application from G into $\mathrm{PI}(A)$ satisfying the above three properties. Of course an action (of a finitely generated group G) is a special case of a partial action. We use the notion of partial action to define an extension property of a given group similar to that of partial isomorphisms.

We say that a group G has the *extension property* for a class of relational structures $\mathcal{C}$ if given any partial action φ of G on a finite element A of $\mathcal{C}$ which is induced by an action $\bar\varphi$ of G on a structure M of $\mathcal{C}$ there exists a finite extension B of A which is in $\mathcal{C}$ and an action $\tilde\varphi$ of G on B which extends φ.

The existence of $\bar\varphi$ and M plays here the role of the first condition in the definition of the extension property in Section 2.2.

The extension property for a class of structures, as defined in the previous section, is the extension property of all free groups for this class of structures. In the sequel, we will try to understand the meaning of the extension properties for different classes of structures. We will indeed try to translate these properties into statements about the profinite topology of groups.

R. Gitik gave the following characterization of LERF groups:

Theorem 3. [11] *A group has the extension property for sets if and only if it is LERF.*

In fact R. Gitik does not use our terminology but that of labeled graphs and covers which are equivalent to our notions of partial actions and extensions.

Going a little further one can translate RZ_n into an extension property:

Theorem 4. [6]

(1) *A group has the extension property for graphs if and only if it is RZ_2.*

(2) *A group has the extension property for n-cycle free n-partitioned structures if and only if it is RZ_n.*

We will not go into the proofs (which have been published) of these three results. Although Theorem 3 and Theorem 4.1 are quite elementary, Theorem 4.2 is technical.

R. Gitik first used her result on LERF groups to give new proofs of results on LERF groups. Using the previous characterisation of groups having the RZ_n property the author was also able to prove some new results:

Theorem 5. [6, 5]
(1) *The free product of two RZ_n groups is RZ_n;*
(2) *Surface groups are RZ.*

5 Various properties

The translation between these two settings, extension properties and properties of the profinite topology of groups, is not a simple matter. We were able to understand what is the extension property for n-cycle free n-partitioned structures (recall that they have been created to prove that free groups are RZ_n), but it appears to be more difficult in other cases.

5.1 RZ_n hierarchy

We first want to stress out that it is not clear that the various RZ_n properties are a strict hierarchy. There are examples of groups which are RF and not LERF (free solvable groups of class greater than 3 [13, 1]), of groups which are LERF and not RZ_2 (free metabelian groups [5]) and of groups which are RZ_2 but not RZ_3 (free nilpotent group of class 3 [16]).

Thereafter we conjecture that this hierarchy is strict, but we lack a proof. Going back to extension properties, the fact that this hierarchy was not strict, for example that RZ_3 and RZ_4 were equivalent would indicate that in a sense that is not clear that one could encode 4-cycle free structures in 3-cycle free structures.

5.2 $\mathcal{T}$-free structures

B. Herwig and D. Lascar [9] obtained other extension properties (for free groups). They defined the classes of $\mathcal{T}$-free structures which contain the class of n-cycle free n-partitioned structures.

We deal with a relational language $\mathcal{L}$ and $\mathcal{L}$-structures.

Let T and M be $\mathcal{L}$-structures. A *weak morphism* from T to M is a mapping f from T to M such that for every symbol R in $\mathcal{L}$ of arity r and every r-tuple $t_1, \ldots, t_r$ we have

$$R_T(t_1, \ldots, t_r) \Rightarrow R_M(f(t_1), \ldots, f(t_r)).$$

Let $\mathcal{T}$ be a finite set of finite structures, a structure M is $\mathcal{T}$-*free* if there is no weak morphism from an element of $\mathcal{T}$ into M.

Theorem 6. [9] *The class of $\mathcal{T}$-free structures has the extension property.*

This result of B. Herwig and D. Lascar [9] gives us information on the profinite topology of free groups. This will be detailed in the next section.

A group has the *maximal extension property* if it has the extension property for all classes of $\mathcal{T}$-free structures. The previous theorem states that free groups have the maximal extension property. B. Herwig and D. Lascar [9] were able to translate this property of free groups into a pure group theoretic setting.

But as before we can also focus on the class of groups which have the extension property for a given class of $\mathcal{T}$-free structures. And then the translation between the two settings is unclear. Moreover it is also unclear whether or not these various extension properties are equivalent. It is possible that an RZ$_3$ group has the extension property for all classes of $\mathcal{T}$-free structures.

5.3 Left systems

To translate extension properties for $\mathcal{T}$-free structures within a pure group theoretic setting, J. Almeida, M. Delgado [2], B. Herwig and D. Lascar [9] introduced left systems of equations in a group.

A *left system* over a group G is a finite set of equations of the following forms:

$$x \equiv_i yc \text{ or } x \equiv_i c$$

where c is an element of G, x, y are variables from a set X and $i = 1, \ldots, n$.

For an n-tuple $\mathcal{H} = (H_1, \ldots, H_n)$ of subgroups of G, a *solution* of the left system $\mathcal{S}$ modulo $\mathcal{H}$ is a family $(v_x)_{x \in X}$ of elements of G such that

$$\begin{cases} v_x H_i = v_y c H_i & \text{for all equation } x \equiv_i yc \text{ in } \mathcal{S} \\ \text{and} \\ v_x H_i = c H_i & \text{for all equation } x \equiv_i c \text{ in } \mathcal{S} \end{cases}$$

A left system is *finitely approximable* in a group G if for all n-tuple $\mathcal{H} = (H_1, \ldots, H_n)$ of finitely generated subgroups of G there exists an n-tuple $\mathcal{K} = (K_1, \ldots, K_n)$ of finite index subgroups of G such that H_i is contained in K_i

and such that $\mathcal{S}$ has a solution modulo $\mathcal{H}$ if and only if it has a solution modulo $\mathcal{K}$.

Details about these notions and the link with $\mathcal{T}$-free structures can be found in the article of B. Herwig and D. Lascar [9]. There it is proved that all left systems are finitely approximable in free groups. We can generalize their work to other groups. The main result that can be proved using the work done by J. Almeida, M. Delgado, B. Herwig and D. Lascar, and inspired by the results of the author about RZ_n groups is the following.

Theorem 7. *A group G has the maximal extension property if and only if all left systems are finitely approximable in G.*

6 Open questions

We already mentioned some problems about extension properties. The first one is to prove that the RZ_n hierarchy is strict, or more generally to understand the relative strength of extension properties for different classes of $\mathcal{T}$-free structures. In particular it would be very interesting to find groups having some extension properties and not others.

Another direction is to study groups having the maximal extension property. The class of groups having the maximal extension property contains free groups, finite groups and finitely generated abelian groups. It is very likely that this class is closed under free products as is the class of RZ_n groups and that it contains surface groups.

We conclude this note with a well-known open question that we state as a conjecture.

A *tournament* is an oriented graph such that to different vertices have exactly one oriented edge linking them. It can be seen as a relational structure in a language $\mathcal{L}$ having only one binary relation symbol with the following requirement:

$$\forall x, y((x \neq y) \to (R(x,y) \lor R(y,x)) \land \neg(R(x,y) \land R(y,x))).$$

Conjecture 1. *The class of tournaments has the extension property.*

The class of tournaments is not one of the classes of $\mathcal{T}$-free structures. But in this case B. Herwig and D. Lascar [9] translated this conjecture within a statement about profinite topology of free groups. Precisely it is equivalent to a conjecture about the oddadic topology of free groups.

The oddadic topology of a group G is the coarsest topology that makes continuous all morphisms from G into finite groups of odd cardinal. The normal subgroups of finite odd index are clopen for this topology. It is not true that all finitely generated subgroups are closed for this topology. A necessary condition for a subgroup H of a free group F to be closed is that for all element x in F, if x^2 is in H then x is in H. When this holds we say that H is closed for square roots.

Conjecture 1 is equivalent to saying that this necessary condition is sufficient:

Conjecture 2. *A finitely generated subgroup of a free group is closed for the oddadic topology of a free group if and only if it is closed for square roots.*

References

[1] S. Agalakov, Finite separability of groups and lie algebras, *Alg. i Log.*, 22:261–268, 1983.

[2] J. Almeida and M. Delgado, Sur certains systèmes d'équations avec contraintes dans un groupe libre, *Portugal. Math.*, 56(4):409–417, 1999.

[3] J. Almeida and M. Delgado, Sur certains systèmes d'équations avec contraintes dans un groupe libre - addenda, *Portugal. Math.*, 58(4):379–387, 2001.

[4] C. Ash, Inevitable graphs : a proof of the type II conjecture and some related decision procedures, *Int. J. of Alg. and Computation*, 1:127–146, 1991.

[5] T. Coulbois, *Propriétés de Ribes-Zalesskiĭ, topologie profinie, produit libre et généralisation*, PhD thesis, Université Paris 7, 2000.

[6] T. Coulbois, Free product, profinite topology and finitely generated subgroups, *Int. J. of Alg. and Computation*, 11:171–184, 2001.

[7] M. Hall, A topology for free groups and related groups, *Annals of Math.*, 52:127–139, 1950.

[8] K. Henckell, S. Margolis, J.-É. Pin and J. Rhodes, Ash's type II theorem, profinite topology and Mal'cev products. I, *Internat. J. Algebra Comput.*, 1(4):411–436, 1991.

[9] B. Herwig and D. Lascar, Extending partial automorphisms and the profinite topology on free groups, *Trans. Amer. Math. Soc.*, 352(5):1985–2021, 2000.

[10] E. Hrushovski, Extending partial automorphisms of graphs, *Combinatorica*, 12:411–416, 1992.

[11] R. Gitik, Graphs and separability properties of groups, *J. Algebra*, 188(1):125–143, 1997.

[12] R. Gitik and E. Rips, On separability properties of groups, *Int. J. of*

Alg. and Computaion, 5:703–711, 1995.

[13] K. Gruenberg, Residual properties of infinite soluble groups, *Proc. Lond. Math. Soc.*, 7:29–62, 1957.

[14] J. Kellendonk and M. Lawson, Partial actions of groups, *Preprint*.

[15] D. Lascar, Autour de la propriété du petit indice, *Proc. Lond. Math. Soc.*, 62:25–53, 1991.

[16] J. Lennox and J. Wilson, On products of subgroups in polycyclic groups, *Arch. der Math.*, 33:305–309, 1979.

[17] S. Margolis, M. Sapir and P. Weil, Closed subgroups in pro-**V** topologies and the extension problem for inverse automata, *Int. J. of Alg. and Computaion*, 11(4):405–445, 2001

[18] J.-É. Pin, A topological approach to a conjecture of Rhodes, *Bull. Australian Math. Soc*, 38:421–431, 1988.

[19] J.-É. Pin and C. Reutenauer, A conjecture on the Hall topology for the free group, *Bull. London Math. Soc.*, 23:356–362, 1991.

[20] L. Ribes and P. Zalesskiĭ, On the profinite topology on a free group, *Bull. Lond. Math. Soc.*, 25:37–43, 1993.

PRESENTATIONS FOR SOME MONOIDS OF PARTIAL TRANSFORMATIONS ON A FINITE CHAIN: A SURVEY

VÍTOR H. FERNANDES*

Departamento de Matemática
Faculdade de Ciências e Tecnologia
Universidade Nova de Lisboa
2829-516 Monte da Caparica, Portugal
AUTHOR'S SECOND ADDRESS:
Centro de Álgebra da Universidade de Lisboa
Av. Prof. Gama Pinto, 2
1649-003 Lisboa, Portugal
E-mail: vhf@mail.fct.unl.pt

We survey some known nice presentations of several monoids of partial transformations on a finite chain: we consider the monoids of all order preserving partial transformations, of all order preserving full transformations and of all order preserving injective partial transformations and also some of its natural extensions.

1 Introduction and Preliminaries

Let X_n be a chain with n elements, say $X_n = \{1 < 2 < \cdots < n\}$, with $n \in \mathbb{N}$. As usual, we denote by $\mathcal{PT}_n$ the monoid of all (partial) transformations of X_n (under composition), by $\mathcal{T}_n$ the submonoid of $\mathcal{PT}_n$ of all full transformations of X_n, by $\mathcal{I}_n$ the symmetric inverse semigroup on X_n, i.e. the submonoid of $\mathcal{PT}_n$ of all injective (partial) transformations of X_n, and by $\mathcal{S}_n$ the symmetric group on X_n, i.e. the subgroup of $\mathcal{PT}_n$ of all injective full transformations (permutations) of X_n.

We say that a transformation s in $\mathcal{PT}_n$ is *order preserving* (respectively, *order reversing*) if, for all $x, y \in \mathrm{Dom}(s)$, $x \leq y$ implies $xs \leq ys$ (respectively,

---

*THE AUTHOR WOULD LIKE TO ACKNOWLEDGE THE FINANCIAL SUPPORT OF FUNDAÇÃO CALOUSTE GULBENKIAN (FCG), FUNDAÇÃO PARA A CIÊNCIA E A TECNOLOGIA (FCT), FACULDADE DE CIÊNCIAS DA UNIVERSIDADE DE LISBOA (FCUL) AND REITORIA DA UNIVERSIDADE DO PORTO. THIS WORK WAS SUPPORTED BY FCT AND FEDER, WITHIN THE PROJECT POCTI/32440/MAT/1999 – "ÁLGEBRA E APLICAÇÕES" OF CAUL, AND IT WAS PREPARED WITHIN THE PROJECT JD: "APRESENTAÇÕES PARA SEMIGRUPOS", FCT-UNL, 1999.

$xs \geq ys$). Clearly, the product of two order preserving transformations or two order reversing transformations is an order preserving transformation and the product of an order preserving transformation with an order reversing transformation, or vice-versa, is an order reversing transformation. We denote by $\mathcal{PO}_n$ the submonoid of $\mathcal{PT}_n$ of all order preserving transformations and by $\mathcal{POD}_n$ the submonoid of $\mathcal{PT}_n$ of all order preserving transformations together with all order reversing transformations. We denote by $\mathcal{O}_n$ the submonoid of $\mathcal{PO}_n$ of all (order preserving) full transformations, by $\mathcal{OD}_n$ the submonoid of $\mathcal{POD}_n$ of all (order preserving or order reversing) full transformations, by $\mathcal{POI}_n$ the inverse submonoid of $\mathcal{I}_n$ of all order preserving transformations and by $\mathcal{PODI}_n$ the inverse submonoid of $\mathcal{I}_n$ whose elements belong to $\mathcal{POD}_n$.

Now, let $a = (a_1, a_2, \ldots, a_t)$ be a sequence of t ($t \geq 0$) elements from the chain X_n. We say that a is *cyclic* (respectively, *anti-cyclic*) if there exists no more than one index $i \in \{1, \ldots, t\}$ such that $a_i > a_{i+1}$ (respectively, $a_i < a_{i+1}$), where a_{t+1} denotes a_1. Let $s \in \mathcal{PT}_n$ and suppose that $\text{Dom}(s) = \{a_1, \ldots, a_t\}$, with $t \geq 0$ and $a_1 < \cdots < a_t$. We say that s is an *orientation preserving* (respectively, *orientation reversing*) transformation if the sequence of its images $(a_1s, \ldots, a_ts)$ is cyclic (respectively, anti-cyclic). The notion of an orientation preserving transformation was introduced by Catarino and Higgins [4], and also by McAlister [17]. Notice that the product of two orientation preserving transformations or of two orientation reversing transformations is an orientation preserving transformation and the product of an orientation preserving transformation with an orientation reversing transformation, or vice-versa, is an orientation reversing transformation. We denote by $\mathcal{POP}_n$ the submonoid of $\mathcal{PT}_n$ of all orientation preserving transformations and by $\mathcal{POR}_n$ the submonoid of $\mathcal{PT}_n$ of all orientation preserving transformations together with all orientation reversing transformations. Denote by $\mathcal{OP}_n$ the submonoid of $\mathcal{POP}_n$ of all (orientation preserving) full transformations, by $\mathcal{OR}_n$ the submonoid of $\mathcal{POR}_n$ of all (orientation preserving or orientation reversing) full transformations, by $\mathcal{POPI}_n$ the inverse submonoid of $\mathcal{I}_n$ whose elements are all orientation preserving transformations and by $\mathcal{PORI}_n$ the inverse submonoid of $\mathcal{I}_n$ whose elements belong to $\mathcal{POR}_n$. Finally, denote by **1** the trivial monoid and by $\mathcal{C}_n$ the cyclic group of order n. The relative inclusions between these semigroups is summarized in Figure 1.

A presentation for the full transformation monoid $\mathcal{T}_n$ was given by Aĭzenštat [1] in terms of a certain type of two generator presentation for the symmetric group S_n, plus an extra generator and seven more relations (see also [21]). A presentation for the partial transformation monoid $\mathcal{PT}_n$ and for the symmetric inverse monoid $\mathcal{I}_n$ was given by Popova [19] in 1961 (see also [21]). Also Popova [20] established in 1962 a presentation for the monoid $\mathcal{PO}_n$.

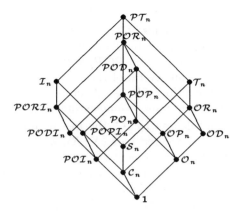

Figure 1. Inclusion relations.

A monoid presentation for $\mathcal{O}_n$, in terms of $2n-2$ generators and n^2 relations, was given in 1962 by Aĭzenštat [2] and a monoid presentation for $\mathcal{POI}_n$, in terms of n generators and $\frac{1}{2}(n^2+5n-4)$ relations, was given by Fernandes [9]. Notice that the rank of these monoids is n, in both cases, and $2n-2$ is the idempotent rank of $\mathcal{O}_n$ (see [9] and [12]). All the presentations known for the monoids $\mathcal{PO}_n$, $\mathcal{OR}_n$, $\mathcal{OP}_n$ and $\mathcal{OD}_n$ are based on Aĭzenštat presentation of $\mathcal{O}_n$ and all presentation known for the monoids $\mathcal{PORI}_n$, $\mathcal{POPI}_n$ and $\mathcal{PODI}_n$ are based on Fernandes presentation of $\mathcal{POI}_n$. First, a presentation for $\mathcal{OP}_n$, on $2n-1$ generators and n^2+n relations, was given by Catarino [5]. Improving Catarino result, Arthur and Ruškuc [3] exhibited a presentation for $\mathcal{OP}_n$, on two generators and $n+2$ relations. Also in [3] a presentation for $\mathcal{OR}_n$ was given, in terms of three generators and $n+6$ relations. The author [8] exhibited a presentation for $\mathcal{POPI}_n$ in terms of $n+1$ (and also on two) generators and $\frac{1}{2}(n^2+7n-2)$ relations. Also, Fernandes, Gomes and Jesus [10] gave presentations for the monoid $\mathcal{PODI}_n$ in terms of n generators and $\frac{1}{2}(n^2+7n-2)$ relations, for the monoid $\mathcal{POPI}_n$ in terms of two generators and $2n$ relations (improving the result presented by the author in [8]) and for the monoid $\mathcal{PORI}_n$ in terms of three generators and $2n+4$ relations.

In 1996 Solomon introduced in [23] two general constructions of transformation monoids associated with a graph: the Catalan monoid $\mathcal{C}(G)$ and the partial Catalan monoid $\mathcal{PC}(G)$ of a directed graph G. The main results of [23] are presentations for $\mathcal{C}(G)$ and $\mathcal{PC}(G)$ when G is a tree. We notice that, in particular, Solomon obtained the results of Aĭzenštat on $\mathcal{O}_n$ and of Popova on $\mathcal{PO}_n$ as corollaries.

Presentations for the monoids $\mathcal{OD}_n$, $\mathcal{POD}_n$ and $\mathcal{POR}_n$ will be exhibited in [11], using the technique given by Theorem 1.3 stated below. Also in this paper a presentation for the monoid $\mathcal{POP}_n$ will be given.

We notice that to obtain several of these presentations, in particular the ones due to the author, considerable use of computational tools (such as D.B. McAlister program *Semigroup for Windows* and GAP [22]) was made.

Let X be a set and denote by X^* the free monoid generated by X. A *monoid presentation* is an ordered pair $\langle X \mid R \rangle$, where X is an alphabet and R is a subset of $X^* \times X^*$. An element (u,v) of $X^* \times X^*$ is called a *relation* and it is usually represented by $u = v$. A monoid M is said to be *defined by a presentation* $\langle X \mid R \rangle$ if M is isomorphic to X^*/ρ_R, where ρ_R denotes the smallest congruence on X^* containing R. For more details see [15] or [21].

Given a finite monoid T, it is clear that we can always exhibit a presentation for it (for example with $|T|$ generators and $|T|^2$ relations). So, by finding a presentation for a finite monoid, we mean in general to find in some sense a nice presentation (e.g. with a small number of generators and relations).

The usual method to find a presentation for a finite monoid is described by the following result.

Theorem 1.1. *Let M be a finite monoid, let X be a generating set for M, let $R \subseteq X^* \times X^*$ be a set of relations, and let $W \subseteq X^*$. Assume that the following conditions are satisfied:*

(1) *The generating set X of M satisfies all the relations from R;*

(2) *For each word $w \in X^*$, there exists a word $w' \in W$ such that the relation $w = w'$ is a consequence of R;*

(3) $|W| \leq |M|$.

Then, M is defined by the presentation $\langle X \mid R \rangle$.

Notice that, if W satisfies the above conditions then, in fact, $|W| = |M|$.

Let X be an alphabet, $R \subseteq X^* \times X^*$ a set of relations and W a subset of X^*. We say that W is a set of *forms* for the presentation $\langle X \mid R \rangle$ if the condition 2 of Theorem 1.1 is satisfied.

Another method to find a new presentation for a monoid from a given one consists of applying Tietze transformations. For a monoid presentation $\langle A \mid R \rangle$, the *elementary Tietze transformations* are:

(T1) Adding a new relation $u = v$ to $\langle A \mid R \rangle$, providing that $u = v$ is a consequence of R;

(T2) Deleting a relation $(u = v) \in R$ from $\langle A \mid R \rangle$, providing that $u = v$ is a consequence of $R \setminus \{u = v\}$;

(T3) Adding a new generating symbol b to A and a new relation $b = w$ to R, where $w \in A^*$;

(T4) If $\langle A \mid R \rangle$ possesses a relation of the form $b = w$, with $b \in A$ and $w \in (A \setminus \{b\})^*$, deleting b from the list of generating symbols, deleting the relation $b = w$, and replacing all remaining appearances of b by w.

The following result is well-known:

Theorem 1.2. *Two finite presentations define the same monoid if and only if one can be obtained from the other by a finite number of applications of elementary Tietze transformations (T1), (T2), (T3) and (T4).*

Recall that the *rank* of a monoid M is, by definition, the minimum of the set $\{|X| : X \subseteq M \text{ and } X \text{ generates } M\}$ (see e.g. [14]).

As for the rest of this section, we will describe a method, due to Fernandes, Gomes and Jesus [10], to obtain a presentation for a finite monoid T given a presentation for a certain submonoid of T.

Let T be a (finite) monoid, S be a submonoid of T and y an element of T such that $y^2 = 1$. Let us suppose that T is generated by S and y. Let $X = \{x_1, \ldots, x_k\}$ ($k \in \mathbb{N}$) be a generating set of S and $\langle X \mid R \rangle$ be a presentation for S. Consider a set of forms W for $\langle X \mid R \rangle$ and suppose there exist two subsets W_α and W_β of W and a word $u_0 \in X^*$ such that $W = W_\alpha \cup W_\beta$ and u_0 is a factor of each word in W_α. Let $Y = X \cup \{y\}$ (notice that Y generates T) and suppose that there exist words $v_0, v_1, \ldots, v_k \in X^*$ such that the following relations over the alphabet Y are satisfied by the generating set Y of T:

(NR$_1$) $yx_i = v_i y$, for all $i \in \{1, \ldots, k\}$;

(NR$_2$) $u_0 y = v_0$.

Observe that the relation (over the alphabet Y)

(NR$_0$) $y^2 = 1$

is also satisfied (by the generating set Y of T), by hypothesis.

Let $\overline{R} = R \cup NR_0 \cup NR_1 \cup NR_2$ and $\overline{W} = W \cup \{wy \mid w \in W_\beta\} \subseteq Y^*$. The following result was proved in [10]:

Theorem 1.3. *If W contains the empty word and $|\overline{W}| \leq |T|$ then the monoid T is defined by the presentation $\langle Y \mid \overline{R} \rangle$.*

2 The monoids $\mathcal{S}_n$, $\mathcal{T}_n$, $\mathcal{I}_n$ and $\mathcal{PT}_n$

The ranks of the monoids $\mathcal{PT}_n$, $\mathcal{T}_n$, $\mathcal{I}_n$ and $\mathcal{S}_n$ are well known results (see [14] or [21]). In this section we consider $n \geq 4$.

First, we consider the monoid $\mathcal{S}_n$, which has rank two and $n!$ elements. Let a be the permutation $(1\,2)$ of order 2 and b be permutation $(1\,2\,\cdots\,n)$ of order n. Then $\{a,b\}$ is a set of generators of $\mathcal{S}_n$. The following group presentation of $\mathcal{S}_n$, in terms of these two generators, due to Moore [18] (see also [6], [21] and [16]), has been known since 1897:

$$\langle a,b \mid a^2 = b^n = (ba)^{n-1} = (ab^{-1}ab)^3 = (ab^{-j}ab^j)^2 = 1 \ (2 \leq j \leq n-2)\rangle.$$

From this group presentation, it can easily be deduced the following monoid presentation for $\mathcal{S}_n$, in terms of the same generators:

$$\langle a,b \mid a^2 = b^n = (ba)^{n-1} = (ab^{n-1}ab)^3 = (ab^{n-j}ab^j)^2 = 1 \ (2 \leq j \leq n-2)\rangle.$$

Notice that this presentation has $n+1$ relations.

In terms of the generators $r_i = (i\,i+1)$, with $1 \leq i \leq n-1$, Moore [18] also gave the following presentation for $\mathcal{S}_n$:

$$\langle r_1, r_2, \ldots, r_{n-1} \mid r_i^2 = 1 \ (1 \leq i \leq n-1), (r_i r_{i+1})^3 = 1 \ (1 \leq i \leq n-2),$$
$$(r_i r_k)^2 = 1 \ (i \leq k-2)\rangle.$$

Next, regarding the monoid $\mathcal{T}_n$, we first notice that it has rank three and n^n elements. A set of generators can be obtained by joining to the permutations a and b any full transformation of rank $n-1$. For instance, if

$$t = \begin{pmatrix} 1 & 2 & 3 & \cdots & n \\ 1 & 1 & 3 & \cdots & n \end{pmatrix},$$

then the set $\{a,b,t\}$ generates the monoid $\mathcal{T}_n$. In 1958, Aĭzenštat [1] (see also [21]) gave a presentation of $\mathcal{T}_n$ that can be established in terms of the generators a, b and t: given a monoid presentation $\langle a,b \mid R \rangle$ of $\mathcal{S}_n$ then

$$\langle a,b,t \mid R, Q \rangle$$

is a monoid presentation for $\mathcal{T}_n$, where Q is the set of relations:

- $at = b^{n-2}ab^2tb^{n-2}ab^2 = bab^{n-1}abtb^{n-1}abab^{n-1} = (tbab^{n-1})^2 = t$;

- $(b^{n-1}abt)^2 = tb^{n-1}abt = (tb^{n-1}ab)^2$;

- $(tbab^{n-2}ab)^2 = (bab^{n-2}abt)^2$.

Thus, considering the Moore monoid presentation of $\mathcal{S}_n$, we obtain the Aĭzenštat presentation of $\mathcal{T}_n$ with $n + 8$ relations.

Now, we focus our attention on the monoid $\mathcal{I}_n$, which has rank three and $\sum_{k=0}^{n} \binom{n}{k}^2 k!$ elements. We obtain a generating set of $\mathcal{I}_n$, with three elements, by joining to the permutations a and b any injective partial transformation of rank $n - 1$. Let (in particular)

$$s = \begin{pmatrix} 2 & 3 & \cdots & n \\ 2 & 3 & \cdots & n \end{pmatrix}.$$

Then, the set $\{a, b, s\}$ generates the monoid $\mathcal{I}_n$.

In 1961 Popova [19] proved that if U is the set of all relations over $\mathcal{S}_n$ holding in $\mathcal{S}_n$ then the monoid $\mathcal{I}_n$ is defined by the presentation

$$\langle \mathcal{S}_n, s \mid U, \ b^{n-1}absb^{n-1}ab = basab^{n-1} = s = s^2, \ (sa)^2 = sas = (as)^2 \rangle$$

(see also [16]). Hence, considering the Moore monoid presentation of $\mathcal{S}_n$, we can deduce the following presentation of $\mathcal{I}_n$ with $n + 6$ relations, in terms of the generators a, b and s:

$\langle\, a, b, s \mid a^2 = b^n = (ba)^{n-1} = (ab^{n-1}ab)^3 = (ab^{n-j}ab^j)^2 = 1 \ (2 \leq j \leq n-2),$
$\quad b^{n-1}absb^{n-1}ab = basab^{n-1} = s = s^2, \ (sa)^2 = sas = (as)^2 \,\rangle.$

Finally, we consider the monoid $\mathcal{PT}_n$. It has rank four and $(n+1)^n$ elements. A generating set with four elements is, for example, the set $\{a, b, s, t\}$. Moreover, joining to any two permutations generating the symmetric group $\mathcal{S}_n$ any full transformation of rank $n - 1$ and any injective partial transformation of rank $n - 1$, we obtain a four elements generating set of $\mathcal{PT}_n$.

Also Popova [19] gave a presentation for the monoid $\mathcal{PT}_n$, considering as letters all elements of $\mathcal{S}_n$ together with t and s and as defining relations all relations from the presentations of $\mathcal{T}_n$ and of $\mathcal{I}_n$, stated above, together with the following four relations:

$$ts = sas, \ st = sa, \ tas = ta, \ tab^{n-1}abas = ab^{n-1}abasab^{n-1}abatab^{n-1}aba.$$

Thus, considering again the Moore monoid presentation of $\mathcal{S}_n$, we can deduce the following presentation of $\mathcal{PT}_n$ with $n + 17$ relations, in terms of the generators a, b, s and t:

$\langle\, a, b, s, t \mid a^2 = b^n = (ba)^{n-1} = (ab^{n-1}ab)^3 = (ab^{n-j}ab^j)^2 = 1 \ (2 \leq j \leq n-2),$
$\quad b^{n-1}absb^{n-1}ab = basab^{n-1} = s = s^2, \ (sa)^2 = sas = (as)^2,$
$\quad at = b^{n-2}ab^2tb^{n-2}ab^2 = bab^{n-1}abtb^{n-1}abab^{n-1} = (tbab^{n-1})^2 = t,$
$\quad (b^{n-1}abt)^2 = tb^{n-1}abt = (tb^{n-1}ab)^2, \ (tbab^{n-2}ab)^2 = (bab^{n-2}abt)^2,$
$\quad ts = sas, \ st = sa, \ tas = ta,$
$\quad tab^{n-1}abas = ab^{n-1}abasab^{n-1}abatab^{n-1}aba \,\rangle.$

3 The monoids $\mathcal{O}_n$, $\mathcal{OP}_n$, $\mathcal{OR}_n$ and $\mathcal{PO}_n$

A presentation $\langle X \mid R \rangle$ for the monoid $\mathcal{O}_n$, of all order preserving full transformations on a chain with n elements, was given by Aĭzenštat [2] in 1962, in terms of the $2n - 2$ element generating set

$$X = \{u_1, u_2, \ldots, u_{n-1}, v_1, v_2, \ldots, v_{n-1}\},$$

where

$$u_i = \begin{pmatrix} 1 & \cdots & i-1 & i & i+1 & \cdots & n \\ 1 & \cdots & i-1 & i+1 & i+1 & \cdots & n \end{pmatrix}$$

and

$$v_i = \begin{pmatrix} 1 & \cdots & n-i & n-i+1 & n-i+2 & \cdots & n \\ 1 & \cdots & n-i & n-i & n-i+2 & \cdots & n \end{pmatrix},$$

for $1 \le i \le n-1$, and with R the following set of n^2 defining relations:

- $v_{n-i} u_i = u_i v_{n-i+1}$, for $2 \le i \le n-1$;
- $u_{n-i} v_i = v_i u_{n-i+1}$, for $2 \le i \le n-1$;
- $v_{n-i} u_i = u_i$, for $1 \le i \le n-1$;
- $u_{n-i} v_i = v_i$, for $1 \le i \le n-1$;
- $u_i v_j = v_j u_i$, for $1 \le i, j \le n-1$, with $j \notin \{n-i, n-i+1\}$;
- $u_1 u_2 u_1 = u_1 u_2$;
- $v_1 v_2 v_1 = v_1 v_2$

(see also [23]).

Notice that all the $2n - 2$ elements of X are idempotents. Also, notice that Gomes and Howie [12] proved that $2n - 2$ is precisely the *idempotent rank* (i.e. the minimum number of idempotent generators) of the monoid $\mathcal{O}_n$. In the same paper, it was also established that, by contrast, the monoid $\mathcal{O}_n$ has rank n. Finally, observe that the number of elements of $\mathcal{O}_n$ is $\binom{2n-1}{n-1}$ (see [13]).

In 1998 Catarino [5] proved that, given any presentation $\langle X \mid Q \rangle$ of $\mathcal{O}_n$, in terms of the generating set X defined above, then the monoid $\mathcal{OP}_n$ admits the presentation $\langle X, b \mid Q, T \rangle$, where b corresponds to the n-cycle $(1\, 2\, \cdots\, n)$ of $\mathcal{S}_n$ (already considered in Section 2) and T is the following set of $2n$ relations:

- $u_i b = b u_{i+1}$, $v_{i+1} b = b v_i$, for $1 \le i \le n-2$;

- $u_{n-1}b = b^2 v_{n-1} v_{n-2} \cdots v_1$, $v_1 b = u_{n-1} u_{n-2} \cdots u_1$;
- $b^n = 1$, $b v_1 v_2 \cdots v_{n-1} = v_1 v_2 \cdots v_{n-1}$.

Hence, considering the previous Aĭzenštat presentation of $\mathcal{O}_n$, we obtain the Catarino presentation for $\mathcal{OP}_n$, in terms of $2n - 1$ generators and $n^2 + 2n$ relations.

It follows quickly from the above relations (also noted in [5]) that $\mathcal{OP}_n$ can be generated by just two elements, namely u_1 and b (from which it is immediate that $\mathcal{OP}_n$ has rank 2). Notice that, Catarino and Higgins [4] proved that $|\mathcal{OP}_n| = n\binom{2n-1}{n-1} - n(n-1)$.

For simplicity, let us denote the transformation u_1 by u. Eliminating redundant generators from the above Catarino presentation and then eliminating redundant relations, Arthur and Ruškuc [3] exhibited the following presentation of $\mathcal{OP}_n$ in terms of the generators u and b, with only $n + 2$ relations:

$$\langle b, u \mid b^n = 1,\ u^2 = u,\ (ub)^n = ub,\ b(ub^{n-1})^{n-1} = (ub^{n-1})^{n-1},$$
$$u\, b^i (ub)^{n-1} b^{n-i} = b^i (ub)^{n-1} b^{n-i} u\ (2 \leq i \leq n-1) \rangle.$$

From this last presentation of $\mathcal{OP}_n$, Arthur and Ruškuc [3] also exhibited the following presentation for the monoid $\mathcal{OR}_n$ of all orientation preserving or reversing full transformations, in terms of the generators u, b and

$$c = \begin{pmatrix} 1 & 2 & 3 & \cdots & n \\ 1 & n & n-1 & \cdots & 2 \end{pmatrix},$$

with $n + 6$ relations:

$$\langle b, u, c \mid b^n = 1,\ u^2 = u,\ (ub)^n = ub,\ b(ub^{n-1})^{n-1} = (ub^{n-1})^{n-1},$$
$$u\, b^i (ub)^{n-1} b^{n-i} = b^i (ub)^{n-1} b^{n-i} u\ (2 \leq i \leq n-1),$$
$$c^2 = 1,\ bc = cb^{n-1},\ uc = c(bu)^{n-1},\ c(ub^{n-1})^{n-2} = b^{n-2}(ub^{n-1})^{n-2} \rangle.$$

Observe that this presentation of $\mathcal{OR}_n$ is obtained from the Arthur and Ruškuc presentation of $\mathcal{OP}_n$ by adding one more generator (which is a permutation of order two) and four more relations, which are (almost trivially) equivalent to the following four relations:

(NR$_0$) $c^2 = 1$;

(NR$_1$) $cb = b^{n-1}c$ and $cu = (bu)^{n-1}c$;

(NR$_2$) $((bu)^{n-1}b)^{n-2}c = b^{n-2}(ub^{n-1})^{n-2}$.

Thus, the method given by Theorem 1.3 can easily be used to give a simple alternative proof of Arthur and Ruškuc's result on the $\mathcal{OR}_n$ presentation: it suffices to show that any element of $\mathcal{OP}_n$ with rank one or two admits a decomposition in $\mathcal{OP}_n$ with $((bu)^{n-1}b)^{n-2}$ as a factor, which is easy to establish.

Notice that the monoid $\mathcal{OR}_n$ has rank three and $n\binom{2n}{n} - \frac{n^2}{2}(n^2 - 2n + 5) + n$ (i.e. $2|\mathcal{OP}_n| - 2\binom{n}{2}^2 - n$) elements (see [17]).

Finally, we focus our attention on the monoid $\mathcal{PO}_n$ of all order preserving partial transformations on a chain with n elements. This monoid has rank $2n - 1$. On the other hand, as $\mathcal{O}_n$, the monoid $\mathcal{PO}_n$ is idempotent generated with idempotent rank $3n - 2$ and it has $\sum_{k=0}^{n} \binom{n}{k}\binom{n+k-1}{k}$ elements (see [12]).

Let $X' = \{u_1, u_2, \ldots, u_{n-1}, v_1, v_2, \ldots, v_{n-1}, c_1, c_2, \ldots, c_n\}$, with

$$c_i = \begin{pmatrix} 1 & \cdots & i-1 & i+1 & \cdots & n \\ 1 & \cdots & i-1 & i+1 & \cdots & n \end{pmatrix},$$

for $1 \le i \le n$. Then X' is a generating set of the monoid $\mathcal{PO}_n$ consisting of $3n - 2$ idempotents.

Consider the following set of relations over the alphabet X':

(A_1) $v_{n-i}u_i = u_i v_{n-i+1}$, for $2 \le i \le n - 1$;

(A_2) $u_{n-i}v_i = v_i u_{n-i+1}$, for $2 \le i \le n - 1$;

(A_3) $v_{n-i}u_i = u_i$, for $1 \le i \le n - 1$;

(A_4) $u_{n-i}v_i = v_i$, for $1 \le i \le n - 1$;

(A_5) $u_i v_j = v_j u_i$, for $1 \le i, j \le n - 1$, with $j \notin \{n - i, n - i + 1\}$;

(A_6) $u_1 u_2 u_1 = u_1 u_2$;

(A_7) $v_1 v_2 v_1 = v_1 v_2$;

(E_1) $c_i c_j = c_j c_i$, for $1 \le i < j \le n$;

(P_1) $u_i c_i = u_i$, for $1 \le i \le n - 1$;

(P_2) $c_i u_i = c_i$, for $1 \le i \le n - 1$;

(P_3) $u_i c_{i+1} = c_i c_{i+1}$, for $1 \le i \le n - 1$;

(P_4) $c_i u_j = u_j c_i$, for $1 \le i, j \le n - 1$ such that $i \notin \{j, j+1\}$;

(P_4') $c_n u_j = u_j c_n$, for $1 \le j \le n - 2$;

(P_5) $v_{n-i}c_{i+1} = v_{n-i}$, for $1 \leq i \leq n-1$;

(P_6) $c_{i+1}v_{n-i} = c_{i+1}$, for $1 \leq i \leq n-1$;

(P_7) $v_{n-i}c_i = c_{i+1}c_i$, for $1 \leq i \leq n-1$;

(P_8) $c_i v_{n-j} = v_{n-j}c_i$, for $1 \leq i, j \leq n-1$ such that $i \notin \{j, j+1\}$;

(P_8') $c_n v_{n-j} = v_{n-j}c_n$, for $1 \leq j \leq n-2$.

Notice that $(A_1) - (A_7)$ is the set of defining relations from the Aĭzenštat presentation of $\mathcal{O}_n$ in terms of the generating set X.

In 1962 Popova [20] exhibited a presentation for the monoid $\mathcal{PO}_n$, in terms of the $3n-2$ elements generating set X', with defining relations consisting of

$$(A_1) - (A_7), (E_1), (P_1) - (P_4), (P_4'), (P_7) - (P_8) \text{ and } (P_8').$$

In this paper the relations $(P_5) - (P_6)$ were considered as being a consequence of $(A_1) - (A_4)$ and $(P_1) - (P_2)$. However, the deduction presented by Popova does not work in the case $i = n-1$. Thus, in view of that proof, the relations

$$v_1 c_n = v_1 \text{ and } c_n v_1 = c_n$$

must be joined to Popova's set of defining relations. Small examples ($n = 3, 4$) tested with GAP [22] showed that, without these relations, the monoids defined by the Popova presentation are probably infinite. Therefore, an accurate presentation for the monoid $\mathcal{PO}_n$, in terms of the $3n-2$ elements generating set X', is for example $\langle X' \mid P \rangle$, with P the set of $\frac{1}{2}(7n^2 - n - 4)$ defining relations consisting of

$$(A_1) - (A_7), (E_1), (P_1) - (P_4), (P_4'), (P_5) - (P_8) \text{ and } (P_8')$$

(see also [23]). Notice that some of these relations can easily be removed. For instance, we can eliminate all relations from (P_5) and (P_6), except for $i = n-1$.

4 The monoids $\mathcal{POI}_n$, $\mathcal{PODI}_n$, $\mathcal{POPI}_n$ and $\mathcal{PORI}_n$

It was proved by Fernandes [7] that $|\mathcal{POI}_n| = \binom{2n}{n}$. The rank of the monoid $\mathcal{POI}_n$ was also discovered by Fernandes [9] as being n. Also in the same paper, it was presented the generating set $\{x_0, x_1, \ldots, x_{n-1}\}$ of $\mathcal{POI}_n$ (with n elements) defined as follows:

$$x_0 = \begin{pmatrix} 2 & \cdots & n-1 & n \\ 1 & \cdots & n-2 & n-1 \end{pmatrix},$$

and, for $i \in \{1, 2, \ldots, n-1\}$,
$$x_i = \begin{pmatrix} 1 & \cdots & n-i-1 & n-i & n-i+2 & \cdots & n \\ 1 & \cdots & n-i-1 & n-i+1 & n-i+2 & \cdots & n \end{pmatrix}.$$

Moreover, Fernandes [9] gave a presentation for the monoid $\mathcal{POI}_n$, in terms of the generators $x_0, x_1, \ldots, x_{n-1}$, with the following $\frac{1}{2}(n^2 + 5n - 4)$ defining relations:

(R_1) $x_i x_0 = x_0 x_{i+1}$, $1 \leq i \leq n-2$;

(R_2) $x_j x_i = x_i x_j$, $2 \leq i+1 < j \leq n-1$;

(R_3) $x_0^2 x_1 = x_0^2 = x_{n-1} x_0^2$;

(R_4) $x_{i+1} x_i x_{i+1} = x_{i+1} x_i = x_i x_{i+1} x_i$, $1 \leq i \leq n-2$;

(R_5) $x_i x_{i+1} \cdots x_{n-1} x_0 x_1 \cdots x_{i-1} x_i = x_i$, $0 \leq i \leq n-1$;

(R_6) $x_{i+1} \cdots x_{n-1} x_0 x_1 \cdots x_{i-1} x_i^2 = x_i^2$, $1 \leq i \leq n-1$.

Using the technique given by Theorem 1.3 and based on this presentation of $\mathcal{POI}_n$, Fernandes, Gomes and Jesus [10] exhibited a presentation for the monoid $\mathcal{PODI}_n$, in terms of the generators $x_1, \ldots, x_{n-1}, d$, with the following $\frac{1}{2}(n^2 + 7n - 2)$ defining relations:

(R_1) $x_i x_0 = x_0 x_{i+1}$, $1 \leq i \leq n-2$;

(R_2) $x_j x_i = x_i x_j$, $2 \leq i+1 < j \leq n-1$;

(R_3) $x_0^2 x_1 = x_0^2 = x_{n-1} x_0^2$;

(R_4) $x_{i+1} x_i x_{i+1} = x_{i+1} x_i = x_i x_{i+1} x_i$, $1 \leq i \leq n-2$;

(R_5) $x_i x_{i+1} \cdots x_{n-1} x_0 x_1 \cdots x_{i-1} x_i = x_i$, $0 \leq i \leq n-1$;

(R_6) $x_{i+1} \cdots x_{n-1} x_0 x_1 \cdots x_{i-1} x_i^2 = x_i^2$, $1 \leq i \leq n-1$;

(NR_0) $d^2 = 1$;

(NR_1) $dx_i = x_{n-i+1} \cdots x_{n-1} x_0 x_1 \cdots x_{n-i-1} d$, $1 \leq i \leq n-1$;

(NR_2) $x_0^{n-1} d = x_{n-1} \cdots x_3 x_2^2$,

where x_0 denotes the word $x_0 = dx_1 \cdots x_{n-1}d$ over $\{x_1, \ldots, x_{n-1}, d\}$. We observe that the letter d corresponds to the following permutation of order two:

$$d = \begin{pmatrix} 1 & 2 & \cdots & n-1 & n \\ n & n-1 & \cdots & 2 & 1 \end{pmatrix}.$$

Notice that the rank of the monoid $\mathcal{PODI}_n$ is $\lfloor \frac{n}{2} \rfloor + 1$ (with $\lfloor \frac{n}{2} \rfloor$ denoting the least integer greater than or equal to $\frac{n}{2}$) and that it has $2\binom{2n}{n} - n^2 - 1$ elements (see [10]).

In [8], the author showed that the number of elements of $\mathcal{POPI}_n$ is $1 + \frac{n}{2}\binom{2n}{n}$. In the same paper it was proved that $\{x_0, x_1, \ldots, x_{n-1}, b\}$, with b denoting again the permutation $(1\,2\,\cdots\,n)$ of order n, is a generating set of the monoid $\mathcal{POPI}_n$. Moreover, still based on the above presentation of $\mathcal{POI}_n$, Fernandes also in [8] gave a presentation for $\mathcal{POPI}_n$, in terms of the generators $x_0, x_1, \ldots, x_{n-1}, b$, and the following $\frac{1}{2}(n^2 + 7n - 2)$ defining relations:

(R_1) $x_i x_0 = x_0 x_{i+1}$, $1 \leq i \leq n-2$;

(R_2) $x_j x_i = x_i x_j$, $2 \leq i+1 < j \leq n-1$;

(R_3) $x_0^2 x_1 = x_0^2 = x_{n-1} x_0^2$;

(R_4) $x_{i+1} x_i x_{i+1} = x_{i+1} x_i = x_i x_{i+1} x_i$, $1 \leq i \leq n-2$;

(R_5) $x_i x_{i+1} \cdots x_{n-1} x_0 x_1 \cdots x_{i-1} x_i = x_i$, $0 \leq i \leq n-1$;

(R_6) $x_{i+1} \cdots x_{n-1} x_0 x_1 \cdots x_{i-1} x_i^2 = x_i^2$, $1 \leq i \leq n-1$;

(R_7) $bx_i = x_{i+1} b$, $1 \leq i \leq n-2$;

(R_8) $bx_0 x_1 = x_1$ and $x_{n-1} x_0 b = x_{n-1}$;

(R_9) $b^n = 1$.

Deduced from this one, the author noticed in the same paper a presentation for the monoid $\mathcal{POPI}_n$ with the same number of relations but just two generators. Observe that it was proved in [8] that the monoid $\mathcal{POPI}_n$ has rank two and that $\{x_1, b\}$ is a generating set of $\mathcal{POPI}_n$. For simplicity, let us denote the transformation x_1 by x. Improving this result, Fernandes, Gomes and Jesus [10] exhibited a presentation for $\mathcal{POPI}_n$, in terms of the generators x and b, with the following $2n$ defining relations:

(R_1') $b^{i-1} x b^{n-i} (xb)^{n-1} = b^{n-1}(xb)^{n-1} b^i x b^{n-i}$, $1 \leq i \leq n-2$;

(R_2'') $b^{i-1}xb^{n-i+1}x = xb^{i-1}xb^{n-i+1}$, $3 \le i \le n-2$;

(R_3') $x^2 = x(xb)^{n-1}$;

(R_4'') $(xb^{n-1}xb)x = xb^{n-1}xb$ and $bxb^{n-1}x = x(bxb^{n-1}x)$;

(R_6') $b(xb)^{n-2}x^2 = x^2$;

(R_8') $(xb)^{n-1}x = x$;

(R_9') $b^n = 1$.

Based on the last presentation of $\mathcal{POPI}_n$ showed above, Fernandes, Gomes and Jesus [10], using the technique given by Theorem 1.3, also exhibited a presentation for $\mathcal{PORI}_n$, in terms of the generators x, b and d, with the following $2n+4$ defining relations:

(R_1') $b^{i-1}xb^{n-i}(xb)^{n-1} = b^{n-1}(xb)^{n-1}b^i xb^{n-i}$, $1 \le i \le n-2$;

(R_2'') $b^{i-1}xb^{n-i+1}x = xb^{i-1}xb^{n-i+1}$, $3 \le i \le n-2$;

(R_3') $x^2 = x(xb)^{n-1}$;

(R_4'') $(xb^{n-1}xb)x = xb^{n-1}xb$ and $bxb^{n-1}x = x(bxb^{n-1}x)$;

(R_6') $b(xb)^{n-2}x^2 = x^2$;

(R_8') $(xb)^{n-1}x = x$;

(R_9') $b^n = 1$;

(NR_0) $d^2 = 1$;

(NR_1) $db = b^{n-1}d$ and $dx = b^{n-1}(xb)^{n-2}b^2 d$;

(NR_2) $(b^{n-1}(xb)^{n-1})^{n-2}d = (b^{n-1}(xb)^{n-1})^{n-2}((xb)^{n-2}b^2)^{n-2}b^{n-1}$.

Notice that the monoid $\mathcal{PORI}_n$ has $1 + n\binom{2n}{n} - \frac{n^2}{2}(n^2 - 2n + 3)$ elements and rank three (see [10]).

Acknowledgments

The author wishes to express many thanks to M.V. Volkov for his helpful comments and translations from the papers by L.M. Popova [19, 20].

References

[1] A.Ya. Aĭzenštat, Defining relations of finite symmetric semigroups, *Mat. Sb. N. S.* **45 (87)** (1958), 261-280 (Russian).

[2] A.Ya. Aĭzenštat, The defining relations of the endomorphism semigroup of a finite linearly ordered set, *Sibirsk. Mat.* **3** (1962), 161-169 (Russian).

[3] R.E. Arthur and N. Ruškuc, Presentations for two extensions of the monoid of order-preserving mappings on a finite chain, *Southeast Asian Bull. Math.* **24** (2000), 1-7.

[4] P.M. Catarino and P.M. Higgins, The monoid of orientation-preserving mappings on a chain, *Semigroup Forum* **58** (1999), 190-206.

[5] P.M. Catarino, Monoids of orientation-preserving transformations of a finite chain and their presentation, *Semigroups and Applications*, 39-46, eds. J.M. Howie & N. Ruškuc, Proceedings of the Conference in St Andrews, Scotland, 1997, World Scientific, 1998.

[6] H.S.M. Coxeter and W.O.J. Moser, *Generators and Relations for Discrete Groups*, (Reprint of Fourth Ed.) Springer-Verlag, 1984.

[7] V.H. Fernandes, Semigroups of order-preserving mappings on a finite chain: a new class of divisors, *Semigroup Forum* **54** (1997), 230-236.

[8] V.H. Fernandes, The monoid of all injective orientation preserving partial transformations on a finite chain, *Comm. Algebra* **28** (2000), 3401-3426.

[9] V.H. Fernandes, The monoid of all injective order preserving partial transformations on a finite chain, *Semigroup Forum* **62** (2001), 178-204.

[10] V.H. Fernandes, G.M.S. Gomes and M.M. Jesus, *Presentations for some monoids of injective partial transformations on a finite chain*, submitted.

[11] V.H. Fernandes, G.M.S. Gomes and M.M. Jesus, *Presentations for some monoids of partial transformations on a finite chain*, work in progress.

[12] G.M.S. Gomes and J.M. Howie, On the ranks of certain semigroups of order-preserving transformations, *Semigroup Forum* **45** (1992), 272-282.

[13] J.M. Howie, Products of idempotents in certain semigroups of transformations, *Proc. Edinburgh Math. Soc.* (2) **17** (1971), 223-236.

[14] J.M. Howie, *Fundamentals of Semigroup Theory*, Oxford University Press, 1995.

[15] G. Lallement, *Semigroups and Combinatorial Applications*, John Wiley & Sons, 1979.

[16] S. Lipscomb, *Symmetric Inverse Semigroups*, Mathematical Surveys and Monographs, Vol. 46, American Mathematical Society, 1996.

[17] D.B. McAlister, Semigroups generated by a group and an idempotent, *Comm. Algebra* **26** (1998), 515-547.

[18] E.H. Moore, Concerning the abstract groups of order $k!$ and $\frac{1}{2}k!$ holohedrically isomorphic with the symmetric and the alternating substitution groups on k letters, *Proc. London Math. Soc. (1)* **28** (1897), 357-366.

[19] L.M. Popova, The defining relations of certain semigroups of partial transformations of a finite set, *Leningradskij gosudarstvennyj pedagogicheskij institut imeni A. I. Gerzena, Uchenye Zapiski* **218** (1961), 191-212 (Russian).

[20] L.M. Popova, The defining relations of the semigroup of partial endomorphisms of a finite linearly ordered set, *Leningradskij gosudarstvennyj pedagogicheskij institut imeni A. I. Gerzena, Uchenye Zapiski* **238** (1962), 78-88 (Russian).

[21] N. Ruškuc, Semigroup Presentations, Ph. D. Thesis, University of St Andrews, 1995.

[22] M. Schönert et al., GAP – Groups, Algorithms, and Programming, Lehrstuhl D für Mathematik, Rheinisch Westfälische Technische Hochschule, Aachen, Germany, fifth edition, 1995.

[23] A. Solomon, Catalan monoids, monoids of local endomorphisms, and their presentations, *Semigroup Forum* **53** (1996), 351-368.

SOME RELATIVES OF AUTOMATIC AND HYPERBOLIC GROUPS

MICHAEL HOFFMANN*[†], DIETRICH KUSKE[2], FRIEDRICH OTTO[3], AND RICHARD M. THOMAS[2]

[1] *Department of Computer Science, Loughborough University, Loughborough LE11 3TU, England. email:* M.Hoffmann@lboro.ac.uk

[2] *Department of Mathematics and Computer Science, University of Leicester, Leicester LE1 7RH, England. email:* {D.Kuske, R.Thomas}@mcs.le.ac.uk

[3] *Fachbereich Mathematik/Informatik, Universität Kassel, 34109 Kassel, Germany. email:* otto@theory.informatik.uni-kassel.de

Automatic and hyperbolic groups are well-studied classes. They share the property that the multiplication can be described in terms of computing devices. The theory of automatic groups has been extended to automatic monoids. In this paper we consider also hyperbolic, asynchronously automatic and rational monoids. We consider the independence of the definitions from the generating set, some closure properties of the classes of asynchronously automatic and of hyperbolic monoids, the complexity of the word problem, and we describe the relation between these classes of monoids.

1 Introduction

The concept of an automatic group was introduced in [2, 10] in order to describe a large class of naturally occurring groups with an easily solvable word problem. This notion was then extended to automatic monoids [5, 22, 25]. In both cases the defining property is the existence of a regular set of normal forms (with respect to some finite generating set A) such that we have a finite automaton with two input tapes that can verify whether two normal forms represent the same monoid element, and, for each generator in A, a finite automaton that can determine whether one normal form represents the monoid element that corresponds to the other normal form multiplied by the

---

*This research was carried out while M Hoffmann worked at the University of Leicester.
*THE AUTHORS WOULD LIKE TO ACKNOWLEDGE THE FINANCIAL SUPPORT OF FUNDAÇÃO CALOUSTE GULBENKIAN (FCG), FUNDAÇÃO PARA A CIÊNCIA E A TECNOLOGIA (FCT), FACULDADE DE CIÊNCIAS DA UNIVERSIDADE DE LISBOA (FCUL) AND REITORIA DA UNIVERSIDADE DO PORTO.

generator in question. According to whether the two input tapes are read synchronously or asynchronously, one distinguishes between automatic and asynchronously automatic monoids.

An important subclass of the class of automatic groups is that of "hyperbolic" groups. These were introduced by Gromov [13] and there are several different equivalent formulations of this notion [1, 13]. Recently, Gilman [12] gave an elegant characterization of hyperbolic groups in terms of pushdown automata. Again, one requires the existence of a regular set of normal forms (over some finite generating set); in this case the pushdown automaton reads three normal forms in sequel and decides whether the third represents the product of the first two. This characterization of hyperbolic groups, unlike many of the others, extends easily to monoids giving rise to the notion of hyperbolic monoids. We will see, however, that, whilst hyperbolic groups are necessarily automatic, this implication no longer holds when we generalize to monoids.

A fourth class of monoids that can be described in this sort of way is the class of "rational" monoids. Here one requires the existence of a set of unique normal forms (over some finite generating set) such that the normal form of any word can be computed by a finite transducer (i.e. a finite automaton with output). So, as in the other notions, we have a regular set of normal forms. Unlike the other concepts considered here, which were first defined for groups and then generalized to monoids, the notion of being rational was first defined (by Sakarovitch in [29]) in the wider class of monoids. As Sakarovitch pointed out, if we restrict ourselves to groups, the concept is not very interesting, since a group is a rational monoid if and only if it is finite; however, as a class of monoids, they are of considerable interest [29, 26, 28].

We study algebraic and computational properties of and the relations between these classes of monoids. In doing so, we consider the independence of the definitions from the generating set, the closure of the classes of asynchronously automatic monoids and of hyperbolic monoids when taking free products, and the complexity of the word problem. Furthermore, the complete inclusion structure of these classes of monoids is described.

2 Definitions, notation and preliminary results

In this section we will survey some notions from the theories of formal languages and of monoids we will use in this paper. The reader is referred to some standard texts (e.g., [3, 9, 14, 20, 21]) and various papers (e.g., [4, 11, 15, 16]) for further details. *Throughout, we will assume that all the monoids considered here are finitely generated.*

For a finite set A, we let A^* denote the free monoid generated by A; its elements are the finite words over the alphabet A and the empty word is denoted by ε. We write $|w|$ for the length of the word w. The reversal of a word $w \in A^*$ will be denoted by $\overline{w}$.

The free product of two monoids M and N is denoted by $M * N$. If M is a monoid and S and T are subsets of M then $S \cdot T$ denotes the set $\{x \cdot y : x \in S, y \in T\}$ of products of elements of S and T. Furthermore, $\langle S \rangle$ is the submonoid of M generated by S. If $M \times N$ is the direct product of two monoids M and N, we let π_1 and π_2 be the projections defined by $(m,n)\pi_1 = m$ and $(m,n)\pi_2 = n$ respectively.

If M is a monoid generated by a finite set A and if $u, v \in A^*$, then we use the notation $u =_M v$ to denote that u and v represent the same element of M, in case where $M = A^*$, we use $u \equiv v$ instead of $u =_{A^*} v$.

A *context-free grammar* over a monoid M is a tuple $G = (M, N, \rightarrow, s)$ where N is a finite set of *non-terminal symbols* disjoint from M, $\rightarrow$ is a finite subset of $N \times (M * N^*)$ and $s \in N$ is the *axiom*. Elements of $\rightarrow$ are called *productions*. Let $\Rightarrow_G \subseteq (M * N^*)^2$ denote the least transitive and reflexive relation on $M * N^*$ satisfying $x\ell y \Rightarrow_G xry$ for all $\ell \rightarrow r$. The set of *sentential forms of G* is the least subset X of $M * N^*$ with $s \in X$ and such that $v \in X$ whenever $u \in X$ and $u \rightarrow v$. The *language $L(G)$ of G* is the set of sentential forms that belong to the monoid M.

In a free monoid A^* it is well known that a language can be generated by a context-free grammar if and only if it is accepted by some (nondeterministic) pushdown automaton; such a language is said to be *context-free*. We can generalize the notion of a context-free language to subsets of arbitrary monoids by saying that a subset S of a monoid M is *context-free* if there is a finite set A and an epimorphism $\varphi : A^* \rightarrow M$ such that $S\varphi^{-1}$ is context-free (one can readily check that this is independent of the choice of A and φ). This is not equivalent to saying that S is generated by a context-free grammar over M; such a language is said to be *algebraic*. An equivalent formulation in that case is to say that a subset S of a monoid M is algebraic if there is a finite set A and an epimorphism $\varphi : A^* \rightarrow M$ such that $S = L\varphi$ for some context-free language $L \subseteq A^*$. From this, we see that any context-free subset of a monoid is necessarily algebraic. In the case of groups, the sets of context-free and algebraic subsets coincide in G if and only if G is either finite or has an infinite cyclic subgroup of finite index [15].

A context-free grammar is said to be *regular* if each of its productions is

381

of the form $x \to my$ with $m \in M$ and $y \in N \cup \{\varepsilon\}$.[a]

A subset of a monoid M that can be generated by a regular grammar is said to be a *rational subset* of M. Rational subsets can also be described set-theoretically as follows:

Definition 2.1. Let M be a monoid. The class $\mathrm{Rat}(M)$ of *rational subsets* of M is the least subset of $\mathcal{P}(M)$ satisfying

- any finite subset of M is rational;
- if $K, L \subseteq M$ are rational, then so are $K \cup L$, $K \cdot L$ and $\langle K \rangle$.

In a free monoid A^* it is well known that a language can be generated by a regular grammar if and only if it is accepted by some (deterministic or nondeterministic) finite automaton; such a language is said to be *regular* or *recognizable*. We can generalize the notion of recognizable to subsets of arbitrary monoids by saying that a subset S of a monoid M is *recognizable* if there is a finite set A and an epimorphism $\varphi : A^* \to M$ such that $S\varphi^{-1}$ is recognizable.

Yet another formulation of the idea of a rational subset is to say that a subset S of a monoid M is rational if there is a finite set A and an epimorphism $\varphi : A^* \to M$ such that $S = L\varphi$ for some recognizable language L; from this we see that any recognizable subset is rational. A monoid in which the sets of recognizable and rational subsets coincide is called a *Kleene monoid*. A group is a Kleene monoid if and only if it is finite.

Next, we establish some preliminary results we will need in this paper. A *rational relation* or *rational transduction* is a rational set in the direct product $M \times N$ of two monoids. A *rational function* is a rational relation in $M \times N$ that happens to be (the graph of) a partial function from the monoid M into the monoid N. Note that, if M is a monoid generated by a finite set A, then any homomorphism $\varphi : M \to N$ into some monoid N is a rational function since its graph $\langle\{(a, a\varphi) : a \in A\}\rangle$ is rational.

In general, the class of rational subsets of a monoid is neither closed under complementation nor under intersection. However, there are some interesting closure properties of rational subsets when we consider free monoids and rational relations between them.

Proposition 2.1. *Let A, B and C be finite sets.*

(1) *The set $\mathrm{Rat}(A^*)$ is closed under complementation and under intersection.*

---

[a]Productions of the form $x \to y$ (or even $x \to \varepsilon$) are allowed since M contains an identity element.

(2) If $R \in \text{Rat}(A^* \times B^*)$, $K \in \text{Rat}(A^*)$ and $L \in \text{Rat}(B^*)$, then R^{-1} and $R \cap (K \times L)$ are elements of $\text{Rat}(A^* \times B^*)$ and $R\pi_1$ is an element of $\text{Rat}(A^*)$.

(3) If $R \in \text{Rat}(A^* \times B^*)$ and $S \in \text{Rat}(B^* \times C^*)$, then $R \circ S \in \text{Rat}(A^* \times C^*)$.

Note that, whilst context-free sets are not closed under intersection, the intersection of a context-free set and a rational set in a free monoid is always context-free. In addition, the classes of rational and context-free languages are closed under taking images via rational relations; in other words, if $R \subseteq A^* \times B^*$ is a rational relation and if L is a rational subset of A^*, then

$$\{\beta \in B^* : (\alpha, \beta) \in R \text{ for some } \alpha \in L\} = [R \cap (L \times A^*)]\pi_2$$

is rational, and similarly for context-free languages.

Let M be a monoid generated by a finite set A and let $\varphi : A^* \to M$ be the natural epimorphism. A subset L of A^* is said to be a *cross-section* of M if φ maps L bijectively onto M; L is then said to be a *rational cross-section* if, in addition, L is a rational subset of A^*. The following result will be useful to us later:

Proposition 2.2. *Let M be a monoid generated by a finite set A and suppose that $e \in A$ (where e is the identity element of M). If $L \subseteq A^*$ is a rational cross-section of M then there exists a constant $k \geqslant 0$ such that no word in L has more than k consecutive occurrences of e.*

Proof. Assume that there are words $u_m e^m v_m \in L$ for any $m \in \mathbb{N}$. By the Pumping Lemma for regular languages, there exist $m, a, b, c \in \mathbb{N}$ with $b > 0$ and $u_m e^a (e^b)^n e^c v_m \in L$ for any $n \in \mathbb{N}$. Since all these words represent the same monoid element, L is not a cross-section, a contradiction. □

Lastly, we have the following result on context-free languages; it states that they are closed under context-free substitutions and under iterated context-free substitutions.

Proposition 2.3. *Let A be a finite set and suppose that $K, L \subseteq A^*$ are context-free languages. Let $\# \in A$; then the following sets are both context-free:*

$$K \cdot_\# L = \{u_0 v_0 u_1 v_1 \ldots u_n v_n u_{n+1} \in A^* :$$
$$u_0 \# u_1 \ldots u_n \# u_{n+1} \in K, \ v_0, v_1, \ldots, v_n \in L\},$$
$$K^\# = \bigcup_{n \geqslant 1} K^n \text{ where } K^1 = K \text{ and } K^{n+1} = K \cdot_\# K^n.$$

Proof. Let $G_K = (A^*, N_K, \to_K, s_K)$ and $G_L = (A^*, N_L, \to_L, s_L)$ be context-free grammars generating the languages K and L respectively. Without loss of generality we may assume that $N_K \cap N_L = \emptyset$. We construct a new context-free grammar $G = (A^*, N_K \cup N_L, \to, s_K)$ as follows:

$$\to \; = \; \to_K \cup \to_L \cup \tag{1}$$
$$\{(x, u_0 s_L u_1 s_L \ldots u_n s_L u_{n+1}) : x \to_K u_0 \# u_1 \ldots \# u_{n+1}\}.$$

One can easily verify that the language of G is $K \cdot_\# L$.

Similarly, the grammar $(A^*, N_K, \to, s_K)$ generates $K^\#$ with

$$\to \; = \; \to_K \cup \{(x, u_0 s_K u_1 s_K \ldots u_n s_K u_{n+1}) : x \to_K u_0 \# u_1 \ldots \# u_{n+1}\}.$$

Note that this is not the same as the grammar given in (1) above with $K = L$ due to our assumption there that $N_K \cap N_L = \emptyset$.

So $K \cdot_\# L$ and $K^\#$ are both context-free as required. □

Similarly to $K \cdot_\# L$ and $K^\#$, we can define $K \cdot_\varepsilon L$ and K^ε: this time, a word from L can be inserted into a word from K at any position (the operators $\cdot_\varepsilon$ and $^\varepsilon$ have been considered in [23] as "sequential insertion" and "insertion closure", respectively.).

Proposition 2.4. *Let B be a finite set and suppose that $K, L \subseteq B^*$ are context-free languages. Then $K \cdot_\varepsilon L$ and K^ε are context-free.*

Proof. Suppose $\# \notin B$ and let $A = B \cup \{\#\}$. Then consider the rational relation $R = \langle \{(b, b) : b \in B\} \cup \{(\varepsilon, \#)\} \rangle$. The image of the context-free set K under this relation, i.e., $H = [R \cap (K \times A^*)]\pi_2$ is context-free. Its elements are obtained from words from K by inserting arbitrary many instances of $\#$. Hence $K \cdot_\varepsilon L = (H \cdot_\# L) \cap B^*$ which is context-free by Proposition 2.3. Similarly, $K^\varepsilon = H^\# \cap B^*$ is context-free. □

3 Classes of monoids

We now consider the various classes of monoids described in Section 1. The connecting themes are that we have a rational set of normal forms in each case and that these are classes of monoids where computation is, in some sense, feasible. We first look at the notion of a "rational monoid":

Definition 3.1. Let M be a monoid. A pair (A, L) is a *rational structure* for M if A is a finite generating set and $L \subseteq A^*$ is a rational cross-section such that the function $\varphi : A^* \to A^*$ defined by $w =_M w\varphi \in L$ for all $w \in A^*$ is rational. The monoid M is *rational* if it has a rational structure. □

Rational monoids were introduced by Sakarovitch in [29]. They have several nice properties; for example, they are Kleene monoids [29] although not all Kleene monoids are rational [26]. Since no infinite group is a Kleene monoid, and since all finite monoids are rational, we immediately have:

Lemma 3.1. *A group is rational if and only if it is finite.*

A finitely generated submonoid of a rational monoid is rational [29]. Furthermore, the presence of a rational monoid allows to concatenate rational relations. Let M_1, M_2, and M_3 be monoids and let $R_i \subseteq M_i \times M_{i+1}$ ($i = 1, 2$) be rational relations. If M_2 is rational, then the relation $R_1 \circ R_2 \subseteq M_1 \times M_3$ is a rational relation as well. In general (i.e. if we remove the hypothesis that M_2 is rational), this is false, and it is an open conjecture as to whether this property characterizes rational monoids.

We now consider "asynchronously automatic" monoids:

Definition 3.2. Let M be a monoid. A pair (A, L) is an *asynchronously automatic structure* for M if A is a finite generating set for M and $L \in \text{Rat}(A^*)$ is mapped surjectively onto M such that the relations

$$L_\varepsilon = \{(u, v) \in L \times L : u =_M v\} \quad \text{and} \quad L_a = \{(u, v) \in L \times L : ua =_M v\}$$

are rational for any $a \in A$. A monoid is *asynchronously automatic* if it has an asynchronously automatic structure. □

There has been a systematic study of asynchronously automatic groups that are often defined in terms of asynchronous automata, see [10]. One can show that any asynchronously automatic monoid (in the sense of Definition 3.2) that happens to be a group satisfies the asynchronous fellow traveler property (cf. [10]) and is therefore an asynchronously automatic group in the usual sense; the converse is also true (and follows fairly immediately) so that the two definitions are equivalent for groups. Up to now there has been less work on asynchronously automatic monoids however.

We now turn to the notion of an "automatic monoid" where we take a synchronous version of Definition 3.2. In order to do this, we need to be able to "pad" two strings together. Let A be a finite set and let $ be a symbol with $ \notin A$. For two words u and v over A, we define the *padded string* $(u,v)\delta$ over $(A \cup \{\$\}) \times (A \cup \{\$\})$ inductively as follows:

- for any $a, b \in A$, we let $(a, b)\delta \equiv (a, b)$.

- for any $a, b \in A$ and for any $u, v \in A^*$, we let $(au, bv)\delta \equiv (a, b)(u, v)\delta$, $(au, \varepsilon)\delta \equiv (a, \$)(u, \varepsilon)\delta$ and $(\varepsilon, bv)\delta \equiv (\$, b)(\varepsilon, v)\delta$.

Given this, we can define the notion of an automatic monoid:

Definition 3.3. Let M be a monoid. A pair (A, L) is an *automatic structure* for M if A is a finite generating set and $L \in \text{Rat}(A^*)$ is mapped surjectively onto M such that the sets

$$L_\varepsilon^\delta = \{(u,v)\delta : u,v \in L, u =_M v\} \quad \text{and} \quad L_a^\delta = \{(u,v)\delta : u,v \in L, ua =_M v\}$$

are rational subsets of the free monoid generated by $(A \cup \{\$\}) \times (A \cup \{\$\})$ for any $a \in A$. A monoid is *automatic* if it has an automatic structure. □

Automatic groups have been the subject of intensive study and there is now quite a comprehensive literature on automatic monoids as well. We should mention (in passing) that there are different possible variations as to precisely how one extends the notion of automaticity from groups to monoids although this will not concern us here (the interested reader is refered to [17, 19]).

Lastly, we consider "hyperbolic monoids":

Definition 3.4. Let M be a monoid, A be a finite generating set for M, and let $\#_1$ and $\#_2$ be symbols with $\#_1, \#_2 \notin A$. Let L be a rational subset of A^* that maps surjectively onto M. Then the pair (A, L) is a *hyperbolic structure* for M if the set

$$\{u\#_1 v\#_2 \overline{w} : u,v,w \in L, uv =_M w\}$$

is a context-free subset of the free monoid $(A \cup \{\#_1, \#_2\})^*$. A monoid is *hyperbolic* if it has a hyperbolic structure. □

This definition has been proposed by Duncan and Gilman [6]. Unlike other definitions of the notion of hyperbolicity in groups, this one does extend naturally to monoids, and a group is hyperbolic (as a monoid) in the sense of Definition 3.4 if and only if it is hyperbolic as a group according to the other (more usual) definitions [12].

The word problem for rational monoids is solvable in linear time [29] since one can compute a normal form of a word w by a transducer. For automatic groups, the word problem is solvable in quadratic time [10] and the same holds for automatic monoids [5]: in order to compute a normal form for a word $a_1 a_2 \ldots a_n$, one computes successively normal forms of the prefix $a_1 a_2 \ldots a_i$ by multiplying the normal form of $a_1 a_2 \ldots a_{i-1}$ with a_i. These multiplications can be carried out in linear time. Since the lengths of two consecutive normal forms differ at most by a constant, the computation of a normal form can be done altogether in quadratic time. Finally, one checks (in linear time) whether the normal forms we have computed represent the same monoid element.

A similar strategy can be applied to compute a normal form in an asynchronously automatic structure. The multiplication by a generator can still be done in linear time, but now the length of the normal form of $a_1 a_2 \ldots a_i$ can be twice that of the normal form of $a_1 a_2 \ldots a_{i-1}$. Since this difference can still be bounded by a multiplicative factor, one gets altogether an exponential upper bound for the time needed to compute a normal form. This shows that the word problem is still decidable if we consider asynchronously automatic monoids.

The remainder of this section is devoted to the proof that the word problem for hyperbolic monoids is decidable in deterministic exponential time.

Lemma 3.2. *Let (Σ, L) be a hyperbolic structure for a monoid M. Then there exists a constant $c \geqslant 0$ such that the following task can be solved in deterministic polynomial time:*

INPUT: *Two words $u, v \in L$.*
OUTPUT: *A word $w \in L$ satisfying $uv =_M w$ and $|w| \leqslant c(|uv| + 2)$.*

Proof. Let $B = A \cup \{\#_1, \#_2\}$. By our hypothesis the language

$$S = \{u\#_1 v \#_2 \overline{w} : u, v, w \in L, uv =_M w\} \subseteq B^*$$

is context-free. Hence there exists a context-free grammar $G = (B^*, N, \rightarrow, s)$ in quadratic Greibach normal form such that $L(G) = S$ (see [8]), that is, each production $(X \rightarrow w)$ of the grammar G satisfies the restriction that $w \in B \cup B \cdot N \cup B \cdot N^2$. Further, we may assume without loss of generality that the set

$$L(G, X) := \{w \in B^* : X \Rightarrow_G w\}$$

is non-empty for each nonterminal $X \in N$. Then we can choose a morphism $\psi : N^* \rightarrow B^*$ as follows. For each $X \in N$, let $w_X \in B^*$ be a (shortest) word from the set $L(G, X)$. Then ψ is obtained by mapping X to w_X for all $X \in N$.

Observe that the grammar G and the mapping ψ can be obtained by preprocessing, that is, for the solution of the problem considered here, they can be considered as given.

Let c be the maximal length of a word $X\psi$ for $X \in N$. We will show that, for $u, v \in L$, we can, in polynomial time, compute some word $\alpha \in N^*$ such that $s \Rightarrow_G u\#_1 v \#_2 \alpha$ and $|\alpha| \leqslant |uv| + 3$. From this word α, the word

$w = \alpha\psi \in A^*$ can be computed in time $O(|\alpha|)$. The length $|w|$ of w is bounded by $c(|uv|+3)$. Then

$$s \Rightarrow_G u\#_1 v\#_2 \alpha \Rightarrow_G u\#_1 v\#_2 w$$

implies $u\#_1 v\#_2 w \in S$ and therefore $\overline{w} \in L$ and $uv =_M \overline{w}$. Thus, indeed, it suffices to show that α can be computed in deterministic polynomial time (we even show that this can be done in linear time).

Let x denote the word $u\#_1 v\#_2$, let $n := |uv|+2 = |x|$, and let $x_1, \ldots, x_n \in B$ be such that $x = x_1 \ldots x_n$. The computation of α proceeds in three main steps.

(1.) First, for each $i, j \in \{1, \ldots, n\}$ with $i \leqslant j$, let

$$V_{i,j} := \{ X \in N : x_i \ldots x_j \in L(G, X) \}.$$

In order to determine the set $V_{i,j}$, we must test membership of the factor $x_i \ldots x_j$ in the context-free language $L(G, X)$ for each nonterminal $X \in N$. As $|x_i \ldots x_j| = j-i+1$, each such test can be executed in time $O((j-i+1)^3)$. As there are $n(n-1)/2$ sets $V_{i,j}$, we see that the computation of all these sets can be realized in time $O(n^5)$.

(2.) For each $i \in \{1, \ldots, n\}$, let

$$N_i := \{ X \in N : \exists \alpha \in N^* : X \Rightarrow_G x_i \ldots x_n \alpha \}.$$

In order to determine these sets effectively, we proceed by induction from n down to 1 as follows. Obviously, $N_n = \{ X \in N : \exists \alpha \in N^* : X \Rightarrow_G x_n \alpha \}$ can be expressed as

$$N_n = \{ X \in N : \exists \beta \in N^* : X \to x_n \beta \},$$

and so N_n can be determined simply by inspecting all the productions of G.

Next let us assume that the sets $N_{i+1}, \ldots, N_n$ have already been computed. For computing the set N_i we use the following observation. If $X \in N_i$, that is, $X \Rightarrow_G x_i \ldots x_n \alpha$ for some $\alpha \in N^*$, then there is a production $X \to x_i \beta$ such that $\beta \Rightarrow_G x_{i+1} \ldots x_n \alpha$ holds. As $i < n$, $x_{i+1} \ldots x_n \neq \varepsilon$ implying that $\beta \neq \varepsilon$, that is, $\beta \in N \cup N^2$.

Thus we must consider all productions of this form. Let $X \to x_i \beta$ be such a production. If $\beta \in N$, then $\beta \in N_{i+1}$ must hold, and this can be checked, since N_{i+1} has already been determined. If $\beta = YZ$ for some $Y, Z \in N$, there are two possibilities. Either $Y \in N_{i+1}$, or else $Y \Rightarrow_G x_{i+1} \ldots x_j$ and $Z \Rightarrow_G x_{j+1} \ldots x_n \alpha$ for some $j \leqslant n-1$. Thus $X \in N_i$ if and only if

(i) there is a production $X \to x_i Y$ such that $Y \in N_{i+1}$, or

(ii) there is a production $X \to x_i YZ$ such that $Y \in N_{i+1}$, or

(iii) there is a production $X \to x_i Y Z$ such that $Y \in V_{i+1,j}$ and $Z \in N_{j+1}$ for some $j \in \{i+1, \ldots, n-1\}$.

Hence the set N_i can be computed effectively. Actually, from the above characterization we see that this computation can be realized in time $O(n-i)$. Hence, the computation of all the sets $N_1, \ldots, N_n$ takes time $O(n^2)$.

(3.) Using the sets N_i, $1 \leqslant i \leqslant n$, and $V_{i,j}$, $1 \leqslant i \leqslant j \leqslant n-1$, we can now determine a word $\alpha \in N^*$ satisfying $s \Rightarrow_G x\alpha$. We use a stack ST that is initialized with s, and that, at the end of the computation, will contain the word α with the first letter of α on the top and the last letter at the bottom.

As $u, v \in L$, there exists some word $\alpha \in N^*$ such that $s \Rightarrow_G x\alpha$, and so $s \in N_1$, and there exists a production $s \to x_1 \beta$ such that

(i) $\beta = Y \in N_2$, or

(ii) $\beta = YZ$ satisfying $Y \in N_2$, or

(iii) $\beta = YZ$ satisfying $Y \in V_{2,j}$ and $Z \in N_{j+1}$ for some $j \in \{2, \ldots, n-1\}$.

First we check whether there exists a production of type (i). If $s \to x_1 Y$ is such a production, we replace s by Y on the top of the stack ST and we move to the letter x_2. If no production satisfies (i), we check whether there is a production of type (ii). If $s \to x_1 YZ$ is such a production, we replace s by YZ on the top of the stack ST and we move to the letter x_2. Finally, if no production satisfies (ii), we choose a production $s \to x_1 YZ$ satisfying (iii), that is, $Y \in V_{2,j}$ and $Z \in N_{j+1}$ for some $j \in \{2, \ldots, n-1\}$. We then replace s by Z on the top of the stack ST and move to the letter x_{j+1}. Actually, we would choose the production in such a way that the index j is as large as possible.

In this way we have reached the following situation. The stack ST contains a word $\gamma \in N^+$, the topmost symbol X is from the set N_i, and we still need to process the suffix $x_i \ldots x_n$ of the input, where $i \geqslant 2$. As $X \in N_i$, we can proceed in the same way as before for s. Thus, after at most n iterations we have completely processed x and thereby created a word $\alpha \in N^*$ as the contents of the stack ST. In each step, we replace one symbol from the stack by at most two symbols, i.e., the length of α exceeds by at most one the number of steps it takes to compute α (since at the beginning, the stack contains the word s). Hence $|\alpha| \leqslant |x| + 1 = |uv| + 3$. This word α is the required output.

Each iteration of the process outlined in (3.) takes linear time in the length of the remaining suffix of x, and so this computation can be performed in time $O(n^2)$. Thus, the word α is obtained from u and v in time $O(|uv|^5)$ as required. □

Proposition 3.3. *Let (A, L) be a hyperbolic structure for a monoid M. Then the following task can be solved in time $2^{O(n)}$:*

INPUT: A word $u \in A^*$.
OUTPUT: A word $v \in L$ such that $u =_M v$.

Proof. For each $a \in A$, let $x_a \in L$ be a representative of the monoid element a, and let $x_e \in L$ be a representative of the identity element of M.

Given the word $u = a_1 a_2 \ldots a_n$, where $a_1, \ldots, a_n \in A$, we successively compute representatives $u_i \in L$ such that $u_i =_M a_1 \ldots a_i$ by applying the procedure from Lemma 3.2 to the inputs u_{i-1} and x_{a_i}, starting with $u_0 := x_e$. Let $c \geqslant 3 + |x_a|$ for all $a \in A$ be a constant that exceeds the constant from Lemma 3.2. Then $|u_i| \leqslant c(|u_{i-1}| + c)$ and therefore $|u_i| \leqslant d^{i+1}$ for some constant d. Hence the computation of u_i from u_{i-1} can be executed in deterministic time polynomial in 2^i, i.e. in $2^{O(i)}$, which implies that $|v|$ can be computed in deterministic time $2^{O(n)}$. □

Theorem 3.4. *The word problem for any hyperbolic monoid can be solved in time $2^{O(n)}$.*

Proof. Given $u, v \in A^*$, representatives $u_0, v_0 \in L$ can be computed in time $2^{O(n)}$ by Proposition 3.3. Now $u =_M v$ if and only if $u_0 =_M v_0$ if and only if $y := u_0 \#_1 x_e \#_2 \overline{v_0}$ belongs to the context free language S, which in turn can be checked in time $O(|y|^3)$. As $|y| \in 2^{O(n)}$, we see that the whole process can be performed in time $2^{O(n)}$. □

4 Asynchronously automatic and hyperbolic structures

With each of our notions the property is defined in terms of a suitable structure; for example, asynchronously automatic monoids require the existence of an asynchronously automatic structure. In this section, we establish some particular extra properties of asynchronously automatic and hyperbolic structures that we can require (in addition to those already stipulated in Definitions 3.2 and 3.4 respectively) without restricting the class of monoids under consideration.

If one has a rational structure with respect to one finite generating set, then one has a rational structure with respect to any finite generating set [29]. The following two propositions show that the same holds for (asynchronously) automatic and for hyperbolic structures:

Proposition 4.1. *If M is an (asynchronously) automatic monoid generated* by the finite set B then there exists an (asynchronously) automatic structure (B, K) for M.

Proof. For automatic structures, this proposition has been established in [7]. So let (A, L) be some asynchronously automatic structure for M. For each $a \in A$ choose $w_a \in B^*$ with $a =_M w_a$. Let R be the rational relation $\langle\{(w_a, a) : a \in A\}\rangle$ and then let $K = (R \cap (B^* \times L))\pi_1$. Then K is a rational subset of B^*. We also have that K is mapped surjectively onto M; this follows from the fact that, for any $m \in M$, there exists $w \equiv a_1 a_2 \ldots a_n \in L$ with $w =_M x$, and then $w_{a_1} w_{a_2} \ldots w_{a_n}$ belongs to K and represents x.

For each $b \in B$, let $v_b \in L$ be some representative of the monoid element b. Suppose that $v_b \equiv a_1 a_2 \ldots a_n$ with $a_i \in A$. Then

$$K_b = R \circ L_{a_1} \circ L_{a_2} \cdots \circ L_{a_n} \circ R^{-1}$$

is rational by Proposition 2.1. Similarly, $K_\epsilon = R \circ L_\epsilon \circ R^{-1}$ is rational. □

Proposition 4.2. *If M is a hyperbolic monoid generated by the finite set B then there exists a hyperbolic structure (B, K) for M.*

Proof. Let (A, L) be some hyperbolic structure for M so that

$$S = \{u' \#_1 v' \#_2 \overline{w'} : u', v', w' \in L, u'v' =_M w'\}$$

is context-free. For each $a \in A$, let $w_a \in B^*$ be a representative of the monoid element a. Consider the rational relation $R \subseteq B^* \times A^*$ defined by $R = \langle\{(w_a, a) : a \in A\}\rangle$. Then $K = (R \cap (B^* \times L))\pi_1$ is a rational subset of B^* that is mapped surjectively onto M (as in the proof of Proposition 4.1).

Let T be the set of all words $u\#_1 v\#_2 \overline{w}$ with $u, v, w \in K$ for which there exist $u', v', w' \in L$ with $(u, u'), (v, v'), (w, w') \in R$ and $u'\#_1 v'\#_2 \overline{w'} \in S$. Note that

$$T = [(R \cdot \{(\#_1, \#_1)\} \cdot R \cdot \{(\#_2, \#_2)\} \cdot \overline{R}) \cap ((B \cup \{\#_1, \#_2\})^* \times S)]\pi_1$$

where $\cdot$ is the multiplication in the monoid $B^* \times A^*$ and

$$\overline{R} = \{(\overline{v}, \overline{w}) : (v, w) \in R\}.$$

This set T is context-free since S is context-free and R is a rational relation.

We show that T describes the multiplication for elements of K. Let $u, v, w \in K$; then there exist $u', v', w' \in L$ with $(u, u'), (v, v'), (w, w') \in R$. Hence, in particular, $u =_M u'$ etc. Thus we have

$$uv =_M w \iff u'v' =_M w' \iff u'\#_1 v'\#_2 \overline{w'} \in S \iff u\#_1 v\#_2 \overline{w} \in T.$$

Hence (B, K) is a hyperbolic structure for M as required. □

We now come to the notion of a structure "with uniqueness":

Definition 4.1. An asynchronously automatic, automatic or hyperbolic structure (A, L) is said to be *with uniqueness* if L is a rational cross-section.

There is no point in introducing a notion of a "rational structure with uniqueness" since L is a rational cross-section for any rational structure (A, L) by definition. In the case of automatic monoids, we always have an automatic structure with uniqueness [5]. Our attempts to exend this result to cover asynchronously automatic monoids is only partially successful. Recall that a rational relation is defined as a rational subset of $A^* \times B^*$. An alternative definition uses two-tape automata (also known as transducer, cf. [3]): a relation $R \subseteq A^* \times B^*$ is rational iff it can be accepted by a finite transducer. Differently from one-tape automata, deterministic transducers give rise to the strictly smaller class of *deterministic rational relations* (cf. [24] or [27] for the precise definitions).

Proposition 4.3. *Let (A, L) be an asynchronously automatic structure for a monoid M such that*

- *the relation L_ε is a deterministic rational relation and*

- *for any word $w \in L$, there exists a lexicographically least word $w\varphi \in L$ with $(w\varphi, w) \in L_\varepsilon$.*

Then there exists a rational cross-section $L' \subseteq L$ such that (A, L') is an asynchronously automatic structure with uniqueness for M.

Proof. Note that $\varphi : L \to L$ is an idempotent function that is rational by [24] (cf. also [27, Thm. 5.2]). Let $L' = L\varphi$ be the image of φ; we will show that (A, L') is an asynchronously automatic structure for M.

Since φ is a rational relation and L a rational set in a free monoid, we have that L' is a rational set. It is mapped bijectively onto M since L' contains a unique element in any equivalence class of L_ε. Furthermore, $L'_a = \varphi^{-1} \circ L_a \circ \varphi$ is a rational relation for $a \in A \cup \{\varepsilon\}$. □

The question as to whether a hyperbolic monoid necessarily has a hyperbolic structure with uniqueness appears to be open. In the case of groups, the answer is "yes". In that case, a hyperbolic structure (A, L) for a group G is necessarily an automatic structure (which is no longer the case for monoids, cf. Example 7.5), and we may obtain an automatic structure (A, L') with uniqueness [5] with $L' \subseteq L$. The pair (A, L') is still a hyperbolic structure for

G, since the set

$$\{u\#_1 v \#_2 \overline{w} : u,v,w \in L', uv =_M w\} = \{u\#_1 v\#_2 \overline{w} : u,v,w \in L, uv =_M w\}$$
$$\cap \; (L' \cdot \{\#_1\} \cdot L' \cdot \{\#_2\} \cdot \overline{L'})$$

is the intersection of a context-free set and a rational set in $(A \cup \{\#_1, \#_2\})^*$, and is hence context-free.

5 Free products

The free product of two automatic monoids is automatic [5]. Before we prove the analogous result for asynchronously automatic monoids, we establish the following:

Lemma 5.1. *Let (A, L) be an asynchronously automatic structure for a monoid M. Let $x \in M$ be represented by $w \in A^*$ and let $X \subseteq L$ be the set of words in L that represent x. Then $(A, (L - X) \cup \{w\})$ is an asynchronously automatic structure for M.*

Proof. We first show that $L' = (L - X) \cup \{w\}$ is rational. Let v be an arbitrary word in X; then $\{v\} \times X = L_\varepsilon \cap (\{v\} \times A^*)$ is rational. Since X is the projection of this set to the second component, X is rational. Since $\{w\}$ is finite, this implies that L' is rational.

Let $R = \{(w, v)\} \cup \langle \{(a, a) : a \in A\} \rangle$; then R is a rational relation. Note that $L'_a = (R \circ L_a \circ R^{-1}) \cap (L' \times L')$ for $a \in A \cup \{\varepsilon\}$ is a rational relation. □

Given this, we now demonstrate that the free product of asynchronously automatic monoids is asynchronously automatic:

Theorem 5.2. *If M_1 and M_2 are asynchronously automatic monoids then the free product $M_1 * M_2$ is asynchronously automatic.*

Proof. Let (A_i, L^i) be an asynchronously automatic structure for M_i ($i = 1, 2$) with $A_1 \cap A_2 = \emptyset$. We can, by Lemma 5.1, assume that the empty word is the only representative in L^i of the identity in M_i.

Let $L \subseteq (A_1 \cup A_2)^*$ consist of all words of the form $w_1 w_2 w_3 \ldots w_n$ with $n \geqslant 0$, $w_i \in (L^1 \cup L^2) - \{\varepsilon\}$ and $w_i \in A_1^*$ if and only if $w_{i+1} \in A_2^*$. We show that $(A_1 \cup A_2, L)$ is an asynchronously automatic structure for $M_1 * M_2$.

Since each L^i is rational, so is L. Then

$$L_\varepsilon = (L_\varepsilon^1 \cup \{(\varepsilon, \varepsilon)\}) \cdot \langle L_\varepsilon^2 L_\varepsilon^1 \cap (A_2^+ A_1^+)^2 \rangle \cdot (L_\varepsilon^2 \cup \{(\varepsilon, \varepsilon)\})$$

is rational.

Now let $a \in A_2$ and $u, v \in L$; then $u \equiv u_1 u_2$ where u_2 is the maximal suffix of the word u that belongs to A_2^*. Similarly, v_2 is the maximal suffix of v with $v_2 \in A_2^*$ and $v \equiv v_1 v_2$. Then $ua =_{M_1 * M_2} v$ if and only if $u_1 \equiv v_1$ and $u_2 a =_{M_2} v_2$, i.e., $(u_2, v_2) \in L_a^2$. Hence

$$L_a = (L_a^1 \cup \langle \{(b,b) : b \in A_1 \cup A_2\} \rangle \cdot \{(b,b) : b \in A_1\} \cdot L_a^2) \cap (L \times L),$$

which is a rational relation. We can obviously argue similarly for $a \in A_1$. □

We will also show that the free product of hyperbolic monoids is hyperbolic under certain assumptions; again, we need a technical result first.

Lemma 5.3. *Let (A, L) be a hyperbolic structure for a monoid M.*

i) If $x \in A^*$ then $(A, L \cup \{x\})$ is a hyperbolic structure for M.

ii) If $W \subseteq A^*$ is rational and $L - W$ maps onto M then $(A, L - W)$ is a hyperbolic structure for M.

Proof. (i) Since L maps onto M, there exists $y \in L$ with $y =_M x$ and the set $L \cup \{x\}$ maps onto M. Let R denote the set of tuples $(u_1 \#_1 u_2 \#_2 u_3, u_1' \#_1 u_2' \#_2 u_3')$ with $u_i \equiv u_i'$ or $u_i \equiv y$ and $u_i' \equiv x$ for $i = 1, 2, 3$, i.e. R replaces occurrences of y by either y or x; this relation is rational. Let $S = \{u \#_1 v \#_2 \overline{w} : u, v, w \in L, uv =_M w\}$; then

$$[R \cap (S \times (A \cup \{\#_1, \#_2\})^*)]\pi_2 = \{u \#_1 v \#_2 \overline{w} : u, v, w \in L \cup \{x\}, uv =_M w\}.$$

Since (A, L) is a hyperbolic structure for M, the set S is context-free. Since context-free languages are closed under rational relations, the set

$$\{u \#_1 v \#_2 \overline{w} : u, v, w \in L \cup \{x\}, uv =_M w\}$$

is also context-free, so that $(A, L \cup \{x\})$ is a hyperbolic structure for M as required.

(ii) Since rational sets in free monoids are closed under set differences, $L - W$ is rational. We have to show that the set

$$S = \{u \#_1 v \#_2 \overline{w} : u, v, w \in L - W, uv =_M w\}$$

is context-free. This follows from the fact that S is the intersection of the context-free set

$$\{u \#_1 v \#_2 \overline{w} : u, v, w \in L, uv =_M w\}$$

and the rational set

$$(L - W)\{\#_1\}(L - W)\{\#_2\}(\overline{L - W})$$

in a free monoid. Hence $(A, L - W)$ is a hyperbolic structure for M. □

Let (A, L) be an (asynchronously) automatic structure for the monoid M and let $x \in M$; then the set of all words $v \in L$ that represent x is rational (see beginning of the proof of Lemma 5.1). We do not know whether this is also the case for hyperbolic structures (we would imagine that it does not hold in general) which is the reason for assuming this fact explicitly in the following lemma. This result proves that the free product of two hyperbolic monoids is hyperbolic in a fairly general case.

Theorem 5.4. *Let (A_i, L_i) be hyperbolic structures for the monoids M_1 and M_2 respectively. If the sets $\{w \in L_i : w =_{M_i} 1\}$ are rational then the free product $M = M_1 * M_2$ is hyperbolic.*

Proof. By our assumption and by Lemma 5.3 we can assume that the only element in L_i that represents the identity element in M_i is the empty word ε; we write L'_i for $L_i - \{\varepsilon\}$. Furthermore, we assume that A_1 and A_2 are disjoint.

Let L be the subset of $(A_1 \cup A_2 \cup \{\$\})^*$ (where $\$ \notin A_1 \cup A_2$) consisting of all words of the form $\$w_1\$w_2\$w_3\ldots\$w_n$ with $n \geqslant 0$, $w_i \in L'_1 \cup L'_2$ and $w_i \in A_1^*$ if and only if $w_{i+1} \in A_2^*$. We will show that $(A_1 \cup A_2 \cup \{\$\}, L)$ is a hyperbolic structure for $M_1 * M_2$ where $\$$ represents the identity.

Since L_1 and L_2 map surjectively onto M_1 and M_2 respectively, and since every element m in M can be expressed as $u_1 v_1 u_2 v_2 \ldots v_{n-1} u_n v_n$ with $u_i \in M_1$ and $v_i \in M_2$, the set L maps surjectively onto M. In addition, L is rational since it equals

$$(\$L'_1 \cup \{\varepsilon\}) \langle (\$L'_2 \$L'_1) \rangle (\$L'_2 \cup \{\varepsilon\}).$$

It remains to show that the set $S = \{u\#_1 v\#_2 \overline{w} : u, v, w \in L, uv =_{M_1 * M_2} w\}$ is context-free.

By our assumption, the set $S_i = \{u\#_1 v\#_2 \overline{w} : u, v, w \in L_i, uv =_{M_i} w\}$ is context-free for $i = 1, 2$. Our first aim is to show that the following sets are also context-free:

$$O = \{u\#_1 v : u, v \in L, uv =_M 1\};$$
$$P = \{u\#_2 \bar{v} : u, v \in L, u =_M v\};$$
$$Q = \{u\#_1 \bar{v} : u, v \in L, u =_M v\}.$$

For $i = 1, 2$, the set $O'_i = \{u\#_1 v\#_2 : u, v \in L'_i, uv =_{M_i} 1\}$ is the intersection of S_i with the rational set of all words ending with $\#_2$; hence O'_i is context-free. Then also

$$O_i = \{\$u\#_1\$v : u, v \in L'_i, uv =_{M_i} 1\}$$

is context-free since it is obtained from O'_i by concatenation with $\{\$\}$ from the left, and substitution of $\#_1\$$ for $\#_1$ and ε for $\#_2$.

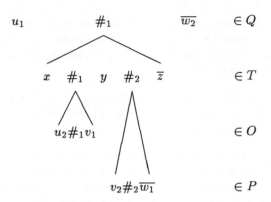

Figure 1. Expressing S using O, T, Q, P (see the proof of Theorem 5.4)

Now consider the set $O_2 \cdot_{\#_1} O_1$. It consists of all words $\$u_2\$u_1\#_1\$v_1\$v_2$ with $u_i, v_i \in L'_i$ and $u_i v_i =_{M_i} 1$. Its iteration $(O_2 \cdot_{\#_1} O_1)^{\#_1}$ is therefore the set of words

$$\$u_1\$u_2\$\ldots\$u_{2n}\#_1\$v_{2n}\$v_{2n-1}\$\ldots\$v_1$$

with $u_i, v_i \in L_1$ for i even, $u_i, v_i \in L_2$ for i odd, and $u_i v_i =_{M_i} 1$ for all i. Hence $(O_2 \cdot_{\#_1} O_1)^{\#_1}$ consists of all words $u\#_1 v$ where $u, v \in L$ start with a block from L'_2, contain an even number of blocks, and satisfy $uv =_M 1$. In order to describe all of O, we therefore first substitute this set into $(O_1 \cup \{\#_1\})$ and then $(O_2 \cup \{\#_1\})$ into the resulting set for $\#_1$. Finally, we add the element $\#_1$. The result is O:

$$O = \{\#_1\} \cup (O_1 \cup \{\#_1\}) \cdot_{\#_1} (O_2 \cdot_{\#_1} O_1)^{\#_1} \cdot_{\#_1} (O_2 \cup \{\#_1\}).$$

Hence, by Proposition 2.3, O is context-free.

To show that P is context-free we follow a similar idea. The sets $P_i = \{\$u\#_2\$\overline{v} : u, v \in L'_i, u =_{M_i} v\}$ are context-free (we intersect S_i with the rational set $A_i^*\{\#_1\#_2\}A_i^*$ and make the appropriate substitutions). Then one obtains

$$P = (P_1 \cup \{\#_2\}) \cdot_{\#_2} (P_2 \cdot_{\#_2} P_1)^{\#_2} \cdot_{\#_2} (P_2 \cup \{\#_2\})$$

which is context-free by Proposition 2.3. Finally, we see that the set Q is context-free, since P is context-free and $Q = P \cdot_{\#_2} \{\#_1\}$.

We will also use the set T consisting of all words $\$u\#_1\$v\#_2\overline{w}\$$ with $u\#_1v\#_2\overline{w} \in S_1 \cup S_2$ that satisfy

$$\varepsilon \in \{u, v, w\} \Rightarrow u \equiv v \equiv w \equiv \varepsilon.$$

The set T is context-free: it is obtained from the intersection of $S_1 \cup S_2$ with a rational set by substituting $\#_1\$$ for $\#_1$ and multiplication with $\$$ from the left and from the right.

Coming back to S, a word in S must be of the form

$$u_1 x u_2 \#_1 v_1 y v_2 \#_2 \overline{w}_1 \overline{z} \overline{w}_2$$

with $u_1, u_2, v_1, v_2, w_1, w_2, x, y, z \in L$, where $u_2 v_1$ collapses to 1, $u_1 =_M w_2$, $v_2 =_M w_1$, $xy =_M z$ and either all of x, y, z are empty or else $x, y, z \in \$L'_i$ for some i. We can express S using the sets O, P, Q and T:

$$\begin{aligned} S = \{u_1 x u_2 \#_1 v_1 y v_2 \#_2 \overline{w}_1 \overline{z} \overline{w}_2 : &\ u_1 \#_1 \overline{w}_2 \in Q,\ x \#_1 y \#_2 \overline{z} \in T, \\ &\ u_2 \#_1 v_1 \in O,\ v_2 \#_2 \overline{w}_1 \in P, \\ &\ u_1 x u_2, v_1 y v_2, w_2 z w_1 \in L\}. \end{aligned}$$

Using substitution we can write S as follows (see Figure 1):

$$S = (((Q \cdot_{\#_1} T) \cdot_{\#_1} O) \cdot_{\#_2} P) \cap (L\#_1 L\#_2 L).$$

Therefore, by Proposition 2.3, S is context-free and M is hyperbolic. □

6 Inclusions

Recall that asynchronously automatic structures are based on rational relations in $A^* \times A^*$, while automatic structures are expressed in terms of rational sets in $[(A \cup \{\$\})^2]^*$. Hence, despite their names, it is not immediately obvious that any automatic monoid is asynchronously automatic. We first establish that this is indeed the case:

Proposition 6.1. *If (A, L) be an automatic structure for the monoid M then (A, L) is an asynchronously automatic structure for M.*

Proof. Consider the homomorphism $\varphi : [(A \cup \{\$\})^2]^* \to A^* \times A^*$ given by $(a, b)\varphi = (a, b)$, $(a, \$)\varphi = (a, \varepsilon)$, $(\$, b)\varphi = (\varepsilon, b)$ and $(\$, \$)\varphi = (\varepsilon, \varepsilon)$ for $a, b \in A$. It is obvious that $L_a^\delta \varphi = L_a$ for $a \in A \cup \{\varepsilon\}$. Now the result follows since rational sets are closed under homomorphic images. □

Recall (Lemma 3.1) that a group is rational if and only if it is finite. Since there are infinite automatic groups, not every automatic monoid is rational. We will see in Example 7.4 that a rational monoid need not be automatic. However, we can show that any rational monoid is, at least, asynchronously automatic.

Theorem 6.2. *If M is a rational monoid generated by the finite set A, then (A, A^*) is an asynchronously automatic structure for M. So any rational monoid is asynchronously automatic.*

Proof. Let (A, L) be a rational structure for M and let $R \subseteq A^* \times L$ be the set of pairs (u, v) with $u =_M v$ and $v \in L$; then R is a rational function. Let $K = A^*$; we will show that (A, K) is an asynchronously automatic structure for M.

For this, it suffices to show that K_a is rational for $a \in A \cup \{\varepsilon\}$. This is clear in the case of K_ε since $K_\varepsilon = R \circ R^{-1}$. For $a \in A$, the relation $R_a = \{(u, ua) : u \in A^*\} = \langle\{(b, b) : b \in A\}\rangle(\varepsilon, a)$ is rational. Hence $K_a = R_a \circ K_\varepsilon$ is rational which completes the proof of this direction. □

Finally, we show that rational monoids are also hyperbolic:

Theorem 6.3. *If (A, L) is a rational structure for the monoid M then (A, L) is a hyperbolic structure for M with uniqueness. So any rational monoid is hyperbolic.*

Proof. Let $B = A \times \{0, 1\}$ and $\varphi : B^* \to A^* \times A^*$ be the homomorphism defined by $(a, 0)\varphi = (a, \varepsilon)$ and $(a, 1)\varphi = (\varepsilon, a)$.

Since (A, L) is a rational structure, the set L is a rational cross-section and $R = \{(u, v) \in A^* \times L : u =_M v\}$ is a rational subset of $A^* \times A^*$. Since the homomorphism φ is surjective, there is a rational set $R' \subseteq B^*$ with $R'\varphi = R$. The set

$$H = \{w\#\overline{w} : w \in R'\} = \{w\#\overline{w} : w \in B^*\} \cap (R'\#_2 B^*)$$

is the intersection of a context-free and a rational language in the free monoid B^* and is therefore context-free.

For $i = 1, 2$, let $\psi_i : B^* \to B^*$ be the projection of a word w onto its letters from $A \times \{i\}$. Since these projections are homomorphisms, the relation

$$S = \{(w, w\psi_0) : w \in B^*\} \cdot (\#_2, \#_2) \cdot \{(w, w\psi_1) : w \in B^*\}$$

is rational. Hence the image of H under S, i.e., the set $K = [S \cap (H \times B^*)]\pi_2$ is context-free. Its elements are words from H obtained by suppressing all letters from $A \times \{1\}$ before $\#_2$ and all letters from $A \times \{0\}$ after $\#_2$. Let β be the homomorphism from B^* onto A^* that simply forgets the number i from $(a, i) \in B$. Then $K\beta$ is the set of words $u\#_2\overline{v} \in A^*$ with $(u, v) \in R$, i.e., with $v \in L$ and $u =_M v$. Hence, by Proposition 2.4, the set

$$\{u_1\#_1 u_2 \#_2 \overline{v} : u_1, u_2, v \in L, u_1 u_2 =_M v\} = (K\beta \cdot_\varepsilon \#_1) \cap (L\#_1 L \#_2 A^*)$$

is context-free. Thus, indeed, (A, L) is a hyperbolic structure for M. Since L is a cross-section, it is a hyperbolic structure with uniqueness. □

7 Examples

The following examples show that all inclusions are proper and verify Figure 2.

Example 7.1. [automatic, rational] Any finite monoid is both rational and automatic and the same holds for finitely generated free monoids [5, 29]. □

Example 7.2. [automatic, not rational, hyperbolic] Let M_2 be an infinite hyperbolic group such as $\mathbb{Z}$. Since M_2 is a group, M_2 is automatic [10]; however M_2 is not rational by Lemma 3.1. □

Example 7.3. [automatic, not rational, not hyperbolic] Let M_3 be the free abelian group $\mathbb{Z} \times \mathbb{Z}$ on two generators; this is the standard example of a non-hyperbolic group which is automatic [10]. Lemma 3.1 implies that M_3 is not rational. □

Example 7.4. [rational, not automatic] Let M_4 be the monoid generated by the set $A = \{a, b, x, y, 0, e\}$ subject to the relations

$$ax = xa = bx, \qquad bx = xb,$$
$$ay = ya = bby, \qquad by = yb,$$
$$xy = yx = xx = yy = 0,$$
$$w0 = 0w = 0 \text{ for any } w \in A,$$
$$we = ew = w \text{ for any } w \in A.$$

M_4 is rational

Since e acts as the identity of M_4, this monoid is generated by $B = A - \{e\}$. Let L be the rational cross-section of M given by

$$L = \{a, b\}^* \cup \{b\}^*\{x, y\} \cup \{0\} \subseteq B^*.$$

Consider the rational relation R defined by

$$\begin{aligned}R = {} & (B^*\{0\}B^* \times \{0\}) \cup (B^*\{x,y\}B^*\{x,y\}B^* \times \{0\}) \\ & \cup \{(a,b),(b,b)\}^*(x,\varepsilon)\{(a,b),(b,b)\}^*(\varepsilon,x) \\ & \cup \{(a,bb),(b,b)\}^*(y,\varepsilon)\{(a,bb),(b,b)\}^*(\varepsilon,y) \\ & \cup \{(a,a),(b,b)\}^*.\end{aligned}$$

Now the fact that M_4 is rational follows from the following two observations (which can be readily checked):

- R is a total and idempotent function from B^* onto L;
- for any $(u,v) \in R$, we have that $u =_{M_4} v$.

M_4 **is not automatic.**

Assume there exists an automatic structure (A, L) for M_4. By [5] we can further assume that (A, L) is an automatic structure with uniqueness. For $i \in \mathbb{N}$, let $w_i \in L$ represent the monoid element a^i; we must have that $w_i \in \{a, e\}^*$. Now consider the set

$$R = \{(u,w)\delta : u,w \in L \text{ and } \exists v \in L \text{ such that } uy =_{M_4} vy \text{ and } vx =_{M_4} wx\}$$

which is rational [18]. Note that

$$(\{a,e\}^* \times \{a,e\}^*)\delta = [(\{a,e\} \times \{a,e\})^* \cdot ((\{a,e\} \times \{\$\})^* \cup (\{\$\} \times \{a,e\})^*)]\delta$$

is rational. Hence

$$H = R \cap (\{a,e\}^* \times \{a,e\}^*)\delta$$

is rational since this is a subset of a finitely generated free monoid. Furthermore, we have

$$H = \{(w_i, w_j)\delta : 0 \leqslant i \leqslant j \leqslant 2i\}.$$

Let k be the constant from the Pumping Lemma for the rational set H. Suppose that $(u,v)\delta \in H$ with $||u| - |v|| > k$. Without loss of generality we may assume that u is shorter than v. Hence $(u,v)\delta \equiv (u,v_1)\delta \, (\varepsilon, v_2)\delta$ with $v \equiv v_1 v_2$ and $|v_2| > k$. Then we can write v_2 as xyz with $|y| > 0$ such that $(u, v_1 xy^n z)\delta \in H$ for any $n > 0$. Hence there are infinitely many words v' such that $(u, v')\delta \in H$, a contradiction. Thus, for any $(u,v)\delta \in H$, we have $||u| - |v|| \leqslant k$. This also gives that $|w_{2j_i}| - |w_i| \leqslant jk$ for all $i, j > 0$. Since there exists (by Proposition 2.2) a constant ℓ such that there are no more than ℓ consecutive occurrences of e in a word in L, we have that

$$i \leqslant |w_i| \leqslant \ell(i+1) + i.$$

Since $|w_{2^j i}| \geqslant 2^j i > \ell(i+1) + i \geqslant |w_i|$ for $j > \ell$, we get $2^j i - \ell(i+1) - i < jk$ for all $i > 0$ and $j > \ell$, a contradiction. Therefore (A, L) is not an automatic structure for M_4. By Proposition 4.1 we have that M_4 is not automatic. □

Example 7.5. [asynchronously automatic and hyperbolic, but not automatic and not rational] Let M_5 be the monoid $M_4 * (\mathbb{Z}, +)$. By Example 7.4, M_4 is rational, which gives, by Theorems 6.2 and 6.3, that M_4 is also asynchronously automatic and hyperbolic. By Theorem 6.3, there exists a hyperbolic structure with uniqueness for M_4. The monoid $(\mathbb{Z}, +)$ is not rational (Lemma 3.1), but it is asynchronously automatic and hyperbolic; and in particular there exists a hyperbolic structure with uniqueness for $(\mathbb{Z}, +)$.

By Theorem 5.2, M_5 is asynchronously automatic and, by Theorem 5.4, M_5 is hyperbolic. However, M_5 is not automatic (if the monoid $M_4 * (\mathbb{Z}, +)$ were automatic, then both M_4 and $(\mathbb{Z}, +)$ would have to be automatic [7]). Since a finitely generated submonoid of a rational monoid is rational [29], M_5 is not rational. □

Example 7.6. [asynchronously automatic, not automatic, not hyperbolic] Let M_6 be a Baumslag-Solitar group represented by

$$\langle a, b : b^{-1} a^i b = a^j \rangle$$

with $i \neq j$. The monoid M_6 is asynchronously automatic but not automatic [10]. Since every hyperbolic group is automatic, M_6 is not hyperbolic. □

Example 7.7. [not asynchronously automatic, but hyperbolic] Let M_7 be the monoid with presentation

$$\langle a, b, c : abc = c = bac = cc, ac = ca, bc = cb \rangle.$$

We will show that M_7 is not asynchronously automatic, but that it is hyperbolic.

M_7 is not asynchronously automatic

Let $A = \{a, b, c, e\}$ where e represents the identity of M_7. Assume that (A, L) is an asynchronously automatic structure for M_7. Let φ be the homomorphism from A^* to itself that suppresses all occurrences of e in a word. Then $L' = L\varphi$ is rational and the natural homomorphism maps L' surjectively onto M_7. Note that, for $x \in \{a, b, c, \varepsilon\}$, we have $L'_x = \varphi^{-1} \circ L_x \circ \varphi$ which is a rational relation; hence, we can assume that no word in L contains an occurrence of e. So, for every $i, j \in \mathbb{N}$ there exists exactly one word in L that represents the monoid element $a^i b^j$, which is the word $a^i b^j$ itself. Note that $B = \{a\}^* \{b\}^*$ is rational. Let w be a word in L representing the monoid

401

element c. By Proposition 2.1 we have that
$$K = L_c \cap (B \times \{w\})$$
is a rational relation. Hence we would also have that $K\pi_1 = \{a^i b^i : i \in \mathbb{N}\}$ is rational, which is not the case. Therefore (A, L) is not an asynchronously automatic structure for M_7. By Proposition 4.1, M_7 is not asynchronously automatic.

M_7 *is hyperbolic*

Let $C = \{a, b, c\}$, $D = \{a, b\}$ and then let $L = D^*\{c, \varepsilon\}$; we will show that (C, L) is a hyperbolic structure for M.

The set $S = \{u\#_1 v\#_2 \overline{w} : u, v, w \in L, uv =_{M_7} w\}$ can be split up into the subsets S_1 and S_2 where
$$S_1 = \{u\#_1 v\#_2 \overline{w} : u, v, w \in D^*, uv =_{M_7} wc\},$$
$$S_2 = \{u\#_1 v\#_2 \overline{wc} : u, v, wc \in L, uv \notin D^*, uv =_{M_7} wc\}.$$

Since each side of any relation in M_7 contains at least one occurrence of c, we can write S_1 as
$$S_1 = \{u\#_1 v\#_2 \overline{w} : u, v, w \in D^*, uv \equiv w\},$$
which is obviously context-free.

We turn our attention to S_2. Note that for $uv \notin D^*$ and $w \in D^\star$, we have
$$uv =_{M_7} wc \iff |uv|_a - |uv|_b = |w|_a - |w|_b,$$
where $|w|_a$ is the number of occurrences of the letter a in the word w and similarly for $|w|_b$. Let $\psi : C^* \to (\mathbb{Z}, +)$ be the homomorphism defined by $a\psi = 1$, $b\psi = -1$ and $c\psi = 0$; then
$$S_2 = \{u\#_1 v\#_2 \overline{wc} : u, v, wc \in L, uv \notin D^*, (uv)\psi = (wc)\psi\}.$$

The sets $K = \{a^n\#_2 a^n, b^n\#_2 b^n : n \in \mathbb{N}\}$ and $L = \{ab, ba\} \cup c^*$ are context-free. Hence $K \cdot_\varepsilon L^\varepsilon$ is context-free by Proposition 2.4. Finally,
$$T = ((K \cdot_\varepsilon L^\varepsilon) \cdot_\varepsilon \{\#_1\}) \cap (C^*\#_1 C^* \#_2 C^*)$$
is context-free as the intersection of a context-free and a rational set in a free monoid. Note that
$$T = \{u\#_1 v\#_2 \overline{w} : u, v, w \in C^*, (uv)\psi = (wc)\psi\}.$$

Hence S_2 is the intersection (in a free monoid) of the context-free set T and the rational set $\{u\#_1 v\#_2 \overline{wc} : u, v, wc \in L, uv \notin D^*\}$, so that S_2 is context-free.

We have shown that S is context-free (as the union of the two context-free sets S_1 and S_2) and therefore that (C, L) is a hyperbolic structure for M_7. □

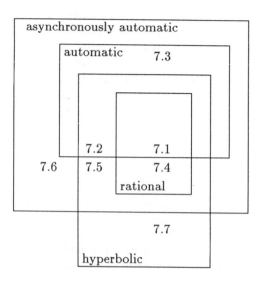

This picture is to be read as follows. The main rectangular boxes denote the four classes of monoids under consideration. The inclusion of these main boxes is shown in Section 6. Note that any intersection of these main boxes contains a number. This is the number of the example in Section 7 that proves that the intersection is not empty.

Figure 2. The complete inclusion structure

8 Conclusion

Recall that, for groups, we have the following implications:

rational $\Rightarrow$ hyperbolic $\Rightarrow$ automatic $\Rightarrow$ asynchronously automatic.

This inclusion structure becomes more complicated when we consider monoids. Figure 2 shows that the classes of rational and automatic monoids are incomparable and also that the classes of hyperbolic and automatic monoids are incomparable. However, if we also consider the concept of an asynchronously automatic monoid, we get a more "connected" picture with the class of rational monoids in the centre.

Besides this inclusion structure, we have shown that the word problem for hyperbolic monoids is decidable in deterministic exponential time. While any automatic monoid has an automatic structure with uniqueness, we proved the corresponding statement for asynchronously automatic monoids only under additional assumptions, and we did not prove or disprove it for hyperbolic monoids. Finally, the closure under free products of hyperbolic monoids could only be shown under an additional assumption; it is not clear whether this assumption is always satisfied, whether one can prove the closure without the assumption, or whether there are free products of hyperbolic monoids that are not hyperbolic.

Acknowledgments

Some of the work on this paper was undertaken during the Workshop on Presentations and Geometry during the Thematic Term on Semigroups, Algorithms, Automata and Languages. The authors are very grateful to the organizers Gracinda Gomes, Jean-Éric Pin and Pedro Silva of the Thematic Term for providing such a wonderful environment, and also to Bob Gilman and Jacques Sakarovitch for helpful conversations about the material in this paper. We thank Mark Kambites and the referees who helped to eliminate some errors in the submitted version of this paper. The first and fourth authors would like to thank Chen-Hui Chiu and Hilary Craig respectively for all their help and encouragement.

References

[1] J.M. Alonso, T. Brady, D. Cooper, V. Ferlini, M. Lustig, M. Mihalik, M. Shapiro and H. Short, Notes on word hyperbolic groups, *in* E. Ghys, A. Haefliger and A. Verjovsky (eds.), *Group Theory from a Geometric Viewpoint*, World Scientific, Singapore (1991), 3–63.
[2] G. Baumslag, S.M. Gersten, M. Shapiro and H. Short, Automatic groups and amalgams, *Journal of Pure and Applied Algebra* **76** (1991), 229–316.
[3] J. Berstel, *Transductions and Context-free Languages*, Teubner Studienbücher, Stuttgart (1979).
[4] J. Berstel and J. Sakarovitch, Recent results in the theory of rational sets, *in* J. Gruska, B. Rovan and J. Wiedermann (eds.), *Mathematical Foundations of Computer Science*, Springer-Verlag, Berlin (1986), 15–28.
[5] C.M. Campbell, E.F. Robertson, N. Ruškuc, and R.M. Thomas. Automatic semigroups, *Theoretical Computer Science* **250** (2001), 365–391.
[6] A.J. Duncan and R. Gilman, *Personal communication*.
[7] A.J. Duncan, E.F. Robertson and N. Ruškuc, Automatic monoids and

change of generators, *Mathematical Proceedings of the Cambridge Philosophical Society* **127** (1999), 403–409.
[8] A. Ehrenfeucht and G. Rozenberg, An easy proof of Greibach normal form, *Information and Control* **63** (1984), 190–199.
[9] S. Eilenberg, *Automata, Languages and Machines. Volume A*, Academic Press, New York (1974).
[10] D.B.A. Epstein, J.W. Cannon, D.F. Holt, S.V.F. Levy, M.S. Paterson and W.P. Thurston, *Word Processing In Groups*, Jones and Bartlett, Boston (1992).
[11] C. Frougny, J. Sakarovitch and P.E. Schupp, Finiteness conditions on subgroups and formal language theory, *Proceeedings of the London Mathematical Society* **58** (1989), 74–88.
[12] R. Gilman, On the definition of word hyperbolic groups, *Mathematische Zeitschrift*, to appear.
[13] M. Gromov, Hyperbolic groups, *in* S.M. Gersten (ed.), *Essays in Group Theory*, Mathematical Sciences Research Institute Publications **8**, Springer-Verlag, Berlin (1987), 75–263.
[14] M.A. Harrison, *Introduction to Formal Language Theory*, Addison-Wesley, 1978.
[15] T. Herbst, On a subclass of context-free groups, *Informatique théorique et Applications* **25** (1991), 255–272.
[16] T. Herbst and R.M. Thomas, Group presentations, formal languages and characterizations of one-counter groups, *Theoretical Computer Science* **112** (1993), 187–213.
[17] M. Hoffmann, *Automatic Semigroups*, Ph.D. thesis, University of Leicester (2000).
[18] M. Hoffmann and R.M. Thomas, Automaticity and commutative semigroups, *Glasgow Journal of Mathematics* **44** (2002), 167–176.
[19] M. Hoffmann and R.M. Thomas, Notions of automaticity in semigroups, *Semigroup Forum*, to appear.
[20] J.E. Hopcroft, R. Motwani and J.D. Ullman, *Introduction to Automata Theory, Languages and Computation*, Addison-Wesley, Boston (2001).
[21] J.M. Howie, *Fundamentals of Semigroup Theory*, Oxford University Press, Oxford (1995).
[22] J.F.P. Hudson, Regular rewrite systems and automatic structures, *in* J. Almeida, G.M.S. Gomes and P.V. Silva (eds.), *Semigroups, Automata and Languages*, World Scientific, Singapore (1998), 145–152.
[23] M. Ito, L. Kari, and G. Thierren, Insertion and deletion closure of languages, *Theoretical Computer Science* **183** (1997), 3–19.
[24] J.H. Johnson, Do rational equivalence relations have regular cross-

sections?, *in ICALP'85*, Lecture Notes in Computer Science **194**, Springer-Verlag, Berlin (1985), 300–309.

[25] F. Otto, A. Sattler-Klein and K. Madlener, Automatic monoids versus monoids with finite convergent presentations, *in* T. Nipkow (ed.), *Rewriting Techniques and Applications – Proceedings RTA '98*, Lecture Notes in Computer Science **1379**, Springer-Verlag, Berlin (1998), 32–46.

[26] M. Pelletier and J. Sakarovitch, Easy multiplications II. Extensions of rational semigroups, *Information and Computation* **88** (1990), 18–59.

[27] M. Pelletier and J. Sakarovitch, On the representation of finite deterministic 2-tape automata, *Theoretical Computer Science* **225** (1999), 1–63.

[28] C.P. Rupert, On commutative Kleene monoids, *Semigroup Forum* **43** (1991), 163-177.

[29] J. Sakarovitch, Easy multiplications I. The realm of Kleene's Theorem, *Information and Computation* **74** (1987), 173–197.

OPERATORS ON CLASSES OF REGULAR LANGUAGES

LIBOR POLÁK*

Department of Mathematics, Masaryk University Brno
Janáčkovo nám 2a, 66295 Brno, Czech Republic
E-mail: polak@math.muni.cz

The consideration of ordered monoids refines the study of classes of regular languages in terms of pseudovarieties of monoids. Here we propose the ordered alternative to the power operator widely discussed for monoids.

1 Introduction

The books by Pin [5] and by Almeida [1] explore the so-called Eilenberg correspondence between certain classes of regular languages, called varieties, and pseudovarieties of monoids and they present numerous sophisticated examples. Crucial role is played there by the so-called *power* operator: for a monoid $(M, \cdot)$ take the set $\mathsf{P}(M)$ of all subsets of M with the operation $X \cdot Y = \{\, xy \mid x \in X, \, y \in Y \,\}$ and for a class $\mathcal{M}$ of finite monoids generate a pseudovariety by the class $\{(\mathsf{P}(M), \cdot) \mid (M, \cdot) \in \mathcal{M}\}$. This leads to a natural operator on varieties of languages. Chapter 11 of [1] is devoted to the topic; a newer survey is Almeida's [2].

The operator P is closely related to the operator P' – we take only nonempty subsets. Also the whole theory is developing simultaneously for both semigroups and monoids.

The syntactic monoid of a language carries implicitly an order. This was used in Pin's generalization [6] of the Eilenberg result: the so-called positive varieties of languages correspond to pseudovarieties of ordered monoids; his exposition [7] uses already the ordered monoid approach. The first valuable application of positive varieties in language theory was probably the work by Pin and Weil [8].

---

*SUPPORTED BY THE MINISTRY OF EDUCATION OF THE CZECH REPUBLIC UNDER THE PROJECT MSM 143100009. THE AUTHOR WOULD LIKE TO ACKNOWLEDGE THE FINANCIAL SUPPORT OF FUNDAÇÃO CALOUSTE GULBENKIAN (FCG), FUNDAÇÃO PARA A CIÊNCIA E A TECNOLOGIA (FCT), FACULDADE DE CIÊNCIAS DA UNIVERSIDADE DE LISBOA (FCUL) AND REITORIA DA UNIVERSIDADE DO PORTO.

In [9] the author introduced the so-called syntactic semiring of a language under the name syntactic semilattice-ordered monoid. The main result of that paper is an Eilenberg-type theorem giving a one-to-one correspondence between the so-called conjunctive varieties of regular languages and pseudovarieties of idempotent semirings. The next author's contribution [10] studies the relationships between the (ordered) syntactic monoid and the syntactic semiring of a given language. We mention here only that the first one is finite if and only if the second one is finite and that these two structures are equationally independent. Also several examples of conjunctive varieties of languages are presented there. It seems that the notion of the syntactic semiring is suitable when discussing equations $r(x_1, \ldots, x_m) = L$ where L is a given regular language over a finite alphabet A and r is a given regular expression over A in variables $x_1, \ldots, x_m$ – see [11].

Here we define the *hereditary* operator on classes of ordered monoids. It should be the right alternative to the power operator for the ordered case. The notion of the syntactic semiring helps us to understand our operator and to complete proofs. In fact, it is more natural to consider the hereditary operator defined on pseudovarieties of ordered monoids with values in the lattice of all pseudovarieties of idempotent semirings – see Section 8.

Our main result is Theorem 4.2 giving the language counterpart of the hereditary operator. In Section 6 we calculate concrete examples. We have chosen the simplest setting: we consider monoids, non-empty subsets and non-empty hereditary subsets. In fact, our P is usually denoted by P′.

2 Syntactic semiring of a language

A structure $(O, \cdot, \leq)$ is called an *ordered monoid* if
 (i) $(O, \cdot)$ is a monoid with the neutral element 1,
 (ii) $(O, \leq)$ is an ordered set,
 (iii) $a, b, c \in O$, $a \leq b$ implies $ac \leq bc$ and $ca \leq cb$.

The structure $(O, \cdot, \geq)$ is called the *dual* of $(O, \cdot, \leq)$. Also for a class $\mathcal{P}$ of ordered monoids we define its dual $\mathcal{P}^d = \{(O, \cdot, \geq) \mid (O, \cdot, \leq) \in \mathcal{P}\}$. The class $\mathcal{P}$ is *self-dual* if $\mathcal{P}^d = \mathcal{P}$.

A structure $(S, \cdot, \vee)$ is called an *idempotent semiring* if
 (i) $(S, \cdot)$ is a monoid,
 (ii) $(S, \vee)$ is a semilattice,
 (iii) $a, b, c \in S$ implies $a(b \vee c) = ab \vee ac$ and $(a \vee b)c = ac \vee bc$.

Note that one usually postulates, in addition, the existence of a neutral element for the join which is absorbing for the multiplication.

Any semilattice $(S, \vee)$ is an ordered set with respect to the relation $\leq$ defined by $a \leq b \iff a \vee b = b$, $a, b \in S$. In this view an idempotent semiring becomes an ordered monoid.

For an ordered set $(O, \leq)$, a subset H of O is *hereditary* if $a \in H$, $b \in O$, $b \leq a$ implies $b \in H$.

Recall that an *ideal* I of a semilattice $(S, \vee)$ is a hereditary subset of $(S, \leq)$ closed with respect to the operation $\vee$.

We also denote

P(B) the set of all non-empty subsets of a set B,

F(B) the set of all non-empty finite subsets of a set B,

$(B] = \{\, a \in O \mid a \leq b \text{ for some } b \in B \,\}$ the hereditary subset of $(O, \leq)$ *generated* by a given $B \subseteq O$,

H$(O, \leq)$ the set of all non-empty hereditary subsets of an ordered set $(O, \leq)$.

Homomorphisms of monoids are semigroup homomorphisms sending neutral elements to neutral elements and homomorphisms of ordered monoids (idempotent semirings) are isotone monoid homomorphisms (monoid homomorphisms which respect also the operation $\vee$). A monoid homomorphism $\lambda : (B^*, \cdot) \to (A^*, \cdot)$ is *literal* if $\lambda(b) \in A$ for all $b \in B$.

Notice that the algebra $(\mathsf{F}(A^*), \cdot, \cup)$ is a free idempotent semiring over A and that any of its ideals is of the form $\mathsf{F}(L)$ for some $L \subseteq A^*$.

A language $L \subseteq A^*$ is *recognizable* by a monoid $(M, \cdot)$ with respect to its subset B (by an ordered monoid $(O, \cdot, \leq)$ with respect to its hereditary subset H, by an idempotent semiring $(S, \cdot, \vee)$ with respect to its ideal I) if there exists a monoid homomorphism $\alpha : (A^*, \cdot) \to (M, \cdot)$ such that $L = \alpha^{-1}(B)$ ($\alpha : (A^*, \cdot) \to (O, \cdot)$ such that $L = \alpha^{-1}(H)$, $\alpha : (A^*, \cdot) \to (S, \cdot)$ such that $L = \alpha^{-1}(I)$).

Clearly, the last notion can be rephrased as follows: A language $L \subseteq A^*$ is recognizable by an idempotent semiring $(S, \cdot, \vee)$ with respect to its ideal I if there exists a semiring homomorphism $\beta : (\mathsf{F}(A^*), \cdot, \cup) \to (S, \cdot, \vee)$ such that $\mathsf{F}(L) = \beta^{-1}(I)$.

We say that a language L is *recognizable* by a monoid $(M, \cdot)$ (by an ordered monoid $(O, \cdot, \leq)$, by an idempotent semiring $(S, \cdot, \vee)$) if it is recognizable with respect to some subset (hereditary subset, ideal). Clearly, recognizability by $(M, \cdot)$ gives recognizability by $(M, \cdot, =)$ and conversely recognizability by $(M, \cdot, \leq)$ yields recognizability by $(M, \cdot)$. Furthermore, *recognizability* means recognizability by a finite monoid. As shown in [10] it is equivalent to recognizability by a finite idempotent semiring.

Following Eilenberg [4, vol. A, pages 61–63] a language L over A defines the so-called *syntactic congruence* on the free monoid $(A^*, \cdot)$ over A by

$$u \approx_L v \text{ if and only if } (\forall p, q \in A^*) (puq \in L \iff pvq \in L).$$

The factor-structure $(A^*, \cdot)/\approx_L$ is called the *syntactic monoid* of L and we denote it by $(\mathsf{O}(L), \cdot)$. As pointed out by Pin [6], it is ordered by

$$v \approx_L \leq_L u \approx_L \text{ if and only if } (\forall p, q \in A^*) (puq \in L \Rightarrow pvq \in L)$$

and we speak about the *ordered syntactic monoid*. We also write $v \preceq_L u$ instead of $v \approx_L \leq_L u \approx_L$.

Also L defines a congruence $\sim_L$ on $(\mathsf{F}(A^*), \cdot, \cup)$ by

$$\{u_1, \ldots, u_k\} \sim_L \{v_1, \ldots, v_l\} \text{ if and only if}$$

$$(\forall p, q \in A^*) (pu_1q, \ldots, pu_kq \in L \iff pv_1q, \ldots, pv_lq \in L).$$

The factor-structure is called the *syntactic semiring* of L; we denote it by $(\mathsf{S}(L), \cdot, \vee)$.

3 Eilenberg-type theorems

Note that a substructure of an ordered monoid is a submonoid with the induced order. A class of finite monoids (ordered monoids, idempotent semirings) is called a *pseudovariety of monoids (ordered monoids, idempotent semirings)* if it is closed under forming of products of finite families, substructures and homomorphic images.

For sets A and B, a semiring homomorphism

$$\psi : (\mathsf{F}(A^*), \cdot, \cup) \to (\mathsf{F}(B^*), \cdot, \cup)$$

and $L \subseteq B^*$ we define

$$\psi^{[-1]}(L) = \{ u \in A^* \mid \psi(\{u\}) \subseteq L \}$$

and

$$\psi^{(-1)}(L) = \{ u \in A^* \mid \psi(\{u\}) \cap L \neq \emptyset \}.$$

Recall that the set of all *regular* languages over a finite alphabet A is the smallest family of subsets of A^* containing the empty set, all singletons $\{u\}$, $u \in A^*$, closed with respect to binary unions and the operations $\cdot$ and $*$. As well-known, the regular languages are exactly the recognizable ones.

For $p, q \in A^*$ and $L \subseteq A^*$ we put
$$p^{-1}L = \{u \in A^* \mid pu \in L\}$$
$$Lq^{-1} = \{u \in A^* \mid uq \in L\}$$
$$p^{-1}Lq^{-1} = \{u \in A^* \mid puq \in L\}$$
We speak about (*left/right*) *quotients* of L.

A *class* of (regular) languages is an operator $\mathcal{L}$ assigning to every finite set A a set $\mathcal{L}(A)$ of regular languages over the alphabet A containing both $\emptyset$ and A^*.

Conditions which such a class of languages can satisfy follow.

($\cap$): for every finite set A, the set $\mathcal{L}(A)$ is closed with respect to finite intersections,

($\cup$): for every finite set A, the set $\mathcal{L}(A)$ is closed with respect to finite unions,

(**Q**): for every finite set A, $a \in A$ and $L \in \mathcal{L}(A)$ we have $a^{-1}L, La^{-1} \in \mathcal{L}(A)$,

($^{-1}$): for every finite sets A and B, monoid homomorphism $\sigma : (A^*, \cdot) \to (B^*, \cdot)$ and $L \in \mathcal{L}(B)$ we have $\sigma^{-1}(L) \in \mathcal{L}(A)$,

($^{[-1]}$): for every finite sets A and B, semiring homomorphism $\psi : (\mathsf{F}(A^*), \cdot, \cup) \to (\mathsf{F}(B^*), \cdot, \cup)$ and $L \in \mathcal{L}(B)$ we have $\psi^{[-1]}(L) \in \mathcal{L}(A)$,

($^{(-1)}$): for every finite sets A and B, semiring homomorphism $\psi : (\mathsf{F}(A^*), \cdot, \cup) \to (\mathsf{F}(B^*), \cdot, \cup)$ and $L \in \mathcal{L}(B)$ we have $\psi^{(-1)}(L) \in \mathcal{L}(A)$,

(**C**): for every finite set A, the set $\mathcal{L}(A)$ is closed with respect to complements.

For a class $\mathcal{L}$ of languages we define its *complement* $\mathsf{C}\mathcal{L}$ by
$$(\mathsf{C}\mathcal{L})(A) = \{\, A^* \setminus L \mid L \in \mathcal{L}(A) \,\} \text{ for every finite set } A \;.$$
Clearly, $\mathcal{L}$ satisfies the condition ($\cap$) if and only if $\mathsf{C}\mathcal{L}$ satisfies ($\cup$). Similarly, $\mathcal{L}$ satisfies the condition ($^{[-1]}$) if and only if $\mathsf{C}\mathcal{L}$ satisfies ($^{(-1)}$). Further, either of the conditions ($^{[-1]}$) and ($^{(-1)}$) implies ($^{-1}$).

A class $\mathcal{L}$ is called a *conjunctive variety of languages* if it satisfies the conditions ($\cap$), ($^{[-1]}$) and (**Q**). Similarly, it is called a *disjunctive variety of languages* if it satisfies ($\cup$), ($^{(-1)}$) and (**Q**).

Further, $\mathcal{L}$ is called *a positive variety of languages* if it satisfies the conditions ($\cap$), ($\cup$), ($^{-1}$), (**Q**) and such a variety is called *a boolean variety of languages* if it satisfies in addition the condition (**C**).

We can assign to any class of languages $\mathcal{L}$ the pseudovarieties of monoids (ordered monoids, idempotent semirings)

$$\mathsf{M}(\mathcal{L}) = \langle\!\{ (\mathsf{O}(L), \cdot) \mid A \text{ a finite set}, L \in \mathcal{L}(A) \}\!\rangle_\mathsf{M} ,$$
$$\mathsf{O}(\mathcal{L}) = \langle\!\{ (\mathsf{O}(L), \cdot, \leq) \mid A \text{ a finite set}, L \in \mathcal{L}(A) \}\!\rangle_\mathsf{O} ,$$
$$\mathsf{S}(\mathcal{L}) = \langle\!\{ (\mathsf{S}(L), \cdot, \vee) \mid A \text{ a finite set}, L \in \mathcal{L}(A) \}\!\rangle_\mathsf{S}$$

generated by all syntactic monoids (ordered syntactic monoids, syntactic semirings) of members of $\mathcal{L}$.

Conversely, for pseudovarieties $\mathcal{M}$ of monoids, $\mathcal{P}$ of ordered monoids and $\mathcal{V}$ of idempotent semirings and a finite set A, we put

$$(\mathsf{L}(\mathcal{M}))(A) = \{ L \subseteq A^* \mid (\mathsf{O}(L), \cdot) \in \mathcal{M} \} ,$$
$$(\mathsf{L}(\mathcal{P}))(A) = \{ L \subseteq A^* \mid (\mathsf{O}(L), \cdot, \leq) \in \mathcal{P} \} ,$$
$$(\mathsf{L}(\mathcal{V}))(A) = \{ L \subseteq A^* \mid (\mathsf{S}(L), \cdot, \vee) \in \mathcal{V} \} .$$

Theorem 3.1 (Eilenberg [4], Pin [6], Polák [9]).

(i) *The operators M and L are mutually inverse bijections between boolean varieties of languages and pseudovarieties of monoids.*

(ii) *The operators O and L are mutually inverse bijections between positive varieties of languages and pseudovarieties of ordered monoids.*

(iii) *The operators S and L are mutually inverse bijections between conjunctive varieties of languages and pseudovarieties of idempotent semirings.*

4 Main results

For a monoid $(M, \cdot)$ we write $\mathsf{P}(M, \cdot) = (\mathsf{P}(M), \cdot)$ where

$$X \cdot Y = \{ xy \mid x \in X, y \in Y \} .$$

Similarly, for an ordered monoid $(O, \cdot, \leq)$ we put

$$\mathsf{H}(O, \cdot, \leq) = (\mathsf{H}(O, \leq), \circ, \subseteq) \text{ and } \mathsf{H}^d(O, \cdot, \leq) = (\mathsf{H}(O, \geq), \circ, \supseteq)$$

where

$$X \circ Y = \{ p \in O \mid \text{there exist } x \in X, y \in Y \text{ such that } p \leq xy \} .$$

For a pseudovariety $\mathcal{M}$ of monoids, let

$$\mathsf{P}(\mathcal{M}) = \langle\!\{ \mathsf{P}(M, \cdot) \mid (M, \cdot) \in \mathcal{M} \}\!\rangle_\mathsf{M}$$

and similarly for a pseudovariety $\mathcal{P}$ of ordered monoids

$$\mathsf{H}(\mathcal{P}) = \langle\!\{ \mathsf{H}(O, \cdot, \leq) \mid (O, \cdot, \leq) \in \mathcal{P} \}\!\rangle_\mathsf{O}$$

and
$$\mathsf{H}^d(\mathcal{P}) = \langle \{\, \mathsf{H}^d(O,\cdot,\leq) \mid (O,\cdot,\leq) \in \mathcal{P}\,\}\rangle_0 \;.$$

For a monoid homomorphism $\lambda : (B^*, \cdot) \to (A^*, \cdot)$ and $L \subseteq B^*$ we put
$$\lambda^{[\,]}(L) = \{u \in A^* \mid \lambda^{-1}(u) \subseteq L\} \text{ and } \lambda^{(\,)}(L) = \{u \in A^* \mid \lambda^{-1}(u) \cap L \neq \emptyset\} \;.$$
Of course, the last set is usually denoted by $\lambda(L)$.

On classes of languages we consider operators
$$\bigcap,\; \bigcup,\; \mathsf{B},\; \mathsf{Q},\; \Sigma,\; \Phi,\; \Psi,\; \Theta,\; \Lambda,\; \Pi \;.$$

Namely, for a given class $\mathcal{L}$ and a given finite set A, we put

$(\bigcap \mathcal{L})(A) = \{\, L_1 \cap \cdots \cap L_k \mid L_1, \ldots, L_k \in \mathcal{L}(A) \,\}$,

$(\bigcup \mathcal{L})(A) = \{\, L_1 \cup \cdots \cup L_k \mid L_1, \ldots, L_k \in \mathcal{L}(A) \,\}$,

$(\mathsf{B}\,\mathcal{L})(A) = \{\, (L_{1,1} \cap \cdots \cap L_{1,l_1}) \cup \cdots \cup (L_{k,1} \cap \cdots \cap L_{k,l_k}) \mid$
$\qquad L_{i,j} \in \mathcal{L}(A) \cup (\mathsf{C}\,\mathcal{L})(A),\; i = 1, \ldots, k,\; j = 1, \ldots, l_i \,\}$,

$(\mathsf{Q}\,\mathcal{L})(A) = \{\, p^{-1}Lq^{-1} \mid p, q \in A^*,\; L \in \mathcal{L}(A) \,\}$,

$(\Sigma \mathcal{L})(A) = \{\, \sigma^{-1}(L) \mid B \text{ is a finite set},\; \sigma : (A^*, \cdot) \to (B^*, \cdot)$
$\qquad \text{is a monoid homomorphism and } L \in \mathcal{L}(B) \,\}$,

$(\Phi \mathcal{L})(A) = \{\, \psi^{[-1]}(L) \mid B \text{ is a finite set},\; \psi : (\mathsf{F}(A^*), \cdot, \cup) \to (\mathsf{F}(B^*), \cdot, \cup)$
$\qquad \text{is a semiring homomorphism},\; L \in \mathcal{L}(B) \,\}$,

$(\Psi \mathcal{L})(A) = \{\, \psi^{(-1)}(L) \mid B \text{ is a finite set},\; \psi : (\mathsf{F}(A^*), \cdot, \cup) \to (\mathsf{F}(B^*), \cdot, \cup)$
$\qquad \text{is a semiring homomorphism},\; L \in \mathcal{L}(B) \,\}$,

$(\Theta \mathcal{L})(A) = \{\, \lambda^{[\,]}(L) \mid B \text{ is a finite set},\; \lambda : (B^*, \cdot) \to (A^*, \cdot)$
$\qquad \text{is a surjective literal monoid homomorphism and } L \in \mathcal{L}(B) \,\}$,

$(\Lambda \mathcal{L})(A) = \{\, \lambda(L) \mid B \text{ is a finite set},\; \lambda : (B^*, \cdot) \to (A^*, \cdot)$
$\qquad \text{is a surjective literal monoid homomorphism and } L \in \mathcal{L}(B) \,\}$.

It is immediate that every of our operators is a closure one and that, for any class $\mathcal{L}$ of languages,
$$\Phi \mathcal{L} = \mathsf{C}\Psi\mathsf{C}\mathcal{L},\; \Psi\mathcal{L} = \mathsf{C}\Phi\mathsf{C}\mathcal{L},\; \Sigma\mathcal{L} \subseteq \Phi\mathcal{L} \cap \Psi\mathcal{L}\,,$$
$$\Theta\mathcal{L} = \mathsf{C}\Lambda\mathsf{C}\mathcal{L},\; \Lambda\mathcal{L} = \mathsf{C}\Theta\mathsf{C}\mathcal{L}\;.$$

Two less trivial statements are formulated in Lemmas 5.2 and 5.3.

According to [5, Theorem 5.1.3] the following is an amalgam of the works by Reutenauer, Straubing and Pin; see also [1, Proposition 11.2.3].

Theorem 4.1 (Reutenauer, Straubing, Pin). *Let $\mathcal{M}$ be a pseudovariety of monoids and let $\mathcal{L}$ be the corresponding boolean variety of languages; that is, $\mathsf{L}\mathcal{M} = \mathcal{L}$ and $\mathsf{M}\mathcal{L} = \mathcal{M}$. Then the corresponding boolean variety for $\mathsf{P}\mathcal{M}$ is $\mathsf{B}\Psi\mathcal{L} = \mathsf{B}\Lambda\mathcal{L}$.*

The following is our main result.

Theorem 4.2. *Let $\mathcal{P}$ be a pseudovariety of ordered monoids and let $\mathcal{L}$ be the corresponding positive variety of languages; that is, $\mathsf{L}\mathcal{P} = \mathcal{L}$ and $\mathsf{O}\mathcal{L} = \mathcal{P}$. Then the corresponding positive variety for $\mathsf{H}\mathcal{P}$ is $\bigcup\bigcap\Phi\mathcal{L} = \bigcup\bigcap\Theta\mathcal{L}$.*

Remark 4.1. The last statement has its dual version; namely the corresponding positive variety for $\mathsf{H}^d\mathcal{P}$ is $\bigcup\bigcap\Psi\mathcal{L} = \bigcup\bigcap\Lambda\mathcal{L}$.

Also it is useful to understand the passage from positive varieties of languages to conjunctive ones and back.

Theorem 4.3.
(i) *If $\mathcal{L}$ is a positive variety of languages, then $\bigcap\Phi\mathcal{L}$ is the smallest conjunctive variety of languages containing $\mathcal{L}$.*
(ii) *If $\mathcal{L}$ is a conjunctive variety of languages, then $\bigcup\mathcal{L}$ is the smallest positive variety of languages containing $\mathcal{L}$.*

5 Proofs of the main theorems

In this section we fix $A = \{a_1, \ldots, a_n\}$ where $a_1, \ldots, a_n$ are pairwise different symbols.

The following lemma is straightforward.

Lemma 5.1. *For any $K, L \subseteq A^*$, $a \in A$, $p, q \in A^*$ we have*
(i) $p^{-1}(K \cap L)q^{-1} = p^{-1}Kq^{-1} \cap p^{-1}Lq^{-1}$,
(ii) $p^{-1}(K \cup L)q^{-1} = p^{-1}Kq^{-1} \cup p^{-1}Lq^{-1}$,
(iii) $p^{-1}(A^*L)q^{-1} = A^*p^{-1}Lq^{-1}$;
if moreover $\sigma : (A^*, \cdot) \to (B^*, \cdot)$ is a monoid homomorphism and $M, N \subseteq B^*$, then
(iv) $a^{-1}\ \sigma^{-1}(M) = \sigma^{-1}((\sigma(a))^{-1}\ M)$,
(v) $\sigma^{-1}(M \cup N) = \sigma^{-1}(M) \cup \sigma^{-1}(N)$;
if moreover $\psi : (\mathsf{F}(A^*), \cdot, \cup) \to (\mathsf{F}(B^*), \cdot, \cup)$ is a semiring homomorphism, then
(vi) $a^{-1}\ \psi^{[-1]}(M)\ =\ \bigcap_{v \in \psi(\{a\})} \psi^{[-1]}(v^{-1}M)$ *and similarly for*

$\psi^{[-1]}(M)\ a^{-1}$,
(vii) $a^{-1}\ \psi^{(-1)}(M) = \bigcup_{v \in \psi(\{a\})} \psi^{(-1)}(v^{-1}M)$ and similarly for $\psi^{(-1)}(M)\ a^{-1}$,
(viii) $\psi^{[-1]}(M \cap N) = \psi^{[-1]}(M) \cap \psi^{[-1]}(N)$.

Also the following lemmas will be useful.

Lemma 5.2. *For any class $\mathcal{L}$ of languages we have $\Lambda\mathcal{L} \subseteq \Psi\mathcal{L}$.*

Proof. Let $\lambda : (B^*, \cdot) \to (A^*, \cdot)$ be a surjective literal monoid homomorphism. Define
$$\psi : (\mathsf{F}(A^*), \cdot, \cup) \to (\mathsf{F}(B^*), \cdot, \cup),\ \{a_i\} \mapsto \lambda^{-1}(a_i),\ i = 1, \ldots, n\ .$$
Then, for $L \subseteq B^*$, we have $\lambda(L) = \psi^{(-1)}(L)$. $\square$

Lemma 5.3. *For any class $\mathcal{L}$ of languages we have $\Psi\mathcal{L} \subseteq \Lambda\Sigma\mathcal{L}$.*

Proof. Let
$$\psi : (\mathsf{F}(A^*), \cdot, \cup) \to (\mathsf{F}(B^*), \cdot, \cup),\ \{a_i\} \mapsto \{v_{i,1}, \ldots, v_{i,k_i}\},\ i = 1, \ldots, n$$
be a semiring homomorphism. Let $C = \{c_{1,1}, \ldots, c_{1,k_1}, \ldots, c_{n,1}, \ldots, c_{n,k_n}\}$ be a new alphabet. Define
$$\lambda : (C^*, \cdot) \to (A^*, \cdot),\ c_{i,j} \mapsto a_i,\ i = 1, \ldots, n,\ j = 1, \ldots, k_i$$
and
$$\sigma : (C^*, \cdot) \to (B^*, \cdot),\ c_{i,j} \mapsto v_{i,j},\ i = 1, \ldots, n,\ j = 1, \ldots, k_i\ .$$
Then, for $L \subseteq B^*$, we have $\psi^{(-1)}(L) = \lambda(\sigma^{-1}(L))$. $\square$

Lemma 5.4. *Let a language $L \subseteq B^*$ be recognizable by an idempotent semiring $(S, \cdot, \vee)$ with respect to its ideal I. Let*
$$\psi : (\mathsf{F}(A^*), \cdot, \cup) \to (\mathsf{F}(B^*), \cdot, \cup)$$
be a semiring homomorphism. Then $K = \psi^{[-1]}(L)$ is again recognizable by $(S, \cdot, \vee)$ with respect to I.

Proof. Let $\beta : (\mathsf{F}(B^*), \cdot, \cup) \to (S, \cdot, \vee)$ be a semiring homomorphism such that $\mathsf{F}(L) = \beta^{-1}(I)$. Then $\mathsf{F}(K) = (\beta\psi)^{-1}(I)$. $\square$

Lemma 5.5. *Let $(O, \cdot, \leq)$ be a finite ordered monoid, let $\alpha : (A^*, \cdot, =) \to \mathsf{H}^d(O, \cdot, \leq)$ be a monoid homomorphism. Further, let H be a hereditary subset of $\mathsf{H}^d(O, \cdot, \leq)$ and let $L = \alpha^{-1}(H)$.*

Then L is a positive boolean combination of languages each of which is an image under a surjective literal homomorphism of a language recognizable by $(O, \cdot, \leq)$.

Proof. Let $\alpha : a_i \mapsto \{s_{i,1}, \ldots, s_{i,k_i}\} \in \mathsf{H}(O, \geq)$, $s_{i,j} \in O$, $i = 1, \ldots, n$, $j = 1, \ldots, k_i$ and let $H = \{P_1, \ldots, P_l\} \subseteq \mathsf{H}(O, \geq)$ with

$$P_m \subseteq P \in \mathsf{H}(O, \geq) \text{ for some } m \Rightarrow P \in H \ .$$

Now $w \in L$ is the same as $\alpha(w) \in \{P_1, \ldots, P_l\}$. Thus $L = \alpha^{-1}(\{P_1\}) \cup \cdots \cup \alpha^{-1}(\{P_l\})$ and it is enough to consider the case

$$L = \alpha^{-1}(\{ Q \in \mathsf{H}(O, \geq) \mid P \subseteq Q \subseteq O \}) \text{ for a given } P \in \mathsf{H}(O, \geq) \ ;$$

that is, $w \in L$ if and only if $\alpha(w) \supseteq P$.

For $w = a_{i_1} \ldots a_{i_m}$ we have

$$\alpha(w) = \{s_{i_1 1}, \ldots, s_{i_1 k_{i_1}}\} \circ \cdots \circ \{s_{i_m 1}, \ldots, s_{i_m k_{i_m}}\}$$

$$= \{s \in O \mid (\exists j_1, \ldots, j_m) \ s \geq s_{i_1 j_1} \ldots s_{i_m j_m}\}.$$

Putting, for $s \in O$, $L_s = \{w \in A^* \mid s \in \alpha(w)\}$, we have $L = \bigcap_{s \in P} L_s$.

Let $B = \{b_{1,1}, \ldots, b_{1,k_1}, \ldots, b_{n,1}, \ldots, b_{n,k_n}\}$ be a new alphabet and let a surjective literal homomorphism $\lambda : (B^*, \cdot) \to (A^*, \cdot)$ be given by $b_{i,j} \mapsto a_i$, $i = 1, \ldots, n$, $j = 1, \ldots, k_i$. Also define $\eta : (B^*, \cdot) \to (O, \cdot)$ by $b_{i,j} \mapsto s_{i,j}$, $i = 1, \ldots, n$, $j = 1, \ldots, k_i$. Then $L_s = \lambda(\eta^{-1}((s]))$. $\square$

We will prove Theorem 4.2 in four steps.

Step 1. Let $\mathcal{L}$ be a positive variety of languages. Then $\bigcup \bigcap \Phi \mathcal{L}$ is again a positive variety of languages. Indeed, it is obvious that $\bigcup \bigcap \Phi \mathcal{L}$ satisfies the conditions $(\bigcap)$, $(\bigcup)$, $(^{-1})$. The condition (Q) follows from Lemma 5.1.

Step 2. For a positive variety of languages $\mathcal{L}$ we have $\Phi \mathcal{L} = \Theta \mathcal{L}$ and $\Psi \mathcal{L} = \Lambda \mathcal{L}$. The equalities follow from Lemmas 5.2 and 5.3.

Step 3. In the notation of Theorem 4.2, we have $\mathsf{O}(\bigcup \bigcap \Phi \mathcal{L}) \subseteq \mathsf{H}\mathcal{P}$. Indeed, let $(O, \cdot, \leq) \in \mathcal{P}$, let $L \subseteq B^*$ be recognizable by $(O, \cdot, \leq)$ with respect to its hereditary subset H. Let $\psi : (\mathsf{F}(A^*), \cdot, \cup) \to (\mathsf{F}(B^*), \cdot, \cup)$ be a semiring homomorphism and let $K = \psi^{[-1]}(L)$.

By [10, Prop. 1], the structure $(O, \cdot, \leq)$ is isomorphic to a substructure of $(\mathsf{H}(O, \leq), \circ, \cup)$ with $s \mapsto (s]$ being the isomorphism. Therefore L is also recognizable by the semiring $(\mathsf{H}(O, \leq), \circ, \cup)$ with respect to its ideal $(H]$ of all hereditary subsets of $(O, \leq)$ which are contained in H.

By Lemma 5.4 the language K is also recognizable by $(H(O, \leq), \circ, \cup)$. By Pin's refinement of the Eilenberg correspondence (Theorem 3.1 (ii)), the ordered syntactic monoids of joins of intersections of such K's remain in H ($\mathcal{P}$).
Step 4. In notation of Theorem 4.2, we have $\mathsf{L}\,(\mathsf{H}\,\mathcal{P}) \subseteq \bigcup \cap \Theta \mathcal{L}$.

Indeed, from Lemma 5.5 we get $\mathsf{L}\,(\mathsf{H}^\mathsf{d}\mathcal{P}) \subseteq \bigcup \cap \Lambda \mathcal{L}$. Taking duals of monoids and complements of languages we get

$$\mathsf{L}\,(\langle\{\,(\mathsf{H}\,(O, \geq), \circ, \subseteq) \mid (O, \cdot, \leq) \in \mathcal{P}\,\}\rangle_\circ) \subseteq \mathsf{C}\bigcup \cap \Lambda \mathcal{L} \ .$$

The last class is $\bigcup \cap \Theta \mathsf{C} \mathcal{L}$. Now take $\mathcal{P}^\mathsf{d}$ instead of $\mathcal{P}$ and $\mathsf{C} \mathcal{L}$ instead of $\mathcal{L}$.
Proof of Remark 1. Use the same tricks as in the proof of Step 4.
Proof of Theorem 4.3. The item (i) follows from Lemma 5.1 (i), (vi), (viii) and the item (ii) from Lemma 5.1 (ii), (v).

6 Hereditary version generalizes the power version

In fact, we will derive here among others Theorem 4.1 from Theorem 4.2.

There are natural operators between classes of (finite) monoids and classes of (finite) ordered monoids; namely for a class $\mathcal{M}$ of monoids

$$\mathcal{M}^\mathsf{o} = \{(O, \cdot, \leq) \mid (O, \cdot) \in \mathcal{M}\}$$

and for a class $\mathcal{P}$ of ordered monoids

$$\mathcal{P}^\mathsf{M} = \{(O, \cdot) \mid (O, \cdot, \leq) \in \mathcal{P}\} \ .$$

We collect now some observations about relationships between the lattices of pseudovarieties of monoids and pseudovarieties of ordered monoids.

Lemma 6.1.
(i) *If $\mathcal{M}$ is a pseudovariety of monoids then $\mathcal{M}^\mathsf{o}$ is a pseudovariety of ordered monoids. Moreover, $\mathsf{L}\,\mathcal{M} = \mathsf{L}\,\mathcal{M}^\mathsf{o}$.*
(ii) *Let $\mathcal{P}$ be a pseudovariety of ordered monoids. Then $\mathcal{P}^\mathsf{M}$ is a class of monoids closed with respect to forming direct products and substructures.*
(iii) *If $\mathcal{M}$ is a pseudovariety of monoids then $\mathcal{M}^{\mathsf{o}\,\mathsf{M}} = \mathcal{M}$.*
(iv) *A pseudovariety $\mathcal{P}$ of ordered monoids is of the form $\mathcal{M}^\mathsf{o}$ for some pseudovariety of monoids $\mathcal{M}$ if and only if it is self-dual.*
(v) *The smallest self-dual pseudovariety of ordered monoids containing a given pseudovariety $\mathcal{P}$ is $\mathcal{P} \vee \mathcal{P}^\mathsf{d}$ (the join is taken in the lattice of all pseudovarieties of ordered monoids) which equals to $\mathcal{P}^{\mathsf{M}\,\mathsf{o}}$ and the largest self-dual pseudovariety contained in $\mathcal{P}$ is $\mathcal{P} \cap \mathcal{P}^\mathsf{d}$ which also equals to $\{(M, \cdot) \mid (M, \cdot, =) \in \mathcal{P}\}^\mathsf{o}$.*

Proof. Almost all is obvious. The only thing to mention is that for a pseudovariety $\mathcal{P}$ of ordered monoids

$$(O,\cdot,\leq),(O,\cdot,\geq)\in\mathcal{P}\iff(O,\cdot,=)\in\mathcal{P} \qquad (*) \, .$$

Indeed, $s\mapsto(s,s)$ $(s\in O)$ is an isomorphism of $(O,\cdot,=)$ onto a substructure of $(O,\cdot,\leq)\times(O,\cdot,\geq)$. □

Lemma 6.2. *For any pseudovariety $\mathcal{M}$ of monoids*
 (i) $\mathsf{H}(\mathcal{M}^o) = \langle\{\,(\mathsf{P}(M),\cdot,\subseteq)\mid(M,\cdot)\in\mathcal{M}\,\}\rangle_o$,
 (ii) $(\mathsf{P}(\mathcal{M}))^o = \mathsf{H}(\mathcal{M}^o)\vee(\mathsf{H}(\mathcal{M}^o))^d$.

Proof. (i) Observe that for an ordered monoid $(O,\cdot,\leq)$, the semiring $(\mathsf{H}(O,\leq),\circ,\cup)$ is a homomorphic image of the semiring $(\mathsf{P}(O),\cdot,\cup)$. Indeed, consider the mapping $\xi:\mathsf{P}(O)\to\mathsf{H}(O,\leq)$, $X\mapsto[X]$.
 (ii) By Lemma 6.1 (v) (or directly from $(*)$),

$$\mathsf{H}(\mathcal{M}^o)\vee(\mathsf{H}(\mathcal{M}^o))^d = \langle\{\,(\mathsf{P}(M),\cdot,=)\mid(M,\cdot)\in\mathcal{M}\,\}\rangle_o$$

which equals the class from the statement since, for any class $\mathcal{N}$ of finite monoids, '

$$(\langle\mathcal{N}\rangle_M)^o = \langle\{\,(M,\cdot,=)\mid(M,\cdot)\in\mathcal{N}\,\}\rangle_o \, .$$

Proof of Theorem 4.1 from Theorem 4.2.
 Let $\mathcal{M}$ be a pseudovariety of monoids and let $\mathcal{L}$ be the corresponding boolean variety of languages; that is $\mathsf{L}\mathcal{M}=\mathcal{L}$ and $\mathsf{M}\mathcal{L}=\mathcal{M}$.
 By Lemma 6.1 we also have $\mathsf{L}\mathcal{M}^o=\mathcal{L}$, $\mathsf{O}\mathcal{L}=\mathcal{M}^o$.
 By Theorem 4.2, the positive variety $\bigcup\bigcap\Phi\mathcal{L}$ corresponds to the pseudovariety $\mathsf{H}(\mathcal{M}^o)$ and similarly $\mathsf{C}\bigcup\bigcap\Phi\mathcal{L}$ corresponds to the pseudovariety $(\mathsf{H}(\mathcal{M}^o))^d$.
 The smallest positive variety containing both $\bigcup\bigcap\Phi\mathcal{L}$ and $\mathsf{C}\bigcup\bigcap\Phi\mathcal{L}$ is the boolean variety $\mathsf{B}\Phi\mathcal{L}=\mathsf{B}\Psi\mathcal{L}$ (use $\mathcal{L}=\mathsf{C}\mathcal{L}$ and Lemma 5.1).
 Using Pin's refinement of the Eilenberg theorem, $\mathsf{B}\Phi\mathcal{L}$ corresponds to the pseudovariety $\mathsf{H}(\mathcal{M}^o)\vee(\mathsf{H}(\mathcal{M}^o))^d$ which is $(\mathsf{P}(\mathcal{M}))^o$ by Lemma 6.2 (ii). Now Lemma 6.1 (i) gives the statement of Theorem 4.1. □

7 Examples

We will calculate here how the operator H acts on the smallest pseudovarieties of ordered monoids.

For a finite monoid $(M, \cdot)$ and $s \in M$, let s^ω denote the idempotent in the subsemigroup of $(M, \cdot)$ generated by $\{s\}$. For a class Π of pseudoidentities, let $[\Pi]$ denote the class of all finite ordered monoids which satisfy all members of Π. Notice that identities for ordered monoids are of the form $f \leq g$ where f and g are usual monoid terms.

We will need the following lemma probably quite often used in the literature.

Lemma 7.1. *Let $\mathcal{U}_1, \mathcal{U}_2, \ldots$ be a sequence of varieties of ordered monoids, let a finite ordered monoid $(S_n, \cdot, \leq)$ generate $\mathcal{U}_n$, $n = 1, 2, \ldots$. Let the pseudovariety $\mathcal{P}$ consist of all finite members of $\bigcup_{n \in \mathbb{N}} \mathcal{U}_n$. Then the sequence $(S_1, \cdot, \leq), (S_2, \cdot, \leq), \ldots$ generates the pseudovariety $\mathcal{P}$.*

Proof. Any $(O, \cdot, \leq) \in \mathcal{P}$ is an isotone monoid homomorphic image (denote this homomorphism by β) of a substructure T of a power $\prod_{\gamma \in \Gamma} (S_n, \cdot, \leq)$ for some n. Choose $(s_{a,\gamma})_{\gamma \in \Gamma} \in \beta^{-1}(a)$ for every $a \in O$. This gives, for any $\gamma \in \Gamma$, a mapping of O to S_n. Take a finite $\Gamma_0 \subseteq \Gamma$ such that all above mappings are represented. Then $(O, \cdot, \leq)$ is an isotone monoid homomorphic image of a substructure of a finite product $\prod_{\gamma \in \Gamma_0} (S_n, \cdot, \leq)$. $\square$

As well-known there are exactly four pseudovarieties of ordered monoids formed by semilattices:

$$\mathcal{J}_1 = [x^2 = x,\ xy = yx],$$
$$\mathcal{J}_1^- = [x^2 = x,\ xy = yx,\ 1 \leq x],$$
$$\mathcal{J}_1^+ = [x^2 = x,\ xy = yx,\ x \leq 1],$$
$$\mathcal{I} = [x = y].$$

Denote also

$$\mathcal{AC}_0 = [x^{\omega+1} = x^\omega,\ xy = yx,\ 1 \leq x],\ \mathcal{AC}_1 = [x^{\omega+1} = x^\omega,\ xy = yx,\ x \leq x^2].$$

Theorem 7.2.
  (i) $\mathsf{H}(\mathcal{I}) = \mathcal{I}$,
  (ii) $\mathsf{H}(\mathcal{J}_1^+) = \mathcal{J}_1^+$,
  (iii) $\mathsf{H}(\mathcal{J}_1^-) = \mathcal{AC}_0$,
  (iv) $\mathsf{H}(\mathcal{J}_1) = \mathcal{AC}_1$.

Proof. The item (i) is clear.
  If $(O, \cdot, \leq)$ satisfies $x \leq x^2$ then also $(\mathsf{H}(O, \leq), \circ, \subseteq)$ satisfies this inequality. The same is true for any of $xy = yx$, $1 \leq x$, $x \leq 1$.

419

Now $x \leq 1$, $x \leq x^2$ gives $x^2 = x$ which yields the statement (ii).

If $H = (\{s_1, \ldots, s_n\}]$ is a hereditary subset of a finite ordered monoid $(O, \cdot, \leq)$ then $H^k = (\{s_{i_1} \ldots s_{i_k} \mid i_1, \ldots, i_k \in \{1, \ldots, n\}\}]$, $k = 1, 2, \ldots$.
If $(O, \cdot, \leq)$ satisfies $x^2 = x$ and $xy = yx$ then $H^{n+1} = H^n$. Thus

$$\mathsf{H}(\mathcal{J}_1^-) \subseteq \mathcal{AC}_0 \text{ and } \mathsf{H}(\mathcal{J}_1) \subseteq \mathcal{AC}_1 \ .$$

Denote by $\mathcal{V}_n$ and $\mathcal{W}_n$ the varieties of all ordered monoids given by

$$x^{n+1} = x^n, \ xy = yx, \ 1 \leq x \quad \text{and} \quad x^{n+1} = x^n, \ xy = yx, \ x \leq x^2 \ ,$$

respectively. Then

$$\mathcal{AC}_0 = \text{ all finite members of } \bigcup_{n \in \mathbb{N}} \mathcal{V}_n, \ \mathcal{AC}_1 = \text{ all finite members of } \bigcup_{n \in \mathbb{N}} \mathcal{W}_n \ .$$

One can easily characterize the identities $f \leq g$ which are valid in $\mathcal{V}_n$: for every variable x, the number of occurrences of x in f is less or equal to the number of occurrences of x in g or both those numbers are $\geq n$.

It is similar for the variety $\mathcal{W}_n$; one only adds the condition that every variable of g is also in f.

Let O_n be the set of all subsets of $\{a_1, \ldots, a_n\}$, $n \in \mathbb{N}$. Then

$$(O_n, \cup, \subseteq) \in \mathcal{J}_1^- \text{ and } (O_n, \cup, =) \in \mathcal{J}_1$$

and consider

$$(S_n, \cdot, \leq) = \mathsf{H}(O_n, \cup, \subseteq), \ (T_n, \cdot, \leq) = \mathsf{H}(O_n, \cup, =) \ .$$

We show that $(S_n, \cdot, \leq)$ generates the variety $\mathcal{V}_n$ and that $(T_n, \cdot, \leq)$ generates the variety $\mathcal{W}_n$.

Indeed, assume that $(S_n, \cdot, \leq)$ satisfies $f \leq g$. If the number of occurrences of a variable x in g is less than both n and the number of occurrences of a variable x in f, substituting $x \mapsto \{\emptyset, \{a_1\}, \ldots, \{a_n\}\}$, $y \mapsto \{\emptyset\}$ for all other variables, we get a contradiction.

Similarly for such an identity for $(T_n, \cdot, \leq)$ and in the case that a variable x is in g but not in f substitute $x \mapsto \{\{a_1\}\}$, $y \mapsto O_n$ for all other variables, to get a contradiction.

Now Lemma 7.1 gives the items (iii) and (iv). □

8 Final remarks

We have seen that the *hereditary* operator corresponds to the power operator when generalizing monoids to the ordered monoids. As well-known and explicitly stated in [2], the power operator on pseudovarieties turns out to be much more complicated than its variety version. So it is natural to study hereditary operator on varieties first. Probably the first non-trivial result in this direction is due to Almeida and Klíma [3] showing that for a given variety $\mathcal{P}$ of ordered monoids the set of all left-linear inequalities valid in $\mathcal{P}$ forms an equational base for $\mathsf{H}\,(\mathcal{P})$ (here, an inequality $f \leq g$ is *left-linear* if every variable has at most one occurrence in f). This is a nice generalization of the well-known result concerning the power operator on varieties of monoids.

Another thing of further interest would be to consider the hereditary operator defined on pseudovarieties of ordered monoids with values in the lattice of all pseudovarieties of idempotent semirings

$$\mathcal{P} \mapsto \langle \{\ (\mathsf{H}\,(O, \leq), \circ, \cup) \mid (O, \cdot, \leq) \in \mathcal{P}\ \}\rangle_\mathsf{S}\ .$$

This will relate the pseudovarieties of ordered monoids and pseudovarieties of idempotent semirings leading to results about comparison of the two independent classifications of regular languages – see Theorem 3.1.

After the author's talk in Coimbra, J.-É. Pin and A. C. Gómez informed the author they were preparing a paper dealing with an ordered version of the power operator.

The present paper would not be never written in time without several discussions with O. Klíma and M. Kunc. The author also wishes to thank to the anonymous referees for numerous suggestions.

References

[1] J. Almeida, *Finite Semigroups and Universal Algebra*, World Scientific, 1994
[2] J. Almeida, Power semigroups: results and problems, *Algebraic Engineering*, M. Ito and C. Nehaniv (eds.), World Scientific, Singapore, 1999, 399–415.
[3] J. Almeida, O. Klíma, Notes from July 2000, unpublished
[4] S. Eilenberg, *Automata, Languages and Machines*, Volumes A and B, Academic Press, 1974 and 1976
[5] J.-É. Pin, *Varieties of Formal Languages*, Plenum, 1986
[6] J.-É. Pin, A variety theorem without complementation, *Izvestiya VUZ Matematika* **39** (1995), 80–90. English version: *Russian Mathem. (Iz. VUZ)* **39** (1995), 74–83
[7] J.-É. Pin, Syntactic semigroups, Chapter 10 in *Handbook of Formal Languages*, G. Rozenberg and A. Salomaa (eds.), Springer, 1997
[8] J.-É. Pin and P. Weil, Polynomial closure and unambiguous product, *Theory Comput. Syst.* **30**, No. 4 (1997), 383–422

[9] L. Polák, A classification of rational languages by semilattice-ordered monoids, http://www.math.muni.cz/~polak
[10] L. Polák, Syntactic semiring of a language, *Proc. Mathematical Foundations of Computer Science 2001*, Springer Lecture Notes in Computer Science, Vol. 2136, 2001, 611–620
[11] L. Polák, Syntactic semiring and language equations, submitted

AUTOMATA IN AUTONOMOUS VARIETIES

OLGA SOKRATOVA*

Institute of Computer Science, University of Tartu, J. Liivi 2, 50409 Tartu, Estonia, E-mail: olga@cs.ut.ee

Department of Mathematics, The University of Iowa, 14 MacLean Hall, Iowa City, IA 52242-1419, USA, E-mail: osokrato@math.uiowa.edu

> Automata and machines in an arbitrary category of autonomous algebras are investigated by using the theory of Ω-rings. These automata include sequential (ordinary) automata and linear automata (dynamical systems). The recognizable subsets of free monoid Ω-rings are investigated. They are defined using Ω-automata as recognizers. For a subset of a free monoid Ω-ring, we define the Myhill and Nerode congruences and prove a generalization of the well-known Myhill-Nerode Theorem.

1 Introduction

Various categorical models of automata have been defined and studied recently. A general definition of machines in an arbitrary category was first given by Arbib and Manes [2]. Their research was motivated by unifying the realization problem for automata theory and control system theory. Their model includes sequential machines, linear control systems, tree automata and stochastic automata. Later, Adámek and Trnková [1] provided further development of the Arbib-Manes model of automata. One more approach can be found in the book by Plotkin et al [8]. Remarkably, the unification of automata and system theory and algebraic analysis of such systems have been fruitful in real-life application [5].

In this paper we deal with automata (machines) in an arbitrary category of autonomous algebras. Although we do not use the categorial framework, that is we do not explore the general categorical definition due to Arbib and Manes, our model is still sufficiently rich to support the main examples, namely sequential and linear machines. On the other hand, it appears that

---

*THE AUTHOR SINCERELY THANKS THE FINANCIAL SUPPORT OF FUNDAÇÃO CALOUSTE GULBENKIAN (FCG), FUNDAÇÃO PARA A CIÊNCIA E A TECNOLOGIA (FCT), FACULDADE DE CIÊNCIAS DA UNIVERSIDADE DE LISBOA (FCUL) AND REITORIA DA UNIVERSIDADE DO PORTO.

in our framework, the proposed machines can be investigated by means of the so-called Ω-rings and acts over them. An Ω-ring is a universal algebra equipped with a binary associative multiplication connected to operations in Ω by two-sided distributivity. Ω-rings provide a natural common generalization of distributive lattices, rings, semirings, and semigroups. Using the theory of Ω-rings, we are able to get new results for this generalized model.

In the second part of the paper we consider recognizable subsets of free Ω-rings. Recognizable subsets of general algebras have been defined by Mezei and Wright [6] in terms of finite congruences. An equivalent definition suggesting the idea of finite algebras as recognizers was given by Steinby [11]. Recognizable subsets of various free objects have been also investigated in the literature. We show that recognizable subsets of free Ω-rings can be defined using Ω-automata as recognizers. This approach allows us to get more explicit characterization of such subsets. For a subset of a free monoid Ω-ring, we define the Myhill and Nerode congruences and prove a generalization of the well-known Myhill-Nerode Theorem.

2 Preliminaries

This section contains concise algebraic background.

Let Ω be a signature. By an Ω-*ring* we mean an Ω-algebra R with multiplication $\cdot$ defined such that

(R1) $(R, \cdot\,)$ is a monoid with the identity element 1;

(R2) $r(s_1 \ldots s_n \omega) = (rs_1) \ldots (rs_n)\omega$,
$(s_1 \ldots s_n \omega)r = (s_1 r) \ldots (s_n r)\omega$,

for any operation $\omega \in \Omega_n$, $n \geq 0$, and arbitrary elements $r, s_1, \ldots, s_n \in R$.

Let an Ω-ring R be given. A *right (unitary) R-act* is an Ω-algebra A equipped with an action $A \times R \to A$ denoted by $(a, r) \mapsto ar$ which satisfies the following conditions for all elements $a, a_1, \ldots, a_n \in A$, $r, s, r_1, \ldots, r_n \in R$ and any operation $\omega \in \Omega_n$ $(n \geq 0)$:

(A1) $(ar)s = a(rs)$;

(A2) $a(r_1 \ldots r_n \omega) = (ar_1) \ldots (ar_n)\omega$;
$(a_1 \ldots a_n \omega)r = (a_1 r) \ldots (a_n r)\omega$;

(A3) $a1 = a$.

It makes sense to consider Ω-rings, and their acts whose Ω-algebras belong to some fixed variety of autonomous Ω-algebras, that is algebras with permuting operations. Then every act over an Ω-ring R can be viewed as a representation of R. Recall that an Ω-algebra A is called *permutative* if any two operations of Ω permute, i.e.,

$$(a_{11}\ldots a_{1n}\omega)(a_{21}\ldots a_{2n}\omega)\ldots(a_{m1}\ldots a_{mn}\omega)\tau =$$
$$= (a_{11}\ldots a_{m1}\tau)(a_{12}\ldots a_{m2}\tau)\ldots(a_{1n}\ldots a_{mn}\tau)\omega,$$

for arbitrary $\omega \in \Omega_n, \tau \in \Omega_m$ and $a_{11}, \ldots, a_{mn} \in A$.

Given a monoid M, we can define the *monoid Ω-ring RM*, as a coproduct of copies of Ω-algebras of R indexed by the elements of M. Recall that RM consists of all polynomials of the form $p(r_1 u_1, \ldots, r_n u_n)$, where $r_i \in R$, $u_i \in M$. The multiplication is defined by extending the multiplication in R and M by the distributivity law. It is well-defined in the case of autonomous underlying Ω-algebras.

In particular, given a set X, the free monoid Ω-ring RX^* consists of all polynomials over R in non-commuting variables of X.

For further information on Ω-rings we refer to the paper [9].

3 Ω-Automata

We follow the notion of automata in an arbitrary category introduced by B.I. Plotkin [8]. For a category of universal algebras this definition can be written down as follows:

Let $\mathcal{K}$ be an arbitrary category of Ω-algebras. Every system $(A, X, B, \circ, *)$ is called an Ω-*machine in* $\mathcal{K}$, where

- $A \in \mathcal{K}$ is an algebra of *states*;
- X is an alphabet of *input signals*;
- $B \in \mathcal{K}$ is an algebra of *output signals*;
- $\circ \colon A \times X \to A$ is the *transition function*; for every $x \in X$, the mapping $a \mapsto a \circ x$ is a morphism in $\mathcal{K}$, i.e.,

$$a_1 \ldots a_n \omega \circ x = (a_1 \circ x) \ldots (a_n \circ x)\, \omega,$$

for all $a_1, \ldots, a_n \in A$, and any operation $\omega \in \Omega_n$ $(n \geq 0)$;

- $* \colon A \times X \to B$ is the *output function*; for every $x \in X$, the mapping $a \mapsto a * x$ is a morphism in $\mathcal{K}$, i.e.,

$$a_1 \ldots a_n \omega * x = (a_1 * x) \ldots (a_n * x)\, \omega,$$

for all $a_1, \ldots, a_n \in A$, and any operation $\omega \in \Omega_n$ $(n \geq 0)$.

As usual, an Ω-machine $(A, \Gamma, B, \circ, *)$ is called a *semigroup* Ω-machine if Γ is a semigroup and
$$a \circ (\gamma_1 \gamma_2) = (a \circ \gamma_1) \circ \gamma_2$$
$$a * (\gamma_1 \gamma_2) = (a \circ \gamma_1) * \gamma_2,$$
for $a \in A$, $\gamma_1, \gamma_2 \in \Gamma$.

Here we deal with Ω-machines in an autonomous variety $\mathcal{A}$ of Ω-algebras with zero. It appears that these machines can be investigated by means of Ω-rings and acts over them, using the following assertion due to Fleischer [4].

Proposition 1. *Every non-trivial autonomous variety $\mathcal{A}$ of Ω-algebras with zero is polynomially equivalent to the variety of acts over a suitable commutative Ω-ring R with zero 0 and identity element 1.*

Let us note that this Ω-ring R was obtained there as the Ω-ring of endomorphisms of the free cyclic Ω-algebra $F_1 \in \mathcal{A}$. Besides, $(R, \Omega) \cong F_1$. In some cases the result holds for autonomous varieties without 0-ary operation, as well.

Thus, throughout we deal with Ω-machines $(A, \Gamma, B, \circ, *)$ in the variety $\mathcal{A}$, where the algebras A, B are considered as R-acts over an Ω-ring R provided by Proposition 1.

Example 3.1. Suppose $\mathcal{A}$ is the category of sets and $R = \{1\}$ is the one-element monoid. Then we get the ordinary (sequential) machines.

Example 3.2. Suppose $\mathcal{A}$ is a category of commutative monoids. Then R is a commutative semiring. In this way we get linear sequential automata, which have been arising in the theory of dynamical systems. For the theory of linear machines we refer to [2, 5, 7, 8].

Given an Ω-machine $(A, \Gamma, B, \circ, *)$ over a semigroup Γ, one can extend it to an Ω-machine $(A, R\Gamma, B, \circ, *)$ over the semigroup Ω-ring $R\Gamma$ by extending the transition and output functions as follows:
$$a \circ p(r_1 \gamma_1, \ldots, r_n \gamma_n) = p(ar_1 \circ \gamma_1, \ldots, ar_n \circ \gamma_n);$$
$$a * p(r_1 \gamma_1, \ldots, r_n \gamma_n) = p(ar_1 * \gamma_1, \ldots, ar_n * \gamma_n).$$

In this way, the Ω-automaton $(A, R\Gamma, \circ)$ consists of the Ω-ring $R\Gamma$ and the $R\Gamma$-act A.

We turn now to universal Ω-machines. First, let A and B be algebras from $\mathcal{A}$. Since the variety $\mathcal{A}$ is commutative, $\text{End}(A)$ and $\text{Hom}(A, B)$ are algebras from $\mathcal{A}$. Then their direct product $\text{End}(A, B) = \text{End}(A) \times \text{Hom}(A, B)$

belongs to $\mathcal{A}$, as well. The multiplication defined by

$$(\phi_1, \psi_1)(\phi_2, \psi_2) = (\phi_1\phi_2, \phi_1\psi_2), \quad \text{for } \phi_i \in \text{End}(A), \psi_i \in \text{Hom}(A, B),$$

makes $\text{End}(A, B)$ an Ω-ring. As usual, the transition and output functions are defined by

$$a \circ (\phi, \psi) = a\phi \quad \text{and} \quad a * (\phi, \psi) = (a\phi)\psi, \quad \text{for } a \in A, (\phi, \psi) \in \text{End}(A, B).$$

The Ω-machine $\text{Atm}(A, B) = (A, \text{End}(A, B), B)$ obtained in this way is a terminal object in the category of all Ω-machines with given (A, B) (see [8]).

Second, let us consider the Ω-machine

$$\text{Atm}(\Gamma, B) = (\text{Hom}_R(R\Gamma, B), \Gamma, B).$$

The transition and output functions are defined by

$$(f \circ \gamma)(u) = f(\gamma u) \text{ and } f * \gamma = f(\gamma), \quad f \in \text{Hom}_R(R\Gamma, B), \gamma \in \Gamma, u \in R\Gamma.$$

Note that these functions are well defined, because

$$(f_1 \ldots f_n \omega \circ \gamma)(u) = (f_1 \ldots f_n \omega)(\gamma u) = f_1(\gamma u) \ldots f_n(\gamma u) \, \omega$$
$$= (f_1 \circ \gamma(u)) \ldots (f_n \circ \gamma(u)) \, \omega$$
$$= ((f_1 \circ \gamma) \ldots (f_n \circ \gamma) \, \omega)(u)$$

and

$$(f_1 \ldots f_n \omega) * \gamma = (f_1 \cdots f_n \omega)(\gamma) = f_1(\gamma) \ldots f_n(\gamma) \, \omega$$
$$= (f_1 * \gamma) \ldots (f_n * \gamma) \, \omega,$$

for all $f_1, \ldots, f_n \in \text{Hom}_R(R\Gamma, B), \gamma \in \Gamma, u \in R\Gamma$, and $\omega \in \Omega_n$. Obviously, $\text{Atm}(\Gamma, B)$ is a semigroup Ω-machine.

Proposition 2. *The Ω-machine $\text{Atm}(\Gamma, B)$ is a terminal object in the category of all Ω-machines with given (Γ, B).*

Proof. Let (A, Γ, B) be an Ω-machine in $\mathcal{A}$. Define the mapping $\mu_1 \colon A \to \text{Hom}_R(R\Gamma, B)$ by

$$\mu_1(a)(u) = a * u \quad \text{for } a \in A, u \in R\Gamma,$$

which is a homomorphism of Ω-algebras. Indeed,

$$\mu_1(a_1 \cdots a_n \omega) = (a_1 \cdots a_n \omega) * u$$
$$= (a_1 * u) \cdots (a_n * u) \, \omega$$
$$= \mu_1(a_1) \cdots \mu_1(a_n)\omega,$$

for all $a_1, \ldots, a_n \in A, u \in R\Gamma$, and $\omega \in \Omega_n$. Furthermore, $(\mu_1, \mathrm{id}_\Gamma, \mathrm{id}_B)$ is a homomorphism of machines:

$$\mu_1(a \circ \gamma)(u) = (a \circ \gamma) * u = a * (\gamma u)$$
$$= \mu_1(a)(\gamma u) = (\mu_1(a) \circ \gamma)(u)$$

and

$$\mathrm{id}_B(a * \gamma) = a * \gamma = \mu_1(a)(\gamma)$$
$$= \mu_1(a) * \gamma,$$

for all $a \in A, \gamma \in \Gamma, u \in R\Gamma$, and $\omega \in \Omega_n$. Moreover, $(\mu_1, \mathrm{id}_\Gamma, \mathrm{id}_B)$ is a unique homomorphism of this sort. Indeed, assume that $(\nu, \mathrm{id}_\Gamma, \mathrm{id}_B) \colon (A, \Gamma, B) \to \mathrm{Atm}(\Gamma, B)$ is also homomorphism. Then

$$\mu_1(a)(\gamma) = a * \gamma = \mathrm{id}_B(a * \gamma)$$
$$= \nu(a) * \gamma = \nu(a)(\gamma),$$

for all $a \in A, \gamma \in \Gamma$. Therefore $\mu_1(a)|_\Gamma = \nu(a)|_\Gamma$, and so $\mu_1(a) = \nu(a)$, for any $a \in A$. Hence $\mu_1 = \nu$. $\square$

The construction of the tensor product of acts over Ω-rings and its properties can be found in [9].

Now, let (A, Γ) be a given Ω-automaton. Consider the Ω-machine

$$\mathrm{Atm}(A, \Gamma) = (A, \Gamma, A \otimes R\Gamma),$$

with the output function defined by

$$a * \gamma = a \otimes 1\gamma, \qquad \text{for } a \in A, \gamma \in \Gamma.$$

Proposition 3. *The Ω-machine $\mathrm{Atm}(A, \Gamma)$ is an initial object in the category of all Ω-machines with given (A, Γ) and $\circ \colon A \times \Gamma \to A$.*

Proof. Let (A, Γ, B) be an Ω-machine in $\mathcal{A}$. Define the homomorphism

$$\mu_3 \colon A \otimes R\Gamma \to B$$

as the extension of a bilinear map $\phi \colon A \times R\Gamma \to B$ such that $\phi(a, u) = a * u$. Then $(\mathrm{id}_A, \mathrm{id}_\Gamma, \mu_3) \colon \mathrm{Atm}(A, \Gamma) \to (A, \Gamma, B)$ is a homomorphism:

$$\mu_3(a * \gamma) = \mu_3(a \otimes 1\gamma) = a * \gamma,$$

for all $a \in A, \gamma \in \Gamma$.

It remains to show that the homomorphism $(\mathrm{id}_A, \mathrm{id}_\Gamma, \mu_3)$ is unique. Assume that $(\mathrm{id}_A, \mathrm{id}_\Gamma, \nu)$ is another homomorphism. Then

$$\mu(a \otimes \gamma) = a * \gamma = \nu(a * \gamma) = \nu(a \otimes \gamma),$$

for any $a \in A, \gamma \in \Gamma$. It follows that $\mu = \nu$, because $A \otimes R\Gamma$ is generated by $\{a \otimes \gamma \mid a \in A, \gamma \in \Gamma\}$. □

4 Recognizable Subsets of Free Ω-Rings

Recognizable subsets were defined in general algebra by Mezei and Wright [6] in terms of finite congruences. An equivalent definition suggesting the idea of finite algebras as recognizers was given by Steinby [11]. In this section we consider recognizable subsets of free Ω-rings and show that they can be defined using Ω-automata as recognizers.

Recall that R is assumed to be the Ω-ring of endomorphisms of the free cyclic Ω-algebra $F_1 \in \mathcal{A}$. We shall investigate recognizable subsets in the monoid Ω-ring RX^* motivated by the following

Proposition 4. *The monoid Ω-ring RX^* is a free Ω-ring over X in the variety of all Ω-rings whose Ω-algebras belong to $\mathcal{A}$.*

Proof. As it was mentioned in the comments to Proposition 1, there is an isomorphism $(R, \Omega) \cong F_1$, where F_1 is the free cyclic Ω-algebra in $\mathcal{A}$. Moreover, this isomorphism takes the identity element 1_R of R to the free generator of F_1. Recall that the Ω-algebra of RX^* is taken to be the free product $\coprod_{u \in X^*} Ru$ of copies Ru of (R, Ω). Hence $Ru \cong (R, \Omega) \cong F_1$.

Let S be an arbitrary Ω-ring whose Ω-algebra belongs to $\mathcal{A}$. Each mapping $f \colon X \to S$ has a unique monoid extension $f' \colon X^* \to (S, \cdot)$. There exists a unique Ω-homomorphism $g \colon R \to S$ such that $g(1_R) = 1_S$. We claim that g is a homomorphism of Ω-rings. Indeed, since 1_R is the only free generator of (R, Ω), any elements $r, s \in R$ can be represented in the form $r = p_r(1_R), s = p_s(1_R)$, for some unary Ω-terms p_r, p_s. Then by using the definition of the multiplication in the monoid Ω-rings [9] we have

$$\begin{aligned} g(rs) &= g(p_r(1_R) \cdot p_s(1_R)) \\ &= g(p_r(p_s(1_R))) \\ &= p_r(p_s(g(1_R))) \\ &= p_r(p_s(1_S)) \\ &= p_r(1_S) \cdot p_s(1_S) \\ &= p_r(g(1_R)) \cdot p_s(g(1_R)) \\ &= g(r)g(c), \end{aligned}$$

as required. For every $u \in X^*$, define an Ω-homomorphism $f_u \colon Ru \to S$ by
$$f_u(ru) = g(r)f'(u), \quad \text{whenever } r \in R.$$
Let $\overline{f} \colon \coprod_{u \in X^*} Ru \to S$ be the unique extension of all f_u. In order to show that $\overline{f}$ is a homomorphism of Ω-rings, we first verify that
$$\overline{f}(ru \cdot sv) = \overline{f}(ru)\overline{f}(sv), \tag{1}$$
for all $r, s \in R, u, v \in X^*$. Representing the elements r, s in the form $r = p_r(1_R), s = p_s(1_R)$, we get

$$\begin{aligned}
\overline{f}(ru)\overline{f}(sv) &= g(r)f'(u)g(s)f'(v) \\
&= g(p_r(1_R))f'(u)g(p_s(1_R))f'(v) \\
&= p_r(1_S)f'(u) \cdot p_s(1_S)f'(v) \\
&= p_r(f'(u))p_s(f'(v)) \\
&= p_r(p_s(f'(u)f'(v))) \\
&= p_r(p_s(f'(uv))) \\
&= p_r(p_s(1_S)f'(uv)) \\
&= p_r(p_s(1_S))f'(uv) \\
&= p_r(1_S)p_s(1_S)f'(uv) \\
&= g(r)g(s)f'(uv) \\
&= g(rs)f'(uv) \\
&= \overline{f}(rsuv) \\
&= \overline{f}(ru \cdot sv)
\end{aligned}$$

Now, applying $\overline{f}$ to the definition of the multiplication in the monoid Ω-rings [9] for any elements from RX^* and using (1) we can easily verify that $\overline{f}$ is a homomorphism of Ω-rings.

Let $h \colon \coprod_{u \in X^*} Ru \to S$ be another extension of f. Then $h(1_R) = 1_S$ implies $h(r) = p_r(1_S) = g(r)$, for any element $r = p_r(1_R) \in R$. Together with $h(u) = f'(u)$, whenever $u \in X^*$, this implies $h = \overline{f}$.

Thus f admits a unique extension $\overline{f} \colon \coprod_{u \in X^*} Ru \to S$, and the result follows. □

Let us give the following definition. An Ω-*recognizer* (A, a_0, F) in the variety $\mathcal{A}$ consists of an Ω-automaton (A, X^*) in $\mathcal{A}$, an initial state $a_0 \in A$

and a set $F \subseteq A$ of final states. An Ω-recognizer (A, a_0, F) is *finite* if A is finite. The *set recognized* by an Ω-recognizer (A, a_0, F) is the set

$$L(A) = \{u \in RX^* \mid a_0 u \in F\}.$$

Recall that $\mathrm{Tr}_\Omega(RX^*)$ is the set of all translations of the Ω-algebra of RX^*. For arbitrary subset $L \subseteq RX^*$, define the following equivalence relations on RX^*:

$$u \; \theta_L \; v \text{ if and only if } (\forall p \in \mathrm{Tr}_\Omega(RX^*))(\forall w, w' \in X^*)$$
$$[p(wuw') \in L \Leftrightarrow p(wvw') \in L]$$

and

$$u \; \rho_L \; v \text{ if and only if } (\forall p \in \mathrm{Tr}_\Omega(RX^*))(\forall w \in X^*)$$
$$[p(uw) \in L \Leftrightarrow p(vw) \in L].$$

We make use of the following technical lemma.

Lemma 5. *For every non-trivial term $t = t(y_1, \ldots, y_n)$ and elements $a, b_1, \ldots, b_{i-1}, b_{i+1}, \ldots, b_n$, there exists a translation $p \in \mathrm{Tr}(A)$ such that*

$$t(b_1, \ldots, b_{i-1}, \xi, b_{i+1}, \ldots, b_n) = p(\xi), \quad \text{for all } \xi.$$

Proposition 6. *θ_L is the greatest congruence on RX^* which saturates L.*

Proof. Given pairs $\langle u_i, v_i \rangle \in \theta_L$ $(i = 1, \ldots, n)$, and an operation $\omega \in \Omega_n$, we have to show that $\langle u_1 \ldots u_n \omega, v_1 \ldots v_n \omega \rangle \in \theta_L$. Take an arbitrary $p \in \mathrm{Tr}_\Omega(RX^*)$ and $w, w' \in X^*$. For any i, define the elementary translation

$$q_i(\xi) = (wv_1w') \ldots (wv_{i-1}w')\xi(wu_{i+1}w') \ldots (wu_nw')\omega.$$

Put $p_i = p \cdot q_i$. Then

$$p(w(u_1 \ldots u_n \omega)w') = p_1(wu_1w') \in L$$
$$\Leftrightarrow p_1(wv_1w') = p_2(wu_2w') \in L$$
$$\Leftrightarrow p_2(wv_2w') \in L$$
$$\Leftrightarrow \ldots$$
$$\Leftrightarrow p_n(wv_nw') = p(w(v_1 \ldots v_n\omega)w') \in L.$$

In order to prove that θ_L is stable with respect to multiplication, we first note that $\langle u, v \rangle \in \theta_L$ implies $\langle ur, vr \rangle \in \theta_L$, for all $r \in R, u, v \in RX^*$. Indeed, since $r = p_r(1_R)$, for some unary Ω-term p_r, we get that either $r = 0$, and then $\langle ur, vr \rangle = \langle 0, 0 \rangle \in \theta_L$, or $p_r(\xi) \in \mathrm{Tr}_\Omega(RX^*)$, and then the definition of θ_L yields $\langle ur, vr \rangle \in \theta_L$, again.

Next, consider an arbitrary nonzero element $c = q(r_1c_1, \ldots, r_nc_n) \in RX^*$, where $r_i \in R, c_i \in X_*$, and q is an Ω-term. We are going to show that $\langle u, v \rangle \in \theta_L$ implies $\langle uc, vc \rangle \in \theta_L$. Take an arbitrary translation $p \in \operatorname{Tr}_\Omega(RX^*)$ and $w, w' \in X^*$. For every $i = 1, \ldots, n$, take the Ω-translation

$$q_i(\xi) = q(wvr_1c_1w', \ldots, wvr_{i-1}c_{i-1}w', \xi, wur_{i+1}c_{i+1}w', \ldots, wur_nc_nw')$$

provided by Lemma 5. Put $p_i = p \cdot q_i$. Then

$$\begin{aligned}
p(w(uc)w') &= p(q(wur_1c_1w', \ldots, wur_nc_nw')) \\
&= p_1(w(ur_1)c_1w') \in L \\
&\Leftrightarrow p_1(w(vr_1)c_1w') = p_2(w(ur_2)c_2w') \in L \\
&\Leftrightarrow p_2(w(vr_2)c_2w') \in L \\
&\Leftrightarrow \ldots \\
&\Leftrightarrow p_n(w(vr_n)c_nw') = p(w(vc)w') \in L.
\end{aligned}$$

Similarly, $\langle uc, vc \rangle \in \theta_L$.

Thus θ_L is a congruence. It is clear that every congruence that saturates L is contained in θ_L. □

The following proposition can be proved similarly.

Proposition 7. *ρ_L is the greatest right congruence on RX^* which saturates L.*

Proposition 8. *$(RX^*/\rho_L, X^*)$ is a minimal Ω-recognizer of L.*

Proof. Transitions are defined as follows:

$$u/\rho_L \cdot w = (uw)/\rho_L, \qquad \text{for } u \in RX^*, w \in X^*.$$

The transition is well defined, since ρ_L is a right congruence. Further, take $1/\rho_L$ as an initial state and $F = \{u/\rho_L \mid u \in L\}$. It is clear that $(RX^*/\rho_L, 1/\rho_L, F)$ recognizes L. Now, let (A, a_0, T) recognize L, as well. Denote by A' the RX^*-subact in A generated by a_0RX^*. Let $\phi: RX^* \to A$ be the RX^*-homomorphism that takes 1 to a_0. By Proposition 7, $\ker \phi \leq \rho_L$. Therefore there exists a surjective RX^*-homomorphism $A' \to RX^*/\rho_L$ which maps a_0 to $1/\rho_L$. Thus $RX^*/\rho_L \preceq A$. □

Given an Ω-automaton (A, Γ) in $\mathcal{A}$, we have a natural homomorphism of Ω-rings $\mu: R\Gamma \to \operatorname{End}_\Omega(A)$ defined by

$$a\mu(u) = au, \qquad \text{for } u \in R\Gamma, a \in A.$$

The Ω-ring $R\Gamma/\ker\mu \cong \mu(R\Gamma)$ is called the Ω-*ring of transitions* of (A, Γ).

Proposition 9. RX^*/θ_L *is isomorphic to the Ω-ring of all transitions of the Ω-automaton $(RX^*/\rho_L, X^*)$.*

Proof. We have to show that $\ker \mu = \theta_L$, for the Ω-automaton $(RX^*/\rho_L, X^*)$. Because of Proposition 6 we show that $\ker \mu$ is the greatest congruence which saturates L. Assume that $\mu(v) = \mu(w)$ and $v \in L$. Then, for every $u \in RX^*$, we have $u/\rho_L \cdot v = u/\rho_L \mu(v) = u/\rho_L \mu(w) = u/\rho_L \cdot w$, and so $(uv)/\rho_L = (uw)/\rho_L$. Therefore $\langle v, w \rangle \in \rho_L$, that together with $v \in L$, implies $w \in L$. Thus $\ker \mu$ saturates L. Suppose now that L is saturated by some congruence σ of RX^*. Given an arbitrary $\langle v, w \rangle \in \sigma$, we show that $\mu(v) = \mu(w)$. It remains to verify that $(uv)/\rho_L = (uw)/\rho_L$, for any $u \in RX^*$. Note that $\langle v, w \rangle \in \sigma$ implies $\langle uv, uw \rangle \in \sigma$. Further, since the congruence σ saturates L, Proposition 7 yields $\sigma \leq \rho_L$. Therefore $\langle uv, uw \rangle \in \rho_L$, which completes the proof. $\square$

The following theorem generalizes the Myhill-Nerode Theorem and gives a few characterizations of recognizable sets.

Theorem 10. *For an arbitrary subset $L \subseteq RX^*$, the following conditions are equivalent:*
  (1) *L is recognized by a finite Ω-recognizer (A, a_0, F) in the variety $\mathcal{A}$;*
  (2) *L is saturated by a congruence of RX^* of a finite index;*
  (3) *θ_L is of a finite index;*
  (4) *L is saturated by a right congruence of RX^* of a finite index;*
  (5) *ρ_L is of a finite index;*
  (6) *there exist a finite Ω-ring A, a homomorphism $\phi \colon RX^* \to A$ and a subset $F \subseteq A$ such that $L = F\phi^{-1}$.*

Proof. The equivalence of conditions (2) and (6) is immediate. The implication (3) $\Rightarrow$ (2) is obvious, since θ_L saturates L. Conversely, if L is saturated by a right congruence σ of RX^* of a finite index, then $\sigma \leq \theta_L$, and so the index of θ_L is lower than the index of σ, which is finite. Thus we have proved (2) $\Leftrightarrow$ (3).

The equivalence of conditions (4) and (5) can be proved similarly.

The implication (2) $\Rightarrow$ (4) is obvious.

The implication (4) $\Rightarrow$ (1) follows from Proposition 8.

Let us prove now (1) $\Rightarrow$ (4). Assume that L is recognized by some finite Ω-recognizer (A, a_0, F). Define the equivalence $\sim$ of RX^* by $u \sim v$ if and only if $a_0 u = a_0 v$. Clearly, $\sim$ is a right congruence and it saturates L.

Finally, let us show (5) $\Rightarrow$ (3). If ρ_L is of a finite index, then the Ω-ring of all transitions of $(RX^*/\rho_L, X^*)$ is finite. Therefore RX^*/θ_L is finite by Proposition 9. □

In this way, we have shown that our definition of recognizable subsets of free Ω-semirings coincides with the definition of Mezei and Wright (see condition (6) of Theorem 10), and the definition of Steinby (see condition (2) of Theorem 10) for general algebras. Thus, among other things, Theorem 10 makes it possible to use their results, as well.

Several examples of recognizable subsets of the semiring $\mathbb{N}_0$ of natural numbers and of the free semiring $\mathbb{N}_0[X]$ are given by the author [10]. For example, it was proven that every finite subset of the semiring $\mathbb{N}_0$ is recognizable, and every ideal of the semiring $\mathbb{N}_0$ is recognizable. Surprisingly, the last result does not hold in general even for the polynomial semiring $\mathbb{N}_0[x]$ over one variable.

References

[1] J. Adámek and V. Trnková, *Automata and algebras in categories.* Mathematics and its Applications (East European Series), 37. Kluwer Academic Publishers Group, Dordrecht, 1990.

[2] M.A. Arbib and E.G. Manes, *Machines in a category: an expository introduction.* SIAM Rev. **16**, 163–192 (1974).

[3] J. Almeida *On pseudovarieties, varieties of languages, filters of congruences, pseudoidentities and related topics*, Algebra Universalis **27**, 333–350 (1990).

[4] V. Fleischer Ω-*rings, over which all acts are n-free*, Acta et Comment. Univ. Tartuensis **390**, 56–83 (1975).

[5] R.E. Kalman, P.L. Falb, and M.A. Arbib, *Topics in Mathematical System Theory* (McGraw-Hill, New York, 1969).

[6] J. Mezei and J.B. Wright *Algebraic automata and context-free sets*, Information and Control **11**, 3–29 (1967).

[7] G. Naude and C. Nolte, *A survey of the realization and duality theories of linear systems over rings* Quaest. Math. **5**, 135–164 (1982).

[8] B. I. Plotkin, L. Ja. Greenglaz and A. A. Gvaramija, *Algebraic structures in automata and databases theory*, World Sci. Publishing, River Edge, NJ, 1992.

[9] O. Sokratova and U. Kaljulaid. Ω-*rings and their flat representations.* Contributions to general algebra **12**, 377-390 (1999).

[10] O. Sokratova, *Linear automata and recognizable subsets in free semirings*, Proceedings of "13th Symposium on Fundamentals of Computa-

tional Theory" FCT 2001, Springer LNCS **2138**, 420-423 (2001).

[11] M. Steinby, *Some algebraic aspects of recognizability and rationality*, Lect. Notes Comput. Sci. **117**, 372–377 (1981).

A SAMPLER OF A TOPOLOGICAL APPROACH TO INVERSE SEMIGROUPS

BENJAMIN STEINBERG[*]

Departamento de Matemática Pura,
Faculdade de Ciências da Universidade do Porto,
R. Campo Alegre, 687
4169-007 Porto, Portugal
E-mail: bsteinbg@agc0.fc.up.pt

Generalizing the usual construction for group presentations, we introduce the standard 2-complex for an inverse semigroup presentation. This complex is a useful tool for understanding both the structure of the maximal subgroups and of the trace product groupoid of the inverse semigroup. Similarly, an inverse semigroup theoretic analog of the Cayley complex of a group presentation is given. Applications are given to the study of certain amalgams of inverse semigroups. Also a Reidemeister-Schreier type theorem is proved for maximal subgroups of inverse semigroups. Quasiconvexity is considered, as well.

1 Introduction

This paper arose out of various talks I have given on [21]. The ideas which will be presented in this paper, had a duplo-genesis: on the one hand, I had in mind the construction of a 2-complex starting with the Schützenberger graphs of an inverse semigroup [22] as its base; on the other hand, an inverse semigroup is a certain type of ordered groupoid [4, 10] and groupoids can be obtained as the fundamental groupoid of a 2-complex so it seemed natural to be able to recover the inverse semigroup from some sort of an ordered 2-complex. What I discovered was that the complex I had been construct-

---

[*]THE AUTHOR WAS SUPPORTED IN PART BY THE FCT AND POCTI APPROVED PROJECTS POCTI/32817/MAT/2000 AND POCTI/MAT/37670/2001 IN PARTICIPATION WITH THE EUROPEAN COMMUNITY FUND FEDER AND BY FCT THROUGH *CENTRO DE MATEMÁTICA DA UNIVERSIDADE DO PORTO*, AS WELL AS BY INTAS PROJECT 99-1224 *COMBINATORIAL AND GEOMETRIC THEORY OF GROUPS AND SEMIGROUPS AND ITS APPLICATIONS TO COMPUTER SCIENCE.* THE AUTHOR ALSO ACKNOWLEDGES THE FINANCIAL SUPPORT OF FUNDAÇÃO CALOUSTE GULBENKIAN (FCG), FUNDAÇÃO PARA A CIÊNCIA E A TECNOLOGIA (FCT), FACULDADE DE CIÊNCIAS DA UNIVERSIDADE DE LISBOA (FCUL) AND THE REITORIA DA UNIVERSIDADE DO PORTO.

ing from standard combinatorial techniques is an (ordered) 2-complex whose fundamental groupoid is the trace product groupoid of the original inverse semigroup. Moreover, the construction generalized the classical construction for groups.

When I wrote [21], I had in mind applications to inverse semigroup amalgams. For these applications it was *essential* to use ordered groupoids. Hence [21] does everything from the viewpoint that inverse semigroups are ordered groupoids. I then reduced the relationship with Schützenberger graphs to a small subsection which showed how to obtain an ordered groupoid presentation of an inverse semigroup from an inverse semigroup presentation.

The purpose of this text is to present the topological approach from a combinatorial inverse semigroup theory point-of-view which does not require experience with ordered groupoids. This forces me to leave out many of the proofs about amalgams, giving instead plausibility arguments. Some of the proofs here are different than those of [21] as I make use of combinatorial techniques instead of groupoid techniques. To avoid redundancy with [21], I omit some proofs which do not require ordered groupoids, but rather are easy manipulations with Green's relations.

The paper is organized as follows. To make the paper accessible to people without a background in combinatorial group theory, we begin with an overview of the classical theory of group presentations. Then we give the construction of the standard 2-complex of an inverse semigroup presentation (generalizing the usual notion from group presentations). The fundamental groupoid of this 2-complex is the trace product groupoid of the inverse semigroup. This complex is closely related to the Munn and right letter mapping representations of the inverse semigroup. Some minor improvements are made here over [21] to handle bizarre defining paths of 2-cells.

The maximal subgroups of the inverse semigroup will turn out to be the fundamental groups of the connected components of the standard 2-complex. This will allow us to obtain a quick topological proof that if a $\mathcal{D}$-class of a finitely presented inverse semigroup has a finite number of idempotents, then the corresponding maximal subgroup is finitely presented — a Reidemeister-Schreier type theorem for inverse semigroups. To obtain a more general result, we introduce here a notion of quasiconvexity for inverse subsemigroups and show that a quasiconvex subgroup of a finitely generated inverse semigroup is finitely generated: this generalizes the classical result for groups [5]. In particular, the maximal subgroup of a $\mathcal{D}$-class with finitely many idempotents is quasiconvex.

We also introduce the Schützenberger complex of an inverse semigroup presentation. This complex is a π_1-trivial covering space of the standard 2-

complex of the presentation and plays a role in the theory similar to that played by the Cayley complex (which it generalizes) in group theory. The 1-skeleton of the Schützenberger complex is the union of all the Schützenberger graphs of the presentation [22].

We end the paper with a survey of the techniques used to prove some results on amalgams given in [21]. These results are variants and generalizations of results of Haataja et al. [7] and Bennett [1] but our approach is conceptually (and technically) simpler.

It should be mentioned that [21] considers the standard 2-complex with an order structure so that the original inverse semigroup can, in fact, be obtained from the standard 2-complex. For simplicity's sake, we have decided to ignore the order component.

We hope the reader will at least be able to obtain a feel for how topology can be used to understand inverse semigroups.

2 Notation and Preliminaries

We assume that the reader is familiar with semigroup theory. Lawson's book [10] is recommended as a reference for inverse semigroups and their relationship with groupoids.

If A is a set, we let $\widetilde{A} = A \cup A^{-1}$. Then $\widetilde{A}^*$ denotes the free monoid with involution on A. If $w \in \widetilde{A}^*$, we use $red(w)$ for the word obtained by freely reducing w [2, 5, 11]. A *Dyck word* is a word w with $red(w) = 1$. A word w is called *cyclically reduced* if w^2 is reduced [2, 5, 11]. Every reduced word (viewed as a free group element) is conjugate to a cyclically reduced word.

If S is a semigroup, $\mathcal{E}(S)$ will denote the idempotents of S.

Following the Serre convention [20], we define a *graph* X to consist of: a set $V(X)$ of *vertices*; a set $E(X)$ of *edges*; an involution $e \mapsto e^{-1}$ on $E(X)$; and a function $\mathbf{d} : E(X) \to V(X)$ which selects the *initial* vertex of an edge (the terminology $\mathbf{d}$ is chosen to suggest the word domain). We define $\mathbf{r} : E(X) \to V(X)$ by $\mathbf{r}(e) = \mathbf{d}(e^{-1})$ (and $\mathbf{r}(e)$ is called the *terminal* vertex of e). Usually we only draw one of the pair $\{e, e^{-1}\}$.

In general we shall require $e^{-1} \neq e$, the exception being when we view groupoids as graphs in which case local group elements of order 2 will have this property. Our reason for choosing this unorthodox notation for graphs is because many of our graphs will, in fact, be groupoids where the terms domain and range have obvious meanings.

Graph morphisms are defined in the obvious way. For $v \in V(X)$, one defines $Star(v) = \mathbf{d}^{-1}(v)$. If $\varphi : X \to Y$ is a surjective graph morphism and, for each $v \in V(X)$, $\varphi : Star(v) \to Star(\varphi(v))$ is bijective, then φ is called a *covering.*

Recall that a groupoid is a category [12] in which each arrow is an isomorphism (i.e. is invertible). A groupoid can be viewed as a graph in an obvious way: the object set is the vertices; the arrow set is the edges; the inverse operation gives rise to the involution. We shall hence use graph notation for groupoids.

If $v \in V(G)$ is a vertex of a groupoid G, then

$$G_v = \{g \in E(G) \mid \mathbf{d}(g) = v = \mathbf{r}(g)\}$$

is a group called the *local group* or *maximal subgroup* of G at v.

An *inverse semigroup* is a semigroup in which each element has a unique inverse. The *trace product groupoid* of an inverse semigroup I, denoted $\mathcal{G}(I)$, is defined as follows: $V(\mathcal{G}(I)) = \mathcal{E}(I)$; $E(\mathcal{G}(I)) = I$; the inverse is the inverse; $\mathbf{d}(s) = ss^{-1}$, $\mathbf{r}(s) = s^{-1}s$; and the product is the usual multiplication. We observe that the $\mathcal{R}$-class R_e of an idempotent $e \in \mathcal{E}(I)$ is $Star(e)$, the $\mathcal{H}$-class H_e of e is the local group $\mathcal{G}(I)_e$ at e, and the $\mathcal{D}$-class D_e of e is the connected component of the groupoid containing e. Inverse semigroup homomorphisms induce morphisms of trace product groupoids.

If G is a group acting on the *left* of a set X, we use $G\backslash X$ to denote quotient; if the action is on the right, we use the customary X/G. We remind the reader that if S is any semigroup and H_e is a maximal subgroup. Then $H_e\backslash R_e$ is isomorphic to the set of $\mathcal{L}$-classes of D_e. When S is inverse this set is isomorphic to the set of idempotents contained in D_e.

Note that our definitions are the opposite of those of Lawson [10]. This is because Lawson defines st in a category if $\mathbf{r}(t) = \mathbf{d}(s)$ while we, to be consistent with the usual path multiplication for graphs, define st if $\mathbf{r}(s) = \mathbf{d}(t)$.

3 2-complexes

We shall be very informal with respect to topological notions; see the literature [2, 5, 11] for a more careful approach.

We define a 2-cell to be a regular n-gon in $\mathbb{R}^2$, $n > 0$ (both the boundary and the interior). By convention, we take a regular 1-gon to be the unit disk with a single vertex $(1,0)$ and one edge pair $\{e, e^{-1}\}$, and we take a regular 2-gon to be the unit disk with vertices $(1,0)$ and $(-1,0)$ with the obvious two edge pairs. If c is a 2-cell, its boundary will be denoted ∂c.

A 2-*complex* X consists of a graph $X^{(1)}$ (the 1-*skeleton*) and a set $C(X)$ consisting of pairs (c, f_c) where c is a 2-cell and $f_c : \partial c \to X^{(1)}$ is a graph morphism called *the attaching map* for c. Topologically, c is glued to $X^{(1)}$ via

f_c. We shall write $V(X)$ and $E(X)$ for the vertex and edge sets of $X^{(1)}$. A *defining path* for $(c, f_c) \in C(X)$ is the image under f_c of a simple, non-empty, closed path in ∂c (a closed path is called simple if the first repeated vertex encountered when tracing the path is the terminal vertex).

Morphisms of 2-complexes are defined in the obvious way. The reader is referred to [2] for the definition of a covering of 2-complexes. Roughly speaking, it is a surjective morphism which is a covering on the 1-skeleton and such that each based 2-cell has a unique based lift.

Let us give some examples of 2-complexes. Of course, any graph is a 2-complex. The two-sphere S^2 can be realized as a 2-complex in various ways. One way is to view the equator as a circle subdivided in two parts with two antipodal points as the vertices; then we attach two 2-cells (which are 2-gons): the northern and southern hemispheres. See Figure 1.

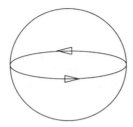

Figure 1. The sphere S^2 as a 2-complex.

The torus T can also be realized as a 2-complex in the following way. Let I denote a unit interval. Recall that T can be obtained from $I \times I$ by identifying opposite sides without any twists; see Figure 2. The four corners then become identified to a single vertex which we take as the unique vertex of our 2-complex. The sides labeled a become a 1-gon in the quotient and similarly for the sides labeled b. We thus obtain the 1-skeleton of the torus. There is also a 2-cell whose defining path is the commutator $aba^{-1}b^{-1}$.

Another example is the projective plane P^2. It can be defined as the quotient of S^2 by the natural action of $\mathbb{Z}/2\mathbb{Z}$ taking each point to its antipodal point. The complex consists of a single vertex, an attached 1-gon (say labeled by a) and an attached 2-cell with defining path a^2.

The plane $\mathbb{R}^2$ can be realized as a 2-complex in an obvious way: the lattice $\mathbb{Z} \times \mathbb{Z}$ of integral points is the vertex set; the edges are the natural grid spanned by these points; the 2-cells fill in the grid. Notice that $\mathbb{Z} \times \mathbb{Z}$ acts on $\mathbb{R}^2$ and that the quotient is T.

Figure 2. The torus T as a 2-complex.

What is of key importance for us is that we can associate a groupoid to each 2-complex.

4 Paths and the Fundamental Groupoid

A *path* in a graph is defined in the usual way; we allow an empty path at each vertex. If p is a path, we shall use $\mathbf{d}(p)$ for its initial vertex, $\mathbf{r}(p)$ for its terminal vertex, and p^{-1} for the reverse path. A path p is said to be *closed* if $\mathbf{d}(p) = \mathbf{r}(p)$. If p is a closed path, we say that a path p' is a *cyclic conjugate* of p if $p = st$ and $p' = ts$ (that is, p and p' are "the same path" starting from different vertices). Connected components of a graph are defined in the usual way. Using terminology suggestive of semigroup theory, connected components will also be called $\mathcal{D}$-*classes*. If v is a vertex, then D_v will denote the $\mathcal{D}$-class of v. Note that if G is a groupoid, then all maximal subgroups at vertices of the same $\mathcal{D}$-class are isomorphic.

We say that a path p is obtained from a path p' by an *elementary homotopy* if p is gotten from p' by inserting or deleting a subpath of the form ee^{-1}. One says that paths p and p' are *homotopic* if p can be turned into p' by a finite sequence of elementary homotopies; this is an equivalence relation and equivalent elements are coterminal. It is well-known that this definition is equivalent to asking that the paths be homotopic (in the topological sense) in the geometric realization of the graph.

The fundamental groupoid of the graph X, denoted $\Pi_1(X)$, is defined by taking $V(\Pi_1(X)) = V(X)$ and $E(\Pi_1(X))$ to be the set of all homotopy classes of paths in X; $\mathbf{d}$, $\mathbf{r}$, and the involution are defined in the obvious way. We can then use path composition to turn $\Pi_1(X)$ into a groupoid. Note that X embeds in $\Pi_1(X)$ in a natural way.

Notions about paths and connectivity are transported to 2-complexes in a natural way via their 1-skeleta. Given a 2-complex X, we can define its fundamental groupoid $\Pi_1(X)$. It is the quotient of $\Pi_1(X^{(1)})$ obtained in the following way. We now identify two paths p, p' if p can be gotten from p' by a finite sequence of elementary homotopies and by insertions or deletions of defining paths of X (recall that a defining path is the image in X of a simple path in ∂c).

If $v \in V(X)$, then $\pi_1(X,v)$ will be used to denote the local group at v and is called the *fundamental group* of X at v. This definition coincides with the usual topological definition. If Y is a connected 2-complex, then $\pi_1(Y)$ will be used to denote the abstract group to which all fundamental groups of Y are isomorphic. Many of the 2-complexes that we shall consider will not be connected, and so we cannot in general speak of the fundamental group of X. A 2-complex will be called π_1-*trivial* if its fundamental group at each vertex is trivial (or, equivalently, each connected component is simply connected).

For example, $\pi_1(S^2) = 1 = \pi_1(\mathbb{R}^2)$. On the other hand, it is simple to see that

$$\pi_1(T) = Gp\langle a,b|aba^{-1}b^{-1}\rangle = \mathbb{Z} \times \mathbb{Z}.$$

The reader can verify that

$$\pi_1(P^2) = Gp\langle a|a^2\rangle = \mathbb{Z}/2\mathbb{Z}.$$

We note if X is a connected 2-complex with finite 1-skeleton, then $\pi_1(X)$ is finitely generated; if X is in fact finite, then $\pi_1(X)$ is finitely presented. For computing the fundamental group of a connected 2-complex, the reader is referred to the literature [2, 11].

5 The Cayley Complex of a Group Presentation

If G is a group generated by A, then the *Cayley graph* [5] of G with respect to A, denoted $\Gamma_A(G)$, is defined as follows.

- $V(\Gamma_A(G)) = G$
- $E(\Gamma_A(G)) = G \times \widetilde{A}$
- $\mathbf{d}(g,a) = g$, $\mathbf{r}(g,a) = ga$
- $(g,a)^{-1} = (ga, a^{-1})$

Normally we only draw the edges labeled by A. For instance, the edge (g,a) will be drawn $g \xrightarrow{a} ga$.

For example, the lattice of integral points in $\mathbb{R}^2$ is the Cayley graph of $\mathbb{Z} \times \mathbb{Z}$ with respect to generators $(1,0)$ and $(0,1)$. The Cayley graph of $\mathbb{Z}/2\mathbb{Z}$

with respect to a single generator is precisely the equator of S^2 under our realization of S^2 as a 2-complex.

The Cayley graph of a free group on generators a, b, $F(a, b)$, is a well-known tree.

Suppose now that $Grp\langle A|R\rangle$ is a group presentation for G. Then the *Cayley complex* [5, 11] (called the Cayley diagram in [2]) is the 2-complex with 1-skeleton $\Gamma_A(G)$ and with 2-cells attached via the following recipe. For each circuit in $\Gamma_A(G)$ labeled by an element w of R, we attach a disk with defining path labeled by w.

For instance, if we consider $\mathbb{Z}/2\mathbb{Z} = Grp\langle a|a^2\rangle$, then the Cayley complex is the two-sphere S^2. Indeed, as we pointed out before, the equator is the Cayley graph. The northern hemisphere corresponds to the path labeled by a^2 from 1; the southern hemisphere corresponds to the path labeled by a^2 from a. Observe that $\mathbb{Z}/2\mathbb{Z}$ acts without fixed points on S^2 as indicated earlier; the quotient complex is the projective plane P^2.

If we consider $\mathbb{Z} \times \mathbb{Z} = Grp\langle a, b|aba^{-1}b^{-1}\rangle$, then the Cayley complex is $\mathbb{R}^2$. Indeed, the Cayley graph is the lattice of integral points and the cells fill in the holes. Again, $\mathbb{Z} \times \mathbb{Z}$ acts without fixed points on $\mathbb{R}^2$ by translations; the quotient complex is the torus T.

The Cayley complex of $F(a, b) = Grp\langle a, b|\emptyset\rangle$ is still a tree, as no cells are added. Once again, $F(a, b)$ acts freely on its Cayley graph and the quotient is a bouquet of two circles joined at a single vertex.

Notice in each of these cases that the Cayley complex is simply connected and that the group acts freely on it without fixed points. This is always the case. The action of g is defined on vertices by $g' \mapsto gg'$ and on edges by $(g', a) \mapsto (gg', a)$; this induces an action on the cells.

6 The Standard 2-complex of a Group Presentation

The *standard* 2-*complex* of a group presentation $G = Grp\langle A|R\rangle$ is the quotient of the Cayley complex by the left action of G. The resulting complex, which we denote $B_{X,R}$, is connected and can be described as follows.

- $V(B_{X,R}) = *$

- $E(B_{X,R}) = \widetilde{A}$

The involution, **d** and **r** are the obvious choices; the two cells are in bijection with the elements of R; $c_r \in R$ has defining path labeled by r.

If we consider $\mathbb{Z} \times \mathbb{Z} = Grp\langle a, b|aba^{-1}b^{-1}\rangle$, then the standard 2-complex is the torus T; see Figure 2. For $\mathbb{Z}/2\mathbb{Z} = Grp\langle a|a^2\rangle$, the standard 2-complex is the projective plane P^2. The standard 2-complex of $FG(a,b) = Grp\langle a,b|\emptyset\rangle$ is a bouquet of two circles.

As another example, if $G = Grp\langle a,b,c,d|aba^{-1}b^{-1}cdc^{-1}d^{-1}\rangle$, then the standard 2-complex of G is the quotient of a regular octagon obtained by identifying opposite sides (without twisting). This is, in fact, the two-holed torus $T\#T$; see Figure 3.

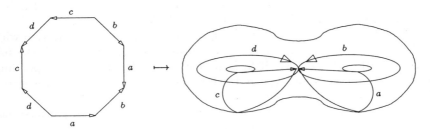

Figure 3. $T\#T$ as the standard 2-complex of $Grp\langle a,b,c,d|aba^{-1}b^{-1}cdc^{-1}d^{-1}\rangle$.

It easily follows from the above description that $\pi_1(B_{X,R}) = G$; alternatively, one can use the result that if a group G acts freely on the left of a simply connected 2-complex X, then $\pi_1(G\backslash X) = G$ where $G\backslash X$ is the quotient complex [2, 11].

Note that the Cayley complex, which we shall denote $\widetilde{B_{X,R}}$, is the universal cover of the standard 2-complex $B_{X,R}$. The covering projection is the quotient map by the action of G.

7 The Schützenberger Complex

In this section, we associate a π_1-trivial 2-complex to each inverse semigroup presentation which plays a role in the theory analogous to that of the Cayley complex in group theory.

Let I be an inverse semigroup generated by A. We first need the notion of a Schützenberger graph, due to Stephen [22]. Stephen, in fact, considers a connected graph for each $\mathcal{R}$-class R; we shall consider one graph which is the disjoint union of these graphs over all the $\mathcal{R}$-classes.

The *Schützenberger graph* of I with respect to A, denoted $\Gamma_A(I)$, is defined as follows.

- $V(\Gamma_A(I)) = I$

- $E(\Gamma_A(I)) = \{(s,a) \mid s \in I, a \in \tilde{A} \text{ and } s \mathcal{R} \, sa\}$ (This latter condition is equivalent to saying $s^{-1}s \leq aa^{-1}$.)

- $\mathbf{d}(s,a) = s$, $\mathbf{r}(s,a) = sa$

- $(s,a)^{-1} = (sa, a^{-1})$

A typical edge is of the form $s \xrightarrow{a} sa$ with $s \mathcal{R} \, sa$. Notice that if G is a group, then the Schützenberger graph is the Cayley graph.

More generally, the connected components of $\Gamma_A(I)$ are in bijection with the $\mathcal{R}$-classes of I; indeed, the connected component of a vertex s is the full subgraph with vertex set R_s. If R_s is an $\mathcal{R}$-class, we denote the corresponding connected component by $\Gamma_A(I, R_s)$, which we call the Schützenberger graph of R_s with respect to A.

Note that $\Gamma_A(I)$ is a disconnected (when I is not a group) inverse automaton [22] with transition inverse monoid I.

The Preston-Vagner representation [10] of I on the *left* of itself induces an action of I on $\Gamma_A(I)$ by partial isomorphisms. It is defined on vertices in the usual way. We extend to edges by defining $t(s,a) = (ts, a)$ if $t^{-1}t \geq ss^{-1}$ (that is, if ts is defined in the left Preston-Vagner representation).

Suppose $I = Inv\langle A|R \rangle$ is an inverse semigroup presentation. Then the *Schützenberger complex* $\widetilde{C_{X,R}}$ is defined as follows. For each $r \in R$ (say r is $u = v$) and each occurrence in $\Gamma_A(I)$ of a path labeled by uv^{-1} we attach a 2-cell. Note that since $u = v$ in I and the transition monoid of the Schützenberger graph is I, all paths labeled by uv^{-1} are closed. The defining path for this 2-cell is chosen in the following manner. Suppose the path begins at q. Let $red(u) = pu's$, $red(v) = pv's$ where u' and v' have neither a common prefix nor a common suffix. Then $u'(v')^{-1}$ labels a closed path at qp which we take as the defining path of the 2-cell. Here qp is the endpoint of the path labeled by p from q. See Figure 4. Notice that if uv^{-1} is a Dyck word, no cell is added. Also observe $u'(v')^{-1}$ is cylically **reduced**.

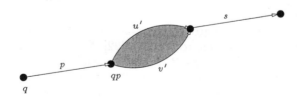

Figure 4. Attaching 2-cells.

Intuitively, each time one sees two paths in $\Gamma_A(I)$ from the same vertex q, one labeled by u and the other by v (and which hence have the same end point), one fills in the essential part of the hole with a 2-cell.

Note that it is not the case that every path labeled $u'(v')^{-1}$ is a loop, nor is every such path a defining path.

Notice that if I is a group, then $\widetilde{C_{X,R}}$ is the Cayley complex of I corresponding to the group presentation obtained by converting R to group relators. The connected component of $\widetilde{C_{X,R}}$ corresponding to an $\mathcal{R}$-class R_s will be denoted $\widetilde{C_{X,R}}(R_s)$.

We remark that this construction differs cosmetically from that of [21]; however it is convenient (especially when I is a group) and allows us prove a more powerful result.

The left action of I on $\Gamma_A(I)$ extends to an action by partial isomorphisms on $\widetilde{C_{X,R}}$. Note that $\widetilde{C_{X,R}}(R_s)$ and $\widetilde{C_{X,R}}(R_t)$ are identified under this action if and only if $s \mathcal{D} t$. Also notice that if e is an idempotent, then the maximal subgroup H_e acts freely on $\widetilde{C_{X,R}}(R_e)$.

We prove one result before giving examples.

Theorem 7.1. *Each connected component of $\widetilde{C_{X,R}}$ is simply connected; that is $\widetilde{C_{X,R}}$ is π_1-trivial.*

Proof. A full proof is given in [21]; we sketch an alternative proof. The reader is assumed to be familiar with Stephen's iterative construction of $\Gamma_A(I, R_s)$ [22]. This construction begins with the Munn tree [10, 22] of a word $w \in \tilde{A}^*$ representing s. For each relation $u = v \in R$, if one side of the relation exists labeling a path from q_1 to q_2, but not the other, then one adds a path with label the other side; this is called an *expansion*. One then folds edges to make the resulting graph an inverse graph [22]. One then iterates the construction (this can only be done effectively if the presentation is finite). The direct limit of the iterations is $\Gamma_A(I, R_s)$.

Let $u = v$ be a relation. Let $red(u) = pu's$, $red(v) = pv's$ where u' and v' have neither a common prefix nor a common suffix. Suppose now that each time we do an expansion corresponding to a relation $u = v$, we make the path at q_1p labeled by $u'(v')^{-1}$ the defining path of a 2-cell (assuming $u'(v')^{-1}$ is not 1; but this happens precisely when uv^{-1} is a Dyck word). The resulting direct limit of the iterates is then $\widetilde{C_{X,R}}(R_s)$. By induction, each iterate is simply connected; indeed, the Munn tree is simply connected and each time we add an element of π_1, we kill it off. It easily follows that the direct limit $\widetilde{C_{X,R}}(R_s)$ is simply connected. □

As a first example, consider the free inverse semigroup on A with presentation $I = Inv\langle A|\emptyset\rangle$. Then the Schützenberger complex is a graph; in fact, it consists of various copies of the Munn trees of non-empty words in $\widetilde{A}^*$. Namely, there is one copy of each Munn tree for each choice of initial and terminal states. More succinctly, we say the Schützenberger complex consists of all doubly pointed Munn trees.

If we consider the bicyclic monoid $I = Inv\langle a|aaa^{-1} = a\rangle$, then the Schützenberger complex is again a graph. Let $\mathbb{R}^+$ be the graph whose vertices are the non-negative integers and whose edges are of the form $n \xrightarrow{a} n+1$ (plus the inverse edges). Then the Schützenberger complex consists of all doubly pointed copies of $\mathbb{R}^+$.

Consider the $n \times n$ Brandt semigroup with structure group $G = Gp\langle A|R\rangle$. Suppose $x_i \mapsto (1,1,i)$ and $a \mapsto (1,a,1)$ $(a \in A)$. Let R' be obtained from R by replacing 1 with aa^{-1} for some $a \in A$. We use the presentation

$$B_n(G) = Inv\langle x_2, \ldots, x_n, A | x_i x_j = x_n^2, (\forall a, a' \in A)\ aa^{-1} = x_i x_i^{-1},$$
$$aa^{-1} = (a')^{-1}a', ax_n^2 = x_n^2, x_n^2 a = x_n^2, R'\rangle. \qquad (7.1.1)$$

The Schützenberger complex of $B_n(G)$ has $n+1$ components. There are n disjoint copies of the Cayley complex of G with an outwardly direct edge labeled by x_i adjoined at each vertex (with distinct endpoints). Also the Cayley complex of G is attached to the vertex 0 as well as a loop labeled by each x_i. All loops at 0 (both those labeled by A and by the x_i) are made defining paths of a 2-cell.

See Figure 5 for the case of $B_2(\mathbb{Z}_2)$ where $x_2 = x$ and where each loop at 0 is the boundary of a disk and a^2 (at 0) is the "equator" of a projective plane.

As another example, consider the inverse semigroup

$$I = Inv\langle a, b|aba^{-1}b^{-1} = 1\rangle. \qquad (7.1.2)$$

(This is of course an abuse of notation). Using Stephen's iterative procedure [22], it is easy to see that this is an aperiodic E-unitary inverse semigroup. The Schützenberger graph consists of a copy of each connected subgraph of the Cayley graph of $\mathbb{Z} \times \mathbb{Z}$ containing the origin which is closed under walking north and east, one copy for each possible choice of terminal state. The Schützenberger complex consists of a copy of each connected subcomplex of $\mathbb{R}^2$ containing the origin which is closed under walking north and east (not necessarily along edge paths), one copy for each possible choice of terminal state.

Let $G = Gp\langle A|R\rangle$ be a group presentation and $M(G) = Inv\langle A|R'\rangle$ where R' consists of all relations $w = w^2$ where $w = 1$ in G. This is the inverse

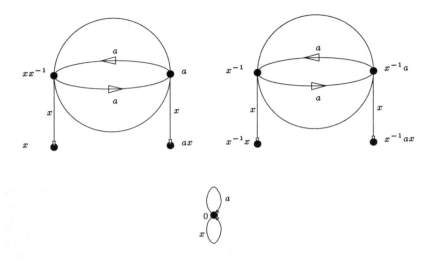

Figure 5. The Schützenberger complex of $B_2(\mathbb{Z}_2)$.

semigroup considered in [14, 3]. The Schützenberger graph of $M(G)$ consists of a copy of all finite connected subgraphs of $\Gamma_A(G)$ containing 1, one for each choice of terminal vertex. The Schützenberger complex is obtained by making each path labeled by a cyclically reduced word which is 1 in G a defining path. If C is such complex, viewed as embedded in the Cayley complex of G with the initial state at 1, then the maximal subgroup of the corresponding $\mathcal{R}$-class of $M(G)$ is the subgroup of G leaving C invariant. As C is finite and the action is free, this subgroup is finite. If G is torsion-free, then $M(G)$ is aperiodic.

8 The Standard 2-complex of an Inverse Semigroup Presentation

Let $I = Inv\langle A|R\rangle$ be an inverse semigroup presentation. We now construct the *standard 2-complex* $C_{A,R}$ of the presentation.

Define $C_{A,R} = I\backslash\widetilde{C_{A,R}}$. As mentioned earlier, two connected components $\widetilde{C_{A,R}}(R_s)$ and $\widetilde{C_{A,R}}(R_t)$ are identified if and only if $s\,\mathcal{D}\,t$. Thus the connected components of $C_{A,R}$ are in bijection with the $\mathcal{D}$-classes of I; if D is a $\mathcal{D}$-class, we use $C_{A,R}(D)$ to denote the corresponding component. Moreover, it is straightforward to verify that if $e \in \mathcal{E}(I)$, then $C_{A,R}(D_e) = H_e\backslash\widetilde{C_{A,R}}(R_e)$ (since H_e is the subsemigroup of I leaving $\widetilde{C_{A,R}}(R_e)$ invariant under the action). Since $V(\widetilde{C_{A,R}}(R_e)) = R_e$ and $H_e \backslash R_e = \mathcal{E}(D_e)$ (the idempotents of the $\mathcal{D}$-class of e), we see $V(C_{A,R}(D_e)) = \mathcal{E}(D_e)$. Since H_e acts freely on the simply connected 2-complex $\widetilde{C_{A,R}}(R_e)$, we have $\pi_1(C_{A,R}(D_e)) = H_e$.

Putting it all together, we obtain the following description of $C_{A,R}$.

- $V(C_{A,R}) = \mathcal{E}(I)$
- $E(C_{A,R}) = \{(e,a) \mid e \mathcal{R} \, ea\}$ (or equivalently, $e \leq aa^{-1}$)
- $\mathbf{d}(e,a) = e$, $\mathbf{r}(e,a) = a^{-1}ea$, $(e,a)^{-1} = (a^{-1}ea, a^{-1})$

A typical edge is of the form $e \xrightarrow{a} a^{-1}ea$ where $e \mathcal{R} \, ea$. For each relation $u = v \in R$ and for every path in $C_{A,R}$ labeled by uv^{-1}, there is a 2-cell. Note that such a path is necessarily closed; indeed if we view $C_{A,R}^{(1)}$ as a (disconnected) inverse automaton [22], then the transition monoid is nothing but the Munn representation [10] (or the direct sum of the right letter mapping representations [9]) of I. Let q be a vertex from which a path labeled uv^{-1} can be read. The defining path for this 2-cell is chosen in the following manner. Let $red(u) = pu's$, $red(v) = pv's$ where u' and v' have neither a common prefix nor a common suffix. Then $u'(v')^{-1}$ labels a closed path at qp which we take as the defining path of the 2-cell. See Figure 4. Notice that if uv^{-1} is a Dyck word, no cell is added.

Intuitively, each time one sees two paths in $C_{A,R}$ from the same vertex q, one labeled by u and the other by v (and which hence have the same end point), one fills in the essential part of the hole with a 2-cell.

Again we remark that the construction in [21] treats the defining paths in a slight different way than we do here.

If I is a group, then $C_{A,R}$ is the standard 2-complex of the corresponding group presentation.

The following result (in fact a stronger result) is proved in [21] and is key to applying our topological theory.

Theorem 8.1. *The map $C_{A,R}^{(1)} \to \mathcal{G}(I)$ which is given by the identity on vertices and which sends an edge (e,a) to ea induces an isomorphism of $\Pi_1(C_{A,R})$ with the trace groupoid $\mathcal{G}(I)$. In particular, $\pi_1(C_{A,R}, e) = H_e$.*

Unlike the group case, $C_{A,R}$ will not necessarily be finite if the presentation is finite; one needs to impose that I has finitely many idempotents to obtain this.

We mention that the projection $\widetilde{C_{A,R}} \to C_{A,R}$ is a covering [21]. This follows since it is built up from the various covering morphisms $\widetilde{C_{A,R}}(R_e) \to$

$C_{A,R}(D_e)$ obtained by quotienting out via the free action of H_e on $\widetilde{C_{A,R}}(R_e)$. Hence the Schützenberger complex is "the" universal cover of the standard 2-complex. The covering map takes an edge (s, x) to the edge $(s^{-1}s, x)$.

We now turn to some examples. Let I be the free inverse semigroup given by the presentation $Inv\langle A|\emptyset\rangle$. Then the standard 2-complex consists of all Munn trees over $\widetilde{A}$. Note that $C_{A,\emptyset}$ is π_1-trivial. This is the case for any aperiodic inverse semigroup.

Consider the bicyclic semigroup $I = Inv\langle a|aaa^{-1} = a\rangle$. The standard 2-complex is just $\mathbb{R}^+$.

Let $G = Gp\langle A|R\rangle$ be a group presentation. We use presentation (7) for $B_n(G)$. Then the standard 2-complex has two connected components. One connected component contains a copy of the standard 2-complex for G plus outwardly directed edges labeled by $x_2, \ldots, x_n$ (with distinct endpoints). The connected component of 0 contains a copy of the standard 2-complex of G plus loops labeled by each x_i. The loop corresponding to each generators is the defining path of a disk.

The 1-skeleton of the standard 2-complex for $B_2(\mathbb{Z}_2)$ is in Figure 6. The 2-cells give rise to two projective planes with "equator" a^2 and to disks attached at 0 with defining paths x and a.

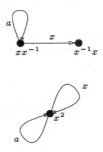

Figure 6. The 1-skeleton of the standard 2-complex of $B_2(\mathbb{Z}_2)$.

Now consider (7.1.2). The standard 2-complex is the disjoint union of all connected subcomplexes of $\mathbb{R}^2$ containing the origin and which are closed under walking north or east (along any path, not just edge paths).

If we consider the case of $M(G)$ (defined earlier) then the standard 2-complex has the following description. We consider the disjoint union of all quotients of finite connected subgraphs of $\Gamma_A(G)$ containing 1 by their stabilizers. We then add in all 2-cells for each path whose label is a cyclically reduced word which is trivial in G.

The following result generalizes a theorem proved in [21] and is a sort of Reidemeister-Schreier type theorem for inverse semigroups. Similar results were obtained in Rŭskuc [19] for semigroup presentations of regular semigroups using Reidemeister-Schreier type rewriting systems.

Theorem 8.2. *Suppose that $I = Inv\langle A|R\rangle$ is an inverse semigroup presentation and H_e is a maximal subgroup of I such that D_e has finitely many idempotents. Then if A is finite, H_e is finitely generated. If, in addition, R has only finitely many relations of the form $u = v$ such that uv^{-1} is not a Dyck word (in particular, if R is finite), then H_e is finitely presented.*

Proof. Consider $C_{A,R}(D_e)$. Then $V(C_{A,R}(D_e)) = \mathcal{E}(D_e)$ and hence is finite. If A is finite, then, by construction, there are at most $2|A|$-edges incident on each vertex of $C_{A,R}$. It follows that $C_{A,R}(D_e)$ is a finite vertex, locally finite graph: that is, $C_{A,R}(D_e)$ is finite. Thus $H_e = \pi_1(C_{A,R}(D_e))$ is finitely generated. If there are only finitely many relations in R of the form $u = v$ with uv^{-1} not a Dyck word, then there are only finitely many labels of defining paths in $C_{A,R}$. Since $C_{A,R}(D_e)^{(1)}$ is finite, we can only find finitely many paths in $C_{A,R}(D_e)^{(1)}$ with any particular label. It follows that $C_{A,R}(D_e)$ is finite, whence $H_e = \pi_1(D_e)$ is finitely presented. □

It is a simple exercise to give an explicit presentation of H_e using the standard techniques for computing the fundamental group of a finite 2-complex [2, 11].

We now generalize this further. Let $K > 0$ be a real number. Let $G = Gp\langle A|R\rangle$ be a group. A subgroup H is said to be K-*quasiconvex* if any geodesic path between elements of H in $\Gamma_A(G)$ belongs to the K-neighborhood of H [3, 5, 6]. Here distance is measured via length of the shortest path between two vertices. It is easy to verify [3, 6] that H is K-quasiconvex if and only if in $H\backslash\Gamma_A(G)$ each circuit at $H \cdot 1$ whose label is a geodesic in $\Gamma_A(G)$ is contained in the K-neighborhood of $H \cdot 1$. A subgroup H is called *quasiconvex* if it is K-quasiconvex for some $K > 0$. It is then clear that any finite index subgroup is quasiconvex. If A is finite (and so the K-neighborhood of any vertex in $H\backslash\Gamma_A(G)$ is finite), then a quasiconvex subgroup H is finitely generated [3, 5, 6]. Note that quasiconvexity depends on the generating set if G is not hyperbolic.

We generalize this to inverse semigroups. Let $I = Inv\langle A|R\rangle$ and let T be an inverse subsemigroup. For $K > 0$, we say T is K-*quasiconvex* if every geodesic in $\Gamma_A(I)$ between vertices of T is contained in the K-neighborhood of T. We say T is *quasiconvex* if it is K-quasiconvex for some $K > 0$.

For example, $E(I)$ is always quasiconvex since distinct idempotents belong to distinct connected components. It follows that quasiconvex inverse subsemigroups of finitely generated inverse semigroups need not be finitely generated. However the situation for subgroups is different.

Suppose H is a subgroup of I with identity e and consider the action of H on $C = \widetilde{C_{A,R}}(R)$ where R is the $\mathcal{R}$-class of H. Then, as was the case above for groups, it is easy to see that H is K-quasiconvex if and only if every circuit at $v = H \cdot e$ in $\overline{C} = H \backslash C$ which is the image of a geodesic path in C is contained in the K-neighborhood of v. In particular if H is a maximal subgroup whose $\mathcal{D}$-class has finitely many idempotents, then H is quasiconvex as $\overline{C}$ is finite vertex.

We then have the following generalization of Theorem 8.2.

Theorem 8.3. *A quasiconvex subgroup of a finitely generated inverse semigroup is finitely generated.*

Proof. We continue with the above notation. Since H acts freely on C which is simply connected, $H = \pi_1(\overline{C}, v)$. Now H is the image of the free group $\pi_1(\overline{C}^{(1)}, v)$. Let C' be the K-neighborhood of v in $\overline{C}^{(1)}$. Then C' is finite since A is finite. Hence $\pi_1(C', v)$ is finitely generated. We show that H is generated by the image of $\pi_1(C', v)$.

Let p be a circuit at v. Then p has a unique lift $\widetilde{p}$ starting at e in C. Let q be the endpoint of $\widetilde{p}$ and let p' be a geodesic path from e to q. Then $\widetilde{p}$ and p' are homotopic as C is simply connected. Hence p is homotopic to the image $\overline{p'}$ of p'. But $\overline{p'}$ is contained in C' since H is K-quasiconvex. It follows that the image of $\pi_1(C', v)$ generates H and so H is finitely generated. □

9 Amalgams

In this section, we study certain amalgams of inverse semigroups. These techniques can also be applied to certain HNN-extensions of inverse semigroups and even to Bass-Serre theory [15].

Suppose S and T are inverse semigroups with intersection a common inverse subsemigroup U. In light of Hall's theorem [8], the *amalgamated product* $S *_U T$ is the universal inverse semigroup containing S and T and, moreover, $S \cap T = U$ (in $S *_U T$). One calls the triple (S, U, T) an *amalgam*.

Let

$$S = Inv\langle A_1|R_1\rangle, \quad T = Inv\langle A_2|R_2\rangle, \quad U = Inv\langle A_3|R_3\rangle \tag{9.0.1}$$

be presentations such that $A_3 = A_1 \cap A_2$ and $R_3 = R_1 \cap R_2$. Then

$$S *_U T = Inv\langle A_1 \cup A_2 | R_1 \cup R_2 \rangle. \tag{9.0.2}$$

9.1 The group case

For the case of a group amalgam (G_1, H, G_2), there is a well-known normal form theorem for $G_1 *_H G_2$, as well as structure theorems for the subgroups [2, 11, 20]. For instance, it is known that if $K \leq G_1 *_H G_2$ is a subgroup, all of whose conjugates intersect trivially with H, then K is free.

Topologically speaking, group amalgams correspond precisely to amalgams of connected topological spaces.

An *amalgam of 2-complexes* consists of a triple (X, Z, Y) of 2-complexes with $Z = X \cap Y$. The *amalgamated product* of X and Y over Z is nothing more that $X \cup Y$. However, one frequently uses the notation $X \cup_Z Y$ to indicate that X and Y are being glued along Z. If X, Y, Z are connected, then so is $X \cup_Z Y$ and $\pi_1(X \cup_Z Y) = \pi_1(X) *_{\pi_1(Z)} \pi_1(Y)$ — this is none other than van Kampen's theorem for 2-complexes [2, 11]. In particular, if Z is simply connected, $\pi_1(X \cup_Z Y) = \pi_1(X) * \pi_1(Y)$ (the free product).

As an example, consider the two-holed torus $T\#T$. For exposition, we will deal with topological spaces instead of 2-complexes. Let T' be a torus minus the interior of a disk. Then $T\#T$ is obtained by gluing two copies of T' along their boundary circles (without twisting); that is, $T\#T = T' \cup_{S^1} T'$. It is well known and easy to show $\pi_1(T') = F_2$ (a free group on two generators) and $\pi_1(S^1) = \mathbb{Z}$. It follows $\pi_1(T\#T) = F_2 *_\mathbb{Z} F_2$. In fact, as we saw earlier,

$$\pi_1(T\#T) = Gp\langle a, b, c, d | aba^{-1}b^{-1}cdc^{-1}d^{-1}\rangle.$$

The relation comes from the fact that we identify a circle homotopic to $aba^{-1}b^{-1}$ with a circle homotopic to $dcd^{-1}c^{-1}$. (See Figure 3 and think about which diagonal of the octagon corresponds to the identified circles.)

Suppose (S, U, T) is an amalgam of groups with presentations as in (9.0.1) and X, Z, and Y are the respective standard 2-complexes of S, U, and T, then (X, Z, Y) is an amalgam of connected 2-complexes and $X \cup_Z Y$ is the standard 2-complex of $S *_U T$.

We shall see shortly that a similar situation arises in the case of certain inverse semigroup amalgams, only we obtain an amalgam of *disconnected 2-complexes*. This leads us to study amalgamations of such and their fundamental groups and groupoids.

Before moving onto amalgams of disconnected 2-complexes, we need to discuss graphs of groups [2, 20] and complexes.

9.2 Graphs of Groups and Complexes

A *graph of groups* $(\mathcal{G}, X)$ consists of a graph X and an assignment of a group G_v to each vertex v, of a group $G_e = G_{e^{-1}}$ to each pair of edges $\{e, e^{-1}\}$, and of a monomorphism $\iota_e : G_e \to G_{\mathbf{d}(e)}$ to each edge e. If $(\mathcal{G}, X)$ is a connected graph of groups and T is a maximal subtree of X, then the *fundamental group* of $(\mathcal{G}, X)$ with respect to the tree T is

$$\pi_1(\mathcal{G}, X) = \langle G_v \ (v \in V(X)), y_e \ (e \in E(X)) | y_t = 1 \ (t \in T),$$

$$y_e^{-1} = y_{e^{-1}} \ (e \in E(X)), y_e^{-1}\iota_e(g)y_e = \iota_{e^{-1}}(g) \ (e \in E(X), g \in G_e)\rangle.$$

This group is independent of the choice of T [2, 20].

For example, let X be the graph consisting of one vertex $*$ and one edge pair $\{e, e^{-1}\}$. Let $G_* = \mathbb{Z}$, $G_e = \mathbb{Z} = G_{e^{-1}}$ and let the monomorphisms be the usual inclusions. Then

$$\pi_1(\mathcal{G}, X) = Grp\langle a, y_e | y_e^{-1} a y_e = a\rangle = \mathbb{Z} \times \mathbb{Z}.$$

A *graph of 2-complexes* $(\mathcal{C}, X)$ consists of a graph X and an assignment to each vertex v of a 2-complex X_v (called a *vertex complex*), to each pair of edges $\{e, e^{-1}\}$ a 2-complex (called an *edge complex*) $X_e = X_{e^{-1}}$, and to each edge e an embedding $\iota_e : X_e \to X_{\mathbf{d}(e)}$. The *realization* of $(\mathcal{C}, X)$ is obtained by replacing each vertex v by X_v and each pair of edges $\{e, e^{-1}\}$ by $X_e \times [0,1]$ (technically speaking, by its 2-skeleton); one then glues, for each pair of edges $\{e, e^{-1}\}$, $X_e \times 0$ to the copy of X_e in $X_{\mathbf{d}(e)}$ and $X_e \times 1$ to the copy in $X_{\mathbf{r}(e)}$.

As an example, let X be the graph consisting of one vertex $*$ and one edge pair $\{e, e^{-1}\}$. Let $X_* = S^1$ (a unit circle) and $X_e = S^1$; the embeddings $\iota_e, \iota_{e^{-1}}$ are just the inclusions. Then the reader is invited to verify that the geometric realization of $(\mathcal{C}, X)$ is the torus T.

A simple application of the Seifert-van Kampen theorem [2, 11] shows that if X is the geometric realization of a connected graph of connected 2-complexes, then $\pi_1(X)$ is isomorphic to the fundamental group of the graph of groups obtained by replacing each vertex and edge complex by its fundamental group. Conversely, the fundamental group of any connected graph of groups can be so realized. See the example of the torus T, above.

9.3 Graphs of Disconnected 2-Complexes and Amalgams

The following simple observation, which we state as a proposition, will be key to studying the structure of maximal subgroups of amalgams. Again, we use $\mathcal{D}$-class as a synonym for connected component.

Proposition 9.1. *Let $(\mathcal{C}, X)$ be a graph of 2-complexes. Then its realization has the following decomposition as a graph $(\mathcal{C}', Y)$ of connected 2-complexes: $V(Y)$ is the set of $\mathcal{D}$-classes of the vertex complexes of X; $E(Y)$ is the set of $\mathcal{D}$-classes of the edge complexes of X; if $e \in E(X)$ and Z is a $\mathcal{D}$-class of X_e, then Z is a subcomplex of a unique $\mathcal{D}$-class of $X_{\mathbf{d}(e)}$ which we define to be the domain of the edge of Y corresponding to Z; the inverse of the edge corresponding to Z is the copy of Z in $X_{e^{-1}}$; and the vertex (edge) complex corresponding to a $\mathcal{D}$-class is the $\mathcal{D}$-class itself.*

We illustrate the above proposition in Figure 7 where we have a segment of disconnected complexes as this is the case of interest.

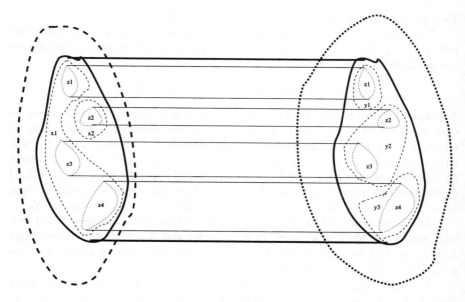

Figure 7. An illustration of Proposition 9.1.

Here, the left vertex complex (surrounded in large thick dashes) has two $\mathcal{D}$-classes $x1$ and $x2$ (dashed lines); the right vertex complex (surrounded in small thick dashes) has three $\mathcal{D}$-classes $y1$, $y2$, and $y3$ (dashed lines); the edge complex (surrounded in thick solid lines) has four $\mathcal{D}$-classes $z1$, $z2$, $z3$, and $z4$ (dotted lines). The resulting decomposition is a connected, bipartite graph of connected 2-complexes with five vertices and four edges (see Figure 8); this will always be the case for a segment of 2-complexes although in general the bipartite graph will not be connected.

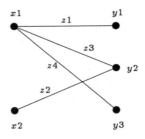

Figure 8. The underlying graph in Figure 7.

Proposition 9.1 gives us a means to obtain a graph of groups decomposition of any local group of the fundamental groupoid of the geometric realization of a graph of disconnected 2-complexes.

Now consider an amalgam of disconnected complexes (X, Z, Y). If we gradually stretch $X \cup_Z Y$ along Z, we see that $X \cup_Z Y$ is homotopy equivalent to the geometric realization of the graph of 2-complexes $(\mathcal{C}, W)$ whose underlying graph is a segment with X as the left vertex complex, Y as the right vertex complex, and Z as the edge complex. Alternatively, one can imagine contracting the subcomplex $Z \times I$ in Figure 7. It follows that $\pi_1(X \cup_Z Y, v)$ has a graph of groups decomposition induced by the graph of complexes decompositions in Proposition 9.1.

If S is an inverse semigroup, we use $D(S)$ for the set of $\mathcal{D}$-classes of S. Let (S, U, T) be an inverse semigroup amalgam. We define a graph of groups $(\mathcal{G}_{(S,U,T)}, \Gamma)$ as follows.

$$V(\Gamma) = D(S) \cup D(T), \ E(\Gamma) = \widetilde{D(U)} \tag{9.1.1}$$

Each $\mathcal{D}$-class of U is contained in a unique $\mathcal{D}$-class of S, which we take as the initial vertex of the corresponding edge, and a unique $\mathcal{D}$-class of T, which we take as its final vertex; we associate to each vertex and edge the maximal subgroup of the corresponding $\mathcal{D}$-class with the obvious inclusion maps [7]). If $e \in \mathcal{E}(S) \cup \mathcal{E}(T)$, we let $(\mathcal{G}_e, \Gamma_e)$ be the connected component of $(\mathcal{G}_{(S,U,T)}, \Gamma)$ containing the $\mathcal{D}$-class of e.

The following is a simplification of what is proved in [21]. Let (S, U, T) be an inverse semigroup amalgam and choose presentations for S, T, and U as per (9.0.1) and choose the presentation (9.0.2) for $S *_U T$.

We aim to generalize the results of [7] where the case of full amalgams is considered (recall that an amalgam is called full if $\mathcal{E}(S) = \mathcal{E}(U) = \mathcal{E}(T)$); see [7] for examples; other generalizations (some of which we shall obtain below) can be found in [1]. We mention that the proof given in [21] is self-contained, while that of [7] depends on the results of Nambooripad and Pastijn [16] and of Ordman [17, 18].

Proposition 9.2. *Suppose (S, U, T) is an inverse semigroup amalgam with $\mathcal{E}(S) = \mathcal{E}(U)$ an order ideal of $\mathcal{E}(T)$. Let X, Z, Y be the respective standard 2-complexes representing S, U, T. Then $X \cup_Z Y$ is the standard 2-complex for $S *_U T$.*

As a consequence we have the following result [21].

Theorem 9.3. *Suppose (S, U, T) is an inverse semigroup amalgam with $\mathcal{E}(S) = \mathcal{E}(U)$ an order ideal of $\mathcal{E}(T)$. Then the maximal subgroup corresponding to an idempotent $e \in \mathcal{E}(S *_U T) = \mathcal{E}(S) \cup \mathcal{E}(T)$ is $\pi_1(\mathcal{G}_e, \Gamma_e)$ (see (9.1.1)). In particular, if S and T only have trivial subgroups, then the subgroups of $S *_U T$ are free.*

Proof. Immediate from Proposition 9.2 and Proposition 9.1. The last statement follows since the fundamental group of a graph of groups with trivial vertex groups is free [2, 20]. □

Moreover, the trace product groupoid of $S *_U T$ is $\Pi_1(X \cup_Z Y)$ which is, by general nonsense, the amalgamated product of $\mathcal{G}(S)$ and $\mathcal{G}(T)$ over $\mathcal{G}(U)$ in the category of groupoids. Work of Ordman [17, 18] gives normal forms for amalgamated products of groupoids. Theorem 9.3 applies, in particular, to the case of full amalgams. Assume (S, U, T) is an inverse semigroup amalgam satisfying the hypotheses of Theorem 9.3. If U has only trivial subgroups, then standard Bass-Serre theory [2, 20] and the Kurosh Theorem [2, 11] imply that the subgroups of $S *_U T$ are isomorphic to free products of free groups and subgroups of the factors S and T.

We now state some further results proved in [21], which coincide in part with results of [1].

Let (S, U, T) be an inverse semigroup amalgam such that one of the following two situations arise: $\mathcal{E}(U)$ is an order ideal in both $\mathcal{E}(S)$ and $\mathcal{E}(T)$, or U is a union of $\mathcal{D}$-classes of S and T (that is, U is $\mathcal{D}$-saturated in S and T). Choose presentations for S, T, and U as per (9.0.1) and choose the presentation (9.0.2) for $S *_U T$. Then we have the following.

Proposition 9.4. *Suppose (S, U, T) is an inverse semigroup amalgam as* above. Let X, Z, Y be the respective standard 2-complexes representing S, U, T. Then $X \cup_Z Y$ embeds as a union of connected components in the standard 2-complex of $S *_U T$.

As an immediate consequence we have

Theorem 9.5. *Let (S, U, T) be an inverse semigroup amalgam such that $E(U)$ is an order ideal of $E(S)$ and $E(T)$, or U is $\mathcal{D}$-saturated in S and T. Let X, Z, Y be the respective standard 2-complexes representing S, U, T (with respect to the presentations (9.0.1)). Then $\Pi_1(X \cup_Z Y)$ embeds as a union of $\mathcal{D}$-classes of $S *_U T$. Thus if $e \in E(S) \cup E(T)$, the maximal subgroup at e is $\pi_1(\mathcal{G}_e, \Gamma_e)$ (see (9.1.1)).*

Proof. Immediate from Proposition 9.4 and Proposition 9.1. □

These results completely describe the structure of the subgroupoid of $\mathcal{G}(S *_U T)$ consisting of the $\mathcal{D}$-classes of idempotents of S or T.

The following corollary is immediate.

Corollary 9.6. *Let (S, U, T) be an amalgam of inverse semigroups such that U is $\mathcal{D}$-saturated in S and T. Then, for $e \in \mathcal{E}(S) \cup \mathcal{E}(T)$, the maximal subgroup at e remains unchanged.*

The above situation occurs, for instance, if S and T have a common ideal U. Another possible application is when S and T have a common $\mathcal{D}$-class U which is a group (for instance, the minimal ideal and the group of units are always $\mathcal{D}$-classes which are subgroups).

We also considered [21] amalgamations in the category of inverse semigroups and prehomomorphisms [10] and obtained similar results.

We mention that Meakin and Yamamura [15, 23] define a notion of a graph of full inverse monoids and the fundamental inverse monoid of such. Full amalgams and full HNN-extensions are special cases of this construction. They obtain a graph of groups decomposition of the maximal subgroups of such a fundamental inverse monoid similar to that for full amalgams. These results can be obtained by our methods as well.

Acknowledgments

I would like to dedicate this article in memory of those who died in the horrible attacks on New York and Washington on September 11, 2001. I was writing this article when I heard the city of my birth was under attack.

References

[1] P. Bennett, *On the structure of inverse semigroup amalgams*, Internat. J. Algebra. Comput. **7** (1998), 577–604.

[2] D. J. Collins, R. I. Grigorchuk, P. F. Kurchanov and H. Zieschang, Combinatorial Group Theory and Applications to Geometry, Springer-Verlag, Berlin, 1998.

[3] M. Delgado, S. W. Margolis, and B. Steinberg, *Combinatorial group theory, inverse monoids, automata, and global semigroup theory*, Internat. J. Algebra and Comput. To appear.

[4] C. Ehresmann, Œuvres Complètes et Commentées, (ed. A. C. Ehresmann) Supplements to Cahiers de Topologie et Géométrie Différentielle, Amiens, 1980-83.

[5] D. B. A. Epstein *with* J. W. Cannon, D. F. Holt, S. V. F Levy, M. S. Paterson, W. P. Thurston, Word Processing in Groups, Jones and Bartlett, Boston-London, 1992.

[6] R. Gitik, *On quasiconvex subgroups of negatively curved groups*, J. Pure Appl. Algebra, **119** (1998), 155–169.

[7] S. Haataja, S. W. Margolis, and J. Meakin, *Bass-Serre theory for groupoids and the structure of full regular semigroup amalgams*, J. Algebra **183** (1996), 38–54.

[8] T. E. Hall, *Free products with amalgamation of inverse semigroups*, J. Algebra **34** (1975), 375–385.

[9] K. Krohn, J. Rhodes, and B. Tilson, *Lectures on the algebraic theory of finite semigroups and finite-state machines*, Chapters 1, 5-9 (Chapter 6 with M. A. Arbib) of The Algebraic Theory of Machines, Languages, and Semigroups, (M. A. Arbib, ed.), Academic Press, New York, 1968.

[10] M. V. Lawson, Inverse Semigroups: The theory of partial symmetries, World Scientific, Singapore, 1998.

[11] R. C. Lyndon and P. E. Schupp, Combinatorial Group Theory, Springer-Verlag, New York, 1977.

[12] S. MacLane, Categories for the Working Mathematician, Springer-Verlag, New York, 1971.

[13] J. C. Meakin, *On the structure of inverse semigroups*, Semigroup Forum **12** (1976), 6–14.

[14] S. W. Margolis and J. C. Meakin, *E-unitary inverse monoids and the Cayley graph of a group presentation*, J. Pure Appl. Algebra **58** (1989), 45–76.

[15] J. C. Meakin and A. Yamamura, *Bass-Serre theory and inverse monoids in* Howie, John M. (ed.) *et al.*, Semigroups and applications. Proceedings

of the conference, St. Andrews, UK, July 2-9, 1997. Singapore: World Scientific. (1998), 125–140.
[16] K. S. S. Nambooripad and F. J. Pastijn, *Amalgamation of regular semigroups*, Houston J. of Math. **15** (1989), 249–254.
[17] E. T. Ordman, *On subgroups of amalgamated free products*, Proc. Cambridge Phil. Soc., **69** (1971), 13–23.
[18] E. T. Ordman, Amalgamated free products of groupoids, Ph. D. Thesis, Princeton University 1969.
[19] N. Rŭskuc, *Presentations for subgroups of monoids*, J. Algebra **220** (1999), 365–380.
[20] J.-P. Serre, Trees, Springer-Verlag, Heidelberg, 1980.
[21] B. Steinberg, *A topological approach to inverse and regular semigroups*, Pacific J. Math. To appear.
[22] J. B. Stephen, *Presentations of inverse monoids*, J. Pure Appl. Algebra **63** (1990), 81–112.
[23] A. Yamamura, *A class of inverse monoids acting on ordered forests*, Preprint 2000.

FINITE SEMIGROUPS AND THE LOGICAL DESCRIPTION OF REGULAR LANGUAGES

HOWARD STRAUBING*

Computer Science Department, Boston College
Chestnut Hill, Massachusetts
USA 02467
E-mail:straubin@cs.bc.edu

We survey the application of finite semigroups to the description of regular languages in first-order logic and generalized first-order logic. The emphasis is on recent results, including formulas with a bounded number of variables, and contacts with computational complexity and universal algebra.

1 Introduction

Formal languages can be defined by formulas of predicate logic and classified according to the types of languages used to define them. This idea was first applied to the languages recognized by finite automata by Büchi [3]. Büchi used formulas of second-order logic. When first-order logic and various modest generalizations thereof are used, finite semigroups become an important tool in this classification.

The present paper is an informal survey of this algebraic approach to what might be called the "descriptive theory of finite automata". I have written extensively about this subject in the monograph [11]. While the fundamentals of the theory will be presented, most of the emphasis will be on research that has appeared since the publication of [11].

2 Defining Regular Languages in Formal Logic

Let Σ be a finite alphabet. Through most of this article our examples will refer to the case where $\Sigma = \{\sigma, \tau\}$, but the results apply to general finite

---

*THE AUTHOR WOULD LIKE TO ACKNOWLEDGE THE FINANCIAL SUPPORT OF FUNDAÇÃO CALOUSTE GULBENKIAN (FCG), FUNDAÇÃO PARA A CIÊNCIA E A TECNOLOGIA (FCT), FACULDADE DE CIÊNCIAS DA UNIVERSIDADE DE LISBOA (FCUL) AND REITORIA DA UNIVERSIDADE DO PORTO.

alphabets. Consider the following first-order sentence:

$$\forall x \forall y((x < y \land \forall z(z \leq x \lor y \leq z)) \to (Q_\sigma x \leftrightarrow Q_\tau y)).$$

Such a sentence is meant to be interpreted in words over the alphabet Σ. The variables denote positions in the word (that is, integers between 1 and the length of the word, inclusive), and the formula $Q_\sigma x$ is interpreted to mean "the letter in position x is σ". Thus the whole sentence says, in effect, that any two consecutive positions contain different letters of Σ. Thus a word w *satisfies* the sentence if the letters σ and τ strictly alternate within w. The sentence consequently *defines* the language L consisting of all such words.

Observe that L is a regular language, and in fact is denoted by the regular expression

$$(\Lambda + \sigma)(\tau\sigma)^*(\Lambda + \tau).$$

It follows from Büchi's results, and is not difficult to prove directly, that any language defined by such a first-order sentence using the binary relation $<$ is a regular language. Let us pose the problem of determining whether a given regular language can be so defined. For instance, is there any regular language that is *not* first-order definable in this sense?

3 The McNaughton-Papert-Schützenberger Theorem

We denote by $FO[<]$ the family of languages over Σ that are so definable. FO means "first-order", and the $<$ in brackets means that this is the only relation on positions—the only *numerical predicate*—that we use in our sentences. (Note that $x \leq y$ is equivalent to $\neg(y < x)$, so we can define both $\leq$ and $=$ in this logic.) We denote by $M(L)$ the syntactic monoid of the language L and by μ_L its syntactic morphism. The following theorem, due to McNaughton and Papert [7], and incorporating an earlier theorem of Schützenberger [9] is a fundamental result.

Theorem 3.1. *Let $L \subseteq \Sigma^*$. $L \in FO[<]$ if and only if $M(L)$ is finite and aperiodic.*

3.1 Using Model-theoretic Games

As a means of showing the reader the kinds of techniques that are used in this subject, we sketch the proof of Theorem 3.1. We begin by proving that every first-order definable language has an aperiodic syntactic monoid.

Let $k \geq 0$. We define the following equivalence reltion on Σ^*: $u \equiv_k v$ if and only if u and v satisfy exactly the same sentences in which the depth of nesting of the quantifiers is no more than k. Since there are only finitely many inequivalent sentences of a given depth, $\equiv_k$ is an equivalence relation on Σ^* of finite index.

Let $u, v \in \Sigma^*$. We define the *k-round Ehrenfeucht-Fraïssé game in (u, v)* as follows. Each player is equipped at the outset with k pebbles, labeled $p_1, \ldots, p_k$. In the ith round, Player I places his pebble labeled p_i on a position in one of the two words; Player II responds by placing her p_i on a position of the other word. Player II wins the game if, after k rounds, the following conditions are satisfied. First, whenever pebble p_i is to the left of pebble p_j in one word, p_i is to the left of p_j in the other word. Second, the letter in the position pebbled by p_i in u is the same as the letter in the position pebbled by p_i in v. Player I wins if Player II does not win.

As an example, let $u = \sigma\tau\sigma$ and $v = \tau\sigma\tau\sigma$. Player II has a winning strategy in the 1-round game in these two words, since the two words contain the same set of letters. But Player I has a winning strategy in the 2-round game: he can place p_1 on the initial τ in v. Player II is forced to respond on the second position of u. Player I then responds by playing p_2 on the first position of u, and Player II now has no safe move in v.

Here is the principal result about these games:

Theorem 3.2. *Let $u, v \in \Sigma^*$, $k \geq 0$. $u \equiv_k v$ if and only if Player II has a winning strategy in the k-round game in (u, v).*

As an illustration of this result, consider our previous example. Since Player I has a winning strategy in the 2-round game in u and v, there must be a sentence of quantifier depth 2 satisfied by one of these words and not the other. Such a sentence says "the first letter is τ". We can write it as

$$\exists x (Q_\tau x \wedge \forall y (y \geq x)).$$

Given this theorem, it is not hard to prove the following facts: First: If $u_1 \equiv_k v_1$, and $u_2 \equiv_k v_2$, then $u_1 u_2 \equiv_k v_1 v_2$. The proof is just the observation that winning strategies for Player II in the k-round games in (u_1, v_1) and (u_2, v_2) can be combined to make a winning strategy for the game in $(u_1 u_2, v_1 v_2)$. Thus $\equiv_k$ is a congruence of finite index on Σ^*.

Second: For all $v \in \Sigma^*$, $k \geq 1$, $v^{2^k} \equiv_k v^{2^k - 1}$. The winning strategy for Player II in these two words is constructed by induction on k. The statement is obvious for $k = 1$. In the inductive step, Player I's first move induces a factorization of the word he plays in as $w\sigma w'$, where $\sigma \in \Sigma$. If $|w| < |w'|$ then Player II factors the other word as $w\sigma w''$, and plays on the position containing σ. In subsequent rounds, if Player I plays in the prefix $w\sigma$ of either word, Player II plays on the corresponding position of the other word. If

465

Player I plays in the factor w' or w'', then Player II responds according to her winning strategy in the $(k-1)$-round game in (w', w''), which exists by the inductive hypothesis. The symmetric argument is made in the case where $|w'| < |w|$.

If L is defined by a sentence of quantifier depth k, then it is a union of $\equiv_k$-classes, and thus $M(L)$ is a homomorphic image of the quotient monoid $\Sigma^*/\equiv_k$. The two facts above imply that this quotient monoid is finite and aperiodic, and thus $M(L)$ is aperiodic. This gives us one direction of the McNaughton-Papert theorem.

3.2   Using the Krohn-Rhodes Theorem

The converse direction of Theorem 3.1 is proved using the following consequence of the Krohn-Rhodes Theorem: If M is aperiodic, then M divides an iterated wreath product

$$U \circ \cdots \circ U,$$

where U is the transformation monoid generated by the automaton over $\Sigma = \{\sigma, \tau\}$, with state set $Q = \{q_1, q_2\}$, such that $Q\sigma = q_1$, $Q\tau = q_2$.

Let L_1, L_2 be languages over Σ. Let $< L_1, L_2 >$ denote the set of all words uv such that $u \in L_1$, and every prefix uv' of uv is in L_2. It is not hard to show that if X is a transformation monoid, then every language recognized by the wreath product $U \circ X$ is a boolean combination of languages $< L_1, L_2 >$, where L_1 and L_2 are recognized by X. Secondly, if one has defining first-order sentences for L_1 and L_2, one can easily construct such a sentence for $< L_1, L_2 >$. It follows that if L is recognized by an aperiodic monoid, then it is first-order definable.

3.3   Non-expressibility and Decidability

An immediate corollary of the McNaughton-Papert Theorem is that any language whose syntactic monoid contains a nontrivial group is not in $FO[<]$. The simplest example is the language consisting of all words of even length.

More generally, we can effectively decide whether a given regular language is in $FO[<]$, since we can effectively compute its syntactic monoid and determine whether the monoid is aperiodic. Further, given a finite aperiodic monoid, we can effectively compute a Krohn-Rhodes decomposition and thus effectively construct a defining first-order sentence for the language.

4 A Gallery of Complementary Results

We give here a summary (not intended to be exhaustive) of results similar to the McNaughton-Papert Theorem. These concern variants of $FO[<]$ obtained by changing the defining sentences along several different dimensions: The kinds of quantifiers used, the numerical predicates allowed, and the number of variables used.

4.1 First-order Sentences with Successor

We denote by $FO[+1]$ the family of languages definable by first-order sentences in which the successor relation $y = x + 1$ is permitted, but the ordering relation $x < y$ is not. The following fundamental result is due to Beauquier and Pin [2], based on earlier work of Thérien and Weiss [15].

Theorem 4.1. *Let $L \in \Sigma^*$ be a regular language. $L \in FO[+1]$ if and only if $M(L)$ is aperiodic, and*

$$esfs'es''f = es''fs'esf$$

for all $e, f, s, s', s'' \in \mu_L(\Sigma^+)$ with e, f idempotent.

Example. Let us return to our original example $L = (\Lambda + \sigma)(\tau\sigma)^*(\Lambda + \tau)$. We have already seen that L is first-order definable, and thus, by Theorem 3.1, $M(L)$ is aperiodic. In fact $M(L) = \{1, 0 = \sigma^2 = \tau^2, \sigma, \tau, \sigma\tau, \tau\sigma\}$, with $\sigma\tau$ and $\tau\sigma$ both idempotent. It follows that $\mu_L(\Sigma^+)$ (that is, the non-identity elements of $M(L)$) satisfy the identity in the statement of Theorem 4.1, and consequently $L \in FO[+1]$. In fact, a defining sentence is

$$\forall x \forall y (y = x + 1 \to (Q_\sigma x \leftrightarrow Q_\tau y)).$$

Example. Theorem 4.1 allows us to give an algebraic proof of the purely model-theoretic fact that $<$ is not first-order definable in terms of successor. Let $\Sigma = \{\rho, \sigma, \tau\}$, and consider the language $\rho^*\sigma\rho^*\tau\rho^*$. One shows easily, either by exhibiting a defining sentence or by computing the syntactic monoid, that $L \in FO[<]$. Let $e = f = \mu_L(\rho)$. (This is the identity of $M(L)$.) Let $s = \mu_L(\sigma)$, $s' = \mu_L(\rho)$, and $s'' = \mu_L(\tau)$. Then $esfs'es''f = \mu_L(\sigma\tau)$ and $es''fs'esf = \mu_L(\tau\sigma)$. The two elements are unequal, since $\sigma\tau \in L$ and $\tau\sigma \notin L$. Thus $L \notin FO[+1]$.

467

4.2 Modular Quantifiers

We introduce a new kind of quantifier, which we call a *modular quantifier*. Let $0 \leq r < n$. We interpret

$$\exists^{(r \bmod n)} x \phi(x)$$

to mean "the number of positions x for which $\phi(x)$ holds is congruent to r modulo n". For example,

$$\exists^{(0 \bmod 2)} x Q_\sigma x$$

defines the set of strings over $\{\sigma, \tau\}$ that contain an even number of occurrences of σ.

We denote by $MOD[<]$ the family of languages over Σ definable by sentences in which $<$ is the only numerical predicate and only modular quantifiers are used. $(FO + MOD)[<]$ denotes the family of languages definable by sentences in which ordinary quantifiers as well as modular quantifiers are allowed. The following theorem is due to Straubing, Thérien and Thomas [14]:

Theorem 4.2. *Let $L \subseteq \Sigma^*$ be a regular language. $L \in MOD[<]$ if and only if $M(L)$ is a solvable group. $L \in (FO + MOD)[<]$ if and only if every group in $M(L)$ is solvable.*

4.3 Regular Numerical Predicates

Let us admit into our defining formulas both the ordering relation and the predicates

$$x \equiv r \pmod{n},$$

where $0 \leq r < n$. We call the numerical predicates that are definable in terms of these *regular numerical predicates*, since any numerical predicate outside this class can be used to define non-regular languages. We denote by $FO[Reg]$ the family of languages over Σ definable by first-order sentences over this base of predicates. (See [11].) The following is due to Barrington, Compton, Straubing and Thérien [1]:

Theorem 4.3. *Let $L \subseteq \Sigma^*$ be a regular language. $L \in FO[Reg]$ if and only if for all $k > 0$, $\mu_L(\Sigma^k)$ contains no nontrivial groups.*

Example. The set of strings over Σ of even length is defined by the sentence

$$\forall x (\forall y (y \leq x) \rightarrow (x \equiv 0 \pmod 2))),$$

and is thus in $FO[Reg]$. Observe that while $M(L)$ is the group of order 2, $\mu_L(\Sigma^k)$ consists of a single element for each k. In contrast, the language consisting of all strings over $\{\sigma,\tau\}$ with an even number of occurrences of σ has the same syntactic monoid. But in this case, for every $k > 0$, $\mu_L(\Sigma^k)$ contains both elements of $M(L)$, and so, by the theorem, is not in $FO[Reg]$.

In [1] it is shown that any regular language definable by a first-order sentence with arbitrary non-regular numerical predicates is in $FO[Reg]$. This theorem depends upon (and is, in fact, a reformulation of) a deep result from circuit complexity.

4.4 Formulas with a Bounded Number of Variables

The language $\Sigma^*\sigma\sigma\Sigma^*\tau\Sigma^*$ is defined by the sentence

$$\exists x \exists y \exists z (Q_\sigma x \wedge Q_\sigma y \wedge Q_\tau z \wedge x < y \wedge y < z \wedge \forall w(w \leq x \vee y \leq w)).$$

Now observe that we can replace the variable w by z without changing the meaning of the formula—the new occurrences of z are bound by the innermost quantification and have nothing to do with the original use of z. Thus, by reusing variables, we have reduced the total number of variables in the sentence to three. Remarkably, this can always be done: every language in $FO[<]$ is definable by a sentence in which only three variables occur. (Immerman and Kozen [6].)

What happens if we allow only two variables? We denote by **DA** the family of finite aperiodic monoids in which every regular $\mathcal{J}$-class is a subsemigroup. (Schützenberger [10].) We denote by $FO^k[<]$ the family of languages over Σ definable by k-variable first-order sentences over $<$. We use the notations $MOD^k[<]$ and $(FO+MOD)^k[<]$ analogously. The following theorem is a combination of results of Thérien and Wilke [16] and Pin and Weil [8]:

Theorem 4.4. *Let $L \subseteq \Sigma^*$ be a regular language. The following are equivalent:*

(a) *L is definable by a first-order sentence over $<$ with only two variables.*

(b) *Both L and its complement are definable by Σ_2 sentences over $<$.*

(c) *$M(L) \in \mathbf{DA}$.*

What happens when we add modular quantifiers to the mix? We extend to this case the notation used above, in which we denote by a superscript the number of variables allowed in our sentences. The following is due to Straubing and Thérien [13]:

Theorem 4.5.

(a) $(FO + MOD)[<] = (FO + MOD)^3[<]$
(b) $MOD[<] = MOD^2[<]$
(c) Let $L \subseteq \Sigma^*$ be a regular language. Then $L \in (FO + MOD)^2[<]$ if and only if $M(L)$ divides a wreath product $M \circ G$, where G is a finite solvable group and $M \in \mathbf{DA}$.
(d) $L \in (FO + MOD)^2[<]$ if and only if both L and its complement are definable by Σ_2-sentences over $MOD[<]$. (That is, sentences whose atomic formulas are formulas quantified with modular quantifiers.)

Example. Let us look once again at our original example $L = (\Lambda + \sigma)(\tau\sigma)^*(\Lambda + \tau)$. $\mu_L(\sigma)$ belongs to the unique nontrivial regular $\mathcal{J}$-class J of $M(L)$, but $\sigma^2 = 0 \notin J$. Thus $M(L) \notin \mathbf{DA}$, and so L is not definable by a first-order sentence with two variables. But L is two-variable definable if we allow modular quantifiers. The sentence

$$\forall x (Q_\sigma x \leftrightarrow \exists^{0 \bmod 2} y (y \leq x)).$$

defines the set of strings $(\sigma\tau)^*(\sigma + \Lambda)$. The disjunction of this with the analogous sentence with σ replaced by τ defines L. The role played by the modular quantifiers in this sentence is rather remarkable. There are no groups in $M(L)$, so we do not need modular counting at all to define the language. Nonetheless, by including them, we are able to define the language more economically than would otherwise be possible.

Up until this point, all of our algebraic characterizations of languages defined by first-order and generalized first-order sentences have been effective in two senses: We have always had an algorithm to determine whether a given regular language is in the class of languages under consideration, and we have always had an algorithm to construct a defining sentence of the required type. However, we know of no algorithm for determining if a given finite monoid divides a wreath product of a monoid in $\mathbf{DA}$ and a solvable group. To see where the difficulty lies, we give an alternative characterization of this family of finite monoids. Let M be a finite monoid, and let J be a regular $\mathcal{J}$-class of M. If M divides a wreath product of a finite group and a monoid in $\mathbf{DA}$, then it is possible to partition the set of $\mathcal{L}$-classes of J in such a manner that whenever s, t belong to the same block of the partition, and $m \in M$, then sm and tm are either both in J or both outside of J, and if they are both in J then they belong to the same block. We thus have a well-defined partial action of M on the set of blocks of this partition. Furthermore, this action is one-to-one. Now it turns out that M divides a wreath product of a solvable group with a monoid in $\mathbf{DA}$ if and only if the partial one-to-one action on the blocks of each J-class can be extended to a solvable permutation group. In fact, the decidability of membership in this class of finite monoids is equivalent to the decidability of the following question: Given a set $\mathcal{F}$ of partial one-to-one maps on a finite set Q, can $\mathcal{F}$ be extended to a solvable group of permutations on a superset of Q? This question remains open.

5 Connections with Computational Complexity

We earlier mentioned some connections with circuit complexity. These are discussed at length in [11]. Here we briefly discuss another contact with computational complexity.

In computational complexity, we usually take the underlying alphabet Σ to be $\{0, 1\}$. Let $\mathcal{C}$ be a class of languages over this alphabet. We define
$L \in \exists \cdot \mathcal{C}$ if and only if there exists a polynomial p, and a language $K \in \mathcal{C}$, such that
$$x \in L \Leftrightarrow \exists y \in \Sigma^{p(|x|)}(xy \in K).$$

We define classes $\forall \cdot \mathcal{C}$ and $\oplus \cdot \mathcal{C}$ analogously, replacing $\exists$ in the definition by, respectively, $\forall$ and $\exists^{0 \bmod 2}$.

For example, if $\mathcal{P}$ denotes the class of polynomial-time languages, then $\exists \cdot \mathcal{P}$ is the class $\mathcal{NP}$, $\forall \cdot \mathcal{P}$ is the class co-$\mathcal{NP}$, and the closure of $\mathcal{P}$ under these two operators is the polynomial-time hierarchy $\mathcal{PH}$.

We also introduce a new computational model for language recognition: Let p be a polynomial, let $f : \Sigma^* \to M$ be a polynomial-time computable function, where M is a finite monoid, and let $X \subseteq M$. We define L to be the set of all strings w such that
$$\prod_{|z|=p(|w|)} f(wz) \in X,$$
where the product in M is taken in lexicographic order of the words z of length $p(|w|)$. We say that L is *polynomially recognized* by the monoid M.

Observe that $L \in \mathcal{NP}$ if and only if L is polynomially recognized by the monoid $\{0, 1\}$ with $X = \{0\}$ as the set of accepting values, and that similarly, L is in co-$\mathcal{NP}$ if and only if it is so recognized with $\{1\}$ as the set of accepting values. It is possible to prove that L is recognized in this sense by a finite monoid if and only if L is in $PSPACE$, and that L is recognized by a finite aperiodic monoid if and only if L is in the polynomial-time hierarchy. (See, for example, Hertrampf, et. al. [5])

The following theorem follows from work of Toda [17]:

Theorem 5.1.

$$\mathcal{PH} \subseteq \exists \cdot \forall \cdot \oplus \mathcal{P} \cap \forall \cdot \exists \cdot \oplus \mathcal{P}.$$

Observe the remarkable similarity between this result and our discussion of formulas with two variables and modular quantifiers. Once again, we can use modular counting to more efficiently express or recognize languages that do not in any intrinsic way require modular counting. The form of the result even suggests our alternative characterization, in terms of Σ_2 formulas, of the two-variable definable languages. We suspect that the underlying algebra is the same; indeed, we conjecture that every language in the polynomial time hierarchy is polynomially recognized by a wreath product of a monoid in **DA** and the cyclic group of order 2.

6 Why Semigroups?

All the classes of regular languages considered in this article (and there are other examples as well) were defined in terms of the kinds of logical formulas used to express the languages, but were characterized in terms of the syntactic monoids of the languages. Why are the answers to these *logical* questions always *algebraic*? Here we outline a general explanation of the phenomenon, based on a generalization of Eilenberg's theorem connecting pseudovarieties of finite semigroups and monoids with varieties of regular languages. ([4].)

Let $\mathcal{C}$ be a class of homomorphisms between finitely generated free monoids such that *(a)* $\mathcal{C}$ is closed under composition, and *(b)* for each finite alphabet Σ, the identity homomorphism on Σ^* is in $\mathcal{C}$. $\mathcal{C}$ is consequently the class of morphisms of a *category* whose objects are the finitely generated free monoids. Examples include: $\mathcal{C}_{all}$, the class of all homomorphisms between finitely generated free monoids, $\mathcal{C}_{ne}$, the class of non-erasing homomorphisms (that is, homomorphisms $\phi : \Sigma^* \to \Gamma^*$ such that $\phi(\Sigma^+) \subseteq \Gamma^+$), and $\mathcal{C}_{lm}$, the class of length-multiplying homomorphisms–those for which there exists $k > 0$ such that $\phi(\Sigma) \subseteq \Gamma^k$.

Given such a class $\mathcal{C}$, we define a $\mathcal{C}$-pseudovariety of homomorphisms to be a family **V** of surjective homomorphisms $\phi : \Sigma^* \to M$, where M is a finite monoid, with the following properties:

(a) Let $\phi : \Sigma^* \to M$ be in **V**, $f : \Gamma^* \to \Sigma^*$ in $\mathcal{C}$, and suppose there is a homomorphism α from $Im(\phi \circ f)$ onto a finite monoid N. Then $\alpha \circ \phi \circ f : \Gamma^* \to N$ is in **V**.

(b) If $\phi : \Sigma^* \to M$ and $\psi : \Sigma^* \to N$ belong to **V**, then so does $\phi \times \psi : \Sigma^* \to Im(\phi \times \psi) \subseteq M \times N$.

In this formalism, the $\mathcal{C}_{all}$ pseudovarieties are in essence identical to pseudovarieties of finite monoids, and the $\mathcal{C}_{ne}$-pseudovarieties to the pseudovarieties of finite semigroups. Given such a $\mathcal{C}$-pseudovariety **V**, we define the corresponding $\mathcal{C}$-variety of languages, which associates to each finite alphabet Σ the family of regular languages L such that the syntactic morphism of L is in **V**. As in the original theory of Eilenberg, the correspondence between $\mathcal{C}$-pseudovarieties and the associated varieties of languages is one-to-one.

The following result is due to the author [12]: Let $\mathcal{Q}$ be a class of quantifiers, either FO, MOD, or $FO+MOD$. Let $k, d > 0$, and let $\mathcal{N}$ be one of the following classes of numerical predicates: equality, equality with successor, ordering, ordering with successor, or all regular numerical predicates. Let

$$\mathcal{Q}[k, d, \mathcal{N}]$$

associate to each finite alphabet Σ the family of languages over Σ^* defined by sentences using the given class of quantifiers, with quantifier depth d, no more than k variables, and numerical predicates in $\mathcal{N}$. Then

Theorem 6.1. $\mathcal{Q}[k, d, \mathcal{N}]$ *is a $\mathcal{C}$-variety of languages, where $\mathcal{C} = \mathcal{C}_{all}$ if $\mathcal{N}$ is equality or ordering, $\mathcal{C} = \mathcal{C}_{ne}$ if $\mathcal{N}$ is one of the classes containing successor, and $\mathcal{C} = \mathcal{C}_{lm}$ if $\mathcal{N}$ is the class of regular numerical predicates.*

It follows that membership in each of these logically defined classes depends only on the syntactic morphism of the language.

References

[1] D. Mix Barrington, K. Compton, H. Straubing, and D. Thérien, "Regular Languages in NC^1", *J. Comp. Syst. Sci.* **44** (1992) 478–499.

[2] D. Beauquier and J. E. Pin, "Factors of Words", *Proc. 16th ICALP,* Springer Lecture Notes in Computer Science **372** (1989) 63–79.

[3] J. R. Büchi, "Weak second-order arithmetic and finite automata", *Zeit. Math. Logik. Grund. Math.* **6** (1960) 66-92.

[4] S. Eilenberg, *Automata, Languages and Machines,* vol. B, Academic Press, New York, 1976.

[5] U. Hertrampf, C. Lautemann, T. Schwentick, H. Vollmer, K. Wagner, "On the Power of Polynomial-Time Bit Reductions", *Proc. 8th IEEE Conference on Structure in Complexity Theory* (1993) 200-207.

[6] N. Immerman and D. Kozen, "Definability with a Bounded Number of Bound Variables", *Information and Computation,* **83**, 121-139 (1989).

[7] R. McNaughton and S. Papert, *Counter-Free Automata,* MIT Press, Cambridge, Massachusetts, 1971.

[8] J.-É. Pin et P. Weil, "Polynomial closure and unambiguous product", *Theory Comput. Systems* **30**, 1-39, (1997).

[9] M. P. Schützenberger, "On Finite Monoids Having Only Trivial Subgroups", *Information and Control* **8** 190-194 (1965).

[10] M. P. Schützenberger, "Sur le Produit de Concatenation Non-ambigu", *Semigroup Forum* **13** (1976), 47-76.

[11] H. Straubing, *Finite Automata, Formal Logic and Circuit Complexity*, Birkhäuser, Boston, 1994.

[12] H. Straubing, "On the logical characterization of regular languages" Under review.

[13] H. Straubing and D. Thérien, "Regular languages defined by generalized first-order formulas with a bounded number of bound variables", *Proc. 2001 STACS*.

[14] H. Straubing, D. Thérien, and W. Thomas, "Regular Languages Defined by Generalized Quantifiers", *Information and Computation* **118** 289-301 (1995).

[15] D. Thérien and A. Weiss, "Graph Congruences and Wreath Products", *J. Pure and Applied Algebra* **35** 205-215 (1985).

[16] D. Thérien and T. Wilke, "Over Words, Two Variables are as Powerful as One Quantifier Alternation", *Proc. 30th ACM Symposium on Theory of Computing*, 256-263 (1988).

[17] S. Toda, "PP is as Hard as the Polynomial-Time Hierarchy", *SIAM J. Computing* **20** (1991) 865-877.

DIAMONDS ARE FOREVER: THE VARIETY DA

PASCAL TESSON AND DENIS THÉRIEN*
*School of Computer Science, McGill University,
3480 rue University, Montréal, Québec,
H3A 2A7, Canada.
E-mail: {ptesso, denis}@ cs.mcgill.ca.*

We survey different characterizations (algebraic, combinatorial, automata-theoretic, logical) of the variety **DA**, the class of monoids whose regular $\mathcal{D}$-classes form aperiodic semigroups. We study the interconnections of such characterizations and outline the importance of this variety in various computational complexity issues.

1 Introduction

The intuition that combinatorial properties of a language should relate to the algebraic structure of its syntactic congruence has been completely validated in the realm of rational languages, i.e. in the case where the syntactic congruence has finite index. Indeed the theory of finite monoids offers an elegant and powerful framework to understand rational languages, and in turn the language approach has proved to be very useful in the study of finite monoids.

A subset $L \subseteq A^*$ is *recognized* by a monoid M if there exists a morphism $\phi : A^* \to M$ and a subset F of M such that $L = \phi^{-1}(F)$. It is easy to show that for any language L, there is a "minimal" recognizer $M(L)$: it has the property that it *divides* (i.e. is a morphic image of a submonoid of) any other monoid M recognizing L. The famous theorem of Kleene states that L is a rational language if and only if $M(L)$ is finite.

The notion of pseudo-variety, which we will simply call variety in this paper, has become central in algebraic automata theory. A class **V** of finite

---

*RESEARCH SUPPORTED BY GRANTS FROM NSERC, FCAR AND THE VON HUMBOLDT FOUNDATION. THE AUTHORS WOULD ALSO LIKE TO ACKNOWLEDGE THE FINANCIAL SUPPORT OF FUNDAÇÃO CALOUSTE GULBENKIAN (FCG), FUNDAÇÃO PARA A CIÊNCIA E A TECNOLOGIA (FCT), FACULDADE DE CIÊNCIAS DA UNIVERSIDADE DE LISBOA (FCUL) AND REITORIA DA UNIVERSIDADE DO PORTO.

monoids forms a variety if and only if it is closed under division and direct product. Certain varieties are real "jewels", in the sense that they admit several different characterizations. Unsurprisingly, these varieties usually turn out to be particularly important. This is certainly the case for the variety **A** of aperiodic (or group-free) monoids: a monoid M is group-free if no subset of it forms a group of cardinality greater than one. For example, the following equivalences can be shown:

- $M(L)$ is aperiodic;
- L is a star-free language [27];
- L can be defined by a first-order sentence [21] (see Section 5);
- L can be defined by a formula in linear temporal logic [15] (see Section 5);
- L can be recognized by a cascade product of set-reset automata [17].

Note that the situation with the variety **G** of finite groups is far from being as satisfactory, essentially because we still understand very little about simple non-Abelian groups.

In recent years, the variety **DA**, a proper subclass of aperiodic monoids, has emerged as a "jewel" in its own right and several beautiful characterizations are now known for it. This paper offers a survey of the most important ones.

For conciseness, we will often refer the reader to the bibliography for additional details. We will first review some basic results, then present congruence relations offering parametrizations of **DA** and algebraic decompositions and parametrizations of the variety. Finally, we will discuss applications of these characterizations to logic and descriptive complexity and to computational complexity related issues.

2 Basic Properties and Notation

All monoids considered in this paper are either free or finite. We denote by $eval_M$ the canonical morphism from M^* into M defined by

$$eval_M(m_1 \ldots m_n) = m_1 \cdot \ldots \cdot m_n$$

and denote by m^ω the unique positive power of m that is an idempotent. Let us recall the fundamental Green equivalences:

- $s \mathcal{R} t$ iff $sM = tM$;

- $s \mathcal{L} t$ iff $Ms = Mt$;
- $s \mathcal{D} t$ iff $s \mathcal{R} \vee \mathcal{L} t$;
- $s \mathcal{J} t$ iff $MsM = MtM$;
- $s \mathcal{H} t$ iff $s \mathcal{R} t$ and $s \mathcal{L} t$.

We also write $s \leq_{\mathcal{R}} t$ if $sM \subseteq tM$ and similarly for the other relations. We also denote by $\mathcal{R}_m$ the $\mathcal{R}$-class of m. For any of the above relations, we say that a given class is regular if and only if it contains an idempotent.

The following facts are elementary:

Lemma 2.1.
(a) In a finite monoid, $\mathcal{D}$ and $\mathcal{J}$ coincide.
(b) If s and t belong to the same $\mathcal{J}$-class, then their product st belongs to $\mathcal{R}_s \cap \mathcal{L}_t$ iff $\mathcal{L}_s \cap \mathcal{R}_t$ contains an idempotent.

The variety **DA** is defined as the class of finite monoids in which every regular $\mathcal{D}$-class forms an aperiodic semigroup. By the lemma above, it follows that M is in **DA** iff $MeM = MfM$ and $e = e^2$ imply $f = f^2$.

The following alternative characterizations are particularly useful.

Theorem 2.2. *The following are equivalent:*
(a) M is in **DA**;
(b) $e = e^2$ and $MeM \subseteq MsM$ imply $ese = e$;
(c) for all x, y, z in M, $(xyz)^\omega y(xyz)^\omega = (xyz)^\omega$.

Proof. (a) $\Rightarrow$ (b) Let $e = usv$: since e is idempotent, we have $e = (usv)^\omega$, which is $\mathcal{J}$-related to $f = (svu)^\omega$. Because M is in **DA**, we have $e \mathcal{J} ef$, so that $e \mathcal{R} es$ and $es = eses$. We conclude that $ese = e$.
(b) $\Leftrightarrow$ (c) Obvious.
(b) $\Rightarrow$ (a) If $e = e^2 \mathcal{J} f$, then $efe = e$. Thus ef is an idempotent, and $effef = ef$: it must be that f^2 is also in the same $\mathcal{J}$-class, hence f is an idempotent. $\square$

If M is not in **DA**, then it contains a regular $\mathcal{J}$-class that has a nonidempotent. Conveniently, this can be used to show that M either contains a group or is divided by one of the aperiodics BA_2 or U, the transformation monoids associated with the automata of Figures 1 and 2 respectively. Throughout this paper, varieties of monoids will be denoted in **boldface**. We'll pay a particular attention to **DA** of course, but will also mention among

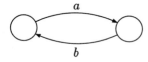
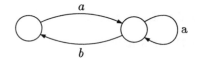

Fig. 1: BA_2 automaton   Fig 2: U automaton

others the variety of aperiodics **A**, commutative aperiodics $\mathbf{A_{com}}$, idempotent monoids or bands $\mathbf{A_1}$, $\mathcal{R}$-trivial monoids **R**, $\mathcal{R}$-trivial bands $\mathbf{R_1}$, $\mathcal{L}$-trivial monoids **L**, $\mathcal{L}$-trivial bands $\mathbf{L_1}$ and solvable groups $\mathbf{G_{sol}}$. Suppose that for every finite alphabet A we have a congruence relation $\sim_A$ over strings in A^*, we will denote by $\mathbf{V}_\sim$ the class of monoids

$$\mathbf{V}_\sim = \{M = A^*/\gamma : \text{ for some } A \text{ with } \sim_A \subseteq \gamma\}.$$

3 Congruences

In this section, we describe various families of congruences, each of which induces a parametrization of the variety **DA**. Let w be a string in A^*. We denote by $\alpha(w)$ the set of letters of A occurring in w and for any $a \in \alpha(w)$, we say that $w = w_0 a w_1$ is the a-left decomposition of w (resp. a-right decomposition of w) if $a \notin \alpha(w_0)$ (resp. $a \notin \alpha(w_1)$).

3.1 A Characterization of **DA** via Congruences

We define a first congruence $\sim_{n,k}$ on A^* where $n = |A|$ by induction on $n+k$. First, we have $x \sim_{n,0} y$ for all $x, y \in A^*$. Next, we let $x \sim_{n,k} y$ when and only when:

(1) $\alpha(x) = \alpha(y)$;
(2) For any $a \in \alpha(x) = \alpha(y)$, if $x = x_0 a x_1$ and $y = y_0 a y_1$ are the a-left decompositions of x and y then $x_0 \sim_{n-1,k} y_0$ and $x_1 \sim_{n,k-1} y_1$;
(3) For any $a \in \alpha(x) = \alpha(y)$, if $x = x_0 a x_1$ and $y = y_0 a y_1$ are the a-right decompositions of x and y then $x_0 \sim_{n,k-1} y_0$ and $x_1 \sim_{n-1,k} y_1$.

This equivalence relation is well-defined since $|\alpha(x_0)| < |\alpha(x)|$ in (2) and $|\alpha(x_1)| < |\alpha(x)|$ in (3). It is easy to check that $\sim_{n,k}$ is in fact a congruence of finite index.

Theorem 3.1. ([35]) *Let $M = A^*/\gamma$, with $|A| = n$. Then $M \in \mathbf{DA}$ iff $\sim_{n,k} \subseteq \gamma$ for some k.*

Proof. For one direction, it suffices to show that $A^*/\sim_{n,k}$ is in **DA**. This will follow if we show that $u \sim_{n,k} v$ whenever u has the form $u = (xyz)^k y (xyz)^k$ and v has the form $v = (xyz)^k$. If $k = 0$, there is nothing to argue. If $k > 0$, we clearly have $\alpha(u) = \alpha(v)$. Let a be a letter in $\alpha(u)$, and let $u = u_0 a u_1$ and $v = v_0 a v_1$ be the a-left decompositions of u and v respectively. Then $u_0 = v_0$: also $u_1 = w(xyz)^{k-1} y(xyz)^k$ and $v_1 = w(xyz)^{k-1}$. Using induction, we see that $(xyz)^{k-1} \sim_{n,k-1} (xyz)^k$ and that $(xyz)^{k-1} yxyz^{k-1} \sim_{n,k-1} (xyz)^{k-1}$, so that $u_1 \sim_{n,k-1} v_1$. By left-right symmetry, we deduce that $u \sim_{n,k} v$.

For the second part of our proof, let us denote as $[u]$, for any $u \in A^*$, the γ equivalence class of u. We define the $\mathcal{R}$-decomposition (with respect to the congruence γ) of a string $u \in A^*$ as the unique factorization $u = u_0 a_1 u_1 \ldots a_t u_t$, with $a_i \in A$ and $u_i \in A^*$ such that:

(1) $[u_0 a_1 \ldots a_s u_s] \, \mathcal{R} \, [u_0 a_1 \ldots a_s u_s a_{s+1}]$ for any $s < t$;

(2) $[u_0 a_1 \ldots a_s] \, \mathcal{R} \, [u_0 a_1 \ldots a_s u_s]$ for any $s \leq t$;

(3) $[u_0] = 1_M$.

It is not hard to see that in the $\mathcal{R}$-decomposition of u, one must have $a_i \notin \alpha(u_{i-1})$. In particular, $u = u_0 \, a_1 \, u_1 a_2 \ldots a_t u_t$ is the a_1-left-decomposition of u. Symmetrically, we can define $\mathcal{L}$-decompositions of strings, which will relate to right-decompositions.

Let now k be the maximum of the number of $\mathcal{R}$-classes and the number of $\mathcal{L}$-classes of M. Let $B \subseteq A$ be a subalphabet with $|B| = m$ and suppose u, v are strings in B^*. We claim that if $u \sim_{m,mk} v$ then $[u] = [v]$ and prove this by induction on m.

For $m = 0$, the claim is trivially true. For the inductive step, assume now $m \geq 1$ and suppose $u \sim_{m,mk} v$, with $\alpha(u) = \alpha(v) \neq \emptyset$. Let $u = u_0 a_1 u_1 \ldots a_t u_t$ be the $\mathcal{R}$-decomposition of u. We have $t \leq k$. Write w_i for $u_i a_{i+1} \ldots a_t u_t$. By our earlier remark, $w_i = u_i a_{i+1} w_{i+1}$ is the a_{i+1}-left-decomposition of w_i, so there must be a decomposition $v = v_0 a_1 \ldots a_t v_t$ such that $u_i \sim_{m-1, mk-i} v_i$ for $i < t$ (note that this actually implies $u_i \sim_{m-1,(m-1)k} v_i$). By the induction hypothesis, we have $[u_i] = [v_i]$ for all $i < t$ so $[u] \, \mathcal{R} \, [u_0 a_1 \ldots u_{t-1} a_t] = [v_0 a_1 \ldots v_{t-1} a_t] \geq_{\mathcal{R}} [v]$. Symmetrically, we get $[v] \geq_{\mathcal{R}} [u]$ and thus $[u] \, \mathcal{R} \, [v]$. The same argument using $\mathcal{L}$-decompositions establishes $[u] \, \mathcal{L} \, [v]$ so, by aperiodicity, we have in fact $[u] = [v]$. □

Corresponding to the definition of $\sim$, we can recursively define $*_{n,k}$ languages in the following way. First, we let A^* be a $*_{|A|,0}$-language for any A. Next, suppose A_0, A_1 are finite alphabets whose difference is a single letter $a \in A_1 - A_0$ and let $L_0 \subseteq A_0^*$, $L_1 \subseteq A_1^*$ be $*_{|A_0|,k}$ and $*_{|A_1|,k}$ languages respectively, then $L_0 a L_1$ and $L_1 a L_0$ are both $*_{|A_1|,k+1}$ languages. From Theorem 3.1, we can immediately conclude:

479

Corollary 3.2. *Let $L \subseteq A^*$. We have $M(L) \in \mathbf{DA}$ if and only if L is a (possibly empty) disjoint union of $*_{n,k}$ languages.*

For any languages $L_0, \ldots, L_t \subseteq A^*$ and letters $a_1, \ldots, a_t \in A$, we say that the language $K = L_0 a_1 L_1 \ldots a_t L_t$ is an *unambiguous concatenation* of the L_i's if for any $w \in K$ there exists a unique factorization $w = w_0 a_1 w_1 \ldots a_t w_t$ with $w_i \in L_i$. For such languages, we refer to the a_i's as *bookmarks*. The results of this subsection provide an alternative proof to a theorem due to Schützenberger [28]:

Theorem 3.3. *Let $L \subseteq A^*$. We have $M(L) \in \mathbf{DA}$ if and only if L is the disjoint union of unambiguous concatenations $K = A_0^* a_1 A_1^* \ldots a_t A_t^*$, with $a_i \in A$ and $A_i \subseteq A$.*

Proof. The $*_{n,k}$ languages are examples of such unambiguous concatenations so the "only if" part of the statement can be obtained from Corollary 3.2.

For the converse, we argue by induction on t that an unambiguous concatenation $K = A_0^* a_1 A_1^* \ldots a_t A_t^*$ is a union of $*_{|A|,t}$ languages. The case $t = 0$ is trivial. For $t \geq 1$, observe that by the unambiguity restriction, it can not be the case that all a_i's belong to the intersection $A_0 \cap A_t$ (or else, the string $(a_1 \ldots a_t)^2$ would admit two different factorizations). Suppose without loss of generality that $a_i \notin A_0$ and consider for any $w \in K$ the a_i-left-decomposition of $w = u a_i v$. Because of unambiguity, exactly one of the following must occur: $u \in K_0 = A_0^* a_1 \ldots a_{i-1} A_{i-1}^*$ while $v \in K_1 = A_i^* a_{i+1} \ldots a_t A_t^*$ or, for some $j < i$ we have $a_i \in A_j$ and $u \in L_j = A_0^* a_1 \ldots a_j A_j^*$ while $v \in L_j' = A_j^* a_{j+1} \ldots a_t A_t^*$. In either case, the languages K_0 and L_j are unambiguous concatenations with at most $t-1$ bookmarks over $(A - \{a_i\})^*$ which, by induction, are disjoint unions of $*_{|A|-1,t-1}$ languages while the languages K_1 and L_j' are unambiguous concatenations with at most $t-1$ bookmarks over A^* and are thus unions of $*_{|A|,t-1}$ languages. □

3.2 An Alternative Parametrization of **DA**

Let $|w|_a$ denote the number of occurrences of the letter a in the word w and let $\alpha_t(w)$ be the vector of dimension $|A|$ giving, for each $a \in A$, the value $|w|_a$ up to the threshold t. We now define a second family of congruences $\approx_{k,t}$ on A^* for any $k, t \in \mathbb{N}$. First, for any t, we let $x \approx_{0,t} y$ for all $x, y \in A^*$. Then recursively, we define $x \approx_{k,t} y$ if and only if

(1) $x \approx_{k-1,t} y$;

(2) $\alpha_t(x) = \alpha_t(y)$;

(3) For all $x = x_0 a x_1$ and $y = y_0 a y_1$ with $|x_0|_a = |y_0|_a \leq t$, we have

$x_0 \approx_{k-1,t} y_0$ and $x_1 \approx_{k-1,t} y_1$;

(4) For all $x = x_0 a x_1$ and $y = y_0 a y_1$ with $|x_1|_a = |y_1|_a \leq t$, we have $x_0 \approx_{k-1,t} y_0$ and $x_1 \approx_{k-1,t} y_1$.

One can check that for all k, t, $\approx_{k,t}$ is a well defined finite index congruence. This congruence is quite close to $\sim_{n,k}$ defined in the previous subsection: where in $\sim_{n,k}$ we were primarily concerned with the first and last occurrence of each letter, we look here at the first and last t occurrences of each letter. Note however that $\approx_{k,1}$ is *not* equal to $\sim_{|A|,k}$ because the recursion on the prefixes and suffixes is slightly different in the two definitions. For example, one can verify that $ab \approx_{1,1} ba$ while $ab \not\approx_{2,1} ba$.

Not so surprisingly, given the similarity to $\sim_{n,k}$, we can obtain from $\approx_{k,t}$ an alternative parametrization of **DA**:

Theorem 3.4. *Let $M = A^*/\gamma$, with $|A| = n$. Then $M \in$ **DA** iff $\approx_{k,t} \subseteq \gamma$ for some k, t.*

Proof. To show that $A^*/\approx_{k,t}$ is in **DA**, we show by induction on k that for any $x, y, z \in A^*$ $u = (xyz)^{kt} y(xyz)^{kt} \approx_{k,t} v = (xyz)^{kt}$. The base case is clear. For $k \geq 1$, it is clear that $\alpha_t(u) = \alpha_t(v)$. Also, let $u = u_0 a u_1$ with $|u_0|_a \leq t$, then clearly this occurrence of the letter a appears in the first $(xyz)^t$ block of u so that in fact we have $u = u_0 a w (xyz)^{(k-1)t} y (xyz)^{kt}$ and $v = u_0 a w (xyz)^{(k-1)t}$. By induction we get $(xyz)^{(k-1)t} y (xyz)^{(k-1)t} \approx_{(k-1),t} (xyz)^{(k-1)t}$ and $(xyz)^{(k-1)t} = xyz^{kt}$ so, since $\approx$ is a congruence, we have $w(xyz)^{(k-1)t} y(xyz)^{kt} \approx_{(k-1),t} w(xyz)^{(k-1)t}$. By left-right symmetry, we can conclude $u \approx_{k,t} v$.

The converse follows from Theorem 3.1 since, from the definitions of $\approx_{k,t}$ and $\sim_{n,k}$, we have $\approx_{nk,1} \subseteq \sim_{n,k}$. □

4 Algebraic decompositions for DA

The variety **DA** admits several interesting algebraic decompositions. We present three different ones here. For two of them, we require the formalism of categories as described in [34].

Let $C = (V, E)$ be a finite directed graph and let C^* be the free category generated by C, i.e. the set of objects of C^* is V and for any $i, j \in V$, $Hom(i, j)$ is the set of paths of finite length from vertex i to vertex j (including, when $i = j$, the path 1_i of length 0). Two paths x, y in C^* are *coterminal* if they belong to the same $Hom(i, j)$ and are *consecutive* if $x \in Hom(i, j)$ and $y \in Hom(j, k)$, for some $i, j, k \in V$. A graph congruence on C^* is an equivalence relation β on the paths such that

1) $x\,\beta\,y$ implies x, y are coterminal;
2) $x_1\beta y_1$ and $x_2\beta y_2$ and x_1, x_2 consecutive implies $(x_1 x_2)\,\beta\,(y_1 y_2)$.

We are interested in the following case. Let γ be a finite-index congruence on the free monoid A^*. We form the graph $C_\gamma = (V, E)$ where $V = A^*/\gamma \times A^*/\gamma$ and for every $(u, a, v) \in A^* \times A \times A^*$ there is an edge $e_{u,a,v}$ from vertex $([u]_\gamma, [av]_\gamma)$ to vertex $([ua]_\gamma, [v]_\gamma)$: of course if $u\,\gamma\,u'$ and $v\,\gamma\,v'$ we do not distinghuish between $e_{u,a,v}$ and $e_{u',a,v'}$ and sometimes use the notation $e_{L,a,K}$ where $L = [u]_\gamma$ and $K = [v]_\gamma$. If $x = x_1 x_2 \ldots x_n \in A^*$ has length n, we will denote by $P_{u,v}(x)$ the path of length n in C^* that goes from vertex $([u]_\gamma, [xv]_\gamma)$ to vertex $([ux]_\gamma, [v]_\gamma)$ where the ith edge is $e_{ux_1\ldots x_{i-1}, x_i, x_{i+1}\ldots x_n v}$. Note that every path in C^* is of that form.

Let β be a finite-index graph congruence on C^*. We form a new congruence $\beta\Box\gamma$ (the *block product* of β and γ) of finite index on A^* by defining $x\,\beta\Box\gamma\,y$ if and only if
1) $x\,\gamma\,y$;
2) for all $u, v \in A^*$, $P_{u,v}(x)\,\beta\,P_{u,v}(y)$.

We define an edge in a graph as *falling* if it connects two distinct strongly connected components. We will also say that two paths p, q are l_1-congruent if they are coterminal and traverse the same set of falling edges. For a finite-index congruence γ on A^* we say that L is a γ-language if it is the union of γ-classes.

Lemma 4.1. *Let L_0, L_1 be γ-languages and suppose $L_0 a L_1$ is an unambiguous concatenation. Then $L_0 a L_1$ is a $l_1\Box\gamma$ language. Conversely, any $l_1\Box\gamma$ language is the disjoint union of unambiguous concatenations $L_0 a_1 L_1 \ldots a_k L_k$ with $a_i \in A$ and where the L_i's are γ-languages.*

Proof. Without loss of generality, we can assume that L_0, L_1 are γ-classes. Clearly, $x \in L_0 a L_1$ if and only if $P_{\epsilon,\epsilon}(x)$ goes through the edge e_{L_0, a, L_1}. But this is a falling edge since the concatenation is unambiguous. Hence the $l_1\Box\gamma$-class of x determines its membership in $L_0 a L_1$. Similarly, the argument for the converse is based on the observation that falling edges in the graph induced by γ correspond to unambiguous concatenation. □

Define on A^* the congruence $x\gamma y$ if and only if $\alpha(x) = \alpha(y)$. It is well known that $M = A^*/\beta$ belongs to the variety $\mathbf{J_1}$ of semilattices if and only if $\beta \supseteq \gamma$. Define now $\delta_1 = \gamma$ and, for every $k > 1$, $\delta_k = l_1\Box\delta_{k-1}$.

Theorem 4.2. $M = A^*/\beta$ *belongs to* **DA** *if and only if* $\beta \supseteq \delta_k$ *for some k.*

Proof. This follows from Lemma 4.1 and the theorem of Schützenberger cited (and re-proved) in Section 3. □

This result is equivalent to saying that **DA** is the smallest variety containing $\mathbf{J_1}$ and closed under block-product with locally trivial categories [23]. Consider the congruence $l_1 \square \gamma$ and suppose $x \gamma y \gamma z \gamma x^2$. We can observe that $P_{u,v}(xyz) \, l_1 \, P_{u,v}(xz)$. Indeed the two paths are certainly coterminal. Moreover for any $x = x_0 a x_1$, the edge e_{ux_0,a,x_1yzv} is identical to the edge e_{ux_0,a,x_1zv} and similarly for any edge coming from a factorization $z = z_0 a z_1$. Finally, for any $y = y_0 a y_1$, the edge e_{uxy_0,a,y_1zv} is not falling and can thus not be taken into account by the l_1 congruence. In the terminology of [23], this means that for any $M = A^*/\beta$ for $\beta \supseteq l_1 \square \gamma$, there is an **LI**-relational morphism (see e.g. [23, 22] for the formal definition) from M to A^*/γ. Since the composition of **LI**-morphisms is also an **LI**-morphism, we can conclude that for any $M \in \mathbf{DA}$, there is an **LI**-relational morphism from M to a semilattice. Conversely, it is not too hard to show that any such monoid will belong to **DA**. This operation, known as the Malcev product, is denoted by Ⓜ . We thus have:

Theorem 4.3. DA = LI Ⓜ $\mathbf{J_1}$.

The last decomposition we present is also using the notion of block product (and is due to [32]). Let γ be again the free $\mathbf{J_1}$-congruence on A^*, and construct the graph $C_\gamma = (V, E)$. We define morphisms from E^* to A^* as follows:

$$\phi_{1,a}(e_{u,b,v}) = \begin{cases} b & \text{if } a \notin \alpha(ub), \\ \epsilon & \text{otherwise;} \end{cases}$$

$$\psi_{1,a}(e_{u,b,v}) = \begin{cases} b & \text{if } a \in \alpha(u), \\ \epsilon & \text{otherwise;} \end{cases}$$

$$\phi_{2,a}(e_{u,b,v}) = \begin{cases} b & \text{if } a \notin \alpha(bv), \\ \epsilon & \text{otherwise;} \end{cases}$$

$$\psi_{2,a}(e_{u,b,v}) = \begin{cases} b & \text{if } a \in \alpha(v), \\ \epsilon & \text{otherwise.} \end{cases}$$

We next set up a congruence on A^* by letting $x \sim_{n,k} y$ if and only if
(1) $x \gamma y$;
(2) For all $a \in A$, for $i = 1, 2$:
   $\phi_{i,a}(P_{\epsilon,\epsilon}(x)) \sim_{n-1,k} \phi_{i,a}(P_{\epsilon,\epsilon}(y))$ and
   $\psi_{i,a}(P_{\epsilon,\epsilon}(x)) \sim_{n,k-1} \psi_{i,a}(P_{\epsilon,\epsilon}(y))$.

It should be clear that this $\sim_{n,k}$ is identical to the one defined in Section 3.

Let $\mathbf{V}$ be the smallest variety containing $\mathbf{J_1}$ and satisfying $\mathbf{V}\Box\mathbf{J_1} = \mathbf{V}$. The above observation establishes that $\mathbf{DA} \subseteq \mathbf{V}$. Since it is easy to show that $\mathbf{DA}$ does satisfy $\mathbf{DA}\Box\mathbf{J_1} = \mathbf{DA}$, we have in fact $\mathbf{V} = \mathbf{DA}$.

5 Logical Descriptions of DA

In this section we will use logical formulas to define properties of strings over a finite alphabet A. First, we consider first-order formulas where variables are integers denoting the positions in a string and predicates are either the binary predicate $<$ or, for each $a \in A$, the unary predicate Q_a, where $Q_a x$ means "position x in the string holds an a". A sentence ϕ in this logic naturally defines a language in A^* consisting of strings over which ϕ holds (see [30] for a detailed introduction). As an example, the sentence $\exists i \exists j (Q_a i \wedge Q_b i \wedge i < j)$ defines the regular language $A^* a A^* b A^*$.

We denote by $\mathbf{FO}[<]$ the class of languages that can be defined by a first-order sentence as above. It is well known that a language belongs to $\mathbf{FO}[<]$ if and only if its syntactic monoid is finite aperiodic [21, 27].

If ϕ is a sentence not containing i as a variable, we denote $\phi_{<i}$ (resp. $\phi_{>i}$) the logical formula obtained from ϕ by requiring that every bound variable in ϕ be strictly less than i (resp. strictly greater than i).

5.1 $\mathbf{\Sigma_2} \cap \mathbf{\Pi_2}[<]$

The most studied subclasses of $\mathbf{FO}[<]$ are the ones defined by bounded quantifier alternation. We will naturally denote by $\mathbf{\Sigma_k}[<]$ (resp. $\mathbf{\Pi_k}[<]$) the class of languages definable by first-order formulas in prenex form having k alternating blocks of existential and universal quantifiers starting with a block of existential quantifiers (resp. a block of universal quantifiers). The following is a corollary of a powerful theorem due to Pin and Weil [24]:

Theorem 5.1. *A language L belongs to both $\mathbf{\Sigma_2}[<]$ and $\mathbf{\Pi_2}[<]$ iff $M(L)$ belongs to $\mathbf{DA}$.*

Proof. The "if" direction of the statement can be obtained from the results of Section 3 without resorting to the result of Pin and Weil. Indeed we noted that if $M(L)$ is in $\mathbf{DA}$, then there exists n, k such that L can be expressed as a disjoint union of $*_{n,k}$ languages. It is clear that $*_{n,0}$ languages are in $\mathbf{\Sigma_2} \cap \mathbf{\Pi_2}[<]$. Now for $n \geq 1$, suppose we want to define the language $L = L_0 a L_1$ where L_0, L_1 are $*_{n,k-1}$ and $*_{n-1,k}$ languages respectively with

L_1 a subset of $(A - \{a\})^*$. Then L can be defined by the Σ_2 sentence

$$\exists i (Q_a i \wedge \phi_{<i} \wedge \psi_{>i})$$

where ϕ and ψ, obtained from the induction, are the Σ_2 sentences defining L_0 and L_1 respectively. Similarly, we can express membership in L by saying that there is a position holding an a and that any position holding an a is either not holding the last a, or defines a suffix belonging to L_1 and a prefix belonging to L_0, i.e. L is defined by the Π_2 sentence:

$$\forall i ((\neg Q_a i \vee (\exists j (Q_a j \wedge i < j)) \vee (Q_a i \wedge \rho_{<i} \wedge \chi_{>i})))$$

where ρ, χ are now the $\Pi_2[<]$ sentences for L_0 and L_1.

For the "only if" part, one first needs a result of Arfi [1] which shows that a language is in $\boldsymbol{\Sigma_2}[<]$ if and only if it is expressible as the disjoint union of languages of the form $A_0^* a_1 A_1^* \ldots a_k A_k^*$ (without any sort of unambiguity restriction). If L is in $\boldsymbol{\Sigma_2} \cap \boldsymbol{\Pi_2}[<]$, both L and its complement can be expressed in that way. Let n be the length of the largest expression needed to obtain L or its complement. Now for any $x, y, z, u, v \in A^*$, if $u(xyz)^{n+1}v$ belongs to L then it belongs to some $A_0^* a_1 A_1^* \ldots a_k A_k^* \subseteq L$ (with $k \leq n$) and it can be factorized as $w_0 a_1 \ldots a_k w_k$. By the pigeon hole principle, some w_j must contain xyz entirely so $\alpha(xyz) \subseteq A_j$. Consequently, $u(xyz)^{n+1} y(xyz)^{n+1} v$ also belongs to $A_0^* a_1 A_1^* \ldots a_k A_k^* \subseteq L$. Similarly, using the representation of L's complement, if $u(xyz)^{n+1} v$ is not in L, then $u(xyz)^{n+1} y(xyz)^{n+1} v$ is not. So $M(L)$ must satisfy $(xyz)^\omega y(xyz)^\omega = (xyz)^\omega$ and thus belongs to **DA**. □

5.2 $\mathbf{FO_2}[<]$ and Unary Temporal Logic

Another meaningful measure of the complexity of these first-order sentences is given by the number of variables used. We will denote by $\mathbf{FO_k}[<]$ the class of sentences using only k different variables (possibly re-used of course). For example, to describe the language of strings in A^* having at least three a's, one would naturally use the three variable sentence

$$\exists i \exists j \exists k ((i < j < k) \wedge Q_a i \wedge Q_a j \wedge Q_a k)$$

which is equivalent to the two-variable sentence:

$$\exists i \exists j ((Q_a i \wedge Q_a j \wedge (i < j) \wedge \exists i ((j < i) \wedge Q_a i)))$$

It is an easy exercise to show that a language L is definable in $\mathbf{FO_1}[<]$ if and only if its syntactic monoid is idempotent and commutative. On the other hand three variables suffice to express *any* first-order definable language [15, 14].

Etessami, Vardi and Wilke showed in [6] that **FO₂** is exactly the class of languages expressible in a fragment of temporal logic called *unary temporal logic* (or **UTL**). Formally, a **UTL** formula is built using ordinary Boolean connectives, elements of the alphabet A and the two unary temporal operators "eventually in the future", denoted $\diamondsuit$ and "eventually in the past": $\diamondsuit\!\!\!\!\leftarrow$. Given a word $w = w_1 w_2 \ldots w_n$ and a **UTL** formula ϕ, we define what it means for ϕ to be true over w at i (and denote $(w, i) \models \phi$) using the following semantics (the semantics of Boolean connectives is clear):

- $(w, i) \models a$ if w_i is a.
- $(w, i) \models \diamondsuit\phi$ if there exists $j > i$ such that $(w, j) \models \phi$.
- $(w, i) \models \diamondsuit\!\!\!\!\leftarrow\phi$ if there exists $j < i$ such that $(w, j) \models \phi$.
- We also say that a formula $\diamondsuit\phi$ (resp. $\diamondsuit\!\!\!\!\leftarrow\phi$) is true over w if $(w, 0) \models \diamondsuit\phi$ (resp. $(w, n+1) \models \diamondsuit\!\!\!\!\leftarrow\phi$).

For a unary temporal logic formula ϕ, we write $L(\phi)$ the set of words over which ϕ is true and denote by **UTL** the class of languages which are $L(\phi)$ for some unary temporal logic formula ϕ.

Once again, the natural classes **FO₂** and **UTL** turn out to have nice algebraic characterizations (obtained in [35]) and one can in fact show:

Theorem 5.2. ([6, 35]) *For a language L, the following are equivalent:*
 (1) $M(L)$ *is in* **DA**;
 (2) L *is in* **UTL**;
 (3) L *is in* **FO₂**.

sketch. (1) $\leftrightarrow$ (3) is one of the main results of [35].

(2) $\leftrightarrow$ (3) is due to [6].

(1) $\leftrightarrow$ (2): we give a direct proof of this last equivalence. For one direction it is sufficient to show by induction that $*_{n,k}$ languages are in **UTL**. It is clear that all $*_{n,0}$ languages are in **UTL**. Next, let $A_1 = A_0 \cup \{a\}$ (with $a \notin A_0$) and let $L_0 \subseteq A_0^*$, $L_1 \subseteq A_1^*$ be $*_{|A_0|,k}$ and $*_{|A_1|,k}$ languages respectively. By induction, we have **UTL** formulas ϕ, ψ defining L_0 and L_1 respectively. It is easy to verify that the language $L_0 a L_1$ will be defined by the formula

$$\diamondsuit a \wedge \phi' \wedge \psi'' \wedge (\neg \diamondsuit\!\!\!\!\leftarrow a)$$

Here, ϕ' is obtained from ϕ, by replacing $\diamondsuit$ operators by $\diamondsuit\!\!\!\!\leftarrow$ and vice-versa and by replacing each subformula $\diamondsuit \rho$ by $\diamondsuit(\rho \wedge \neg \diamondsuit\!\!\!\!\leftarrow a)$ and similarly for subformulas

$\Diamond \rho$. Similarly, ψ'' is obtained from ψ by replacing each subformula $\Diamond \rho$ by $\Diamond(\rho \wedge \Diamond a)$.

For the converse, let u, v, x, y, z be strings in A^* and define $s_n = u(xyz)^n y(xyz)^n v$ and $t_n = u(xyz)^{2n} v$. We want to prove by induction on n that if ϕ is a temporal logic formula of operator depth at most $d \leq n$, then $(s_n, i) \models \phi$ if and only if $(t_n, f_d(i)) \models \phi$ where $f_d(i)$ maps a position in the $u(xyz)^d$ prefix or $(xyz)^d v$ suffix of s_n to the same position in t_n. Positions in the middle $(xyz)^{n-d} y(xyz)^{n-d}$ are mapped to the corresponding position in the dth copy of xyz from the left in t_n if they lie within $(xyz)^{n-d}$ and mapped to the dth copy of xyz in t_n *from the right* otherwise.

The claim is trivially true for $n = 0, 1$. Next, suppose $(s_{n+1}, i) \models \Diamond \phi$ where ϕ has operator depth $d \leq n$. By definition, there exists $j > i$ such that $(s_{n+1}, j) \models \phi$ and so, by induction we have $(t_{n+1}, f_d(j)) \models \phi$. Clearly, $f_d(j) > f_{d+1}(i)$ so we do have $(t_{n+1}, f_{d+1}(i)) \models \Diamond \phi$. Conversely, suppose $(t_{n+1}, f_{d+1}(i)) \models \Diamond \phi$. There is $k > f_{d+1}(i)$ with $(t_{n+1}, k) \models \phi$. For any j such that $f_d(j) = k$, we have by induction $(s_{n+1}, j) \models \phi$ and one can choose such a j to be greater than i so $(s_{n+1}, i) \models \Diamond \phi$. This proves our claim and shows that the syntactic monoid of $L(\phi)$ satisfies $(xyz)^\omega y(xyz)^\omega = (xyz)^\omega$, i.e. is in **DA**. □

It is possible to extend further the expressive power of **FO** by allowing the use of so-called modular quantifiers (see e.g. [30] for relevant definitions). One denotes by $(\mathbf{FO} + \mathbf{MOD})_\mathbf{k}[<]$ the class of languages definable by such generalized first-order sentences using at most k variables. These were studied by Straubing and Thérien in [31] who showed $(\mathbf{FO} + \mathbf{MOD})_3[<] = \mathbf{FO} + \mathbf{MOD}[<]$ and, most notably, that $(\mathbf{FO} + \mathbf{MOD})_2[<] = \mathbf{DA} * \mathbf{G_{sol}}$, where $\mathbf{DA} * \mathbf{G_{sol}}$ is the class of languages whose syntactic monoid divides the wreath product of a monoid in **DA** and a solvable group. Another generalization of Theorem 5.2 has recently yielded algebraic characterizations and decidability results for the levels of the so-called Until/Since hierarchy in temporal logic [36].

To conclude this section, we would like to point out that the above results imply the existence of polynomial space algorithms for deciding whether a regular language L (input, say, as a DFA) is in $\mathbf{\Sigma_2} \cap \mathbf{\Pi_2}[<]$, $\mathbf{FO_2}[<]$ or **UTL**.

6 Monoids as Machines and Complexity Problems

The notion of recognition of a language by a finite monoid can be extended in various ways to provide a meaningful algebraic point of view on computation and complexity classes. In these contexts, the monoid is viewed as a machine using some relatively simple mechanism to process input. Quite naturally, the power of this machine is related to the algebraic properties of the underlying monoid. This approach has been particularly successful in providing algebraic characterizations of classes of languages recognized by small depth boolean circuits via the notion of programs over monoids (see Subsection 6.2 for definitions and basic results) and of subclasses of PSPACE via the leaf-language formalism [12].

This, in turn, has given rise to questions concerning the computational complexity of certain problems whose hardness is parametrized by an underlying monoid. Typically, these are studied first in the restricted cases of groups and aperiodic monoids. In the latter case, **DA** almost always turns up at the center of these investigations and we will survey such examples in this section.

6.1 2-way Partially Ordered DFA's

A finite automaton (FA) is said to be partially ordered (or p.o.) if all its strongly connected components are trivial, i.e. if there is a partial order $\prec$ on the states of the automaton such that whenever there exists a transition from state s to state s', we have $s \prec s'$. It is a classical result that a language is recognized by a p.o. deterministic FA if and only if its syntactic monoid is $\mathcal{R}$-trivial. It is also quite simple to show, by using the result of Arfi already cited in the proof of Theorem 5.1, that a language can be recognized by a p.o. non-deterministic FA if and only if it belongs to $\Sigma_2[<]$.

Two-way finite automata (2-DFA's) are known to have the same computational power as DFA's and have thus received limited attention since their introduction [13]. They can, however, provide more concise descriptions of certain languages and can be studied as a way to compute translations on strings [20].

Formally, a 2-DFA is a tuple[a] $T = (S, A, \delta, l_0, F)$ where S, the set of states, is the disjoint union of two sets L (the states "reached" from the left) and R (the one "reached" from the right), A is the input alphabet, $l_0 \in L$ is the initial state, $F \subseteq S$ is the set of accepting states, $\delta : (S \times A) \cup (L \times \{\triangleleft\}) \cup (R \times \{\triangleright\}) \to S$ is a total transition function and $\triangleright, \triangleleft \notin A$ are the left- and right markers respectively. If T is in state s reading the symbol a, it enters state $\delta(s, a)$ and moves its head to the right if $\delta(s, a) \in L$ and to the left if $\delta(s, a) \in R$. On input $w = w_1 w_2 \ldots w_n$, the machine is initially in state l_0 scanning the symbol w_1 within the string $\triangleright w_1 w_2 \ldots w_n \triangleleft$. For technical reasons, we require that a 2-DFA halts for any given input by "falling off" its input tape, i.e. by encountering a transition $\delta(r, \triangleright) \in R$ or a transition $\delta(l, \triangleleft) \in L$.

---

[a]We present here the definition given in [20].

Recently, it was proved that a language can be recognized by a partially ordered 2-DFA if and only if its syntactic monoid is in **DA**. In order to establish this result, [29] introduces the notion of *turtle programs*, of which we give here a brief description. A turtle program P over an alphabet A is a sequence of turtle instructions, i.e. pairs (d,t) where $d \in \{\rightarrow, \leftarrow\}$ is a direction and $t \in A$ is a target symbol. On an input $w = w_1 w_2 \ldots w_n$, a turtle program whose first instruction is a pair $(\rightarrow, a)$ (resp. $(\leftarrow, a)$) moves a read-only head (the turtle) from w_1 right (resp. from w_n left) until the head encounters a position w_i holding an a. If no such position exists, we say that the instruction has failed. Each subsequent instruction similarly moves the turtle right or left from its current position until it encounters its target symbol or fails by reaching either end of the input. We define as $L(P)$, the language recognized by the turtle program P, as the set of words w such that the sequence of all instructions of the program succeed on input w.

Theorem 6.1. ([29]) *Let K be a language. The following are equivalent:*

(a) $M(K)$ *is in* **DA**;

(b) K *is the Boolean combination of turtle languages, i.e. languages recognized by a turtle program;*

(c) K *can be recognized by a p.o. 2-DFA.*

sketch. (a) $\rightarrow$ (b) follows from the $\sim_{n,k}$ characterization of **DA**.

For (b) $\rightarrow$ (c), note first that a turtle language can trivially be recognized by a p.o. 2-DFA and that the class of languages recognizable by p.o. 2-DFA's is closed under complementation. It is a simple exercise to show that this class is also closed under union, which completes the argument.

(c) $\rightarrow$ (a) is the hardest implication. Let T be 2-DFA over the alphabet A, with states $S = L \dot\cup R$ partially ordered by $\prec$ and recognizing a language K. Without loss of generality, we will assume that T always halts by exiting on the right. We will show how to obtain from T a p.o. NFA (non-deterministic finite automaton) N for K. Together with Theorem 5.1, this will yield our result since K's complement is also recognized by a p.o. 2-DFA. States of our NFA N are odd length sequences of T states $(s_0, s_1, \ldots, s_{2k})$ with $s_i \in L$ for even i and $s_i \in R$ for odd i and such that $s_i \prec s_j$ for $i < j$. This last restriction shows that N has a bounded number of states. We think of these "crossing sequences" (following the terminology of [13]) as the sequence of states in which T might possibly cross the border between two input tape cells and add a transition in N from state $p = (l_0, r_1, \ldots, l_i)$ to state $q = (l'_0, r'_1, \ldots, l'_j)$ while reading a if and only if they are locally consistent, i.e. if there is some word $w = uav$ such that execution of T on w yields crossing sequences p and q

on the left and right borders of the cell containing the a. Similarly, N's initial states are any sequences $p = (l_0, r_1, \ldots, l_k)$ such that l_0 is T's initial state and such that $\delta(r_i, \triangleright) = l_{i+1}$ and the final states are sequences $p = (l_0, r_1, \ldots, l_k)$ such that $\delta(l_i, \triangleleft) = r_{i+1}$ for $i < k$ and $\delta(l_k, \triangleleft) = l$ for some accepting state l of T. One can easily verify that N recognizes K.

We define a relation $\sqsubset$ on the states of N as follows: let $p = (l_0, r_1, \ldots, l_t) \sqsubset q = (l'_0, r'_1, \ldots, l'_u)$ if and only if there exist indices $p_1 < \ldots < p_k = t$ and $0 = q_1 < \ldots < q_k$ such that for all $1 \leq i \leq k$ we have $l_{p_i} \prec l'_{q_i} \prec r'_{q_{i+1}-1} \prec r_{p_i+1}$. In other words $p \sqsubset q$ if and only if the two chains $l_0 \prec r_1 \prec \ldots \prec l_t$ and $l'_0 \prec r'_1 \prec \ldots \prec l'_u$ can be interleaved to form a new chain starting in l_0, ending in l'_u and such that each *new* pair in the chain is of the form $l_i \prec l'_j$ or $r'_i \prec r_j$. It can be shown that $\sqsubset$ is a partial-order and that N has a transition from p to q only if $p \sqsubset q$. □

Because we know that languages recognized by **DA** lie in $\Sigma_2 \cap \Pi_2[<]$, we can conclude that $K \subseteq A^*$ is recognized by a p.o. 2-DFA if and only if both K and its complement $\overline{K}$ can be recognized by p.o. NFA's. This somewhat innocent looking automata-theoretic fact has, to the best of our knowledge, no proof simpler than the one implicitly presented in this survey.

6.2 Equation and Program Satisfiability

We briefly mentioned earlier the notion of programs over monoids. Formally, given an input alphabet A, an n-input program of length s over a monoid M is a sequence of instructions

$$(i_1, f_1), (i_2, f_2), \ldots, (i_s, f_s)$$

with $i_j \in \{1, \ldots, n\}$ and $f_j : A \to M$, together with an accepting subset $F \subseteq M$. The program accepts the input $x_1 x_2 \ldots x_n$ iff the sequence $f_1(x_{i_1}) \ldots f_l(x_{i_s})$ evaluates to an element of F.

We say that $L \subseteq A^*$ is P-recognized by M if there is a sequence $(\varphi_0, \varphi_1, \ldots)$ where φ_n is an n-input program over M, of length $l(n)$ with $l(n)$ polynomial in n, such that the program φ_n accepts $L \cap A^n$. This formalism gives rise to algebraic characterizations of many well-known boolean circuit classes (see [4] and [19] for a thorough survey). In particular, a language L belongs to the class NC^1 (resp. AC^0, CC^0, ACC^0) if and only if it is recognized by a program over a finite monoid (resp. aperiodic monoid, solvable monoid, solvable group).

For any fixed M we define the two following similar problems:

- EQUATION SATISFIABILITY over M (EQNSAT$_M$): we are given an equation over a monoid M of the form
$$c_0 X_{i_1} c_1 \ldots c_{n-1} X_{i_n} c_n = m$$
where the c_i's and m are constants in M and the X_i's are variables which might be repeated. The problem is to determine whether one can assign values in M to the variables so that the equation is satisfied.

- PROGRAM SATISFIABILITY over M (PROGSAT$_M$): we are given an n-input program ϕ of length l over M and a target set $F \subseteq M$. The problem is to determine whether there exists $x \in \{0,1\}^*$ such that $\phi(x) \in F$.

Both problems trivially lie in NP. Also, it is a simple exercise to show that EQNSAT$_M$ reduces to PROGSAT$_M$ for any M.

The hardness of these problems is of course highly dependent of the underlying monoid. In the case where M is aperiodic, the following dichotomy results involving **DA** can be obtained [3]:

Theorem 6.2.
(a) EQNSAT$_M$ and PROGSAT$_M$ lie in P for any M in **DA**.
(b) If M is not in **DA** then PROGSAT$_M$ is NP-complete.
(c) If M is not in **DA**, there exists a monoid N in the variety $\langle M \rangle$ generated by M such that EQNSAT$_M$ is NP-complete.

Proof. For (a), it is sufficient to establish the upper bound for PROGSAT$_M$. Recall from Section 3 that the set $M_F = \{w : eval_M(w) \in F\}$ is the disjoint union of unambiguous concatenations $M_0^* a_1 M_1^* \ldots a_k M_k^*$. It is thus sufficient to determine whether there is some $x \in \{0,1\}^*$ with $\phi(x) \in M_0^* a_1 M_1^* \ldots a_k M_k^*$.

This can only happen if there are k instructions $i_1, \ldots, i_k$ in the program that can, on some input, output the bookmarks $a_1, \ldots, a_k$. So, it suffices to consider the at most $\binom{l}{k}$ k-tuples of instructions possibly responsible for the presence of the bookmarks and, for each of them, check in linear time whether there is an x such that $\phi(x)$ is of the form $M_0^* a_1 M_1^* \ldots a_k M_k^*$.

For (b) and (c), it is sufficient to show that PROGSAT$_U$ and EQNSAT$_{BA_2}$ are both NP-complete. These lower bounds follow from fairly simple reductions from 3-SAT and 1-3SAT respectively. □

6.3 Communication Complexity

Over the past twenty years, communication complexity has become an essential and versatile tool in theoretical computer science (see [18] for a complete introduction). The communication complexity of monoids has been studied [26, 33] to understand the power (and limitations) of communication complexity based lower bounds on small depth circuits (e.g. [8, 9, 10, 11] among many others) and to provide an algebraic point of view on some of these circuit and communication complexity bounds.

We consider the following communication game involving k parties: the players $P_1, \ldots, P_k$ want to collaborate to evaluate, in a given monoid M, a string of n elements $x_1, \ldots, x_n$. Each player P_j, has access to all the x_i except the ones such that $i \equiv j \pmod{k}$. One can imagine, for instance that P_j has these x_i's written on his forehead, available to all players but himself. The players are allowed to communicate by broadcasting bits according to some fixed protocol. The k-party communication complexity of M, denoted $C^{(k)}(M)$, is the least number of bits that they need to communicate in the worst case such that all players know the value $eval_M(x_1, \ldots, x_n)$. (Note that $C^{(k)}(M)$ is in fact a function from $\mathbb{N}$ to $\mathbb{N}$ as the players may have a specific protocol for each input length.)

To argue for the relevance of this definition, we note that the communication complexity of a regular language[b] L is $\Theta(C^{(k)}(M(L)))$ and that for any $f : \mathbb{N} \to \mathbb{N}$, the class of monoids M such that $C^{(k)}(M) = O(f)$ forms a variety (see [26]).

For groups, the question is completely solved: a group G has k-party communication complexity $O(1)$ if and only if it is nilpotent of class $(k-1)$ and has complexity $\Theta(n)$ otherwise. If we turn to aperiodics, **DA** appears as the key variety.

Theorem 6.3. ([26, 25]) *Let M be an aperiodic monoid.*

*(1) In the 2-party case, $C^{(2)}(M) = O(\log n)$ if and only if M belongs to **DA** and $C^{(2)}(M) = \Theta(n)$ otherwise.*

(2) If $M \in \mathbf{V}_{\approx_{k,t}}$ then $C^{(k+1)}(M) = O(1)$.

*(3) In fact, there exists a k such that $C^{(k)}(M) = O(1)$ only if M belongs to **DA**.*

Proof. Both of the upper bounds for **DA** are easy to derive from the parametrization of **DA** induced by the congruence $\approx_{k,t}$ defined in Section 3. It is sufficient to establish the upper bound for $A^*/\approx_{k,t}$ for any alphabet A and any k, t. Let us first describe the constant cost $(k+1)$-party protocol for $A^*/\approx_{k,t}$ which we will build by induction on k. The case $k = 0$ is obvious

---

[b]This can be defined in a way similar to that of a monoid, with the x_i's being words in A^*

since $\approx_{0,t}$ is trivial. Suppose $k \geq 1$ and take a set of representatives $[u_i]$ of $A^*/\approx_{k,t}$. For each u_i, the players will check whether $w \approx_{k,t} u_i$. First, it is clear that by exchanging $O(t) = O(1)$ bits, the players can insure that $\alpha_t(w) = \alpha_t(u_i)$. If $k = 1$, we are done because there is no recursive condition to check. If $k \geq 2$, suppose $u_i = v_0 a v_1$ with $|v_0|_a \leq t$. There exists a factorization $w = w_0 a w_1$, with $|v_0|_a = |w_0|_a$ but in order to handle the recursion, the players need to identify the location of the jth a (where $|v_0|_a = j - 1$).

To achieve this, each player sends a list of identities of the players they think *ignore* the first j a's. This requires only $O(k \log t) = O(1)$ communication. Of course, only the player who has the first a on his forehead will incorrectly identify the first member in that list (say this player is Player l), while all others will agree on designating him as the one ignoring the first a. We are assuming that there are at least 3 players, so Player l can indeed be identified. We can correct his list by adding an l in the first position and shifting the rest of his list right. Now, the second positions in the lists of all but one of the players agree and we can repeat this procedure for j rounds. In the end, all players know which player ignores the jth a and all except that player know the location of that a. Hence, there are k players available to check whether $v_0 \approx_{k-1,t} w_0$ and $v_1 \approx_{k-1,t} w_1$. By induction this is doable with $O(1)$ communication. Left-right symmetry completes the protocol.

For the 2-party $O(\log n)$ upper bound, it suffices to note that by exchanging the positions of the first t a's that they see in w, *both* players can correctly identify the location of the jth a of w for any $j \leq t$ and apply the induction step. $\square$

Naturally, we can ask whether the k-player protocol just described is optimal. We conjecture in fact:

Conjecture 6.4. *If M is aperiodic, we have $M \in \mathbf{V}_{\approx_{k,t}}$ for some t if and only if $C^{(k+1)}(M) = O(1)$.*

For $k = 0$, this is obviously true since only the trivial monoid has 1-party communication complexity $O(1)$ (since there is no access to the input). For $k = 1$, one can easily prove that $\bigcup_t \mathbf{V}_{\approx_{1,t}} = \mathbf{A}_{\text{com}}$ and it is shown in [26] that this variety also corresponds exactly to the aperiodics having constant 2-party communication complexity.

Only partial results are known when $k \geq 2$ [25]. We should stress that this conjecture would imply that the regular language $A^* a_1 A^* \ldots A^* a_k A^*$ can not be recognized by k players in constant communication. Of course, any set of $(k-1)$ input positions is seen by at least one player in the k-party model and the former remark would strengthen the intuition according to which this is the only real building block of the k-party model's computational power. Already, the existing results on the communication complexity of groups can be seen as stating that a group has high k-party complexity if and only if its computing power is not limited to "counting" subwords of length $(k-1)$ (see [26] for a formal presentation).

6.4 Learning

We have seen in the communication complexity setting how an algebraic approach can help in outlining some fundamentally hard aspects of a computational problem. In [7], an algebraic point of view on computational learning theory is developed.

The model considered is Angluin's query-based model of exact-learning. The setup is as follows. We fix a monoid M and consider functions $f : M^n \to M$, representable by an expression over M, i.e. $f(X_1, \ldots, X_n) = c_0 X_{i_1} c_1 \ldots c_{s-1} X_{i_s} c_s$. The goal of a *Learner* (i.e. of a learning algorithm) is to identify such an unknown *target function* f by interacting, according to some protocol, with a *Teacher* who knows f. We assume that the Learner knows n and s that he is allowed to make two types of queries: Evaluation and Equivalence. For an *Evaluation* query, the Learner provides $w \in M^n$ and the Teacher simply returns $f(w)$. For an *Equivalence* query, the Learner provides an hypothesis h (in our case, we consider that h is also given as an expression over M) and the Teacher either returns YES if $h(w) = f(w)$ for all $w \in M^n$ or returns a counterexample $(w, f(w))$ where $w \in M^n$ and $f(w) \neq h(w)$.

The two measures of complexity we are concerned with are the number of queries made by the Learner and the computation time he requires. We say that expressions over a monoid M are learnable with $q_1(n, s)$ Evaluation queries, $q_2(n, s)$ Equivalence queries and $t(n, s)$ time if there exists a learning algorithm that can produce, within these bounds, an M-expression E such that $E(w) = f(w)$ for all $w \in M^n$.

Theorem 6.5. *Expressions over M in* **DA** *are learnable from a polynomial number of Evaluation queries and unbounded computation time.*

Proof. As in the PROGSAT$_M$ algorithm, we will use the fact that when M is in **DA**, the subsets of M^* defined as $M_m = \{w : eval_M(w) = m\}$ can be expressed as the disjoint union of unambiguous concatenations $M_0^* a_1 M_1^* \ldots a_k M_k^*$. Suppose we have an expression $E(X_1, \ldots, X_n)$ over some M in **DA** such that for each m in M, the unambiguous concatenations needed to express M_m have at most k bookmarks.

The Learner asks the Teacher to evaluate $E(w)$ for the $\binom{|M|^{2k}}{n2k}$ words of M^n having at most $2k$ non-identity components. By exhaustive search, one can construct an expression $F(X_1, \ldots, X_n)$ agreeing with E on all these inputs.

Consider now any $x \in M^n$. We claim that $E(x) = F(x)$. Indeed, construct $z \in M^n$ by replacing in x each letter not occurring as a bookmark in either $E(x)$ or $F(x)$ (seen as words in M^*). By the definition of F, we have $E(z) = F(z)$, but also $E(z) = E(x)$ and $F(x) = F(z)$ since the bookmarks of $E(x)$ and $F(x)$ still appear in $E(z)$ and $F(z)$. Thus, E and F are equivalent expressions. □

In the special cases where M is aperiodic and commutative, $\mathcal{R}$-trivial and idempotent or $\mathcal{L}$-trivial and idempotent, it is possible to actually learn the expressions in polynomial time, but it is not known for a general $M \in \mathbf{DA}$ whether an expression F as above can be obtained more efficiently than by exhaustive search.

In contrast to Theorem 6.5, it is also established in [7] that expressions over BA_2 and U, the minimal aperiodics outside $\mathbf{DA}$, are not learnable with a subexponential number of Evaluation queries.

If one considers Equivalence queries alone, then it can be shown that expressions over a commutative idempotent monoid can be learned using a linear number of Equivalence queries alone, while expressions over any non-idempotent M can not be learned with a sub-exponential number of queries.

6.5 Membership

Consider the problem MEMB of determining whether a function g from a set X to itself can be written as the composition of some functions $f_1, \ldots, f_t$, i.e. of deciding whether g is a member of the transformation monoid $\langle f_1, \ldots, f_t \rangle$ generated by the f_i's.

MEMB is known to be complete for polynomial space in general [16]. A series of technically involved papers (culminating in [2]) was necessary to show that when the f_i's are permutations on X, i.e. when the monoid $\langle f_1, \ldots, f_m \rangle$ is a group, MEMB lies in the class NC (polylogarithmic parallel time using a polynomial number of processors).

For any variety $\mathbf{V}$, we denote by $MEMB(\mathbf{V})$ the restriction of MEMB to instances where the monoid $\langle f_1, \ldots, f_t \rangle$ lies in $\mathbf{V}$. The proof of [16] shows in fact that $MEMB(\mathbf{A})$ is already PSPACE-complete. The complexity of MEMB for subvarieties of $\mathbf{A}$ was investigated in [5] who proved:

- $MEMB(\mathbf{R_1} \vee \mathbf{L_1})$ lies in P.

- $MEMB(\mathbf{A_1})$ lies in NP.

- $MEMB(\mathbf{R})$ is NP-complete.

- $MEMB(\mathbf{V})$ is PSPACE-complete for any aperiodic variety $\mathbf{V}$ not contained in $\mathbf{DA}$ and NP-hard for any aperiodic variety $\mathbf{V}$ not contained in $\mathbf{R_1} \vee \mathbf{L_1}$.

It is further conjectured that $MEMB(\mathbf{DA})$ in fact lies in NP. The missing upper bound would follow if one could show the following refinement of Schützenberger's theorem: a language recognizable by an n-state automaton whose transformation monoid lies in $\mathbf{DA}$ can be expressed as the disjoint union of concatenations $A_0^* a_1 \ldots a_k A_k^*$ where k is bounded by a polynomial in n.

7 Summary

We have shown that languages whose syntactic monoid lies in $\mathbf{DA}$ have useful combinatorial, logical, and automata-theoretic characterizations. For a language $L \subseteq A^*$, the following equivalences are known:

- $M(L)$ is in $\mathbf{DA}$;

- L is the union of $\sim_{n,k}$ classes for $n = |A|$ and some k;

- L is the union of $\approx_{k,t}$ classes for some k, t;

- L is the disjoint union of unambiguous concatenations $A_0^* a_1 A_1^* \ldots a_k A_k^*$ with $A_i \subseteq A$, $a_i \in A$;

- L is both $\mathbf{\Sigma_2}[<]$ and $\mathbf{\Pi_2}[<]$ definable;

- L is in $\mathbf{FO_2}[<]$, i.e. is definable by a two variable first-order formula;

- L is in $\mathbf{UTL}$, i.e. definable by a unary temporal logic formula;

- L can be recognized by a partially-ordered 2-way DFA.

We have also shown how these characterizations can be used to answer questions motivated by computation and complexity theory: checking the satisfiability of an equation or a program, evaluating a product using limited communication, learning expressions can always be done efficiently when the underlying monoid lies in $\mathbf{DA}$ whereas these problems are, typically, provably much harder when the monoid is aperiodic but not in $\mathbf{DA}$.

Moreover, the variety **DA** has a number of useful algebraic characterizations. In particular, **DA** is the smallest variety **V** containing $\mathbf{J_1}$ satisfying $\mathbf{V} \Box \mathbf{J_1} = \mathbf{V}$ and the smallest variety **V** containing $\mathbf{J_1}$ satisfying $l_1 \Box \mathbf{V} = \mathbf{V}$.

References

[1] M. Arfi. Opérations polynomiales et hiérarchies de concaténation. *Theoretical Computer Science*, 91:71–84, 1991.

[2] L. Babai, E. Luks, and A. Seress. Permutation groups in NC. In *Proc. ACM STOC*, pages 409–420, 1987.

[3] D. A. Mix Barrington, P. McKenzie, C. Moore, P. Tesson, and D. Thérien. Equation satisfiability and program satisfiability for finite monoids. In *Proc. MFCS'00*, pages 172–181, 2000.

[4] D. A. Mix Barrington and D. Thérien. Finite monoids and the fine structure of NC^1. *Journal of the ACM*, 35(4):941–952, 1988.

[5] M. Beaudry, P. McKenzie, and D. Thérien. The membership problem in aperiodic transformation monoids. *Journal of the ACM*, 39(3):599–616, 1992.

[6] K. Etessami, M. Vardi, and T. Wilke. First-order logic with two variables and unary temporal logic. In *Proc. IEEE LICS*, pages 228–235, 1997.

[7] R. Gavaldà, P. Tesson, and D. Thérien. Learning expressions and programs over monoids. Submitted to Information and Computation. Extended abstract appears in Proc. STACS 2001. Available from www.lsi.upc.es/~gavalda/papers.html, 2001.

[8] V. Grolmusz. Separating the communication complexities of MOD m and MOD p circuits. In *Proc. 33rd IEEE FOCS*, pages 278–287, 1992.

[9] V. Grolmusz. A weight-size trade-off for circuits and MOD m gates. In *Proc. 26th ACM STOC*, pages 68–74, 1994.

[10] V. Grolmusz. Circuits and multi-party protocols. *Computational Complexity*, 7(1):1–18, 1998.

[11] J. Håstad and M. Goldmann. On the power of small-depth threshold circuits. In *Proc. 31st IEEE FOCS*, pages 610–618, 1990.

[12] U. Hertrampf, C. Lautemann, T. Schwentick, H. Vollmer, and K. Wagner. On the power of polynomial time bit-reductions. In *Conf. on Structure in Complexity Theory*, pages 200–207, 1993.

[13] J. E. Hopcroft and J. D. Ullman. *Introduction to Automata Theory, Languages and Computation*. Addison-Wesley, 1979.

[14] N. Immerman and D. Kozen, Definability with bounded number of bound variables. *Information and Computation*, 83(2):121–139, 1989.

[15] J. A. W. Kamp. *Tense Logic and the Theory of Linear Order*. PhD

thesis, University of California, Berkeley, 1968.
[16] D. Kozen. Lower bounds for natural proof systems. In *Proc. IEEE FOCS*, pages 254–266, 1977.
[17] K. Krohn and J. L. Rhodes, *Algebraic Theory of Machines, I Principles of Finite Semigroups and Machines, Transactions of the American Mathematical Society*, 116:450–464, 1965.
[18] E. Kushilevitz and N. Nisan. *Communication Complexity.* Cambridge University Press, 1997.
[19] P. McKenzie, P. Péladeau, and D. Thérien. NC^1: The automata theoretic viewpoint. *Computational Complexity*, 1:330–359, 1991.
[20] P. McKenzie, T. Schwentick, D. Thérien, and H. Vollmer. The many faces of a translation. In *Proc. ICALP'00, LNCS 1853*, pages 890–901, 2000.
[21] R. McNaughton and S. Papert. *Counter-Free Automata.* MIT Press, Cambridge, Mass., 1971.
[22] J.-É. Pin. *Varieties of formal languages.* North Oxford Academic Publishers Ltd, London, 1986.
[23] J.-É. Pin, H. Straubing, and D. Thérien. Locally trivial categories and unambiguous concatenation. *J. Pure Applied Algebra*, 52:297–311, 1988.
[24] J.-É. Pin and P. Weil. Polynomial closure and unambiguous product. *Theory Comput. Systems*, 30:383–422, 1997.
[25] P. Pudlák, K. Rheinhardt, P. Tesson, and D. Thérien. On the multiparty communication complexity of regular languages. Draft available at www.cs.mcgill.ca/~ptesso, 2002.
[26] J.-F. Raymond, P. Tesson, and D. Thérien. An algebraic approach to communication complexity. *Proc. ICALP'98, LNCS 1443*, pages 29–40, 1998.
[27] M. P. Schützenberger. On finite monoids having only trivial subgroups. *Information and Control*, 8(2):190–194, April 1965.
[28] M. P. Schützenberger. Sur le produit de concaténation non ambigu. *Semigroup Forum*, 13:47–75, 1976.
[29] T. Schwentick, D. Thérien, and H. Vollmer. Partially-ordered two-way automata: a new characterization of **DA**. In *Proc. DLT'01*, 37–56, 2001.
[30] H. Straubing. *Finite Automata, Formal Logic and Circuit Complexity.* Boston: Birkhäuser, 1994.
[31] H. Straubing and D. Thérien. Regular languages defined by generalized first-order formulas with a bounded number of bound variables. In *Proc. STACS'01*, pages 551–562, 2001.
[32] H. Straubing and D. Thérien. Weakly iterated block products of finite monoids. In *Proc. of the 5th Latin American Theoretical Informatics*

Conference (LATIN'02), 2002.
[33] P. Tesson. An algebraic approach to communication complexity. Master's thesis, School of Computer Science, McGill University, 1999.
[34] D. Thérien. Two-sided wreath product of categories. *J. Pure and Applied Algebra*, 74:307–315, 1991.
[35] D. Thérien and T. Wilke. Over words, two variables are as powerful as one quantifier alternation. In *Proc. ACM STOC'98*, pages 256–263, 1998.
[36] D. Thérien and T. Wilke. Nesting until and since in linear temporal logic. Submitted to STACS'02, 2002.

DECIDABILITY PROBLEMS IN FINITE SEMIGROUPS

PETER G. TROTTER*

School of Mathematics and Physics, The University of Tasmania, Hobart, Tasmania, Australia
E-mail: Peter.Trotter@utas.edu.au

This article begins with a brief description of the beginnings of the study of decision problems. There follows a discussion of important decision problems for semigroups. Attention is then restricted to finite semigroups and in particular to the membership problem for classes of finite semigroups.

1 Decision problems: decidability and undecidability

Suppose we are given a collection of instances, a statement about the instances and we wish to know for each instance whether the statement is true or false. A *decision procedure* is an algorithm that provides the answer for each instance. In other words, it is a procedure that can be described in advance and will provide the answer for each instance in a finite number of steps. The problem of discovering the algorithm is a *decision problem*. The problem is said to be *solvable*, or *decidable*, if such an algorithm exists; otherwise it is *undecidable*. It is *effectively decidable* if an algorithm is actually found. The following decision problem is effectively decidable by Euclid's Algorithm.

INSTANCE: $p, q \in \mathbb{Z}$
STATEMENT: p and q are relatively prime.

The above notions, especially of decision procedures were formalised by the 1930's. At the 1900 International Congress of Mathematicians, D.Hilbert proposed 23 problems that profoundly influenced the direction of mathematical research in the succeeding century. The 10th problem was to devise what amounted to a decision procedure for determining whether or not an arbitrary Diophantine equation is solvable in integers. It was finally shown in 1970 by Y.Matijasevich that there is no such procedure; that is, the associated decision

---

*THE AUTHOR WOULD LIKE TO ACKNOWLEDGE THE FINANCIAL SUPPORT OF FUNDAÇÃO CALOUSTE GULBENKIAN (FCG), FUNDAÇÃO PARA A CIÊNCIA E A TECNOLOGIA (FCT), FACULDADE DE CIÊNCIAS DA UNIVERSIDADE DE LISBOA (FCUL) AND REITORIA DA UNIVERSIDADE DO PORTO.

problem is undecidable. By 1917 Hilbert had expanded his 10th problem to the problem of decidability, in a finite number of predetermined steps, of the truth or falsity of a given statement in a given mathematical theory. In 1937, A.Turing showed that this problem is not solvable; his procedure is briefly outlined below.

Let us be more specific with the notions of instance, statement and algorithm in our definition of a decision procedure. A theory in which a decision procedure is to be applied should involve just a finite number of symbols for operations, logical connectives, punctuation, variables and so on; formulas should be finite in length and algorithms should involve just a finite number of instructions. The theory can then be encoded by words from A^+ for some finite alphabet A. Turing modelled algorithms with his *Turing machines*.

A Turing machine T consists of a finite set Q of states (including an initial state q_0 and a terminal state q_t), a finite alphabet A and the symbol 1 (the identity element of A^*) and a partial function $\phi : (A \cup 1) \times Q \to (A \cup 1) \times Q \times \{L, R, S\}$. It has a tape made up of infinitely many tape squares with an identified central square. The consecutive letters from A that make up an input are inserted in the consecutive tape squares from the right of the centre, while all other tape squares contain 1. The machine starts in state q_0 reading the central square. Suppose at some step the machine is in state q and is reading the letter x in some tape square; then the components of $\phi(x, q)$ are respectively a letter that replaces x in its square, the next state of the machine and the next tape square that is to be read (L, R, S respectively denote one square to the left, one square to the right, or the same square). The machine stops when it achieves state q_t, or when it has insufficient instructions to continue; we say it *halts* when it achieves state q_t. It might not stop at all.

It is easy to imagine how such a machine models an algorithm and how truth or falsity of statements about instances can be tested. A Turing machine can also be regarded as a function $T : A^+ \to A^* \cup \infty$ with an input as above and the corresponding output being whatever is on the tape when the machine is in state q_t or, if the machine does not halt, the output is ∞.

The Halting Problem for Turing Machines over an alphabet A.

INSTANCE: A Turing machine T over A and an input I
STATEMENT: T halts for I.

There are only countably many Turing machines and inputs over the alphabet A. Turing assumed that the Halting Problem is decidable and consequently, for each Turing machine T and each input $x \in A^+$, the output $T(x) \in A^* \cup \infty$ could be computed. Turing then obtained a contradiction by an argument analogous to Cantor's diagonalisation argument that shows $\mathbb{R}$ is uncountable. Therefore *the Halting Problem is undecidable*.

In the mid 1930's, A.Church showed that if elementary Number Theory is consistent then it includes undecidable problems; from this he showed that the first order predicate calculus has undecidable problems. The discovery of the undecidability of the Halting Problem, or of any other decision problem, is significant because it potentially allows the undecidability of a given problem to be shown by reducing the given problem to the known undecidable problem. There are now many decision problems in mathematics that are known to be undecidable.

2 Some decision problems for semigroups

There are well known decision problems in algebraic theory and more specifically in semigroup theory that have been quite important to their respective development. We mention some of them below.

The Word Problem for a finitely presented semigroup $S = \langle A, R \rangle$.

INSTANCE: $u, v \in A^+$
STATEMENT: $u = v$ in S.

Of course any finitely generated free semigroup and any finite semigroup have decidable word problem, as do the members of various classes of semigroups. However there exist examples of semigroups and of groups with undecidable word problems (see Post [38] and Markov [32] for semigroups and Novikov [36] and Boone [14] for groups). It is usual to say that a class of semigroups has solvable word problem if each of its members has that property. So the class **S** of all finite semigroups has solvable word problem but the class $\mathcal{S}$ of all semigroups has not. There are useful (but non-algorithmic) criterion that ensure decidability of a word problem such as the following due to Boone and Higman [15].

Theorem 2.1. *A finitely presented semigroup has decidable word problem if and only if it is embeddable in a simple semigroup that in turn embeds in a finitely presented semigroup.*

The Uniform Word Problem for a class $\mathcal{C}$ of semigroups requires for its solution one algorithm that solves the word problem for each member of $\mathcal{C}$.

INSTANCE: $S = \langle A, R \rangle$ is a finite presentation, $S \in \mathcal{C}$, $u, v \in A^+$
STATEMENT: $u = v$ in S.

Obviously this is undecidable for the class $\mathcal{S}$ of all semigroups but as well, by Gurevich [21], the uniform word problem is undecidable for the class **S** of all finite semigroups. The same is true for finite groups by Slobodskoii [45]. But there are classes of semigroups with decidable uniform word problem, such as commutative semigroups (Mal'cev [30]).

Recall that a class of semigroups is a *variety* if and only if it is closed under subsemigroups, homomorphic images and direct products, or equivalently if and only if it is an equationally defined class. A *pseudovariety* of finite semigroups is a class of finite semigroups closed under subsemigroups, homomorphic images and finitary direct products.

Problem 1. *Which pseudovarieties of finite semigroups have solvable uniform word problem?*

The Finiteness Problem for semigroups is very relevant to this survey.

INSTANCE: S is a finitely presented semigroup
STATEMENT: S is finite.

Burnside's bounded problem of 1902 is a refinement of this in that it has the same statement but an instance is a finitely generated semigroup that satisfies the identity $x^{p+n} = x^n$ for some $p, n \in \mathbb{N}$. By Morse and Hedlund [34] there is an infinite 3-generator semigroup satisfying $x^2 = 0$, and an infinite 2-generator semigroup satisfying $x^3 = 0$. Also, Adian and Novikov [37] have shown that finitely generated relatively free groups of odd exponent > 665 are infinite.

Problem 2. *Is there a semigroup with undecidable finiteness problem? Is there a finitely presented infinite semigroup that is either periodic or nil?*

The 2nd of these is from Kharlampovich and Sapir [28].

A variety $\mathcal{V}$ has a *free object* $F_A(\mathcal{V})$ on any non-empty set A. For $u, v \in A^+$, u and v are equal in $F_A(\mathcal{V})$ under the natural projection $A^+ \to F_A(\mathcal{V})$ if and only if $u = v$ is an *identity* in $\mathcal{V}$. The countably generated free objects A^+ and $F_A(\mathcal{V})$ are amenable to algorithmic approaches.

It is unusual for a pseudovariety to include free objects and so pseudovarieties are often not equationally defined. Reiterman [39] introduced the notion of a pseudoidentity for a pseudovariety. These are defined in free profinite objects (a notion that has been much developed by Almeida and Weil [8]).

For the pseudovariety **S** of all finite semigroups the free profinite semigroup $\overline{\Omega}_A \mathbf{S}$ is the projective limit of all A-generated finite semigroups partially ordered by being related by surjective morphisms that uniquely extend the identity map on A. Since each finite semigroup is endowed with the discrete topology then as a projective limit, $\overline{\Omega}_A \mathbf{S}$ carries a topology under which it is compact and totally disconnected. We likewise define a relatively free profinite semigroup $\overline{\Omega}_A \mathbf{V}$ by restricting to semigroups in a subpseudovariety $\mathbf{V}$ of $\mathbf{S}$. The universal property for free objects generalises to free profinite objects; that is, for any A-generated $S \in \mathbf{V}$ there is a unique continuous morphism $\overline{\Omega}_A \mathbf{S} \to S$ that extends the identity map on A. For any $u, v \in \overline{\Omega}_A \mathbf{S}$, $u = v$ is a *pseudoidentity* in $\mathbf{V}$ if and only if u and v are equal under the natural projection $\overline{\Omega}_A \mathbf{S} \to \overline{\Omega}_A \mathbf{V}$.

Two types of elements in $\overline{\Omega}_A \mathbf{S}$ are especially significant. For $z \in \overline{\Omega}_A \mathbf{S}$, let $z^\omega \in \overline{\Omega}_A \mathbf{S}$ be such that the natural projection ϕ of $\overline{\Omega}_A \mathbf{S}$ onto an A-generated finite semigroup S takes z^ω to the (unique) idempotent power of $z\phi$. Let $z^{\omega-1}$ be such that $z^{\omega-1}\phi$ is the inverse of $(z^\omega z)\phi$ in S. We define $\Omega_A^\kappa \mathbf{S}$ to be the (countable) closure under the operation $\omega - 1$ of the A-generated subsemigroup of (the uncountable) $\overline{\Omega}_A \mathbf{S}$. The pseudoidentities of the bulk of the well studied pseudovarieties are generated by pseudoidentities that involve only terms from $\Omega_A^\kappa \mathbf{S}$.

Example 2.1. The pseudovariety of finite groups $\mathbf{G} = [\![x^\omega y = y = yx^\omega]\!]$. The pseudovariety of finite aperiodic semigroups $\mathbf{A} = [\![x^{\omega+1} = x^\omega]\!]$.

There are pseudovarieties not having a basis of pseudoidentities that are only from $\Omega_A^\kappa \mathbf{S}$ (see Weil [49]). This complication in some ways encourages the study of pseudovarieties that are subclasses of varieties. For a variety $\mathcal{V}$, denote by $\mathcal{V}_{fin}$ the pseudovariety of finite members of $\mathcal{V}$. Results for varieties can then sometimes be applied to pseudovarieties. An example is the next theorem, due to Sapir [43].

Theorem 2.2. *For any finitely based variety $\mathcal{V}$ of semigroups the uniform word problem is decidable in $\mathcal{V}_{fin}$ if and only if the nil semigroups of $\mathcal{V}$ are locally finite and either $\mathcal{V}$ is periodic or it excludes $\overleftarrow{P} \times P^1$, $\overleftarrow{P}^1 \times P$ or $P \cup \overleftarrow{P}$ where*

$$P = \left\{ \begin{pmatrix} 1 & 0 \\ 0 & 0 \end{pmatrix}, \begin{pmatrix} 0 & 1 \\ 0 & 0 \end{pmatrix}, \begin{pmatrix} 0 & 0 \\ 0 & 0 \end{pmatrix} \right\}, \overleftarrow{P} = \left\{ \begin{pmatrix} 1 & 0 \\ 0 & 0 \end{pmatrix}, \begin{pmatrix} 0 & 0 \\ 1 & 0 \end{pmatrix}, \begin{pmatrix} 0 & 0 \\ 0 & 0 \end{pmatrix} \right\}.$$

We note that, by [28], the condition on nil semigroups in this theorem is decidable, so the theorem provides an algorithm for solving the uniform word problem in $\mathcal{V}_{fin}$. This observation also has relevance to the finiteness problem and its connection with the corresponding problem for groups. The next result is by Sapir [42].

Theorem 2.3. *A finitely based periodic semigroup variety is locally finite if and only if its groups and its nil semigroups are locally finite.*

The Equational Problem for a variety $\mathcal{V}$ of semigroups with a finite basis of identities over a countable set A of variables.

    INSTANCE: $u, v \in A^+$
    STATEMENT: $u = v$ is an identity for $\mathcal{V}$.

This is solvable if and only if the word problem in $F_A(\mathcal{V})$ is solvable. The word problem for free objects in varieties of semigroups has attracted considerable attention over the last 30 years. For example it is solvable for varieties such as $\mathcal{S}$, and for the unary semigroup varieties of completely regular semigroups and of inverse semigroups along with various of their subvarieties. The word problem for free Burnside semigroups of index ≥ 3 was shown to be solvable early in the last decade (see de Luca and Varricchio [16] for index ≥ 5 and period 1, McCammond [33] for index ≥ 6, do Lago [17] for index ≥ 4 and Guba [20] for index ≥ 3). By Murskii [35] there is a variety of semigroups that has undecidable equational problem.

At first sight the equational problem seems of no relevance to pseudovarieties because a pseudoidentity relates terms from the uncountable free profinite semigroups. However the problem becomes sensible if we restrict our attention to pseudoidentities that are also identities. Let **The Weak** (or **Weak-κ**) **Equational Problem** for a pseudovariety **V** with a finite basis of pseudoidentities over a countable set of variables A be given as

    INSTANCE: $u, v \in A^+$ (or $u, v \in \Omega_A^\kappa \mathbf{S}$ respectively)
    STATEMENT: $u = v$ is a pseudoidentity for **V**.

For any variety $\mathcal{V}$ the equational problem and the weak equational problem coincide for $\mathcal{V}_{fin}$. By Almeida's [2] description of $\overline{\Omega}_A \mathbf{J}$, where **J** is the pseudovariety of $\mathcal{J}$-trivial finite semigroups, there is a natural projection from $\Omega_A^\kappa \mathbf{S}$ onto $\overline{\Omega}_A \mathbf{J}$ and the weak-κ equational problem is solvable for **J**. By Albert, Baldinger and Rhodes [1] there is a subpseudovariety of **S** that has undecidable weak equational problem.

The Identity problem for a finitely based variety $\mathcal{V}$.

    INSTANCE: A finite set of identities Σ over A and $u, v \in A^+$
    STATEMENT: $u = v$ is an identity for $\mathcal{V} \cap [\Sigma]$.

This problem requires the equational problem to be uniformly solved for all finitely based subvarieties of $\mathcal{V}$. As with the equational problem, we can formulate **The Weak** (or **Weak-κ**) **Identity Problem** for a pseudovariety. The variety $\mathcal{S}$ has undecidable identity problem (see [35]) and the pseudovariety **S** has undecidable weak identity problem (see [1]). The following is the pseudovariety version of a problem from [28].

Problem 3. *Is there a finitely based pseudovariety with undecidable weak (or weak-κ) identity problem such that every finitely based subpseudovariety has decidable weak (or weak-κ) equational problem?*

Let us now look at a quite different type of decision problem; namely **The Embeddability Problem** for a class C of semigroups.

INSTANCE: A finite semigroup S
STATEMENT: S is embeddable in a member of C.

The next theorem is the group theoretic form of a result of Evans [19].

Theorem 2.4. *Let **H** be a pseudovariety of groups. The uniform word problem for **H** is decidable if and only if the membership of the set of finite partial groups that are embeddable in elements of **H** is decidable.*

Kublanovskii [22] used this result and the fact that **G** has undecidable uniform word problem to obtain

Theorem 2.5. *The embeddability problem for the class of (finite) completely 0-simple semigroups is undecidable.*

The construction in the proof of this theorem has been used to prove some decidability results that at first appear to have little in common. From [22] we have

Theorem 2.6. *For a pseudovariety **H** of finite groups the decidability of the following are equivalent:*
*the uniform word problem for **H**;*
*the embeddability problem for finite 0-simple semigroups over **H**;*
*the embeddability problem for finite Brandt semigroups over **H**;*
*the embeddability problem of degree 4 nilpotent semigroups into finite 0-simple semigroups over **H**;*
*the embeddability problem of degree 3 nilpotent semigroups into finite Brandt semigroups over **H**.*

Suppose $S \in \mathbf{S}$ and $A \subseteq S \times S$. Then the pairs in A are *eventually $\mathcal{L}$-related* if and only if there exists an over-semigroup $T \geq S$ such that A is a subset of Green's $\mathcal{L}$-relation on T. If T can be chosen from a particular class $\mathcal{K}$ of semigroups then we say that A is *eventually $\mathcal{L}$-related in $\mathcal{K}$*. We can similarly define eventually $\mathcal{R}$, $\mathcal{H}$, $\mathcal{D}$ and $\mathcal{J}$-relations on S. Let $\mathcal{L}^*$ be the relation on S be given by

$$(a,b) \in \mathcal{L}^* \text{ iff } ax = ay \Leftrightarrow bx = by \; \forall \; x,y \in S.$$

Dually define $\mathcal{R}^*$. It is well known that $\mathcal{L}^*$ and $\mathcal{R}^*$ are respectively the eventually $\mathcal{L}$ and eventually $\mathcal{R}$-relations on S. However it is not always the case that $\mathcal{L}^* \cap \mathcal{R}^*$ is the eventually $\mathcal{H}$-relation. Jackson [26] has extended a result of Sapir that is based on Kublanovskii's construction to show

Theorem 2.7. *The problem of determining if a subset of a finite semigroup lies within an eventually $\mathcal{H}$-related class is undecidable in both the class* **S** *of finite semigroups and the class $\mathcal{S}$ of all semigroups.*

As well, Kublanovskii and Sapir [29] have shown that the problem of determining if a pair of elements of a finite semigroup is eventually $\mathcal{J}$-related is undecidable in **S**.

Kublanovskii's construction has also been used by Jackson [27] to show the undecidability of the problem of determining if an amalgam of finite semigroups is embeddable in a member of **S**, or of $\mathcal{S}$; this has also been independently proved by Sapir [44].

3 Membership problem for classes of finite semigroups

The study of finite semigroups has been dominated by **The Membership Problem** for a pseudovariety **V**.

    INSTANCE: A finite semigroup S
    STATEMENT: $S \in \mathbf{V}$.

If this is solvable for **V** it is usual to simply say **V** *is decidable*. The problem is dominant partly because of the pseudovariety version of the Krohn-Rhodes theorem and the variety theorem of Eilenberg and Schützenberger [18]. By Almeida [2] there do exist finitely based pseudovarieties of finite semigroups that are not decidable. If **V** is a pseudovariety with a finite basis of pseudoidentities, each of which can be tested for satisfiability in any finite semigroup, then **V** is decidable. So a natural task when given a pseudovariety is to attempt to find a finite basis for its pseudoidentities. However, the condition is not necessary; the pseudovariety $\mathbf{V}(B_2^1)$ generated by B_2^1, the five element aperiodic Brandt semigroup with adjoined identity, is not finitely based but it is decidable [2].

There are many examples in [2] of finitely based pseudovarieties; often they are subpseudovarieties of **DS**, the pseudovariety of finite semigroups whose regular $\mathcal{D}$-classes are completely simple semigroups. It is easy to see

that
$$\mathbf{DS} = [\![((xy)^\omega(yx)^\omega(xy)^\omega)^\omega = (xy)^\omega]\!].$$

Pseudovarieties can be defined in a less direct way than the way in which we defined **DS**. Suppose we generate a pseudovariety from a class of regular semigroups. For example, let $\mathbf{CS^0}$ be the pseudovariety generated by finite 0-simple semigroups. By [22]

$$\mathbf{CS^0} = [\![x^{\omega+2} = x^2, (xy)^{\omega+1}x = xyx, xyx(zx)^\omega = x(zx)^\omega yx]\!].$$

It follows that the subpseudovariety of semigroups of $\mathbf{CS^0}$ whose subgroups are in a decidable pseudovariety of groups is also decidable.

Let $\mathcal{R}(\mathbf{V})$ be the class of finite regular semigroups whose idempotent generated subsemigroups form the pseudovariety $\mathbf{V}$. In the case where $\mathbf{V}$ is a pseudovariety of monoidal bands, or of all finite completely regular semigroups then the pseudovariety generated by $\mathcal{R}(\mathbf{V})$ has been shown to be finitely based and decidable (see [11] and [24]). Furthermore, for such $\mathbf{V}$, the pseudovariety generated by the the class $L_r(\mathcal{R}(\mathbf{V}))$ is finitely based, where $S \in L_r(\mathcal{R}(\mathbf{V}))$ if and only if S is regular and $eSe \in \mathcal{R}(\mathbf{V})$ for each idempotent e of S (see [13]).

Some pseudovarieties have been intensively studied because of their connection with varieties of languages. In particular **A** consists of precisely the semigroups that recognise star-free languages and **J** is made up of the finite semigroups that recognise piecewise testable regular languages. Both of these are decidable pseudovarieties. However there are long standing unresolved membership problems in Language theory.

3.1 Operations on pseudovarieties

The membership problem for pseudovarieties obtained by combining decidable pseudovarieties via the operations like join, Malcev product and semidirect product has been intensively studied for about 25 years. This is because of the complexity problem associated with the Krohn-Rhodes theorem.

Let $\mathbf{U}$ and $\mathbf{V}$ be pseudovarieties of finite semigroups. Then $\mathbf{U} \vee \mathbf{V}$ is the least pseudovariety containing $\mathbf{U}$ and $\mathbf{V}$. The Malcev product $\mathbf{U}m\mathbf{V}$ and the semidirect product $\mathbf{U} * \mathbf{V}$ are respectively the pseudovarieties generated by the Malcev products and semidirect products of members of $\mathbf{U}$ by members of $\mathbf{V}$. These can be more precisely defined in terms of *relational morphisms*.

A relational morphism $\tau : S \to T$ for $S, T \in \mathbf{S}$ is a binary relation $\tau \subseteq S \times T$ such that $s\tau \subseteq T$ and $(s_1\tau)(s_2\tau) \subseteq (s_1s_2)\tau \ \forall \ s, s_1, s_2 \in S$. Then

$$\mathbf{U}m\mathbf{V} = \{S \in \mathbf{S}; \quad \exists \text{ a relational morphism } \tau : S \to T \in \mathbf{V}$$
$$\text{such that } (e)\tau^{-1} \in \mathbf{U} \ \forall e \in E(T)\},$$
$$\mathbf{U} * \mathbf{V} = \{S \in \mathbf{S} : \quad \exists \text{ a relational morphism } \tau : S \to T \in \mathbf{V}$$
$$\text{such that } D(\tau) \in g\mathbf{U}\}.$$

Here $D(\tau)$ is Tilson's derived semigroupoid of τ (see [47]) and $g\mathbf{U}$ is the variety of semigroupoids generated by the members of $\mathbf{U}$ (considered as one vertex semigroupoids).

A variant of the Krohn-Rhodes theorem is that a finite semigroup S is in $\mathbf{A} * \mathbf{G} * \mathbf{A} * \mathbf{G} * \ldots * \mathbf{A} * \mathbf{G} * \mathbf{A}$ for some number n of occurences of $\mathbf{G}$; the least number n is the complexity of S. **The Complexity Problem** for a pseudovariety $\mathbf{V}$ is as follows.

INSTANCE: $S \in \mathbf{V}, n \in \mathbb{Z}, n \geq 0$
STATEMENT: S has complexity n.

This is decidable when $\mathbf{V}$ consists of either the finite completely regular semigroups [10] or of the finite orthodox semigroups.

It can be shown that $\mathbf{U}m\mathbf{G} = \mathbf{U} * \mathbf{G}$ if $l\mathbf{U} = g\mathbf{U}$ where $l\mathbf{U}$ is the variety of semigroupoids whose one vertex subsemigroups are in $\mathbf{U}$. As well (see [31]) if $\mathbf{V}$ is a pseudovariety closed under Rhodes expansions (as $\mathbf{G}$ is) then $\mathbf{A}m\mathbf{V} = \mathbf{A} * \mathbf{V}$. Hence membership of Malcev products of pseudovarieties is very relevant. A major breakthrough is due to Ash [12]

Theorem 3.1. *For a decidable pseudovariety $\mathbf{U}$ of finite semigroups $\mathbf{U}m\mathbf{G}$ is decidable.*

More specifically, it is well known that for any $S \in \mathbf{S}$ there exists a finite group $G \in \mathbf{G}$ and a relational morphism $\lambda : S \to G$ such that

$$(1)\lambda^{-1} = \bigcap \{(1)\tau^{-1}; \ \tau : S \to T \in \mathbf{G} \text{ is a relational morphism}\}.$$

Ash showed that $(1)\lambda^{-1}$ is the least weakly self conjugate subsemigroup of S that contains $E(S)$. It follows that whenever $\mathbf{U}m\mathbf{G} = \mathbf{U} * \mathbf{G}$ and $\mathbf{U}$ is decidable then $\mathbf{U} * \mathbf{G}$ is decidable. From above the equation holds when $l\mathbf{U} = g\mathbf{U}$; it also holds (as noted) when $\mathbf{U} = \mathbf{A}$ or when $\mathbf{U} = \mathbf{J}$ (see [25]). For various decidable pseudovarieties $\mathbf{U}$, it has been shown that $\mathbf{U}*\mathbf{G}$ is decidable because it consists of the finite semigroups whose idempotent generated subsemigroups are in $\mathbf{U}$ [6]. On the negative side, Rhodes [40] showed

Theorem 3.2. *There exists a finite set of identities Σ such that $\mathbf{A} * \mathbf{G} * (\mathbf{A} \cap [\![\Sigma]\!])$ is undecidable. Also, there exists a decidable pseudovariety $\mathbf{U}$ such that $\mathbf{A}m\mathbf{U}$ is undecidable.*

There are various joins of decidable pseudovarieties that are known to be decidable because they are finitely based (see [2]). The join $\mathbf{J} \vee \mathbf{B}$, where $\mathbf{B}$ is the pseudovariety of finite bands, is not finitely based but is decidable [51]. The join $\mathbf{J} \vee \mathbf{G}$ is not finitely based [48] but is decidable [5] [46]. As well, by [1], there is a decidable pseudovariety $\mathbf{U}$ such that $\mathbf{U} \vee \mathbf{Com}$ is undecidable, where $\mathbf{Com}$ consists of all finite commutative semigroups.

The Strong Decidability Problem for a pseudovariety $\mathbf{V}$.

INSTANCE: $S \in \mathbf{S}$ and $A \subseteq S$
STATEMENT: for every relational morphism $\tau : S \to T \in \mathbf{V}$ $\exists\, t \in T$ such that $A \subseteq (t)\tau^{-1}$.

This is stronger than the decidability problem for $\mathbf{V}$ in that a strongly decidable pseudovariety is decidable. In his proof of Theorem 8, Ash showed that $\mathbf{G}$ is strongly decidable; so is $\mathbf{A}$ (see [23]). Rhodes and Steinberg [41] have shown that if the weak identity problem in $\mathbf{V}$ is undecidable then $\mathbf{V}$ is not strongly decidable; they used this to demonstrate a decidable but not strongly decidable pseudovariety.

Almeida and Weil explored the bases of semidirect products of decidable pseudovarieties in [9]. This led Almeida to a notion that is stronger even than strong decidability. For convenience in describing the next problem we identify a relational morphism $\tau : S \to T$ with its canonical extension $\tau : S^1 \to T^1$.

The Hyperdecidability Problem for a pseudovariety $\mathbf{V}$.

INSTANCE: A finite graph Γ, $S \in \mathbf{S}$ and a map $\gamma : \Gamma \to S^1$
STATEMENT: for each relational morphism $\tau : S \to T \in \mathbf{V}$ $\exists$ a map $\beta : \Gamma \to T^1$ such that $(x\gamma, x\beta) \in \tau$ and for each edge $e \in \Gamma$, with $e\alpha$ and $e\omega$ as the initial and final vertex respectively, $(e\alpha)\beta(e)\beta = (e\omega)\beta$.

The pseudovariety $\mathbf{V}$ is called *hyperdecidable* if the above problem is decidable. In [3] Almeida showed that for pseudovarieties $\mathbf{U}$ and $\mathbf{V}$ of finite semigroups, if $g\mathbf{U}$ has a defining set of pseudoidentities written on graphs with at most integer n vertices, if $g\mathbf{U}$ is decidable and if $\mathbf{V}$ is hyperdecidable then $\mathbf{U} * \mathbf{V}$ is decidable. Almeida and Steinberg [7] have developed a hierarchy of increasingly specialised decidability criterion with the aim not only of providing algorithms for deciding the membership of some pseudovariety semidirect products whose first components satisfy the criterion but are such that the semidirect product inherits the criterion; in that way the membership of iterated semidirect products, such as those in the definition of complexity, might be decided. It was announced in the initial session of the thematic term that the aim has not yet been achieved. However, there are recent surveys by Almeida [4] and by Weil [50] that detail the hierarchy and the obstacles to achieving the aim. Various pseudovarieties have already been investigated with respect to them satisfying the specialised decidability criteria.

Acknowledgements

The author would like to acknowledge the financial support of Fundação Calouste Gulbenkian (FCG), Fundação para a Ciência e a Technologia (FCT), Faculdade de Ciências da Universidade de Lisboa (FCUL) and Reitoria da Universidade do Porto.

References

[1] D. Albert, R. Baldinger and J. Rhodes, *Undecidability of the identity problem for finite semigroups*, J. Symbolic Logic, **57**, 179-192 (1992).

[2] J. Almeida, *Finite Semigroups and Universal Algebra*, World Scientific, Singapore, 1994.

[3] J. Almeida, *Hyperdecidable pseudovarieties and the calculation of semidirect products*, Int. J. Algebra and Comp. **9**, 241-261 (1999).

[4] J. Almeida, *Some key problems on finite semigroups*, Semigroup Forum **64**, 159-179 (2002).

[5] J. Almeida, A. Azevedo and M. Zeitoun, *On the joins of $\mathcal{J}$-trivial and completely regular pseudovarieties*, Int. J. Algebra and Comp. **9**, 99-112 (1999).

[6] J. Almeida and A. Escada, *On the equation $V*G=EV$*, J. Pure Appl. Algebra , (to appear).

[7] J. Almeida and B. Steinberg, *On the decidability of iterated semidirect products with applications to complexity*, Proc London Math. Soc. **80**, 50-74 (2000).

[8] J. Almeida and P. Weil, *Relatively free profinite monoids: An introduction and examples*, ed J. Fountain, (Semigroups, Formal Languages and Groups, Kluwer, Netherlands, 1995).

[9] J. Almeida and P. Weil, *Profinite categories and semidirect products*, J. Pure Appl. Algebra **123**, 1-50 (1998).

[10] M.A. Arbib, *The algebraic theory of machines, languages and semigroups*, Academic Press, New York, 1968.

[11] C. Ash, *Finite semigroups with commuting idempotents*, J. Austral. Math. Soc. Series A **43**, 81-90 (1987).

[12] C. Ash, *Inevitable graphs: a proof of the type II conjecture and some related decision procedures*, Int. J. Algebra and Comp. **1**, 127-146 (1991).

[13] K. Auinger and P.G. Trotter, *Pseudovarieties, regular semigroups and semidirect products*, J. London Math. Soc. **58**, 284-296 (1998).

[14] W.W. Boone, *Certain simple unsolvable problems in group theory I*, Proc. Kon. Ned. Akad. Wetensch. A **57**, 231-237 (1954).

[15] W.W. Boone and G. Higman, *An algebraic characterisation of groups with soluble word problem*, J. Austral. Math. Soc. **18**, 41-53 (1974).

[16] A. de Luca and S. Varricchio, *On non-counting regular classes*, ed M. Peterson, (Automata, Languages and Programming, Lecture Notes in Computer Science 443, Springer-Verlag, Berlin 1990).

[17] A. do Lago, *On the Burnside semigroups $x^n = x^{n+m}$*, ed I. Simon, (1st. Latin American Symposium on Theoretical Informatics, Lecture Notes in Computer Science 583, Springer-Verlag, Berlin 1992).

[18] S. Eilenberg, *Automata, languages and machines Vol. B*, Academic Press, New York, 1976.

[19] T. Evans, *The word problem for abstract algebras*, J. London Math. Soc. **26**, 64-71 (1951).

[20] V. Guba, *The word problem for the relatively free semigroup satisfying $t^m = t^{m+n}$, $m \geq 3$*, Int. J. Algebra and Comp. **3**, 335-347 (1993).

[21] Y. Gurevich, *The problem of equality of words for certain classes of semigroups*, Algebra i Logika **5**, 25-35 (1966).

[22] T.E. Hall, S.I. Kublanovskii, S. Margolis and M.V. Sapir and P.G. Trotter, *Algorithmic problems for finite groups and finite 0-simple semigroups*, J. Pure Appl. Algebra **119**, 75-96 (1997).

[23] K. Henkell, *Pointlike sets: the finest aperiodic cover of a finite semigroup*, J. Pure Appl. Algebra **55**, 85-126 (1988).

[24] K. Henkell, S. Margolis, J-E. Pin and J. Rhodes, *Ash's type II theorem, profinite topology and Mal'cev Products I*, Int. J. Algebra and Comp. **1**, 411-436 (1991).

[25] K. Henkell and J. Rhodes, *The theorem of Knast, the PG=BG and the type II conjectures*, ed J. Rhodes, (Monoids and Semigroups with Applications, World Scientific, Singapore 1991).

[26] M. Jackson, *Some undecidable embedding problems for finite semigroups*, Proc. Edinburgh Math. Soc. **42**, 113-125 (1999).

[27] M. Jackson, *The embeddability of ring and semigroup amalgams is undecidable*, J. Austral Math. Soc. Series A **69**, 272-286 (2000).

[28] O.G. Kharlampovich and M. Sapir, *Algorithmic problems in Varieties*, Int. J. Algebra and Comp. **5**, 379-602 (1995).

[29] S. Kublanovsky and M. Sapir, *Potential divisibility in finite semigroups*

is undecidable, Int. J. Algebra and Comp. **8**, 671-679 (1998).
[30] A.I. Mal'cev, *On the homomorphisms onto finite groups*, Uchen. Zapiski Ivanovsk. Ped. Instituta **18**, 49-60 (1958).
[31] S.W. Margolis, *Kernels and expansion: an historical and technical perspective*, ed J. Rhodes (Monoids and Semigroups with Applications, World Scientific, Singapore 1991).
[32] A. Markov, *On the impossibility of certain algorithms in the theory of associative systems*, Dokl. Akad. Nauk. SSSR **55**, 583-586 (1947).
[33] J. McCammond, *The solution of the word problem for the relatively free semigroups satisfying $x^a = x^{a+b}$ with $a \geq 6$*, Int. J. Algebra and Comp. **1**, 1-32 (1991).
[34] M. Morse and G. Hedlund, *Unending chess, symbolic dynamics and a problem in semigroups*, Duke Maths. J **11**, 1-7 (1944).
[35] V.L. Murskii, *Some examples of varieties of semigroups*, Mat. Zametki, **3**, 663-670 (1968).
[36] P.S. Novikov, *On algorithmic unsolvability of the problem of identity*, Dokl. Akad. Nauk. SSSR **85**, 709-719 (1952).
[37] P.S. Novikov and S.I. Adian, *On infinite periodic groups, I, II, III*, Izv. Akad. Nauk. SSSR **32**, 212-244,251-524,709-731 (1968).
[38] E.L. Post, *Recursive unsolvability of a problem of Thue*, J. Symbolic Logic **12**, 1-11 (1947).
[39] J. Reiterman, *The Birkhoff theorem for finite algebras*, Algebra Universalis **14**, 1-10 (1982).
[40] J. Rhodes, *Undecidability, automata and pseudovarieties of finite semigroups*, Int. J. Algebra and Comp. **9**, 455-473 (1999).
[41] J. Rhodes and B. Steinberg, *Pointlike sets, hyperdecidability and the identity problem for finite semigroups*, Int. J. Algebra and Comp. **9**, 475-481 (1999).
[42] M. Sapir, *Problems of Burnside type and the finite basis property in varieties of semigroups*, Izv. Akad. Nauk. SSSR **51**, 319-340 (1987).
[43] M. Sapir, *Weak word problem for finite semigroups*, ed J. Rhodes (Monoids and Semigroups with Applications, World Scientific, Singapore 1991).
[44] M.V. Sapir, *Algorithmic problems for amalgams of finite semigroups*, J. Algebra **229**, 514-531 (2000).
[45] A.M. Slobodskoii, *Undecidability of the universal theory of finite groups*, Algebra i Logika **20**, 207-230 (1981).
[46] B. Steinberg, *On pointlike sets and joins of pseudovarieties*, Int. J. Algebra and Comp. **8**, 203-234 (1998).
[47] B. Tilson, *Categories as algebra: an essential ingredient in the theory of*

monoids, J. Pure Appl. Algebra **48**, 83-198 (1987).
[48] P.G. Trotter and M.V. Volkov, *The finite basis problem in the pseudovariety joins of aperiodic semigroups with groups*, Semigroup Forum **52**, 83-91 (1996).
[49] P. Weil, *Implicit operations on pseudo-varieties: an introduction*, ed J. Rhodes (Monoids and Semigroups with Applications, World Scientific, Singapore 1991).
[50] P. Weil, *Profinite methods in semigroup theory*, Int. J. Algebra and Comp. , (to appear).
[51] M.Zeitoun, *On the decidability of the membership problem of the pseudovariety* $\mathbf{J} \vee \mathbf{B}$, Int. J. Algebra and Comp. **5**, 47-64 (1995).